Heterogeneous Catalysis

Heterogeneous Catalysis

A Versatile Tool for the Synthesis of Bioactive Heterocycles

Edited by

K.L. Ameta

Andrea Penoni

CRC Press
Taylor & Francis Group
Boca Raton London New York

CRC Press is an imprint of the
Taylor & Francis Group, an **informa** business

CRC Press
Taylor & Francis Group
6000 Broken Sound Parkway NW, Suite 300
Boca Raton, FL 33487-2742

Printed on acid-free paper
Version Date: 20140624

International Standard Book Number-13: 978-1-4665-9482-1 (Hardback)

Library of Congress Cataloging-in-Publication Data

Heterogeneous catalysis : a versatile tool for the synthesis of bioactive heterocycles / editors, K.L. Ameta, Andrea Penoni.
pages cm
"A CRC title."
Includes bibliographical references and index.
ISBN 978-1-4665-9482-1 (hardcover : alk. paper) 1. Heterogeneous catalysis. 2. Heterocyclic compounds--Synthesis. I. Ameta, K. L., editor. II. Penoni, Andrea, editor.

QD505.H462 2015
547'.215--dc23 2014024209

Visit the Taylor & Francis Web site at
http://www.taylorandfrancis.com

and the CRC Press Web site at
http://www.crcpress.com

Contents

Preface

Heterocyclic compounds are among the most important and valuable products synthesized and studied by organic chemists and by many other scientists. Heterocycles belong to what is probably the largest and most varied family of organic compounds. They have a relevant role and different applications in various research fields. The importance of heterocyclic molecules has continued to grow in recent decades and they currently occupy a primary and crucial role in biochemistry, medicine, agriculture, physics, materials science, and many other subjects. Heterocyclic frameworks are ubiquitously present in the huge number of bioactive molecules that have been afforded and are continuously produced for the treatment of many diseases. The majority of the top ten drugs sold by pharmaceutical companies in recent years belong to the field of heterocyclic substances, and many different synthetic methodologies and techniques have been developed in the last century to achieve bioactive heterocyclic compounds.

For many decades, research on bioactive heterocycles has been one of the main topics of interest for medicinal chemists because of the numerous pharmacological activities of this class of compounds. Nitrogen-, sulfur-, and oxygen-containing five- and six-membered heterocyclic compounds, and smaller and larger cyclic structures, have had enormous significance in the field of medicinal chemistry. The majority of pharmaceuticals and biologically active agrochemicals are heterocycles, while countless additives and modifiers used in industrial applications such as cosmetics, reprography, information storage, and plastics are heterocyclic in nature. One striking structural feature inherent to heterocycles, which continues to be exploited to great advantage by the drug industry, is their ability to manifest substituents around a core scaffold in defined three-dimensional representations. For more than a century, heterocycles have constituted one of the largest areas of research in organic chemistry. The presence of heterocycles in all kinds of organic compounds of interest in electronics, biology, optics, pharmacology, materials science, and so on is widely known and recognized. However, specifically, sulfur- and nitrogen-containing heterocyclic compounds have maintained the interest of researchers through decades of historical development of organic synthesis.

Much time has passed since the "first organic synthesis" of urea by Wohler in 1828, and the extraordinary ability of synthetic organic chemists has reached levels of real excellence in the research field of total syntheses of highly complex molecules with dramatic selectivities (e.g., palytoxin by Kishi in 1994 and brevetoxin B by Nicolaou in 1998).

> Organic chemists are now able to synthesize small quantities of almost any known natural product, given sufficient time, resources and effort. (Jones et al. 2011)

Very high levels of sophistication were obtained, particularly in the study of catalytic processes. Catalysis is one of the topics in the "Twelve Principles of Green

Chemistry" (Anastas and Warner 1998). In recent years, a renewed sensitivity to the environmental problems connected with organic syntheses was the driving force that motivated chemists to introduce novel techniques and particularly eco-friendly procedures. Atom economy, minimization of side product formation, solventless conditions, and use of unconventional techniques to run reactions (microwave, ultrasound, ball-milling, ionic liquids, etc.) have been introduced and play relevant roles in today's chemistry laboratories. These techniques will probably be used more intensively and in larger amounts in synthetic laboratories in the future.

Recently, intensive studies have focused on the development of catalytic systems owing to their importance in synthetic organic chemistry. One of the most attractive synthetic strategies favored by organic chemists is the use of heterogeneous catalysts in increasing the efficiency of a wide range of organic syntheses. Heterogeneous catalysis is being used in the fine chemicals industry because of the need for more environmentally friendly production technology. This tendency is assisted by the availability of catalytic materials and modern techniques in creating and investigating specific active sites on catalyst surfaces.

Heterogeneous catalysis is one of the most powerful tools for the preparation of building blocks and fine chemicals. Reuse and recycling of heterogeneous catalysts is a milestone and a fundamental topic in the aim to introduce a green approach or an environmental friendly synthesis of scaffolds and target molecules.

Many of these techniques are reported in the reviews that you can read in the chapters of this book. Thus, the purpose of the present book is to provide a succinct summary of protocols for the synthesis of various bioactive heterocycles of different sizes using heterogeneous catalytic approaches. We would like to thank all the authors that gave robust contributions to this book, providing extraordinary reviews on the syntheses of different and highly valuable bioactive heterocyclic compounds through the use of heterogeneous catalysis. We trust that both undergraduate and graduate students, researchers, and faculties will find in these pages an interesting guide to and an update on the state of the art of the enormous versatility of heterogeneous catalysis in the synthesis of bioactive heterocycles. We think that even experts in the research field will find this book inspiration and motivation for future projects.

K. L. Ameta
Andrea Penoni

REFERENCES

Jones, S. B., Simmons, B., Mastracchio, A., MacMillan, D. W. C. 2011. Collective synthesis of natural products by means of organocascade catalysis. *Nature* 475: 183–188.

Anastas, P. T., Warner, J. C. 1998. *Green Chemistry: Theory and Practice*, Oxford University Press: New York, p. 30.

Editors

K. L. Ameta received his doctorate degree in organic chemistry from M. L. Sukhadia University, Udaipur, India, in 2002. Presently, he is working as assistant professor of chemistry at the Faculty of Arts, Science and Commerce, Mody Institute of Technology and Science, Deemed University, Lakshmangarh, Rajasthan, India. Dr. Ameta has vast experience of teaching both graduate and postgraduate students. His research area is the synthesis, characterization, and biological evaluation of bioactive heterocyclic systems of different sizes. In addition, he has keen interest in heterogeneous catalyzed organic synthesis and photocatalysis. He has published a number of research articles in various journals of national and international repute.

Andrea Penoni is assistant professor in organic chemistry at the Department of Science and High Technology at the University of Insubria, Italy. He graduated in chemistry (1996, laurea degree) and received a PhD (2000, doctoral research) at the University of Milan under the supervision of Professor Sergio Cenini working on a project on organometallic chemistry and homogeneous catalysis. He did his postdoctoral training at the University of Milan, and at the University of Oklahoma with Professor Kenneth M. Nicholas (2000–2001), where he studied annulation reactions between nitro- and nitrosoaromatics with alkynes. Since 2003, Dr. Penoni has been a faculty member at the University of Insubria (Como) and works in the research group of Professor Giovanni Palmisano. His research is particularly focused on the synthesis of heterocycles, naturally occurring compounds, and potentially bioactive molecules. Further, he is involved in research projects on the carbon–nitrogen bond formation mediated by metals and metal complexes and in the synthesis of nitrogen-containing heterocycles by metathesis reactions.

Contributors

Chetna Ameta
Department of Chemistry
M. L. Sukhadia University
Udaipur, India

K. L. Ameta
Department of Chemistry
Mody Institute of Technology and Science
Lakshmangarh, India

Laishram Ronibala Devi
Department of Chemistry
Manipur University
Imphal, India

Essam M. Hussein
Department of Chemistry
Assiut University
Assiut, Egypt

Renuka Jain
Department of Chemistry
University of Rajasthan
Jaipur, India

Bahador Karami
Department of Chemistry
Yasouj University
Yasouj, Iran

Sushma S. Kauthale
Department of Chemistry
Deogiri College
Aurangabad, India

Dharma Kishore
Department of Chemistry
Banasthali University
Banasthali, India

Manohar V. Kulkarni
Department of Chemistry
Karnatak University
Dharwad, India

Deepak Kumar
Department of Chemistry
University of Rajasthan
Jaipur, India

Sudesh Kumar
Department of Chemistry
Banasthali University
Banasthali, India

Daniela Lanari
Department of Chemistry and Technology of Medicine
Università degli Studi
Perugia, Italy

Rujirat Longloilert
Center for Petroleum, Petrochemicals, and Advanced Materials
Chulalongkorn University
Bangkok, Thailand

Xin Lv
College of Chemistry and Life Sciences
Zhejiang Normal University
Jinhua, People's Republic of China

Rajendra P. Pawar
Department of Chemistry
Deogiri College
Aurangabad, India

Prachi Rathi
Department of Chemistry
Banasthali University
Banasthali, India

Ambadas B. Rode
Department of Chemistry
Kongju National University
Kongju City, Republic of Korea

Ornelio Rosati
Department of Chemistry and Technology of Medicine
Università degli Studi
Perugia, Italy

Kanti Sharma
Department of Chemistry
R.L. Saharia Govt. P.G. College
Jaipur, India

Lokesh A. Shastri
Department of Chemistry
Karnatak University
Dharwad, India

Sandeep V. Shinde
Department of Chemistry
Pratibha Niketan College
Nanded, India

Okram Mukherjee Singh
Department of Chemistry
Manipur University
Imphal, India

Sunil U. Tekale
Department of Chemistry
Deogiri College
Aurangabad, India

Sujitra Wongkasemjit
Center for Petroleum, Petrochemicals, and Advanced Materials
Chulalongkorn University
Bangkok, Thailand

Guodong Yuan
College of Chemistry and Life Sciences
Zhejiang Normal University
Jinhua, People's Republic of China

1 Synthesis of Bioactive Heterocyclic Systems Promoted by Silica-Supported Catalysts

Daniela Lanari and Ornelio Rosati

CONTENTS

1.1 INTRODUCTION

Heterocyclic compounds are probably the largest and most varied family of organic compounds. Even if only considering the most common heterocyclic elements (oxygen, nitrogen, and sulfur) numerous heterocyclic system combinations are possible (Figure 1.1). This class of compounds is not limited to monocyclic derivatives as there are numerous well-known polycyclic compounds incorporating one or more heterocyclic rings.

Many natural and nonnatural bioactive compounds are based on a heterocyclic skeleton. Heterocycles are still the target of many industrial and academic chemists. From the discovery of the first synthesis of heterocycles, much effort has been put into the development of new synthetic approaches to compounds such as Fischer indoles (Fischer and Jourdan, 1883), Knorr pyrroles (Knorr, 1884), and Biginelli dihydropyrimidinones (Biginelli, 1891a,b). Unfortunately, many of these synthetic approaches cannot be applied to industrial production due to several drawbacks. To resolve this problem, green chemistry and sustainable chemistry have become a priority for the development of new and more environmentally acceptable processes in the chemical industry. From this point of view, the traditional concept of process efficiency, which is focused largely on chemical yield, must be revised to one that assigns economic value to the elimination of waste at source and avoidance of the use of toxic and hazardous substances (Sheldon and Downing, 1999). In particular, heterogeneous

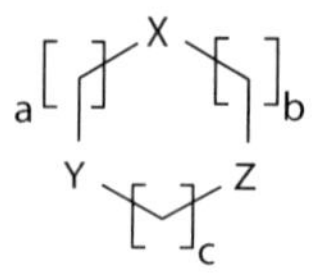

FIGURE 1.1 Generic formula of possible combinations of C, O, N, and S in a heterocycle.

solid catalysts have gained great importance as a sustainable alternative to classical homogenous catalysts (Gladysz, 2002). One obvious advantage of this choice is the ease of separation from the reaction mixture, which allows the recovery and reuse of the solid, thus minimizing waste and disposal issues. Furthermore, solid catalysts allow the design of continuous flow processes that are economically attractive on an industrial scale (Strappaveccia et al., 2013; Bonollo et al., 2012). The preparation and characterization of silica-bound catalysts are indeed a very active area of research (Corma and Garcia, 2006). Much effort has been directed at finding a convenient methodology to covalently bind the catalyst (most commonly), taking advantage of the silanol groups on the silica surface. This chapter will illustrate a variety of synthetic protocols for the construction of heterocyclic systems employing silica-bound catalysts. Specifically, silica-supported acids and bases will be taken into account as they are convenient alternatives to typical acid and base catalysts in order to avoid environmental pollution and corrosion problems.

1.2 SILICA-SUPPORTED ACIDS

Silica sulfuric acid (SSA) has recently attracted the interest of many chemists (Ali et al., 2003; Azizian et al., 2005; Rapolu et al., 2013) thanks to its ability to enhance the reactivity and selectivity of many types of reactions such as oxidation (Mirjalili et al., 2003), carbon–carbon bond formation (Chen and Lu, 2005), cycloaddition (Saheli et al., 2005), protection–deprotection steps (Hajipour et al., 2005), esterification (Chakraborti et al., 2009), and the synthesis of heterocyclic compounds (Maleki et al., 2012). SSA, which is prepared from the reaction of silica gel with chlorosulfonic acid (Zolfigol, 2001), is easier to handle than sulfuric acid and can be readily separated from the products by simple filtration. Additionally, it is recyclable and may have applicability in the large-scale industrial synthesis of fine chemicals and pharmaceuticals.

Maleki and coauthors have recently reported the use of SSA for the solvent-free synthesis of 2,4,5-trisubstituted imidazoles (**1a–l**) from the condensation reaction of benzil or benzoin, functionalized aromatic aldehydes, and ammonium acetate (Scheme 1.1) (Maleki et al., 2012).

Reaction conditions were optimized by employing 4-chlorobenzaldehyde as the test compound. A higher yield (94%) was obtained when equimolar amounts of benzil (1 mmol) and aldehyde (1 mmol) along with ammonium acetate (5 mmol) were allowed to react in the presence of 0.01 g of the catalyst at 110°C. The protocol was then extended to different aromatic aldehydes bearing ortho-, meta-, and para-substitutions under solvent-free conditions; the corresponding results are given in

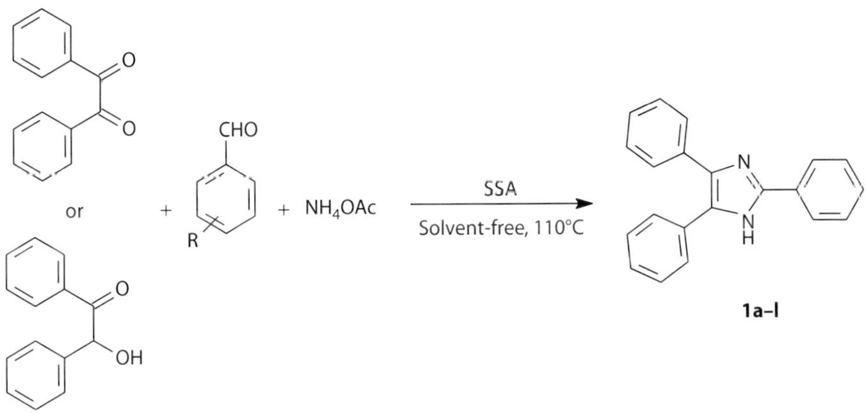

1a–l

SCHEME 1.1

Table 1.1. The reaction proceeded smoothly when aldehydes with either electron-withdrawing or electron-donating groups were employed. Furthermore, the recyclability of the catalyst was studied. At the end of each reaction, SSA was filtered, washed with diethyl ether, dried at 120°C for 3 h, and reused in a subsequent reaction cycle. The recycled catalyst was employed for three consecutive runs and no significant loss in its efficiency was observed.

Zolfigol et al. (2011) have synthesized a modified silica sulfuric acid (MSSA) as a new type of heterogeneous catalyst for the conjugated addition of pyrrole, indole, and thiols to Michael acceptors. This catalyst has the advantages of being a stable, nonvolatile, inexpensive, and safe for handling source of acid (Figure 1.2).

TABLE 1.1
Synthesis of 2,4,5-Triaryl-1*H*-Midazoles (2a–l) Using SSA (0.01 g) under Solvent-Free Conditions

		Time (min)		Yield (%)	
Aldehyde	**Product**	**Benzil**	**Benzoin**	**Benzil**	**Benzoin**
Benzaldehyde	**1a**	50	60	90	85
4-Cl-benzaldehyde	**1b**	45	60	94	87
4-Me-benzaldeide	**1c**	50	65	85	80
4-OMe-benzaldehyde	**1d**	60	75	80	75
4-OH-benzaldehyde	**1e**	50	55	85	78
4-F-benzaldehyde	**1f**	50	60	86	80
2-OMe-benzaldehyde	**1g**	60	65	85	81
3-Br-benzaldehyde	**1h**	45	50	90	85
3-NO_2-benzaldehyde	**1i**	60	70	85	80
2-Me-benzaldehyde	**1j**	55	60	90	80
4-Br-benzaldehyde	**1k**	45	65	92	85
3,4-DiOMe-benzaldehyde	**1l**	55	60	85	80

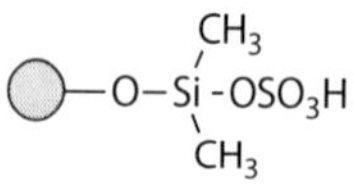

FIGURE 1.2 Modified silica sulfuric acid (MSSA).

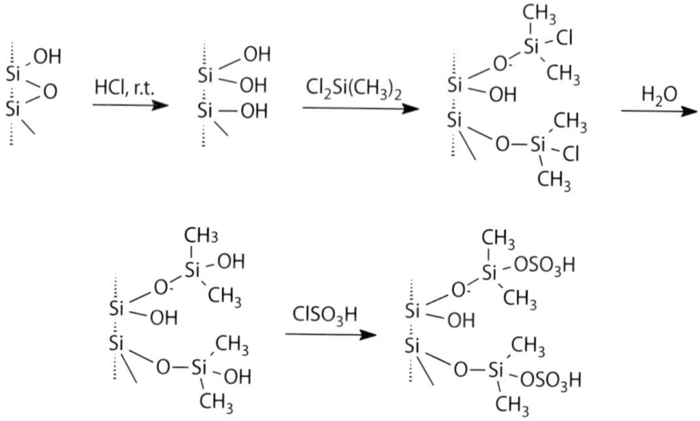

SCHEME 1.2

The synthesis of the new MSSA is depicted in Scheme 1.2. The first step was the activation of the silica gel surface with HCl, then an excess of dimethylchlorosilane (DMCS) was used to accomplish the silanization process. Conversion of Si–Cl to Si–OH occurred when water was added to the solid product after 10 min of stirring at room temperature. After completion of all reactions, samples were dried in a nitrogen stream at 120°C for 24 h. Finally, the silanol groups were allowed to react with chlorosulfonic acid to yield the desired MSSA, without any side products as HCl gas evolved from the reaction vessel immediately.

The reaction of chalcone (**2**) with indole (**3**) (1:1 molar ratios in a 1 mmol scale) was performed to optimize the experimental conditions with respect to temperature, time, and ratio of MSSA to the substrate. It was found that 200 mg of MSSA was sufficient to obtain the desired Michael adduct in 98% yield in 90 min at room temperature in CH_3CN. A variety of α,β-unsaturated ketones and electron-deficient olefins were tested under the optimized reaction conditions (Scheme 1.3a). All the reactions proceeded easily and the products **4** were isolated with comparable yields (85%–98%) in short reaction times. No side products such as N-alkylation products were detected.

It was also possible to perform the pyrrole **5** Michael addition with a variety of α,β-unsaturated ketones **2** catalyzed by MSSA in the same fashion, affording 2-substituted pyrroles **6** in good yields (Scheme 1.3b). Furthermore, MSSA was used as a solid acid heterogeneous catalyst to synthesize 1,1,3-triindolyl compounds through the tandem Michael addition and Friedel–Craft reaction of α,β-unsaturated aldehydes or ketones and indole (**3**) (Scheme 1.4). The authors found two protocols of general applicability employing 1 mmol of α,β-unsaturated compound and 4 mmol

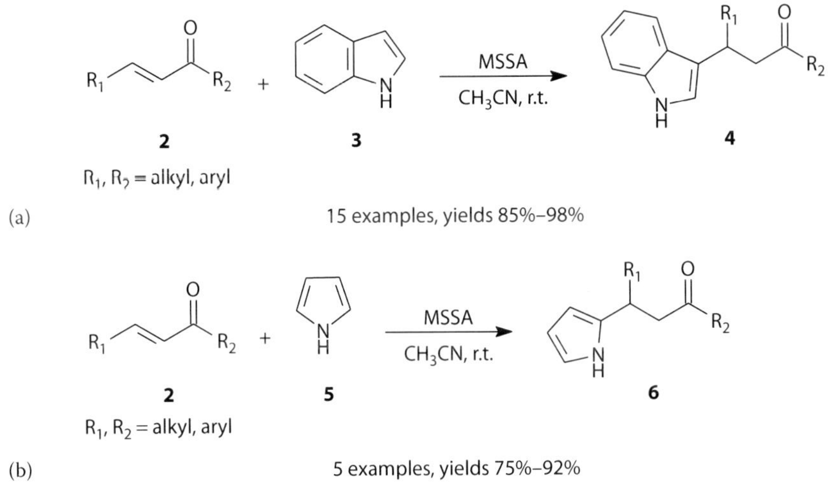

SCHEME 1.3

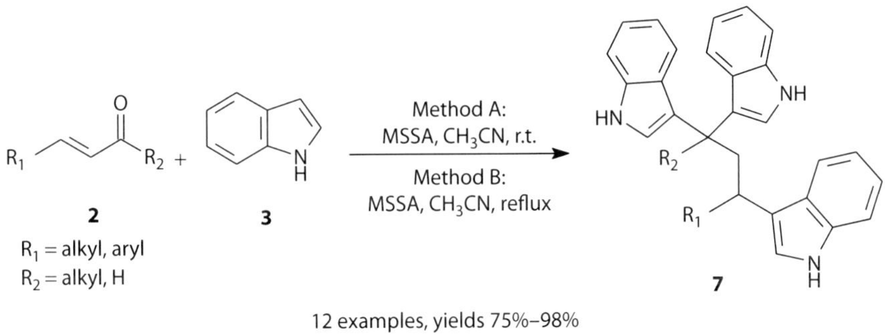

SCHEME 1.4

of indole in 5 ml of CH_3CN (Method A: 250 mg MSSA, 10 h, room temperature; or Method B: 250 mg MSSA, 20 min at reflux condition) to obtain products **7** in good yields (75%–98%).

Davoodnia and coworkers set up a simple and efficient procedure for the synthesis of 14-aryl-14*H*-dibenzo[*a*,*j*]xanthenes using a one-pot condensation reaction of β-naphthol and aryl aldehydes catalyzed by silica gel–supported polyphosphoric acid (PPA/SiO_2) (Khojastehnezhad et al., 2011). Xanthene derivatives occupy a prominent position in medicinal chemistry (Wang et al., 1997) and have been investigated for their agricultural bactericide activity (Handique and Baruah, 2002), anti-inflammatory effect (Sirkecioglu et al., 1995), and antiviral activity.

PPA/SiO_2 was prepared according to the literature procedure (Aoyama et al., 2004). The reaction between β-naphthol (**8**) (2 mmol) and benzaldehyde (1 mmol) to yield compound **9a** was chosen as a test reaction for the optimization of the protocol (Scheme 1.5).

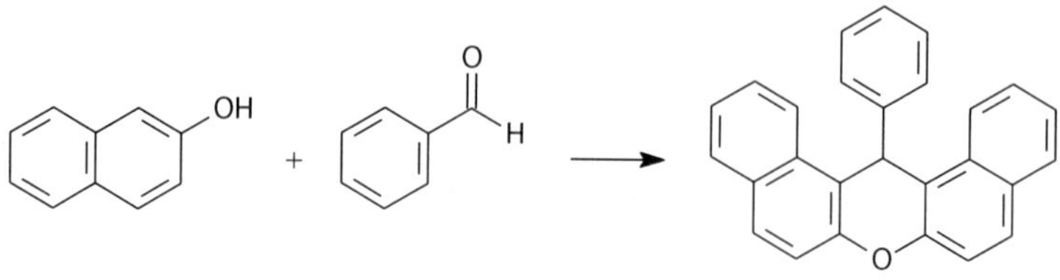

8 9a

SCHEME 1.5

It was found that the efficiency of the reaction is mainly affected by the amount of PPA/SiO_2, with 0.03 g being the optimal amount to catalyze the reaction. Furthermore, the transformation was carried out in different solvents and under solvent-free conditions, as shown in Table 1.2. In the latter experimental conditions, higher yield and the shortest reaction time were obtained.

Furthermore, the effect of temperature under solvent-free conditions on the reaction yield was studied. The best result was obtained at 120°C, when product **9a** was recovered in excellent yield. In order to evaluate the generality of this model, a variety of 14-aryl-14*H*-dibenzo[*a*,*j*]xanthenes were prepared using the optimized reaction conditions. Results are shown in Table 1.3. All aromatic aldehydes (Scheme 1.6), carrying either electron-withdrawing or electron-donating groups, consistently afforded the desired xanthenes **9a–i** in excellent yields with short reaction times.

Reusability of the catalyst was also investigated; it was found that it could be reused for at least three consecutive runs without significant loss of its activity.

Shaterian and coauthors also investigated the use of PPA/SiO_2 for the synthesis of heterocycles containing a phthalazine moiety (Shaterian et al., 2009). Such compounds are of great interest because they show pharmacological and biological anticonvulsant (Grasso et al., 2000), cardiotonic (Nomoto et al., 1990), and vasorelaxant (Watanabe et al., 1998) activities.

TABLE 1.2
Synthesis of 14-Phenyl-14*H*-Dibenzo[*a*,*j*]xanthene (9a) in the Presence of PPA/SiO_2 (0.03 g) at Different Temperatures and in Different Reaction Media

Entry	Solvent	Temperature (°C)	Time (h)	Yield (%)[a]
1	MeOH	64	5	Trace
2	EtOH	78	5	65
3	CH_3CN	81	5	82
4	H_2O	100	5	None
5	Solvent-free	120	0.5	90

[a] The yields were calculated based on benzaldehyde and refer to the pure isolated products.

TABLE 1.3
Preparation of 14-Aryl-14*H*-Dibenzo[*a,j*] xanthenes Using PPA/SiO_2 (0.03 g) as Catalyst

Entry	R_1	R_2	Product	Time (min)	Yield (%)
1	H	H	**9a**	30	90
2	H	Br	**9b**	35	86
3	H	Cl	**9c**	40	89
4	H	F	**9d**	35	90
5	OH	H	**9e**	30	88
6	H	OH	**9f**	35	90
7	H	OMe	**9g**	40	89
8	H	Me	**9h**	40	88
9	NO_2	H	**9i**	40	86

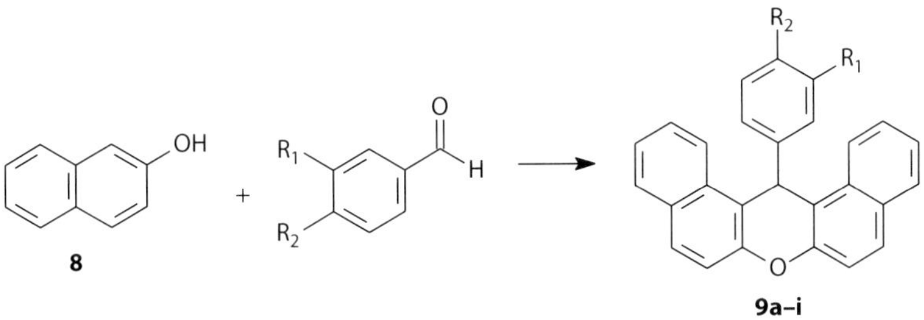

SCHEME 1.6

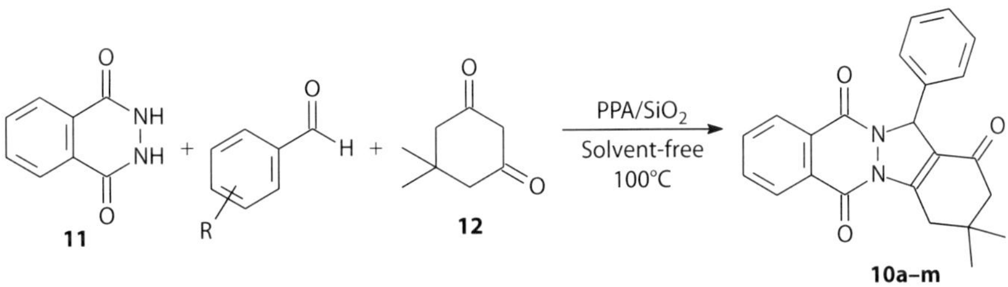

SCHEME 1.7

Specifically, an efficient protocol to afford 2*H*-indazolo[2,1-*b*]phthalazine-1,6,11(13*H*)-triones (**10**) from a condensation reaction of phthalhydrazide (**11**), 5,5-dimethyl-1,3-cyclohexanedione (**12**) (dimedone), and aromatic aldehydes was studied (Scheme 1.7).

The best yields and shortest reaction times were obtained when the transformation was performed at 100°C in solvent-free conditions using a molar ratio of **11**, aldehyde, and **12** of 1/1/1.2 at a 1 mmol scale employing 0.1 g of PPA/SiO_2, which corresponds to 0.05 mmol of H^+. A further increase in the amount of catalyst did not have any significant effect on the product yield. The generality of the reaction was tested using different aromatic aldehydes. In all cases the yields of the corresponding products **10a–m** were consistently good to excellent (Table 1.4); five new analogs of phthalazine were also prepared (Table 1.4, entries 9–13).

The catalyst was recovered after each run, washed 3 times with acetone, dried at 100°C for 30 min prior to use and tested for its activity in five consecutive runs without any significant loss in activity.

The synthesis of xanthene derivatives using silica-supported acid was also investigated by Karimi et al. (2011). They reported a facile procedure for the synthesis of 12-aryl-8,9,10,12-tetrahydrobenzo[*a*]xanthen-11-one derivatives through a multicomponent reaction using Caro's acid–silica gel (CA–SiO_2) as a nontoxic, inexpensive, and easily obtained catalyst. Benzaldehyde, β-naphthol **8**, and dimedone (**12**) were allowed to react at different temperatures and with different amounts of catalyst to find the optimal reaction conditions for the preparation of xanthenone (**11**) (Scheme 1.8).

When the reaction was run on a 1 mmol scale, 0.1 g of CA–SiO_2 at 60°C under solvent-free conditions delivered the best result for the synthesis of product **11** (80% yield). When the protocol was applied to other multicomponent reactions employing groups bearing substituted benzaldehydes and electron-donating groups, excellent results were obtained in all cases (yields 75%–90%).

The cyclization of 2′-amino- and 2′-hydroxy-chalcones (**12a,b**) (Scheme 1.9) using silica-supported sodium hydrogen sulfate ($NaHSO_4$–SiO_2) was accomplished by Kumar and Perumal (2006). Preparation of tetrahydroquinolones and flavonones is a matter of great interest as these compounds display various pharmacological

TABLE 1.4
Preparation of 2*H*-Indazolo[2,1-*b*]phthalazine-Trione (10a–m) Derivatives Using PPA–SiO_2 as Catalyst under Solvent-Free Conditions

Entry	Aldehyde	Product	Time (min)	Yield (%)[a]
1	Benzaldehyde	**10a**	8	92
2	4-Cl-benzaldehyde	**10b**	6	93
3	4-Br-benzaldehyde	**10c**	8	86
4	4-F-benzaldehyde	**10d**	12	82
5	4-NO_2-benzaldehyde	**10e**	8	89
6	2-Cl-benzaldehyde	**10f**	12	81
7	3-NO_2-benzaldehyde	**10g**	13	86
8	4-Me-benzaldehyde	**10h**	12	79
9	3,4,5-TriOMe-benzaldehyde	**10i**	4	90
10	2-Me-benzaldehyde	**10j**	24	87
11	2,4-DiCl-benzaldehyde	**10k**	9	86
12	3-Cl-benzaldehyde	**10l**	16	78
13	4-OH-3-OMe-benzaldehyde	**10m**	10	92

[a] Yields refer to isolated pure products.

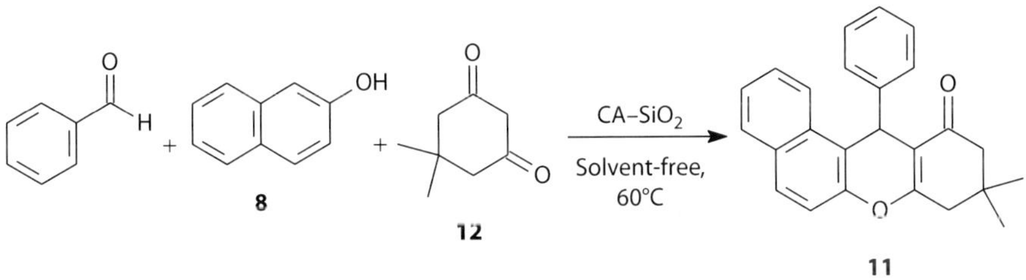

SCHEME 1.8

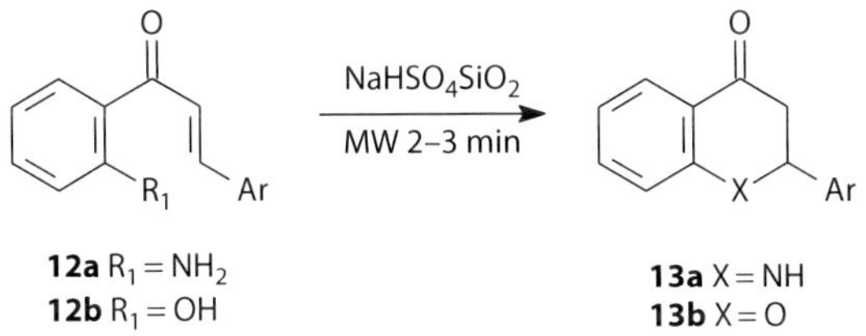

SCHEME 1.9

activities (Ito et al., 1986; De Meyer et al., 1991; Singh and Kapil, 1993). The desired products **13a,b** were obtained through a cyclization reaction of the corresponding chalcones (1 mmol scale) using 2 g of silica gel preimpregnated with $NaHSO_4$; the reaction mixture was then irradiated by microwaves until completion. Some representative examples are reported in Table 1.5.

Substituted 2′-amino-chalcones were obtained in good yields and short reaction times especially if compared to the corresponding reactions run employing conventional heating (Table 1.5, entries 1–4). Also, 2′-hydroxy-chalcones were yielded in a similar fashion (Table 1.5, entries 5–6). The catalyst can be easily prepared (Nishiguchi and Kamio, 1989), safely handled, and removed from the reaction mixture by filtration. The advantages of this protocol are operational simplicity, fast and clean reactions, and high reaction yields.

Chari et al. (2007) reported the use of $NaHSO_4$–SiO_2 to catalyze the condensation of *o*-phenylenediamines (**14**) with ketones (**15**) under microwave irradiation and in solvent-free conditions to afford the corresponding 1,5-benzo-diazepine derivatives (**16**) in high yields. Benzodiazepines and their derivatives are an important class of bioactive molecules and are widely used as anticonvulsant, antianxiety, analgesic, sedative, anti-inflammatory, and hypnotic agents (Schutz, 1982; Landquist, 1984). Taking into account their importance, a variety of synthetic procedures have been developed (Yadav et al., 2002; Jarikote et al., 2003; Kumar et al., 2006b), however many methodologies suffer from drawbacks such as long reaction times, expensive reagents, harsh conditions, low product yields, and difficult recovery and reusability of the catalysts. Treatment of substituted *o*-phenylenediamine (**14**) (1 mmol) with various ketones (**15**) (2.5 mmol) in the presence of $NaHSO_4$–SiO_2 (5 mol%) under solvent-free conditions yielded the corresponding 2,3-dihydro-1,5-benzodiazepines (**16**) in 85%–98% yield, as reported in Scheme 1.10.

TABLE 1.5
$NaHSO_4$–SiO_2-Catalyzed Cyclization of 2′-Aminochalcones under Solventless Conditions

Entry	Chalcone	Product	MW Time (min) (Yield %)	Conventional Heating Time (h) (Yield %)
1	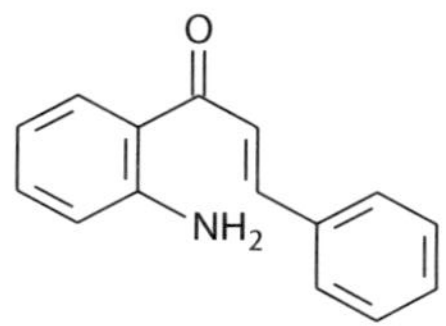	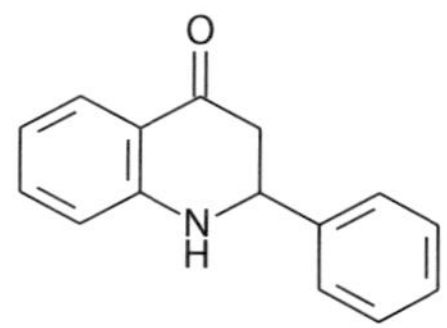	2 (95)	2 (72)
2	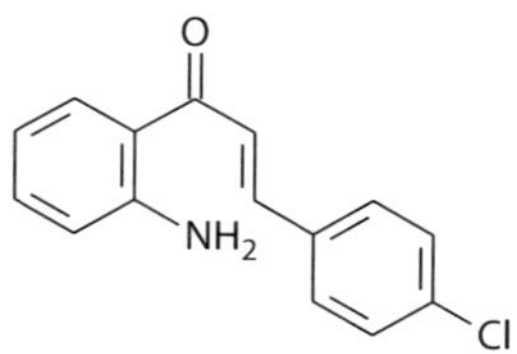	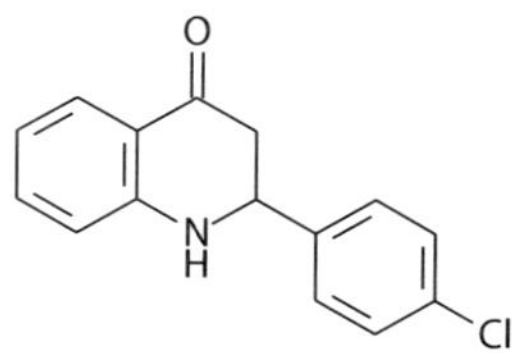	2 (88)	2 (70)
3	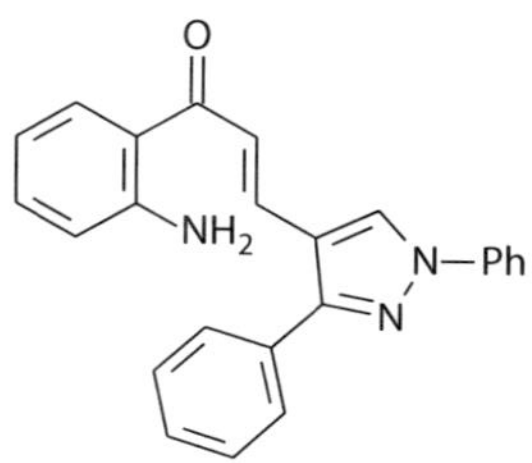	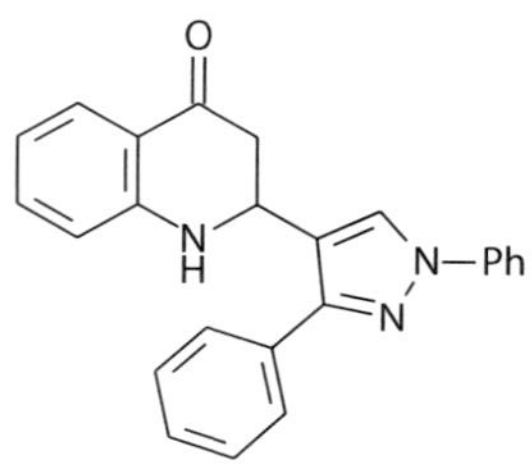	3 (80)	4 (65)

4	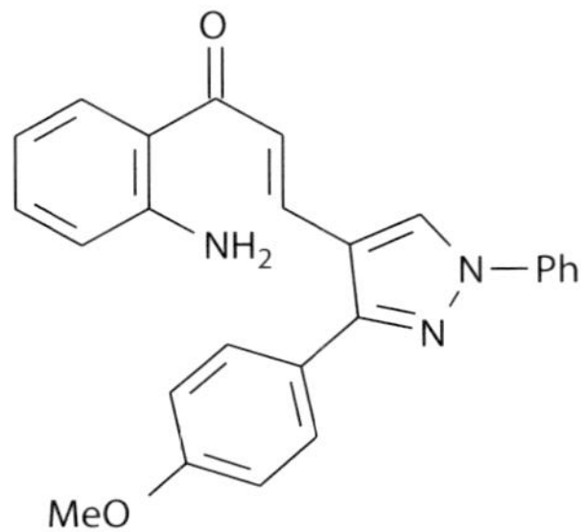	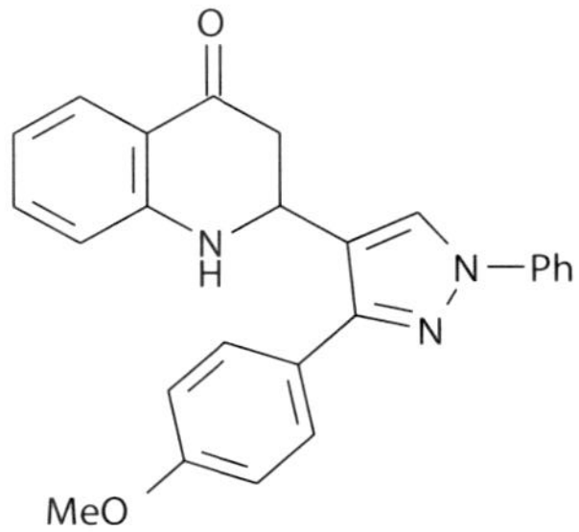	3 (92)	4 (74)
5	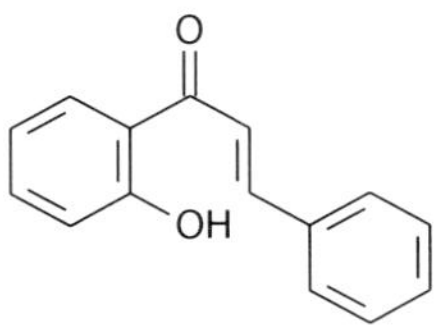	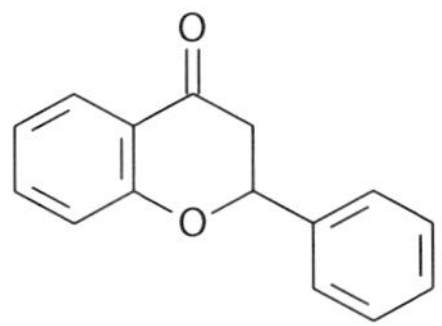	6 (75)	6 (60)
6	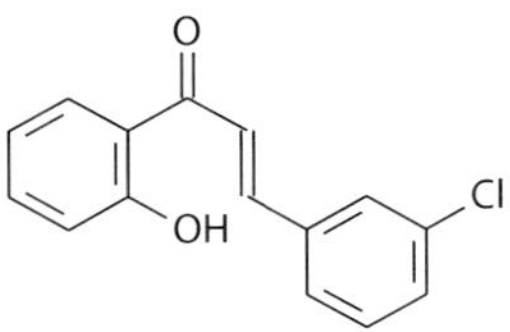	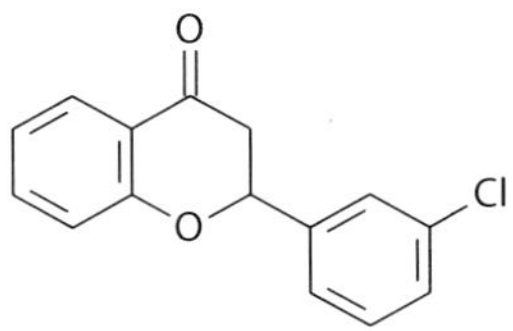	8 (68)	7 (52)

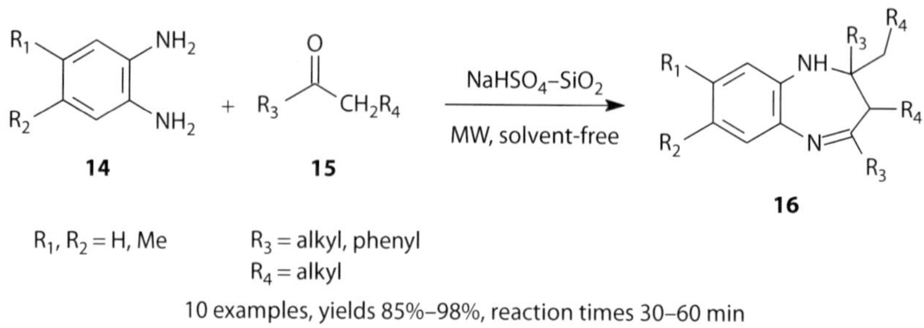

SCHEME 1.10

The use of microwave irradiation combined with the heterogeneous catalyst accelerated the reaction rates and improved the yields of the desired products. The reusability of $NaHSO_4$–SiO_2 was tested in the reaction of *o*-phenylenediamine with acetone for the synthesis of 2,3-dihydro-,2,4-trimethyl-*1H*-1,5-benzo[*b*][1,4]diazepine; it was found that the catalyst remained active for up to four cycles.

El Maatougui et al. (2011) described the synthesis of three different libraries of pyridazin-3-ones using silica-supported aluminum trichloride (Si–$AlCl_3$) as a heterogeneous and reusable catalyst. In previous work, the same authors discovered a novel family of potent platelet aggregation inhibitors derived from the pyridazin-3-one scaffold and incorporated diverse α,β-unsaturated residues at position 5 of the heterocycles, namely chemotypes A, B, and C (Figure 1.3) (Coelho et al., 2007).

The synthetic pathways developed to access the target pyridazinone arrays are based on the assembly of the exocyclic α,β-unsaturated framework through either Knoevenagel (chemotype A) or Claisen–Schmidt (chemotypes B and C) condensations (Scheme 1.11) with silica-supported aluminum chloride employed as a catalyst. In comparison with conventional aluminum chloride, the silica-supported equivalent offers several advantages over the free catalyst (milder acidity, superior shelf life, and the ability to conduct nonaqueous workups) and the polystyrene-supported version (no swelling and ability to carry out reaction in polar solvents).

For the synthesis of chemotype A, the best yields were obtained when 0.4 equivalents of catalyst were employed with the use of ethanol as solvent. The same reaction conditions gave the best results also for the preparation of compounds belonging to chemotype B, whilst the synthesis of the library for chemotype C worked better when dioxane was employed instead of ethanol at a higher temperature (110°C). The

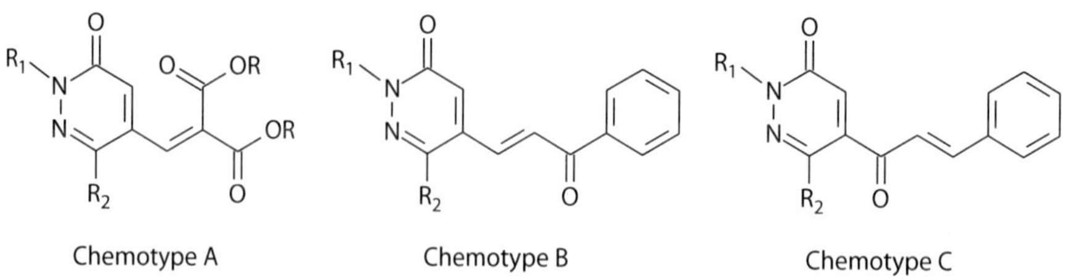

FIGURE 1.3 Pyridazin-3-one derivatives with potent platelet aggregation inhibition.

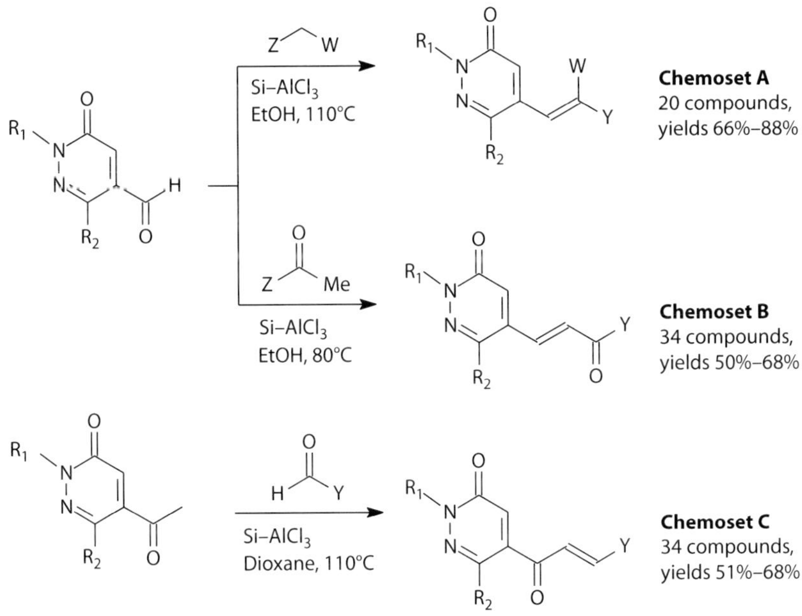

SCHEME 1.11

recyclability of silica-supported aluminum chloride was also investigated. Once the reaction had finished, the heterogeneous catalyst was separated from the reaction mixture by filtration, submitted to a washing protocol, and then dried under vacuum for 12 h. The recovered catalyst was routinely used for at least three consecutive runs without significant loss of activity. Preliminary biological data from thrombin-induced aggregation studies on human platelets performed on representative compounds were also discussed in the paper, demonstrating that antiplatelet activity can be improved through structural manipulation of model chemotypes.

1.3 SILICA-SUPPORTED HETEROPOLY ACIDS

It is important to note that the synthesis of many heterocycles is often carried out under acid-catalyzed reactions, so much effort has been put into the search for solid acid catalysts (Rosati et al., 2007; Dhakshinamoorthy et al., 2011; Sreekumar and Padmakumar, 1998; Kandarpa et al., 2011; Krishnakumar and Swaminathan, 2011; Huang et al., 2008). From this point of view, catalysis by heteropoly acids (HPAs) and related compounds is a field of increasing importance worldwide. To avoid the use of conventional acid catalysts (sulfuric, phosphoric, and hydrofluoric acids and boron trifluoride) and the related environmental pollution and corrosion problems (Vázquez et al., 2002), insoluble solid acid catalysts such as HPAs can be used. HPAs are mixed oxides composed of a central ion or "heteroatom," generally P, As, Si or Ge, bonded to an appropriate number of oxygen atoms and surrounded by a shell of octahedral MO_6 units. HPAs with Keggin structure and related polyoxometalates are quite common and are represented by the formula $H_{8-x}[XM_{12}O_{40}]$, where X is the

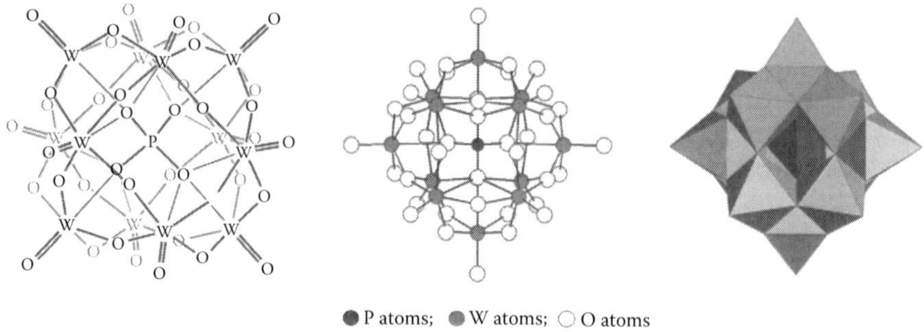

FIGURE 1.4 Some representations of Keggin structure of $[PW_{12}O_{40}]^{3-}$.

heteroatom (e.g., P^{5+} or Si^{4+}), x is its oxidation state, and M is the addendum atom (usually W^{6+} or Mo^{6+}) (Figure 1.4). Despite their rather daunting formulae, they are easy to prepare, simply by mixing phosphate and tungstate in the required amounts at the appropriate pH (Pope, 1983).

This kind of compound shows very strong acidity (Kozhevnikov, 1998). For example, $H_3PW_{12}O_{40}$ has a higher acid strength (Ho = −13.16) than CF_3SO_3H or H_2SO_4. For this reason, HPAs, such as tungstophosphoric acid ($H_3PW_{12}O_{40}$), tungstosilicic acid ($H_4SiW_{12}O_{40}$), and molybdatophosphoric acid ($H_3PMo_{12}O_{40}$) (Moffat, 2001), have attracted much interest as catalysts in both academic and industrial applications (Kozhevnikov, 1998; Okuhara et al., 1996; Da Silva Rocha et al., 2007; Misono, 2001; Da Silva Rocha et al., 2005). They are insoluble in nonpolar solvents. On the other hand, the main drawback of HPAs is their solubility in polar solvents and reactants, such as water and ethanol, which severely limits their application as recyclable solid acid catalysts in the liquid phase. However, they exhibit high thermal stability in the solid state and have been applied in a variety of vapor-phase processes for the production of petrochemicals, for example, olefin hydration and reaction of acetic acid with ethylene (Mizuno and Misono, 1998). From a synthetic point of view, a variety of useful transformations (Izumi et al., 1992) such as dehydration (Turek et al., 2005; Dias et al., 2006), cyclization (Bennardi et al., 2007), esterification (Pizzio et al., 2003), amine oxidation, olefin epoxidation (Ding et al., 2005), oxidation of alcohols (Firouzabadi et al., 2003), Friedel–Crafts reactions (Kaur et al., 2002; Lan et al., 2007), Mannich reactions (Azizi et al., 2006; Rasalkar et al., 2007; Wang et al., 2007), cyanosilylation (Firouzabadi et al., 2005), and ring-opening of epoxides (Azizi and Saidi, 2007) have been developed using HPAs as catalysts, offering a strong option for efficient and cleaner processing compared to conventional mineral acids. HPAs have the advantages of being noncorrosive, environmentally friendly because of their reusability, economically feasible solid acid catalysts compared to conventional homogeneous acids, highly flexible in modification of acid strength, easy to handle, nontoxic, and experimentally simple (Kozhevnikov, 2002; Romanelli et al., 2004). In order to overcome drawbacks such as solubility in polar media and low specific surface (1–5 m^2 g^{-1}), HPAs have been immobilized by occlusion in different materials with high surface areas (Kozhevnikov et al., 1995, 1996) using the sol-gel technique (Izumi et al., 1995, 1997; Izumi, 1997). Various supports such as silica

(Misono, 1987; Rocchiccioli-Deltcheff et al., 1990; Vàzquez et al., 1999), alumina (Shikata et al., 1997), activated carbon (Schwegler et al., 1992; Dupont et al., 1995), and MCM-41 (Dupont and Lefebvre, 1996) have been used for supporting HPAs. Some of the goals of using supported HPAs are surface area enhancement, higher dispersion of acidic protons, heterogenization, and acid strength control. These supported catalysts have been widely studied and found useful in many reactions such as the synthesis of 2,4-dihydropyrimidones (Rafiee and Shahbazi, 2006), metanethole (Torviso et al., 2006), and α-aminonitriles (Rafiee et al., 2007) and esterification (Zhang et al., 2005; Izumi et al., 1999).

Supported HPAs were found to be useful also in the synthesis of several heterocyclic compounds. A representation of synthetic applications of this class of catalysts is reported.

Chromone skeletons (4*H*-benzopyran-4-one) (flavones and chromones) are widely represented in naturally occurring compounds such as those of the flavonoid family (Martens and Mithöfer, 2005). These compounds have been reported to exhibit multiple biological properties, for example, antibacterial, antifungal (Alam, 2004; Göker et al., 2005), anticancer, antioxidant (Chu et al., 2004), anti-HIV (Wu et al., 2003), inhibition of histamine release (Yano et al., 2005), and antifeedant (Morimoto et al., 2003; Ohmura et al., 2000). Several methods for the synthesis of these compounds have been reported, for example, from chalcones, via an intramolecular Wittig strategy (Barton and Ollis, 1979; Ganguly et al., 2005), and from cyclocondensation of 1,3-diaryl 1,3-diketone obtained by rearrangement of *o*-hydroxyacetophenone aryl ester (Varma et al., 1998). These methodologies are carried out under (1) conventional catalysis, such as sulfuric acid in glacial acetic acid (Wheeler, 1952); and cation-exchange resins in isopropanol (Hoshino and Takeno, 1987), glacial acetic acid–anhydrous sodium acetate, or aqueous potassium carbonate (Saxena et al., 1985); and (2) more environmentally friendly conditions such as the use of $CuCl_2$ in ethanol (Kabalka and Mereddy, 2005) and ionic liquid under microwave irradiation (Sarda et al., 2006). Bennardi and colleagues described the use of the HPA $H_6P_2W_{18}O_{62}{\cdot}24H_2O$ with a Wells–Dawson structure as a reusable heterogeneous catalyst in the synthesis of substituted flavones and chromones (**18**) by cyclization of 1-(2-hydroxyphenyl)-3-aryl-1,3-propanediones (**17**) (Scheme 1.12) (Bennardi et al., 2008).

The reactions were carried out in the presence of bulk Wells–Dawson acid (WD, 1 mmol%, solvent-free) or supported Wells–Dawson acid containing 40% by weight of the acid on silica (WD_{40}/SiO_2, 1 mmol%, toluene) at the optimized temperature of 110°C (Table 1.6).

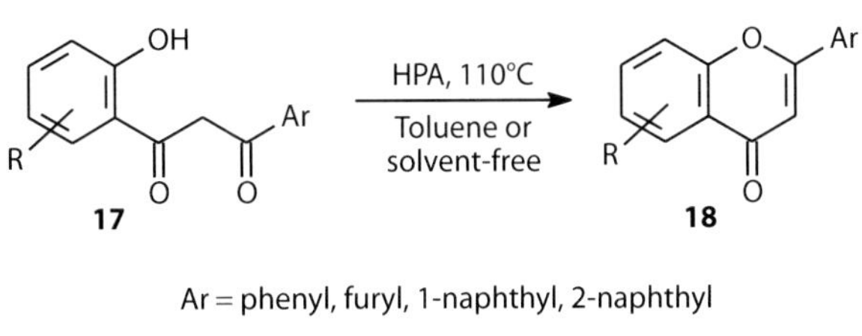

SCHEME 1.12

TABLE 1.6
Preparation of Flavones and Chromones Using 1 mmol% of Bulk and Silica-Supported $H_6P_2W_{18}O_{62}·24H_2O$

Entry	Flavone/Chromone (18) R	Ar	Time (h) (WD)[a]	Yield (%) (WD)[a]	Time (h) (WD_{40}/SiO_2)[b]	Yield (%) (WD_{40}/SiO_2)[b]
1	H	Ph	0.5	87	4.5	91
2	6-Cl	Ph	0.7	87	4.5	86
3	7-Cl	Ph	0.5	86	4.5	85
4	6-Me	Ph	0.5	86	5	86
5	7-Me	Ph	0.5	86	5	87
6	7-MeO	Ph	0.5	82	4	83
7	6-Br	Ph	0.6	85	4.5	84
8	H	2-Naphthyl	0.7	88	5	87
9	7-Me	1-Naphthyl	0.7	88	5	88
10	7-Cl	1-Naphthyl	0.8	85	5	87
11	7-Cl	2-Furyl	0.6	87	4.5	87

[a] Solvent-free.
[b] In toluene, 110°C.

In all the cases, the desired products were obtained with high selectivity, almost free of secondary products, and the unchanged starting materials were recovered nearly quantitatively. The yields of flavones were similar to those of chromones, however, the experiments performed in solvent-free conditions showed a substantial reduction of the reaction times. No stereoelectronic effects owing to the substituent were observed on the yield for any of the catalysts (Table 1.6). Both recycled catalysts showed almost constant activity when used in two consecutive batches after the first one.

2-Oxazolines, 2-imidazolines, and 2-thiazolines are very important moieties due to their extensive applications in chemistry, biochemistry, and pharmacology (Fan et al., 2007; Saxena et al., 2001). These heterocycles are found in the structures of many biologically active natural products (Roy et al., 1999; Greenhill and Lue, 1993). Imidazoline derivatives are of great interest and importance because of their pharmaceutical and synthetic material applications. They exhibit significant biological and pharmacological activities including antihypertensive (Bousquet and Feldman, 1999), antihyperglycemic (Le Bihan et al., 1999), antidepressive (Vizi, 1986), antihypercholesterolemic (Li et al., 1996), and anti-inflammatory (Ueno et al., 1995) activities. They are also known as important intermediates in organic transformations (Yang et al., 1997; Puntener et al., 2000). Optically active mono- and bis-derivatives of these heterocycles have been widely used as both auxiliaries and ligands in asymmetric synthesis (Desimoni et al., 2006; Matsumoto et al., 2005; Hao et al., 2006; Abrunhosa et al., 2001; Ghosh et al., 1998; Lee et al., 2004). Although the synthesis of 2-substituted oxazolines, imidazolines, and thiazolines can be achieved by several methods such as from carboxylic acids (Vorbruggen and Krolikiewicz, 1993;

Cwik et al., 2002), esters (Neef et al., 1981), nitriles (Mohammadpoor-Baltork and Abdollahi-Alibeik, 2003; Mirkhani et al., 2006, 2007; Pathan et al., 2006; Clarke and Wood, 1996; Jnaneshwara et al., 1998; Mohammadpoor-Baltork et al., 2005), hydroximoylchlorides (Salgado-Zamora et al., 1998), hydroxy amides (Boland et al., 2002), aziridines (Ghorai et al., 2006), *N-tert*-butoxycarbonyl-protected α-amino acids (Ramalingam et al., 2007), and aldehydes (Fujioka et al., 2007), they suffer from some disadvantages such as long reaction times, low yields, difficult preparation of starting materials, and tedious workup. Mohammadpoor-Baltork et al. (2008) reported the use of supported 12-tungstophosphoric acid (TPA) as a highly efficient catalyst for the clean, suitable, and chemoselective synthesis of 2-substituted oxazolines (**19**), imidazolines (**20**), and thiazolines (**21**) (Scheme 1.13).

The TPA was supported on inorganic (high-surface-area silica gel and active carbon) and organic (poly-(4-styrylmethyl)pyridinium chloride: PMP) materials by the method of incipient wetness (Misono et al., 1982; Chang, 1995; Fournier et al., 1991; Yadav and Kirthivasan, 1995). The content of TPA on all supports, which were calculated from the W content in the supported TPA catalysts, was determined by neutron activation analysis (NAA). The amounts of W loading on the supported catalysts were 29.75 ± 0.9, 28.65 ± 0.9, and 29.53 ± 0.8 for SiO_2–TPA, C–TPA, and PMP–TPA, respectively. The specific surface area for SiO_2–TPA was 180 m^2/g and in the case of C–TPA it was 720 m^2/g. The ability of the TPA supported on inorganic (high-surface-area silica gel and active carbon) and organic (PMP) supports was investigated with regard to the synthesis of 2-substituted oxazolines starting from benzonitrile and 2-aminoethanol in the presence of 0.15–0.25 mol% of supported catalysts. The reactions were performed in different solvents such as chloroform, dichloromethane, 1,2-dichloroethane, and methanol, and in the absence of solvent at room temperature and under reflux conditions. The best results were obtained with molar ratios of benzonitrile/2-aminoethanol/supported catalysts of 1:4:0.002 at 100°C under solvent-free conditions (Table 1.7, entries 1–9).

The results showed that the supports did not present significant catalytic activity (Table 1.7, entry 11–13). On the other hand, the catalytic activity of TPA supported on high-surface-area SiO_2, active carbon, and PMP has been increased with respect to the use of TPA alone (Table 1.7, entry 10). The optimized reaction conditions were used for the synthesis of several 2-oxazoline derivatives (**19**) starting from various aromatic (or heteroaromatic) nitriles and 2-aminoethanol (or 2-amino-2-methylpropanol). As reported in Table 1.8, in all entries, high to excellent yields were obtained. Notably, the authors reported that both mono- and bis-oxazolines can be obtained by this

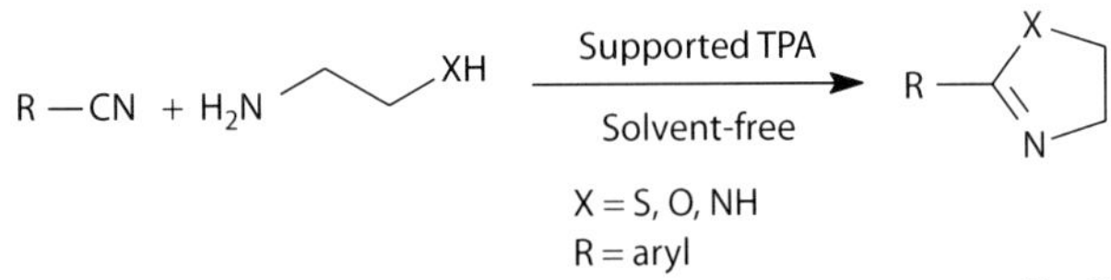

X = O **19**
X = NH **20**
X = S **21**

SCHEME 1.13

TABLE 1.7
Investigation of Catalyst Effects in the Synthesis of 2-Phenyloxazoline (19) from Benzonitrile (1 mmol) and 2-Aminoethanol (4 mmol) under Solvent-Free Conditions at 100°C

Entry	Catalyst	TPA (mol%)	Time (h)	Yield (%)
1	SiO_2–TPA	0.15	3	78
2	SiO_2–TPA	0.2	3	90
3	SiO_2–TPA	0.25	3	90
4	C–TPA	0.15	3	80
5	C–TPA	0.2	3	92
6	C–TPA	0.25	3	93
7	PMP–TPA	0.15	2	75
8	PMP–TPA	0.2	2	90
9	PMP–TPA	0.25	2	92
10	TPA	0.2	3	35
11	SiO_2		3	20
12	Active carbon		3	12
13	PMP		3	10

catalytic system and the selectivity depends on the reaction time. The bis-oxazoline has been synthesized from dinitrile simply by increasing the reaction time to 6 h (Table 1.8, entry 9). The preparation of mono-oxazolines from dinitrile compounds is of great interest because the remaining nitrile group can be converted to other functional groups (Jnaneshwara et al., 1999).

Mohammadpoor-Baltork and colleagues also investigated the scope and generality of this catalytic system by the synthesis of 2-imidazolines (**20**) (Table 1.9, entries 1–6) and 2-thiazolines (**21**) (Table 1.9, entries 7–12). The reactions were carried out starting from different types of arylnitriles and ethylenediamine or 2-aminoethanethiol, respectively, under the same reaction conditions used in the synthesis of 2-oxazolines, obtaining the corresponding derivatives in good to excellent yields.

It is important to note that the synthesis of bis-imidazolines from the reaction of dinitriles with ethylenediamine catalyzed by supported TPAs failed (Table 1.9, entry 6). On the other hand, the treatment of dinitriles with 2-aminoethanethiol produced bis-thiazolines exclusively (Table 1.9, entry 12).

Synthesis of 2-alkyloxazolines, 2-alkylimidazolines, and 2-alkylthiazolines with this catalytic system also failed. Therefore, this process was chemoselective for the conversion of arylnitriles to their corresponding 2-aryloxazolines, 2-arylimidazolines, and 2-arylthiazolines. The catalysts have been recovered by simple filtration after dilution of the reaction mixture with EtOH and reused after drying at 100°C and weighing. The reusability of the catalysts was examined in the reaction of benzonitrile with 2-aminoethanol. The results showed that the catalyst could be reused five consecutive times. The determination of the amount of catalyst loading after each run by NAA showed a catalyst leaching effect.

TABLE 1.8
Representative Examples of Synthesis of 2-Oxazolines and Bis-Oxazolines Catalyzed by Supported TPA under Solvent-Free Conditions at 100°C

			SiO_2–TPA		C–TPA		PMP–TPA	
Entry	Nitrile	Oxazoline	Time (min)	Yield (%)	Time (h)	Yield (%)	Time (h)	Yield (%)
1	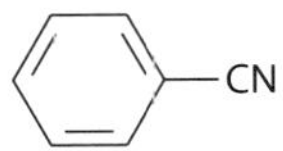	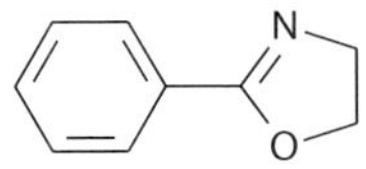	3	90	3	92	2	90
2	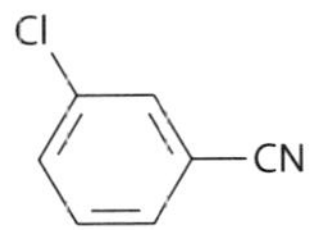	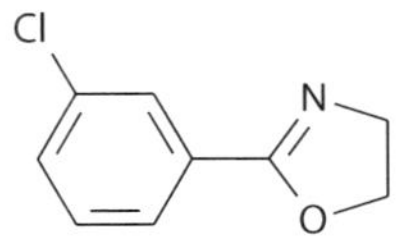	3	93	1.5	95	1.5	93
3	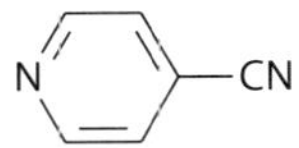	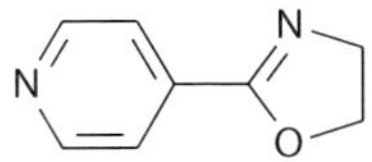	0.5	94	0.75	95	0.75	96
4	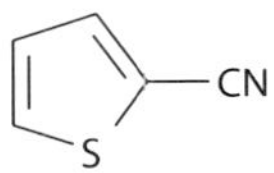	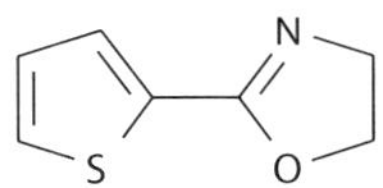	3	88	2.5	90	3	94
5		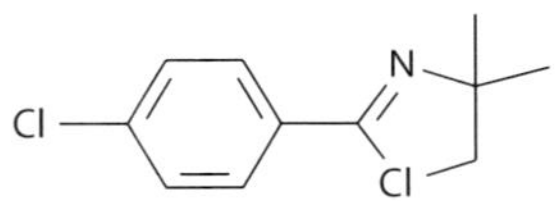	5.5	70	5.5	75	5	76

(continued)

TABLE 1.8 (Continued)
Representative Examples of Synthesis of 2-Oxazolines and Bis-Oxazolines Catalyzed by Supported TPA under Solvent-Free Conditions at 100°C

			SiO_2–TPA		C–TPA		PMP–TPA	
Entry	Nitrile	Oxazoline	Time (min)	Yield (%)	Time (h)	Yield (%)	Time (h)	Yield (%)
6	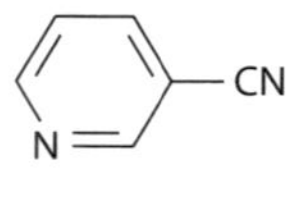	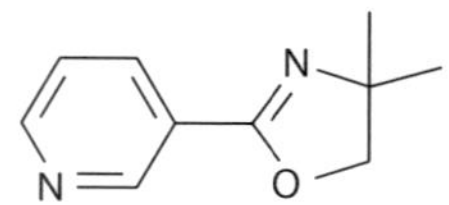	4	65	5	67	5	75
7	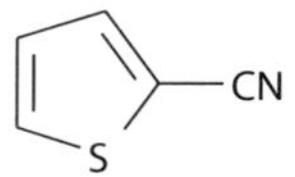	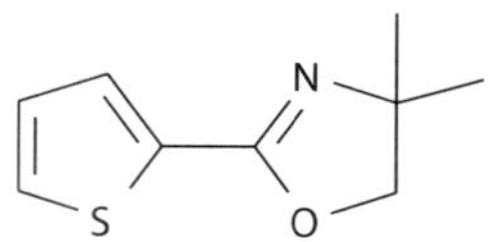	4.5	75	4.5	75	4.5	77
8	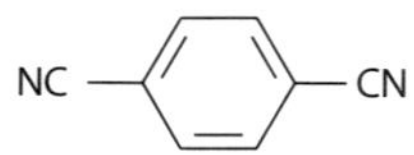	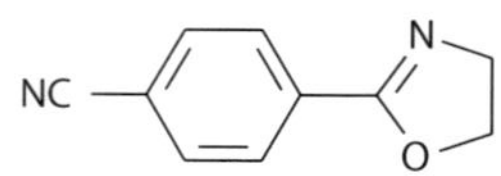	0.5	90	0.3	90	0.3	93
9	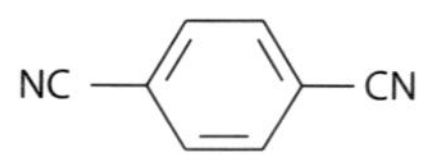	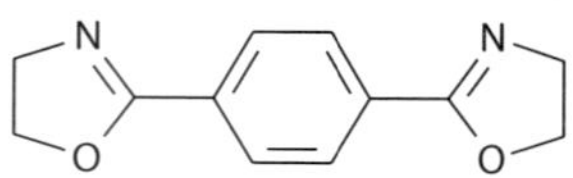	6	75	6	75	6	77

TABLE 1.9
Representative Examples of Synthesis of 2-Imidazolines and 2-Thiazolines Catalyzed by Supported TPA under Solvent-Free Conditions at 110°C

			SiO_2–TPA		C–TPA		PMP–TPA	
Entry	Nitrile	Imidazoline/Thiazoline	Time (min)	Yield (%)	Time (h)	Yield (%)	Time (h)	Yield (%)
1	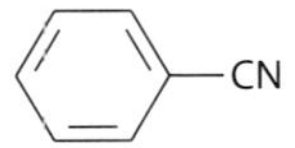	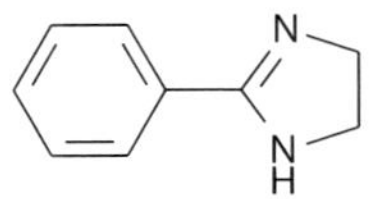	4	80	4	88	3.5	80
2	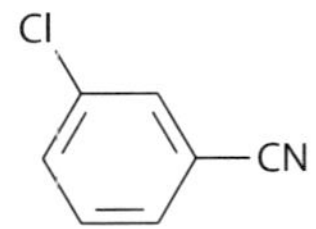	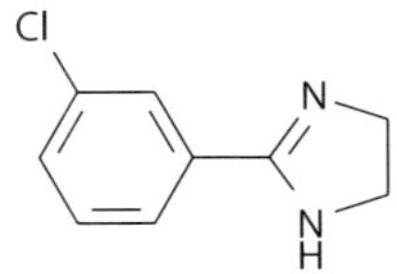	3	90	2	95	2	95
3	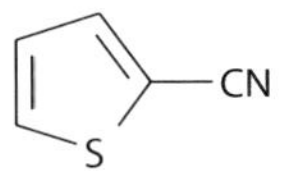	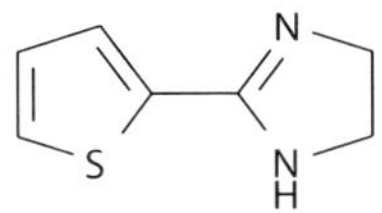	3	90	3	90	2	90
4	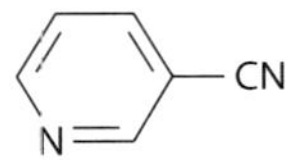	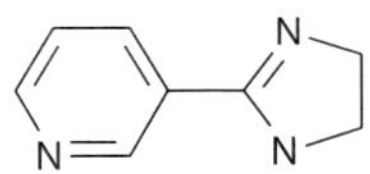	3	90	4	85	3	91
5	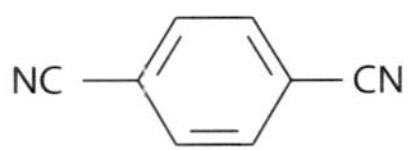	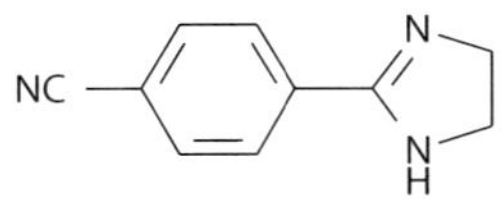	0.8	90	0.8	90	0.3	92

(continued)

TABLE 1.9 (Continued)
Representative Examples of Synthesis of 2-Imidazolines and 2-Thiazolines Catalyzed by Supported TPA under Solvent-Free Conditions at 110°C

Entry	Nitrile	Imidazoline/Thiazoline	SiO_2–TPA Time (min)	SiO_2–TPA Yield (%)	C–TPA Time (h)	C–TPA Yield (%)	PMP–TPA Time (h)	PMP–TPA Yield (%)
6		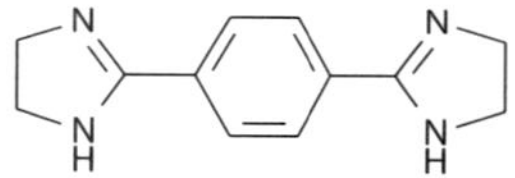	12	7	12	4	12	7
7	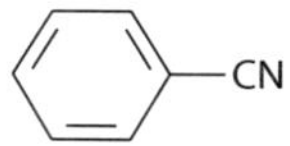	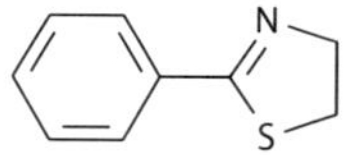	5	95	5	95	5	95
8	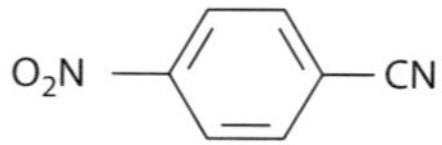	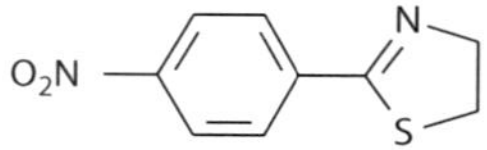	5	80	8	85	5	96
9	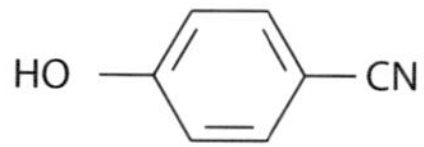	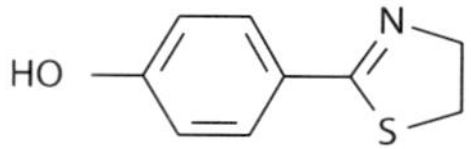	5	60	9	65	5	80
10	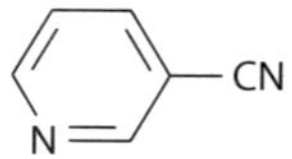	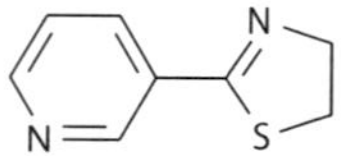	8	96	7	93	5	97
11	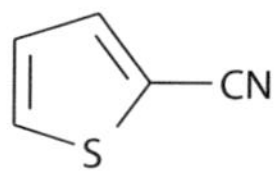	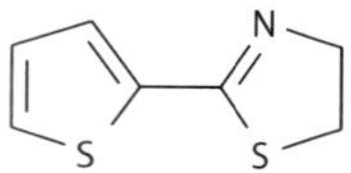	8	95	5	90	6	90
12		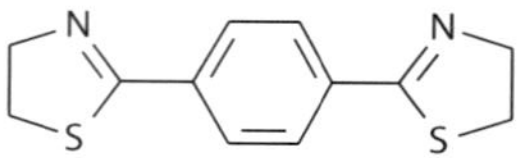	12	92	12	87	12	93

Another approach to the synthesis of 2-imidazolines (**22**) mediated by silica-supported tungstosilicic acid, starting from 1,2-diamines and nitrile derivatives (Scheme 1.14), was reported later by Nasr-Esfahani et al. (2010). In contrast to the previously described method, this approach has proved to be efficient also for the synthesis of *bis*-imidazolines (**23**).

The supported tungstosilicic acid catalysts were prepared by the method of incipient wetness. Preliminary experiments established the best $H_4SiW_{12}O_{40}$ loading on SiO_2 (20 wt.%) and the best mole fraction of catalyst (0.4 mol% of tungstosilicic acid) to use in the test reaction of benzonitrile and 1,2-ethylenediamine. The results of some representative examples of reactions between ethylenediamine or 1,2-propanediamine and various nitriles in the presence of $H_4SiW_{12}O_{40}$–SiO_2 under reflux conditions are shown in Table 1.10. As shown in all entries, the reactions of various aromatic nitriles gave the corresponding 2-imidazolines (**22**) and bis-imidazolines (**23**) in good to excellent yields. It is noteworthy that by increasing the reaction times to 130–135 min, bis-imidazolines can also be obtained from dinitriles using this catalytic system (Table 1.10, entry 6). This method was also applied to large-scale synthesis of imidazolines and bis-imidazolines (100 mmol) and the results were comparable to those of small-scale experiments.

A poor catalytic effect was observed for the use of silica gel as catalyst only in the test reaction of benzonitrile and 1,2-ethylenediamine (12 h, 25% yield). This result show that supported and unsupported $H_4SiW_{12}O_{40}$ accelerates the rate of reaction versus the action of SiO_2 alone. The improvement of the catalytic activity by supporting $H_4SiW_{12}O_{40}$ on SiO_2 was also proved. A comparison of the supported and unsupported $H_4SiW_{12}O_{40}$ catalytic activity is reported in Table 1.10 (entries 1–3). The reusability of the catalyst is important for large-scale operation and from an industrial point of view. Therefore, the reusability of the catalysts was examined in the reaction of benzonitrile with ethylenediamine. Since the reaction medium is heterogeneous, the catalysts can be recovered by simple filtration after dilution of the reaction mixture with EtOH. The recovered catalyst was dried at 100°C and reused four consecutive times in the reaction of nitrile and 1,2-ethylenediamine with a minimal loss of catalytic activity (60 min/90%, 60 min/90%, 65 min/87%, 65 min/85%).

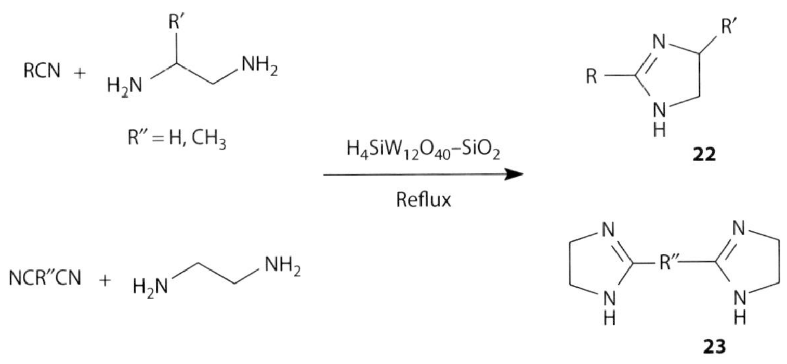

SCHEME 1.14

TABLE 1.10
Representative Examples of Synthesis of Imidazolines (22) and Bis-Imidazolines (23) in the Presence of $H_4SiW_{12}O_{40}$–SiO_2 and Bulk $H_4SiW_{12}O_{40}$

				$H_4SiW_{12}O_{40}$–SiO_2		$H_4SiW_{12}O_{40}$	
Entry	Nitrile	1,2-Diamine	Product	Time (min)	Yield (%)	Time (h)	Yield (%)
1	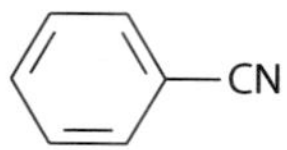	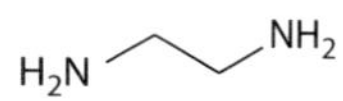	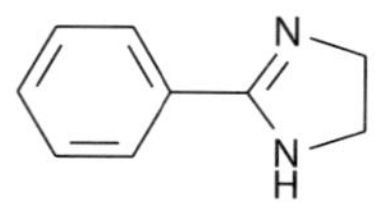	45	90	8	80
2	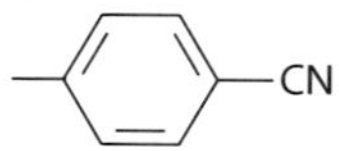	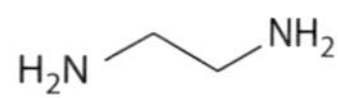	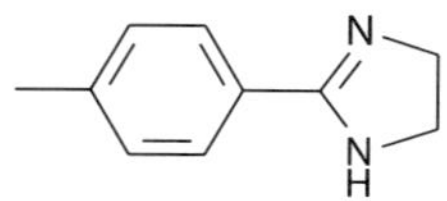	105	95	15	88
3	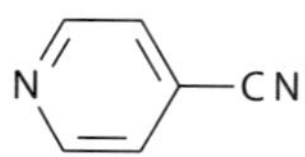	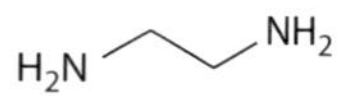	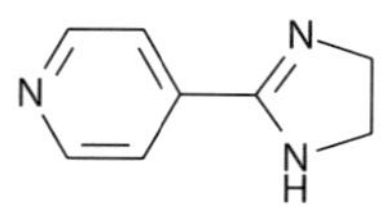	135	78	18	70
4	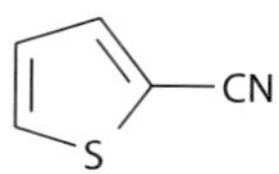	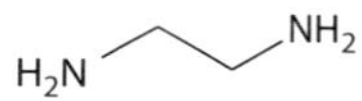	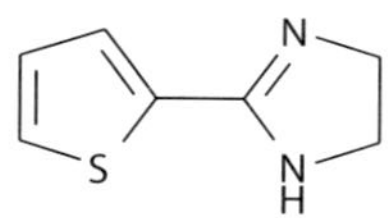	135	85	—	—

5		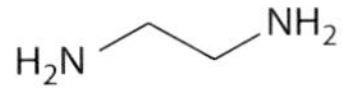	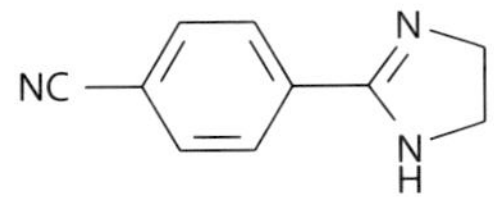	60	90	—	—
6		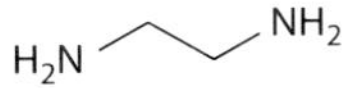	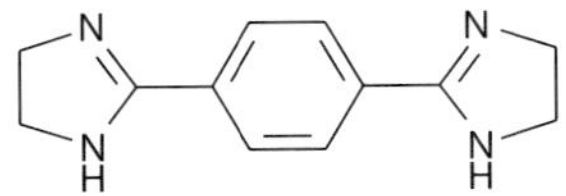	130	87	—	—
7	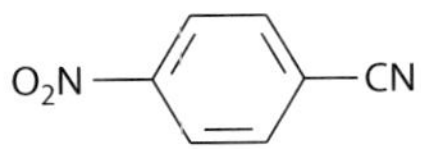	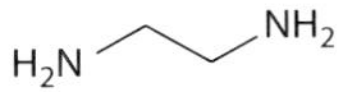	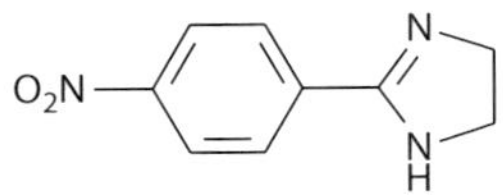	220	60	—	—
8	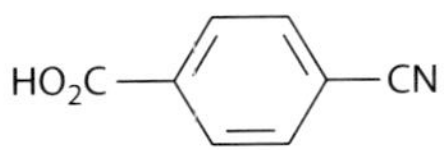	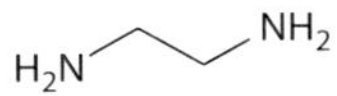	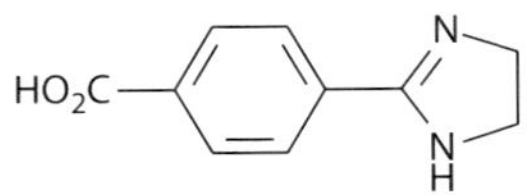	120	80	—	—
9		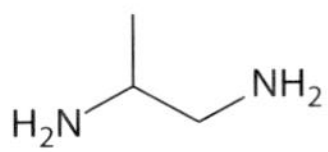	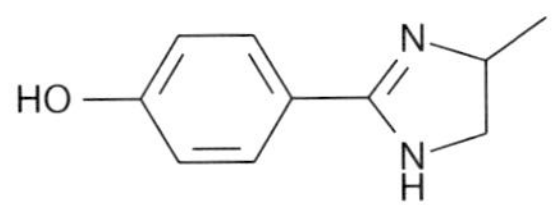	40	82	—	—
10	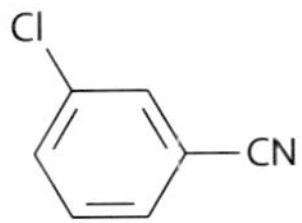	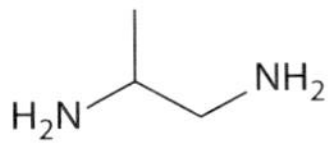	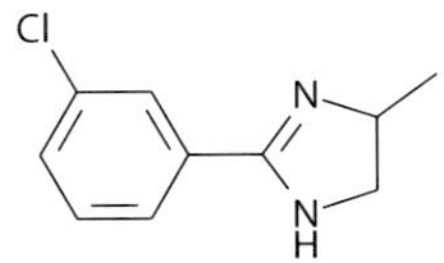	90	75	—	—

Among a large variety of *N*-containing heterocyclic compounds, heterocycles containing hydrazine moieties as "fusion sites" have received considerable attention because of their pharmacological properties (Turk et al., 2001; Clark et al., 2004).

Fused phthalazines are heterocycles containing hydrazine moieties as "fusion sites" that exhibit significant biological and pharmacological activities such as inhibition of p38 MAP kinase (Mavel et al., 2002) and selective binding of GABA receptor (Carling et al., 2004; Street et al., 2004). They are used as antianxiety drugs (Imamura et al., 2003), antitumor agents (Kim et al., 2004), and high-affinity ligands to the a2d-1 subunit of the calcium channel (Lebsack et al., 2004). Fused phthalazine derivatives also possess some biological activities such as anticonvulsant (Grasso et al., 2000), cardiotonic (Nomoto et al., 1990), and vasorelaxant (Watanabe et al., 1998) activities. Although many reports have been published about solution-phase synthesis of fused phthalazine derivatives (Potts and Lovelette, 1969), reports on the synthesis of phthalazines fused with indazole are limited (Sayyafi et al, 2008; Shaterian et al., 2009; Nagarapu et al., 2009). To achieve a more environmentally friendly approach to the synthesis of fused phthalazine derivatives, Sabitha et al. (2010) proposed the use of phosphomolybdic acid ($H_3PMo_{12}O_{40}$ = PMA), supported on silica (PMA–SiO_2) as catalyst in a one-pot three-component condensation of 1,3-diones (**12** and **24**), phthalhydrazide **11**, and aldehydes (Schemes 1.15 and 1.16). In a typical procedure, 1,3-dione derivative (1 mmol), aldehyde (1 mmol), and **11** (1 mmol) were reacted with 1.0 g of PMA–SiO_2 (0.05 mmol). All reactions were carried out under solvent-free conditions at 80°C, affording 2,3,4,13-tetrahydro-1*H*-indazolo[1,2-*b*]phthalazine-1,6,11-trione derivatives (**25**) with good yields and short reaction times (Table 1.11). No side products or decomposition of the products were observed.

Under similar conditions, acetylacetone (**26**) also reacted efficiently with **11** and aldehydes to give the corresponding products **27a–c** in 80%–84% yield within

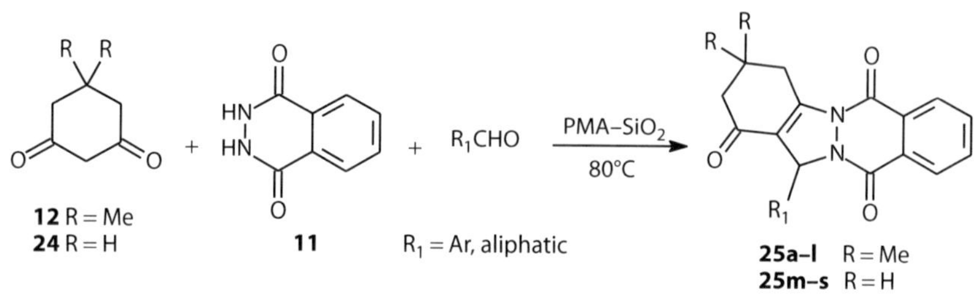

SCHEME 1.15

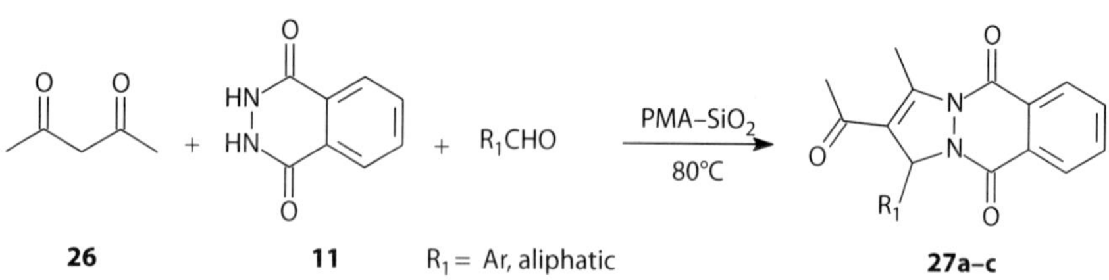

SCHEME 1.16

TABLE 1.11
PMA–SiO_2-Catalyzed Synthesis of Indazolo[1,2-*b*] phthalazine-Triones (25) and Pyrazolo[1,2-*b*] phthalazin-Triones (27)

Entry	Aldehyde	Product	Time (min)	Yield (%)
1	Benzaldehyde	**25a**	30	85
2	4-Chlorobenzaldehyde	**25b**	30	90
3	4-Fluorobenzaldehyde	**25c**	30	85
4	2-Bromobenzaldehyde	**25d**	40	85
5	3-Chlorobenzaldehyde	**25e**	35	86
6	4-Methoxybenzaldehyde	**25f**	40	85
7	4-Nitrobenzaldehyde	**25g**	30	90
8	4-Isopropylbenzaldehyde	**25h**	40	87
9	4-*tert*-Butylylbenzaldehyde	**25i**	40	87
10	Butanal	**25j**	20	92
11	Hexanal	**25k**	20	91
12	2-Methylpropanal	**25l**	25	90
13	Benzaldehyde	**25m**	30	84
14	4-Isopropylbenzaldehyde	**25n**	35	82
15	4-Methoxybenzaldehyde	**25o**	40	81
16	4-Nitrobenzaldehyde	**25p**	40	80
17	Butanal	**25q**	25	86
18	Pentanal	**25r**	25	84
19	2-Methylpropanal	**25s**	30	84
20	Benzaldehyde	**27a**	40	82
21	4-Methoxybenzaldehyde	**27b**	45	80
22	Butanal	**27c**	40	84

40–45 min (Scheme 1.16, Table 1.11, entries 20–22). These results clearly show the generality of the reaction using PMA–SiO_2 under solvent-free conditions. However, heteroaromatic aldehydes such as furan- and thiophene carbaldehydes did not undergo this condensation reaction; instead, the starting materials were recovered.

It is assumed that the reaction proceeded by Knovenagel condensation of 1,3-dione and aldehyde, followed by Michael addition of the phthalhydrazide (**11**), and subsequent cyclization to afford the product **25** or **27**.

Also, in this synthesis the catalyst has been recovered by filtration and reused. Using benzaldehyde as a model reaction, the recovered catalyst has been recycled up to three times with no decrease in the conversion.

1.4 SILICA-SUPPORTED BASES

An interesting basic solid catalyst was obtained for the first time by Sharma, immobilizing zirconium complex onto the functionalized support and thereby combining the properties such as catalyst selectivity and activity with the ease of separation and catalyst reuse (Sharma and Sharma, 2011). Due to its high surface area, excellent stability (chemical and thermal), good accessibility, and ease of functionalization of the surface groups, silica gel has been chosen as the support material (Price et al., 2000; Sharma et al., 2003). The catalyst zirconium(IV)-modified silica gel (Zr-CAP-SG) has been investigated for its catalytic activity in significant organic heterocycle synthesis such as Pechmann condensation of phenols with β-ketoesters to yield coumarins, condensation of various 1,2-diamines with 1,2-diketone to give quinoxalines, and one-pot multicomponent synthesis of 2,4,5-trisubstituted imidazoles. The biological, pharmaceutical, medicinal, and therapeutic properties of coumarins (Kennedy and Tharnes, 1997), quinoxaline derivatives (Hui et al., 2006), and substituted imidazoles (Kühl, 2007) have been well established in the literature.

The grafted catalyst was prepared by following the slightly modified method of Prado and Airoldi (2000) The functionalized aminopropyl silica gel (APSG), prepared according to a reported procedure, and ligand **28** were refluxed in a diglyme solution for 3 h at 140°C under nitrogen (Scheme 1.17). The obtained solid material **29** was filtered off and washed thoroughly with acetonitrile and acetone and then was dried overnight at 90°C. Finally, the grafted silica gel **29** was stirred with zirconium(IV) oxychloride in acetonitrile for 2 h at 80°C. The solid material **30** (Zr-CAP-SG) was filtered off and washed thoroughly with acetone, then dried in a vacuum oven overnight at 90°C. The decrease in the surface area of the catalyst (Karandikar et al., 2004) was found to be 40%, which is indicative of the grafting of ligand and hence zirconium oxychloride onto the silica gel. The metal loading of Zr-CAP-SG was confirmed by ICP-MS and found to be 0.17 mmol/g.

The catalytic activity of the organic–inorganic hybrid material (Zr-CAP-SG) was tested in the Pechmann condensation of phenols with β-ketoesters to yield coumarins (Scheme 1.18, Table 1.12), condensation of various 1,2-diamines with benzil (Scheme 1.19a, Table 1.13) or phenanthrene-9,10-dione (Scheme 1.19b, Table 1.13)

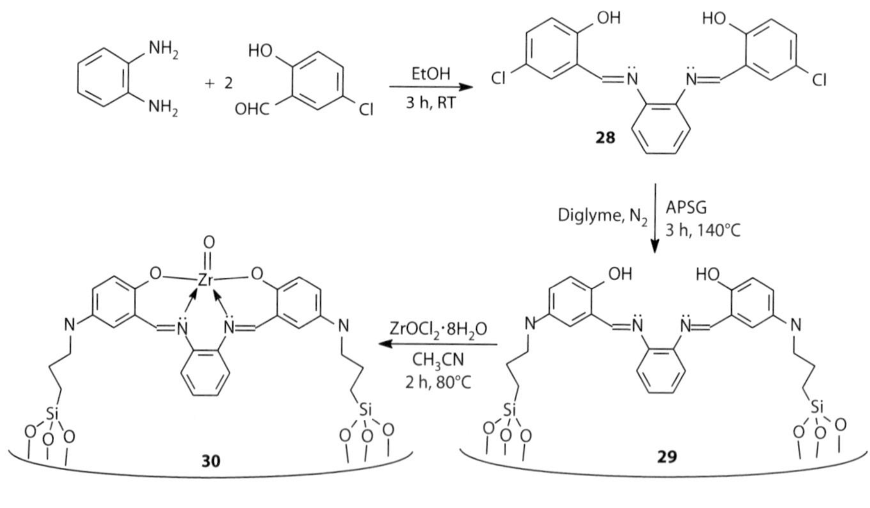

SCHEME 1.17

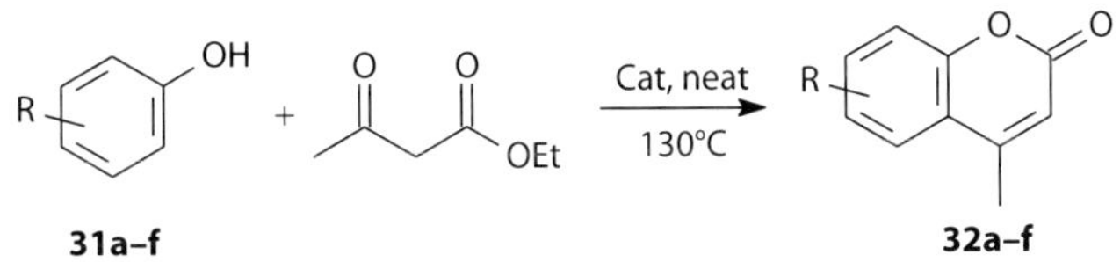

SCHEME 1.18

TABLE 1.12
Pechmann Condensation of Various Phenols with β-Keto Ester Using Zr-CAP-SG as the Catalyst

Entry	R	Phenol	Coumarin	Time (min)	Yield (%)
1	3-OH; 5-OH	**31a**	**32a**	5	>99
2	3-OH; 5-OH	**31b**	**32b**	5	98, 96, 96, 94, 93[a]
3	2-OH; 3-OH	**31c**	**32c**	5	>99
4	3-OH	**31d**	**32d**	30	94
5	3-Me	**31e**	**32e**	30	96
6	4-NO2	**31f**	**32f**	480	0

[a] Recycling experiment.

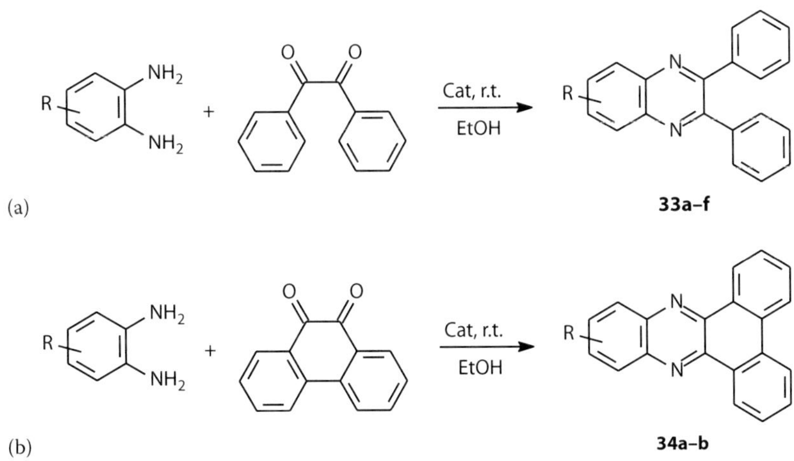

SCHEME 1.19

TABLE 1.13
Condensation of Various 1,2-Diamines with Benzil or Phenanthrene-9,10-Dione in Ethanol at Room Temperature Using Zr-CAP-SG as the Catalyst

Entry	R	Product	Time (h)	Yield (%)
1	H	**33a**	1.5	92
2	H	**33b**	1.5	90, 90, 89, 90, 88[a]
3	4-Me	**33c**	1.0	95
4	Ethylene diamine	**33d**	4.5	86
5	4-PhCO	**33e**	3.0	83
6	4-Cl	**33f**	2.5	85
7	H	**34a**	1.7	89
8	4-Me	**34b**	2.0	87

[a] Recycling experiment.

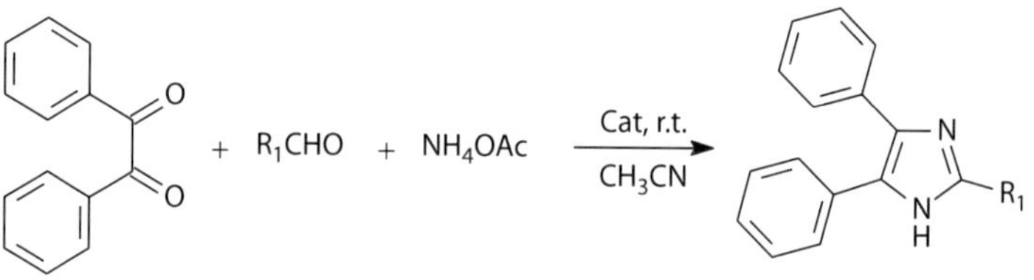

35a–h

SCHEME 1.20

to give quinoxalines, and one-pot multicomponent synthesis of 2,4,5-trisubstituted imidazoles (Scheme 1.20, Table 1.14), under different optimized reaction conditions.

For example, ethanol clearly stands out as the solvent of choice for the synthesis of quinoxalines derivatives. However, acetonitrile proved to be a better solvent for the preparation of substituted imidazoles, since the reaction time period is shortest in this case. Otherwise, the Pechmann condensation of coumarins performed better when the reaction was carried out under neat conditions and at higher temperatures. None of the reactions proceeded in the absence of the catalyst or with only silica as the catalyst.

The recyclable nature of the catalyst has been proved for each type of synthesis at microscale (Table 1.12, entry 2; Table 1.13, entry 2; and Table 1.14, entry 2).

Mukhopadhyay and Ray (2011) described the synthesis of 2-amino-5-alkylidenethiazol-4-ones from ketones, rhodanine, and secondary amines promoted by a silica–pyridine-based catalyst. Compounds containing a 2-amino-5-arylidene-1,3-thiazol-4(5H)-one moiety display a wide range of interesting pharmaceutical activities such as antiviral (Abdel-Ghani, 1999), antimicrobial (Soltero-Higgin et al., 2004), cardiotonic (Andreani et al., 1996), and anti-inflammatory (Nasr and Said, 2003) effects. The one-pot synthesis of thiazol-4-ones was optimized on a model

TABLE 1.14
One-Pot Condensation of Various Aromatic Aldehydes with Benzil and Ammonium Acetate in Acetonitrile at Room Temperature Using Zr-CAP-SG as the Catalyst

Entry	Product	R_1	Time (h)	Yield (%)
1	**35a**	C_6H_5	4.45	87
2	**35b**	C_6H_5	4.45	87, 85, 86, 84, 84[a]
3	**35c**	4-Me	8.45	71
4	**35d**	4-Cl	6.0	79
5	**35e**	4-NO_2	7.0	73
6	**35f**	4-Br	6.5	78
7	**35g**	Furan-2-yl	5.5	86
8	**35h**	Piridin-2-yl	5.0	89

[a] Recycling experiment.

reaction involving 2-acetylthiophene, rhodanine, and morpholine under a variety of experimental conditions. A screening employing several bases and the silica as cocatalyst (due to the well-documented evidence that the silanol groups present on the silica coordinate with the carbonyl oxygen increase its electrophilicity) (Ribeiro et al., 2011) has been performed. Pyridine is the base with the best catalytic activity; therefore, in order to achieve better yields and an easier workup, the authors synthesized a new silica-based substituted pyridine catalyst according to Scheme 1.21.

The novel silica-supported catalyst (40 mg, loading 0.67 mmol/g) was employed to synthesize a variety of 2-amino-5-alkylidenethiazol-4-ones from rhodanine (1 mmol), amines (1 mmol), and ketones (1 mmol) in good to excellent yields, as depicted in Scheme 1.22.

Recycling of the catalyst was monitored during the reaction of 4-chloroacetophenone with rhodanine and morpholine under optimized conditions. The recycled catalyst was used at least eight times without any further treatment.

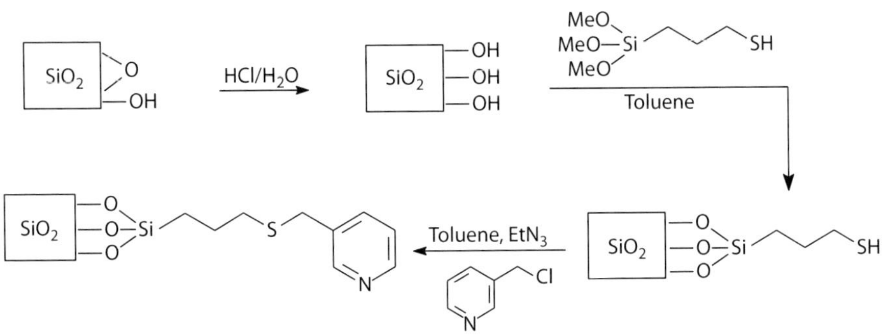

SCHEME 1.21

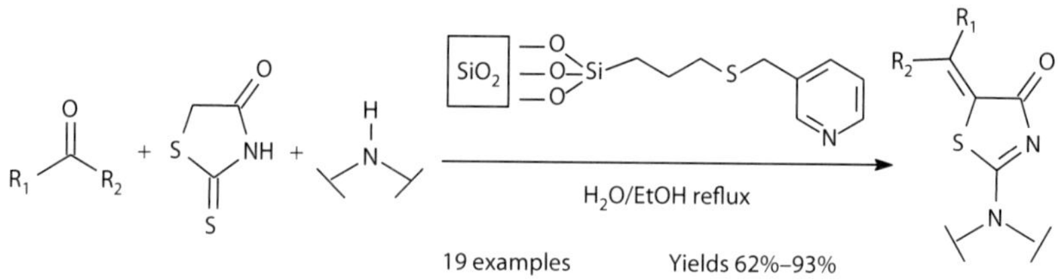

R_1, R_2 = alkyl, aryl

36 (*Z*-isomer)

SCHEME 1.22

The development of sustainable and environmentally friendly chemical synthesis processes has also affected the design of highly desirable novel organocatalysts (Ishida and Haruta, 2007; Huang et al., 2011). Although the success of organocatalysts in homogeneous reactions is hugely acknowledged, there are several drawbacks, like high catalyst loading, and the recovery and reuse of the expensive catalysts. In this context, a successful strategy for the design of a heterogeneous organocatalyst is to heterogenize the active homogeneous counterpart via an immobilization technique. The covalent attachment of the species containing the active sites to solid supports is of course the most useful strategy for the immobilization of homogeneous organocatalysts (Ray et al., 2007; Luts et al., 2007), due to the improved catalyst recycling efficiency and minimum possibility of leaching of the organic fragment.

Mesoporous materials present exceptionally high surface area, and thermal and mechanical stability; thus, mesoporous hybrid organosilica represents a suitable solid support for the design of organocatalysts (Nandi et al., 2011; Jain et al., 2011).

Mondal et al. reported the synthesis of a new robust, non-air-sensitive, metal-free triazine-functionalized mesoporous organocatalyst TFMO-1 (Figure 1.5) and successfully used this material as an excellent catalyst in the one-pot three-component condensation reactions for the syntheses of 2-amino-chromene derivatives under solvent-free conditions (Mondal et al., 2012). TFMO-1 is an easily separable heterogeneous organocatalyst devoid of metal ions and thus is highly desirable to address industrial and environmental concerns.

TFMO-1 with a 2-D hexagonal mesoporous structure has been designed via consecutive surface functionalization of SBA-15 with 3-mercaptopropyl triethoxysilane followed by a thiol–ene click reaction with 2,4,6-triallyloxy-1,3,5-triazine in the presence of azobisisobutylronitrile (AIBN) initiator.

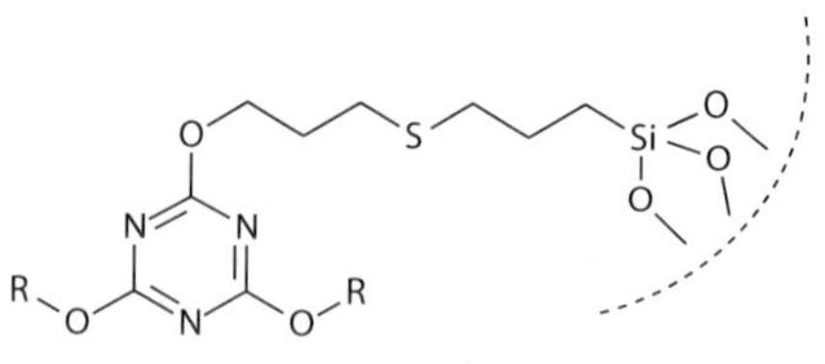

$R = -(CH_2)_3-S-(CH_2)_3-Si(O)_nSi$

FIGURE 1.5 Mesoporous organocatalyst TFMO-1.

The BET surface areas for the pure SBA-15 and TFMO-1 were 610 and 405 m^2/g, respectively. A considerable decrease in the BET surface area upon covalent grafting of the organic triazine groups through the thiol–ene click reaction at the surface of the thiol-functionalized SBA-15 material suggests that the organic groups have been anchored at the surface of the mesopores.

The organic group loading of the mesoporous organocatalyst (TFMO-1) is found to be 0.96 mmol/g, as determined by the nitrogen content (4.06%) in the elemental analysis.

In recent years, interest in substituted 2-amino-4*H*-chromene derivatives has been continuously increasing: these molecules have several important pharmacological properties, such as antimicrobial (Khafagy et al., 2002), antiviral (Martínez-Grau and Marco, 1997; Smith et al., 1998), mutagenic (Hiramoto et al., 1997), antiproliferative (Dell and Smith, 1993a,b), antitumor, and central nervous system activities (Mohr et al., 1975; Eiden and Denk, 1991).

These heterocyclic compounds can be synthesized through several routes: traditional base-catalyzed reaction of salicylaldehydes and cyano derivatives with activated methylenes (Niefang et al., 2000; Costantino et al., 2008); coumarin scaffold modifications under basic conditions (Rosati et al., 2010); multicomponent reactions starting from resorcinol derivatives, aldehydes, and cyano derivatives with activated methylenes via electrochemical synthesis (Makarem et al., 2008); HPA catalysis (Heravi et al., 2007); microwave-assisted synthesis mediated by K_2CO_3 (Kidwai et al., 2005) and basic catalysts such as piperidine (Elagamey et al., 1993; Bloxham et al., 1994), NaOH (Zhang et al., 2007), I_2/K_2CO_3 (Ren and Cai, 2008), $TiCl_4$ (Kumar et al., 2006a), $InCl_3$ (Shanthi and Perumal, 2007), and Et_3N (Shaabani et al., 2009).

A few heterogeneous catalysts, such as nanosized magnesium oxide, Mg/Al hydrotalcite, and nanostructured diphosphate ($Na_2CaP_2O_7$), have been developed for the synthesis of 2-amino-chromene (Solhy et al., 2010a,b).

The condensation reaction of aromatic aldehyde, malononitrile, and 1-naphthol mediated by the mesoporous organocatalyst TFMO-1 was carried out under solvent-free conditions (Scheme 1.23). To determine the best weight of the catalyst required for the condensation reaction, the synthesis was performed in the presence of varying amounts of catalyst (from 0.01 to 0.08 g) at 110°C. It was found that 0.04 g of TFMO-1 is the optimum amount of catalyst required to carry out the reactions.

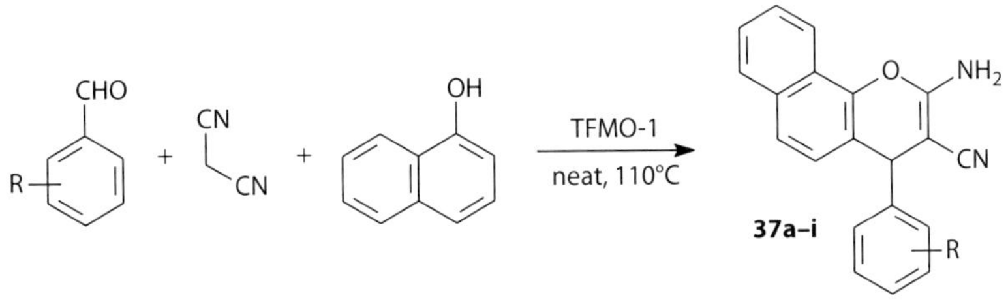

SCHEME 1.23

TABLE 1.15
Synthesis of 2-Amino-4*H*-Chromenes over TFMO-1 under Solvent-Free Conditions[a]

Entry	Aldehyde	Time (h)	Yield[b] (%)	TON[c]	Product
1	4-NO_2-benzaldehyde	4	92	24.0	**37a**
2	4-Br-benzaldehyde	5	92	24.0	**37b**
3	4-Cl-benzaldehyde	4	90	23.4	**37c**
4	4-Me-benzaldehyde	4	90	23.4	**37d**
5	4-OH-benzaldehyde	6	88	23.0	**37e**
6	Thiophene-2-carbaldehyde	5	86	22.3	**37f**
7	3-Br-benzaldehyde	5	86	22.3	**37g**
8	2-NO_2-benzaldehyde	6	88	23.0	**37h**
9	Cinnamaldehyde	7	87	22.6	**37i**

[a] Reaction conditions: aromatic aldehyde (1 mmol), 1-naphthol (1 mmol), malononitrile (1 mmol), solvent-free conditions, reaction temperature 110°C, 40 mg catalyst TFMO-1.

[b] Isolated yield of the pure product.

[c] Turn over number (TON) = moles of substrate converted per mole of active site.

The condensation reaction for aldehydes with both electron-donating (Table 1.15, entries 4 and 5) and electron-withdrawing groups (Table 1.15, entries 1–3 and 7–8) proceeded smoothly and very efficiently. Thus, the nature and position of the substitution in the aromatic ring did not have a great effect on the reactions. The meta-substituted aromatic aldehyde (Table 1.15, entry 7) and the α,β-unsaturated aromatic aldehyde (Table 1.15, entry 9) undergo the condensation reaction without any difficulty. Several sensitive functional groups, such as –Cl, –Br, and –NO_2, attached to the aromatic ring are also compatible in this reaction. A heterocyclic aldehyde, such as thiophene-2-adehyde, participated in the condensation reaction with equal efficiency (Table 1.15, entry 6). The turnover number (TON) for different reactions varies from 24 to 22.6 for different aldehydes, suggesting high catalytic efficiency of TFMO-1 in these reactions.

In order to extend the scope of the reaction, the reaction was performed with various activated phenols (Table 1.16) using 4-chlorobenzaldehyde as the representative case for 4 h. It was found that 1-naphthol showed the higher reactivity among the activated phenols in this condensation reaction.

Among different active methylene compounds (malononitrile, ethyl cyanoacetate, and diethyl malonate), only malononitrile showed a high product yield, suggesting that it was the best-suited active methylene compound for the synthesis of 2-amino-chromene derivatives over TFMO-1.

The effect of reaction temperature in the condensation of 4-chlorobenzaldehyde, malononitrile, and 1-naphtol under solvent-free conditions mediated by TFMO-1

TABLE 1.16
Synthesis of 2-Amino-4*H*-Chromene Derivatives Catalyzed by TFMO-1[a]

Entry	Activated Phenol	Time (h)	Yield[b] (%)
1	1-Naphthol	4	90
2	2-Naphthol	4	45
3	Resorcinol	4	64

[a] Reaction conditions: 4-chlorobenzaldehyde (1 mmol)-activated phenolic compounds (1 mmol), malononitrile (1 mmol), solvent-free, reaction temperature 110°C, 40 mg catalyst TFMO-1.
[b] Isolated yield of pure product.

was also observed. At room temperature (25°C), only a trace amount of product (5%) was obtained. A temperature of 110°C resulted in the maximum yield and this is considered the optimum temperature for the completion of this reaction (Table 1.15, entry 3, yield 90%).

Notably, the mesoporosity plays a crucial role in this condensation reaction. Indeed, the yield of the condensation product over mesoporous organocatalyst TFMO-1 is much higher than the homogeneous 2,4,6-triallyloxy-1,3,5-triazine form.

The recyclability of the catalyst (TFMO-1) was tested in the condensation reaction between 4-chlorobenzaldehyde, malononitrile, and 1-naphtol (Table 1.17). Only a very small drop in the product yield in each catalytic cycle was observed. It is clear from Table 1.17 that the TON of the catalyst is retained from fresh to the sixth reaction cycle, suggesting the high catalytic efficiency of TFMO-1.

TABLE 1.17
Recycling Potential of the Mesoporous Organocatalyst (TFMO-1)

No. of Cycles[a]	Fresh	Run 1	Run 2	Run 3	Run 4	Run 5	Run 6
Yield[b] (%)	90	90	90	88	88	85	84
Time (h)	4	4	4	4	4	4	4
TON[c]	23.4	23.4	23.4	23.0	23.0	22.1	21.8

[a] Reaction conditions: 4-chlorobenzaldehyde (1 mmol), 1-naphthol (1 mmol), malononitrile (1 mmol), solvent-free, 110°C, 40 mg catalyst TFMO-1.
[b] Isolated yield of pure product.
[c] Turnover number (TON) = moles of substrate converted per mole of active site.

ACKNOWLEDGMENTS

The authors gratefully acknowledge Prof. Massimo Curini for helpful discussions.

REFERENCES

Abdel-Ghani, E. 1999. Regioselective base-induced condensations of acrylic acid derivatives. *Journal of Chemical Research* (3):174–175A.

Abrunhosa, I., Gulea, M., Levillain, J., Masson, S. 2001. Synthesis of new chiral thiazoline-containing ligands. *Tetrahedron: Asymmetry* 12(20):2851–2859.

Alam, S. 2004. Synthesis, antibacterial and antifungal activity of some derivatives of 2-phenyl-chromen-4-one. *Journal of Chemical Sciences* 116(6):325–331.

Ali, M., Zolfigol, M.A., Mohammadpoor, B., Mirjalili, I., Bamoniri, A. 2003. Silica sulfuric acid: An efficient catalyst for the direct conversion of primary and secondary trimethylsilyl ethers to their corresponding ethers under mild and heterogeneous conditions. *Synlett* 12:1877–1879.

Andreani, A., Rambaldi, M., Leoni, A., Locatelli, A., Bossa, R., Chiericozzi, M., Galatulas, I., Salvatore, G. 1996. Synthesis and cardiotonic activity of imidazo[2,1-b]thiazoles bearing a lactam ring. *European Journal of Medicinal Chemistry* 31(5):383–387.

Aoyama, T., Takido, T., Kodomari, M. 2004. Silica gel-supported polyphosphoric acid (PPA/SiO_2) as an efficient and reusable catalyst for conversion of carbonyl compounds into oxathioacetals and dithioacetals. *Synlett* 13:2307–2310.

Azizi, N., Saidi, M.R. 2007. Highly efficient ring opening reactions of epoxides with deactivated aromatic amines catalyzed by heteropoly acids in water. *Tetrahedron* 63(4):888–891.

Azizi, N., Torkiyan, L., Saidi, M.R. 2006. Highly efficient one-pot three-component Mannich reaction in water catalyzed by heteropoly acids. *Organic Letters* 8(10):2079–2082.

Azizian, J., Karimi, A.R., Kazemizadeh, Z., Mohammadi, A.A., Mohammadizadeh, M.R. 2005. Silica sulfuric acid-catalyzed reaction of 4-hydroxy proline with 11H-indeno[1,2-b]quinoxalin-11-one and isatin derivatives: A novel synthesis of new pyrrole compounds. *Synthesis* 1095–1097.

Barton, D., Ollis, W.D. 1979. *Comprehensive Organic Chemistry: The Synthesis and Reactions of Organic Compounds*, 1st edn. Oxford, NY: Pergamon Press.

Bennardi, D.O., Romanelli, G.P., Jios, J.L., Autino, J.C., Baronetti, G.T., Thomas, H.J. 2008. Synthesis of substituted flavones and chromones using a Wells-Dawson heteropolyacid as catalyst. *ARKIVOC* (xi):123–130.

Bennardi, D.O., Romanelli, G.P., Jíos, J.L., Vázquez, P., Cáceres, C., Autino, J.C. 2007. Synthesis of substituted flavones and arylchromones using P and Si keggin heteropolyacids as catalysts. *Heterocyclic Communications* 13:77.

Biginelli, P. 1891a. Ueber Aldehyduramide des Acetessigäthers. *Berichte der deutschen chemischen Gesellschaft* 24:1317–1319.

Biginelli, P. 1891b. Ueber Aldehyduramide des Acetessigäthers II. *Berichte der deutschen chemischen Gesellschaft* 24:2962–2967.

Bloxham, J., Dell, C.P., Smith, C.W. 1994. Preparation of some new benzylidenemalononitriles by an SNAr reaction—Application to naphtho[1,2-b]pyran synthesis. *Heterocycles* 38(2):399–408.

Boland, N.A., Casey, M., Hynes, S.J., Matthews, J.W., Smyth, M.P. 2002. A novel general route for the preparation of enantiopure imidazolines. *Journal of Organic Chemistry* 67(11):3919–3922.

Bonollo, S., Lanari, D., Longo, J.M., Vaccaro, L. 2012. E-factor minimized protocols for the polystyryl-BEMP catalyzed conjugate additions of various nucleophiles to α,β-unsaturated carbonyl compounds. *Green Chemistry* 14:164–169.

Bousquet, P., Feldman, J. 1999. Drugs acting on imidazoline receptors: A review of their pharmacology, their use in blood pressure control and their potential interest in cardioprotection. *Drugs* 58(5):799–812.

Carling, R.W., Moore, K.W., Street, L.J., Wild, D., Isted, C., Leeson, P.D., Thomas, S., et al. 2004. 3-phenyl-6-(2-pyridyl)methyloxy-1,2,4-triazolo[3,4-α]phthalazines and analogues: High-affinity γ-aminobutyric acid-A benzodiazepine receptor ligands with α 2, α 3, and α 5 subtype binding selectivity over α 1. *Journal of Medicinal Chemistry* 47(7):1807–1822.

Chakraborti, A.K., Singh, B., Chankeshwara, S.V., Patel, A.R. 2009. Protic acid immobilized on solid support as an extremely efficient recyclable catalyst system for a direct and atom economical esterification of carboxylic acids with alcohols. *Journal of Organic Chemistry* 74:5967–5974.

Chang, T.H. 1995. NMR characterization of the supported 12-heteropoly acids. *Journal of the Chemical Society-Faraday Transactions* 91(2):375–379.

Chari, M.A., Shobha, D., Syamasundar, K. 2007. Silica gel/$NaHSO_4$: An efficient and recyclable heterogeneous catalyst for high yield synthesis of 1,5-benzodiazepine derivatives under microwave irradiation. *Journal of Heterocyclic Chemistry* 44:929–932.

Chen, W.Y., Lu, J. 2005. Silica sulfuric acid catalyzed one-pot synthesis of α-aminonitriles. *Synlett* 15:2293–2296.

Chu, H.W., Wu, H.T., Lee, Y.J. 2004. Regioselective hydroxylation of 2-hydroxychalcones by dimethyldioxirane towards polymethoxylated flavonoids. *Tetrahedron* 60(11):2647–2655.

Clark, M.P., Laughlin, S.K., Laufersweiler, M.J., Bookland, R.G., Brugel, T.A., Golebiowski, A., Sabat, M.P., et al. 2004. Development of orally bioavailable bicyclic pyrazolones as inhibitors of tumor necrosis factor-α production. *Journal of Medicinal Chemistry* 47(11):2724–2727.

Clarke, D.S., Wood, R. 1996. A facile one stage synthesis of oxazolines under microwave irradiation. *Synthetic Communications* 26(7):1335–1340.

Coelho, A., Sotelo, E., Fraitz, N., Yáñez, M., Laguna, R., Cano, E., Raviña, E. 2007. Design, synthesis, and structure-activity relationships of a novel series of 5-alkylidenepyridazin-3(2H)-ones with a non-cAMP-based antiplatelet activity. *Journal of Medicinal Chemistry* 50:6476–6484.

Corma, A., Garcia, H. 2006. Silica-bound homogenous catalysts as recoverable and reusable catalysts in organic synthesis and citations therein. *Advanced Synthesis and Catalysis* 348:1391–1412.

Costantino, U., Curini, M., Montanari, F., Nocchetti, M., Rosati, O. 2008. Hydrotalcite-like compounds as heterogeneous catalysts in liquid phase organic synthesis. II. Preparation of 4H-chromenes promoted by hydrotalcite doped with hydrous tin(IV) oxide. *Microporous and Mesoporous Materials* 107(1–2):16–22.

Cwik, A., Hell, Z., Hegedus, A., Finta, Z., Horvath, Z. 2002. A simple synthesis of 2-substituted oxazolines and oxazines. *Tetrahedron Letters* 43(22):3985–3987.

Da Silva Rocha, K.A., Kozhevnikov, I.V., Gusevskaya, E.V. 2005. Isomerisation of α-pinene oxide over silica supported heteropoly acid $H_3PW_{12}O_{40}$. *Applied Catalysis A: General* 294:106–110.

Da Silva Rocha, K.A., Robles-Dutenhefner, P.A., Sousa, E.M.B., Kozhevnikova, E.F., Kozhevnikov, I.V., Gusevskaya, E.V. 2007. Pd–heteropoly acid as a bifunctional heterogeneous catalyst for one-pot conversion of citronellal to menthol. *Applied Catalysis A: General* 317:171–174.

Dell, C.P., Smith, C.W. 1993a. Antiproliferative derivatives of 4H-naphtho 1,2-b pyran. *European Patent Application, EP537949, Chemical Abstracts*, 119, 139102d.1.

Dell, C.P., Smith, C.W. 1993b. *Antiproliferative Derivatives of 4H-Naphto 1,2-b Pyran*. Lilly Industries LTD, GB.

De Meyer, N., Haemers, A., Mishra, L., Pandey, I.A.C., Pieters, D.A.V., Berghe, A.J., Viletinck, A.J. 1991. 4′-Hydroxy-3-methoxyflavones with potent antipicornavirus activity. *Journal of Medicinal Chemistry* 34:736–746.

Desimoni, G., Faita, G., Jorgensen, K.A. 2006. C-2-symmetric chiral bis(oxazoline) ligands in asymmetric catalysis. *Chemical Reviews* 106(9):3561–3651.

Dhakshinamoorthy, A., Kanagaraj, K., Pitchumani, K. 2011. Zn^{2+}-K10-clay (clayzic) as an efficient water-tolerant, solid acid catalyst for the synthesis of benzimidazoles and quinoxalines at room temperature. *Tetrahedron Letters* 52:69–73.

Dias, A.S., Lima, S., Pillinger, M., Valente, A.A. 2006. Acidic cesium salts of 12-tungstophosphoric acid as catalysts for the dehydration of xylose into furfural. *Carbohydrate Research* 341:2946–2953.

Ding, Y., Ma, B., Gao, Q., Li, G., Yan, L., Suo, J. 2005. A spectroscopic study on the 12-heteropolyacids of molybdenum and tungsten ($H_3PMo_{12-n}W_nO_{40}$) combined with cetylpyridinium bromide in the epoxidation of cyclopentene. *Journal of Molecular Catalysis A: Chemical* 230:121–130.

Dupont, P., Lefebvre, F. 1996. Esterification of propanoic acid by butanol and 2-ethylhexanol catalyzed by heteropolyacids pure or supported on carbon. *Journal of Molecular Catalysis A: Chemical* 114(1–3):299–307.

Dupont, P., Vedrine, J.C., Paumard, E., Hequet, G., Lefebrvre, F. 1995. Heteropolyacids supported on activated carbon as catalysts for the esterification of acrylic acid by butanol. *Applied Catalysis A: General* 129(2):217–227.

Eiden, F., Denk, F. 1991. Synthesis and CNS-activity of pyran derivatives: 6,8-dioxabicyclo[3,2,1] octanes. Part 133. *Archiv Der Pharmazie* 324(6):353–354.

Elagamey, A.G.A., Eltaweel, F.M.A., Khodeir, M.N.M., Elnagdi, M.H. 1993. Nitriles in heterocyclic synthesis—The reaction of polyhydric naphthalenes, 4-methylcoumarin-3-carbonitrile, and alkylidenemalononitrile with methylenemalononitrile. *Bulletin of the Chemical Society of Japan* 66(2):464–468.

El Maatougui, A., Azuaje, J., Sotelo, E., Caamaño, O., Coelho, A. 2011. Silica-supported aluminum chloride-assisted solution phase synthesis of pyridazinone-based antiplatelet agents. *ACS Combinatorial Science* 13:7–12.

Fan, L.J., Lobkovsky, E., Ganem, B. 2007. Bioactive 2-oxazolines: A new approach via one-pot, four-component reaction. *Organic Letters* 9(10):2015–2017.

Firouzabadi, H., Iranpoor, N., Amani, K. 2003. Keggin-type heteropoly acids revealed high catalytic activity for swift and selective oxidation of various hydroxy functionalities to the corresponding carbonyl groups using ferric nitrate as an oxidant under mild and solvent-free conditions. *Synthesis* 408–412.

Firouzabadi, H., Iranpoor, N., Wan, A.A. 2005. Dodecatungestophosphoric acid ($H_3PW_{12}O_{40}$) as a solid green Bronsted acid catalyzes high yielding and efficient trimethylcyanosylilation reactions of aldehydes and ketones by trimethylsilyl cyanide. *Journal of Organometallic Chemistry* 690(6):1556–1559.

Fischer, E., Jourdan, F. 1883. Ueber die Hydrazine der Brenztraubensäure. *Berichte der Deutschen Chemischen Gesellschaft* 16(2):2241–2245.

Fournier, M., Thouvenot, R., Rocchiccioli-Deltcheff, C. 1991. Catalysis by polyoxometalates. Part 1. Supported polyoxoanions of the keggin structure: Spectroscopic study (IR, Raman, UV) of solutions used for impregnation. *Journal of the Chemical Society-Faraday Transactions* 87(2):349–356.

Fujioka, H., Murai, K., Kubo, O., Ohba, Y., Kita, Y. 2007. One-pot synthesis of imidazolines from aldehydes: Detailed study about solvents and substrates. *Tetrahedron* 63(3):638–643.

Ganguly, A.K., Kaur, S., Mahata, P.K., Biswas, D., Pramanik, B.N., Chan, T.M. 2005. Synthesis and properties of 3-acyl-γ-pyrones, a novel class of flavones and chromones. *Tetrahedron Letters* 46(23):4119–4121.

Ghorai, M.K., Das, K., Kumar, A., Das, A. 2006. A convenient synthetic route to 2-aryl-*N*-tosylazetidines and their ZnX2 (X = I, OTf) mediated regioselective nucleophilic ring opening reactions: Synthesis of γ-iodoamines and tetrahydropyrimidines. *Tetrahedron Letters* 47(30):5393–5397.

Ghosh, A.K., Cho, H., Cappiello, J. 1998. Bis(oxazoline) derived cationic aqua complexes: Highly effective catalysts for enantioselective Diels–Alder reactions. *Tetrahedron Asymmetry* 9(20):3687 3691.

Gladysz, J.A. 2002. Introduction: Recoverable catalysts and reagents, perspective and prospective. *Chemical Reviews* 102:3215–3892.

Göker, H., Boykin, D.W., Yildiz, S. 2005. Synthesis and potent antimicrobial activity of some novel 2-phenyl or methyl-4H-1-benzopyran-4-ones carrying amidinobenzimidazoles. *Bioorganic and Medicinal Chemistry* 13(5):1707–1714.

Grasso, S., De Sarro, G., Micale, N., Zappala, M., Puia, G., Baraldi, M., Demicheli, C. 2000. Synthesis and anticonvulsant activity of novel and potent 6,7-methylenedioxyphthalazin-1(2H)-ones. *Journal of Medicinal Chemistry* 43:2851–2859.

Greenhill, J.V., Lue, P. 1993. 5 Amidines and guanidines in medicinal chemistry. In: Ellis, G.P., Luscombe, D.K. (eds), *Progress in Medicinal Chemistry*, pp. 203–326. Amsterdam: Elsevier.

Hajipour, A.R., Zarei, A., Khazdooz, L., Pourmousavi, S.A., Ruoho, A.E. 2005. Silicasulfuric acid/$NaNO_2$ as a new reagent for deprotection of *S,S*-acetals under solvent-free conditions. *Bulletin of the Korean Chemical Society* 26:808–810.

Handique, J.G., Baruah, J.B. 2002. Polyphenolic compounds: An overview. *Reactive and Functional Polymers* 52:163–188.

Hao, X.Q., Gong, J.F., Du, C.X., Wu, L.Y., Wu, Y.J., Song, M.P. 2006. Synthesis, characterization and photoluminescent properties of platinum complexes with novel bis(imidazoline) pincer ligands. *Tetrahedron Letters* 47(29):5033–5036.

Heravi, M.M., Bakhtiari, K., Zadsirjan, V., Bamoharram, F.F., Heravi, O.M. 2007. Aqua mediated synthesis of substituted 2-amino-4*H*-chromenes catalyzed by green and reusable Preyssler heteropolyacid. *Bioorganic and Medicinal Chemistry Letters* 17(15):4262–4265.

Hiramoto, K., Nasuhara, A., Michikoshi, K., Kato, T., Kikugawa, K. 1997. DNA strand-breaking activity and mutagenicity of 2,3-dihydro-3,5-dihydroxy-6-methyl-4H-pyran-4-one (DDMP), a Maillard reaction product of glucose and glycine. *Mutation Research Genetic Toxicology and Environmental Mutagenesis* 395(1):47–56.

Hoshino, Y., Takeno, N. 1987. A facile preparation of flavones using nonaqueous cation-exchange resin. *Bulletin of the Chemical Society of Japan* 60(5):1919–1920.

Huang, T.K., Wang, R., Shi, L., Lu, X.X. 2008. Montmorillonite K-10: An efficient and reusable catalyst for the synthesis of quinoxaline derivatives in water. *Catalysis Communications* 9:1143–1147.

Huang, X., Liu, X.M., Luo, Q.A., Liu, J.Q., Shen, J.C. 2011. Artificial selenoenzymes: Designed and redesigned. *Chemical Society Reviews* 40(3):1171–1184.

Hui, X., Desrivot, J., Bories, C., Loiseau, P.M., Franck, X., Hocquemiller, R., Figadere, B. 2006. Synthesis and antiprotozoal activity of some new synthetic substituted quinoxalines. *Bioorganic and Medicinal Chemistry Letters* 16(4):815–820.

Imamura, Y., Noda, A., Imamura, T., Ono, Y., Okawara, T., Noda, H. 2003. A novel methylthio metabolite of s-triazolo[3,4-a]phthalazine, a lead compound for the development of antianxiety drugs, in rats. *Life Sciences* 74(1):29–36.

Ishida, T., Haruta, M. 2007. Gold catalysts: Towards sustainable chemistry. *Angewandte Chemie-International Edition* 46(38):7154–7156.

Ito, M., Ishimoto, S., Nishida, Y., Shiramizu, T., Yunoki, H. 1986. Effects of baicalein, a flavonoid, and other anti-inflammatory agents on glyoxalase-I activity. *Agricultural and Biological Chemistry* 50:1063–1065.

Izumi, Y. 1997. Hydration/hydrolysis by solid acids. *Catalysis Today* 33(4):371–409.

Izumi, Y., Hisano, K., Hida, T. 1999. Acid catalysis of silica-included heteropolyacid in polar reaction media. *Applied Catalysis A: General* 181(2):277–282.

Izumi, Y., Ono, M., Kitagawa, M., Yoshida, M., Urabe, K. 1995. Silica-included heteropoly compounds as solid acid catalysts. *Microporous Materials* 5(4):255–262.

Izumi, Y., Urabe, K., Onaka, M. 1992. *Zeolites Clay and Heteropolyacid in Organic Reactions*, vol. 99. Tokyo: Kodansha.

Izumi, Y., Urabe, K., Onaka, M. 1997. Development of catalyst materials for acid-catalyzed reactions in the liquid phase. *Catalysis Today* 35(1–2):183–188.

Jain, S.L., Modak, A., Bhaumik, A. 2011. A novel mesoporous silica-grafted organocatalyst for the Michael addition reaction, synthesized via the click method. *Green Chemistry* 13(3):586–590.

Jarikote, D.V., Siddiqui, S., Rajagopal, R., Daniel, T., Lahoti, R., Srinivasan, K.V. 2003. Room temperature ionic liquid promoted synthesis of 1,5-benzodiazepine derivatives under ambient conditions. *Tetrahedron Letters* 44:1835–1838.

Jnaneshwara, G.K., Deshpande, V.H., Bedekar, A.V. 1999. Clay-catalyzed conversion of 2,2-disubstituted malononitriles to 2-oxazolines: Towards unnatural amino acids. *Journal of Chemical Research, Synopses* (4):252–253.

Jnaneshwara, G.K., Deshpande, V.H., Lalithambika, M., Ravindranathan, T., Bedekar, A.V. 1998. Natural kaolinitic clay catalyzed conversion of nitriles to 2-oxazolines. *Tetrahedron Letters* 39(5–6):459–462.

Kabalka, G.W., Mereddy, A.R. 2005. Microwave-assisted synthesis of functionalized flavones and chromones. *Tetrahedron Letters* 46(37):6315–6317.

Kandarpa, P., Anamika, J., Nirada, D. 2011. Characterization of an iron-rich kaolinite clay and its application as heterogeneous catalyst for the microwave-mediated dry synthesis of *N*-containing heterocycles. *Research Journal of Chemistry and Environment* 15:86–91.

Karandikar, P., Agashe, M., Vijayamohanan, K., Chandwadkar, A.J. 2004. Cu2+-perchlorophthalocyanine immobilized MCM-41: Catalyst for oxidation of alkenes. *Applied Catalysis A: General* 257(2):133–143.

Karimi, N., Oskooie, H.A., Heravi, M.M., Tahershamsi, L. 2011. Caro's acid-silica gel-catalyzed one-pot synthesis of 12-aryl-8,9,10,12-tetrahydrobenzo[a] xanthen-11-ones. *Synthetic Communications* 41:307–312.

Kaur, J., Griffin, K., Harrison, B., Kozhevnikov, I.V. 2002. Friedel–Crafts acylation catalysed by heteropoly acids. *Journal of Catalysis* 208(2):448–455.

Kennedy, R.O., Tharnes, R.D, (eds). 1997. *Coumarins: Biology, Application and Mode of Action*. Chichester: Wiley.

Khafagy, M.M., Abd El-Wahab, A.H.F., Eid, F.A., El-Agrody, A.M. 2002. Synthesis of halogen derivatives of benzo[*h*]chromene and benzo[*a*]anthracene with promising antimicrobial activities. *Farmaco* 57(9):715–722.

Khojastehnezhad, A., Davoodnia, A., Bakavoli, M., Tavakoli-Hoseini, N., Zeinali-Dastmalbaf, M. 2011. Silica gel-supported polyphosphoric acid (PPA/SiO_2): An efficient and reusable heterogeneous catalyst for facile synthesis of 14-Aryl-14*H*-dibenzo[*a,j*]xanthenes under solvent-free conditions. *Chinese Journal of Chemistry* 29:297–303.

Kidwai, M., Saxena, S., Khan, M.K.R., Thukral, S.S. 2005. Aqua mediated synthesis of substituted 2-amino-4*H*-chromenes and *in vitro* study as antibacterial agents. *Bioorganic and Medicinal Chemistry Letters* 15(19):4295–4298.

Kim, J.S., Lee, H.J., Suh, M.E., Choo, H.Y.P., Lee, S.K., Park, H.J., Kim, C., Park, S.W., Lee, C.O. 2004. Synthesis and cytotoxicity of 1-substituted 2-methyl-1*H*-imidazo[4,5-*g*]phthalazine-4,9-dione derivatives. *Bioorganic and Medicinal Chemistry* 12(13):3683–3686.

Knorr, L. 1884. Synthese von Pyrrolderivaten. *Berichte der Deutschen Chemischen Gesellschaft* 17:1635–1642.

Kozhevnikov, I.V. 1998. Catalysis by heteropoly acids and multicomponent polyoxometalates in liquid-phase reactions. *Chemical Reviews* 98:171–198.
Kozhevnikov, I.V. 2002. *Catalysis by Polyoxometalates: Catalysts for Fine Chemical Synthesis*. New York: Wiley.
Kozhevnikov, I.V., Kloetstra, K.R., Sinnema, A., Zandbergen, H.W., van Bekkum, H. 1996. Study of catalysts comprising heteropoly acid $H_3PW_{12}O_{40}$ supported on MCM-41 molecular sieve and amorphous silica. *Journal of Molecular Catalysis A: Chemical* 114(1–3):287–298.
Kozhevnikov, I.V., Sinnema, A., Jansen, R.J.J., Pamin, K., Van Bekkum, H. 1995. New acid catalyst comprising heteropoly acid on a mesoporous molecular sieve. *Catalysis Letters* 30(1–4):241–252.
Krishnakumar, B., Swaminathan, M. 2011. Solvent free synthesis of quinoxalines, dipyridophenazines and chalcones under microwave irradiation with sulfated Degussa titania as a novel solid acid catalyst. *Journal of the Molecular Catalysis A: Chemical* 350:16.
Kühl, O. 2007. The chemistry of functionalised *N*-heterocyclic carbenes. *Chemical Society Reviews* 36(4):592–607.
Kumar, B.S., Srinivasulu, N., Udupi, R.H., Rajitha, B., Reddy, Y.T., Reddy, P.N., Kumar, P.S. 2006a. An efficient approach towards three component coupling of one pot reaction for synthesis of functionalized benzopyrans. *Journal of Heterocyclic Chemistry* 43(6):1691–1693.
Kumar, K.H., Perumal, P.T. 2006. A simple and facile solventless procedure for the cyclization of 2′-amino- and 2′-hydroxy-chalcones using silica-supported sodium hydrogen sulphate as heterogenous catalyst. *Canadian Journal of Chemistry* 84:1079–1086.
Kumar, R., Chaudhary, P., Nimesh, S., Verma, A.K., Chandra, R. 2006b. An efficient synthesis of 1,5-benzadiazepine derivatives catalyzed by silver nitrate. *Green Chemistry* 8:519–521.
Lan, K., Fen, S., Shan, Z.X. 2007. Synthesis of aromatic cycloketones via intramolecular Friedel–Crafts acylation catalyzed by heteropoly acids. *Australian Journal of Chemistry* 60(1):80–82.
Landquist, J.K. 1984. Application as pharmaceuticals. In: Katritzky, A.R., Rees, C.W. (eds), *Comprehensive Heterocyclic Chemistry*, vol. 1, p. 166. Oxford: Pergamon.
Le Bihan, G., Rondu, F., Pelé-Tounian, F., Wang, X., Lidy, S., Touboul, E., Lamouri, A., et al. 1999. Design and synthesis of imidazoline derivatives active on glucose homeostasis in a rat model of type II diabetes. 2. Syntheses and biological activities of 1,4-dialkyl-, 1,4-dibenzyl, and 1-benzyl-4-alkyl-2-(4′,5′-dihydro-1′H-imidazol-2′-yl) piperazines and isosteric analogues of imidazoline. *Journal of Medicinal Chemistry* 42(9):1587–1603.
Lebsack, A.D., Gunzner, J., Wang, B.W., Pracitto, R., Schaffhauser, H., Santini, A., Aiyar, J., et al. 2004. Identification and synthesis of [1,2,4]triazolo[3,4-*a*]phthalazine derivatives as high-affinity ligands to the $\alpha(2)\delta$-1 subunit of voltage gated calcium channel. *Bioorganic and Medicinal Chemistry Letters* 14(10):2463–2467.
Lee, A., Kim, W., Lee, J., Hyeon, T., Kim, B.M. 2004. Heterogeneous asymmetric nitro-Mannich reaction using a bis(oxazoline) ligand grafted on mesoporous silica. *Tetrahedron Asymmetry* 15(17):2595–2598.
Li, H.Y., Drummond, S., DeLucca, I., Boswell, G.A. 1996. Singlet oxygen oxidation of pyrroles: Synthesis and chemical transformations of novel 4,4-bis(trifluoromethyl)imidazoline analogs. *Tetrahedron* 52(34):11153–11162.
Luts, T., Suprun, W., Hofmann, D., Klepel, O., Papp, H. 2007. Epoxidation of olefins catalyzed by novel Mn(III) and Mo(IV) Salen complexes immobilized on mesoporous silica gel: Part I. Synthesis and characterization of homogeneous and immobilized Mn(III) and Mo(IV) Salen complexes. *Journal of Molecular Catalysis A: Chemical* 261(1):16–23.
Makarem, S., Mohammadi, A.A., Fakhari, A.R. 2008. A multi-component electro-organic synthesis of 2-amino-4H-chromenes. *Tetrahedron Letters* 49(50):7194–7196.

Maleki, B., Keshvari-Shirvan, H., Taimazi, F., Akbarzadeh, E. 2012. Sulfuric acid immobilized on silica gel as highly efficient and heterogeneous catalyst for the one-pot synthesis of 2,4,5-triaryl-1H-imidazoles. *International Journal of Organic Chemistry* 2:93–101.

Martens, S., Mithöfer, A. 2005. Flavones and flavone syntheses. *Phytochemistry* 66(20):2399–2407.

Martínez-Grau, A., Marco, J.L. 1997. Friedländer reaction on 2-amino-3-cyano-4*H*-pyrans: Synthesis of derivatives of 4*H*-pyran[2,3-*b*]quinoline, new tacrine analogues. *Bioorganic and Medicinal Chemistry Letters* 7(24):3165–3170.

Matsumoto, K., Jitsukawa, K., Masuda, H. 2005. Preparation of new bis(oxazoline) ligand bearing non-covalent interaction sites and an application in the highly asymmetric Diels–Alder reaction. *Tetrahedron Letters* 46(34):5687–5690.

Mavel, S., Thery, I., Gueiffier, A. 2002. Synthesis of imidazo[2,1-*a*]phthalazines, potential inhibitors of p38 MAP kinase. Prediction of binding affinities of protein ligands. *Archiv Der Pharmazie* 335(1):7–14.

Mirjalili, B.F., Zolfigol, M.A., Bamoniri, A., Zaghaghi, Z. 2003. Silica sulfuric acid/$KClO_3$/wet SiO_2 as an efficient heterogeneous method for the oxidation of alcohols under mild conditions. *Journal of Chemical Research* (5):273–277.

Mirkhani, V., Moghadam, M., Tangestaninejad, S., Kargar, H. 2006. Rapid and efficient synthesis of 2-imidazolines and bis-imidazolines under ultrasonic irradiation. *Tetrahedron Letters* 47(13):2129–2132.

Mirkhani, V., Mohammadpoor-Baltork, I., Moghadam, M., Tangestaninejad, S., Abdollahi-Alibeik, M., Kargar, H. 2007. $ZrOCl_2$ center dot AH(2)O: An efficient and reusable catalyst for the synthesis of imidazolines and bis-imidazolines under various reaction conditions. *Applied Catalysis A: General* 325(1):99–104.

Misono, M. 1987. Heterogeneous catalysis by heteropoly compounds of molybdenum and tungsten. *Catalysis Reviews Science and Engineering* 29(2–3):269–321.

Misono, M. 2001. Unique acid catalysis of heteropoly compounds(heteropolyoxometalates) in the solid state. *Chemical Communications* 1141–1142.

Misono, M., Mizuno, N., Katamura, K., Kasai, A., Konishi, Y., Sakata, K., Okuhara, T., Yoneda, Y. 1982. Catalysis by heteropoly compounds. III. The structure and properties of 12-heteropolyacids of molybdenum and tungsten and their salts pertinent to heterogeneous catalysis. *Bulletin of the Chemical Society of Japan* 55(2):400–406.

Mizuno, N., Misono, M. 1998. Heterogeneous catalysis. *Chemical Reviews* 98:199–218.

Moffat, J.B. 2001. *The Surface and Catalytic Properties of Heteropolyoxometalates*. New York: Kluwer.

Mohammadpoor-Baltork, I., Abdollahi-Alibeik, M. 2003. Microwave-assisted facile and convenient synthesis of imidazolines. *Bulletin of the Korean Chemical Society* 24(9):1354–1356.

Mohammadpoor-Baltork, I., Khosropour, A.R., Hojati, S.F. 2005. A novel and chemoselective synthesis of 2-aryloxazolines and bis-oxazolines catalyzed by Bi(III) salts. *Synlett* (18):2747–2750.

Mohammadpoor-Baltork, I., Moghadam, M., Tangestaninejad, S., Mirkhani, V., Hojati, S.F. 2008. Supported 12-tungstophosphoric acid as heterogeneous and recoverable catalysts for the synthesis of oxazolines, imidazolines and thiazolines under solvent-free conditions. *Polyhedron* 27(2):750–758.

Mohr, S.J., Chirigos, M.A., Fuhrman, F.S., Pryor, J.W. 1975. Pyran copolymer as an effective adjuvant to chemotherapy against a murine leukemia and solid tumor. *Cancer Research* 35(12):3750–3754.

Mondal, J., Modak, A., Nandi, M., Uyama, H., Bhaumik, A. 2012. Triazine functionalized ordered mesoporous organosilica as a novel organocatalyst for the facile one-pot synthesis of 2-amino-4*H*-chromenes under solvent-free conditions. *RSC Advances* 2(30):11306–11317.

Morimoto, M., Tanimoto, K., Nakano, S., Ozaki, T., Nakano, A., Komai, K. 2003. Insect antifeedant activity of flavones and chromones against Spodoptera litura. *Journal of Agricultural and Food Chemistry* 51(2):389–393.

Mukhopadhyay, C., Ray, S. 2011. Synthesis of 2-amino-5-alkylidenethiazol-4-ones from ketones, rhodanine, and amines with the aid of re-usable heterogeneous silica-pyridine based catalyst. *Tetrahedron* 67(41):7936–7945.

Nagarapu, L., Bantu, R., Mereyala, H.B. 2009. TMSCl-mediated one-pot, three-component synthesis of 2*H*-indazolo[2,1-*b*]phthalazine-triones. *Journal of Heterocyclic Chemistry* 46:728–731.

Nandi, M., Mondal, J., Sarkar, K., Yamauchi, Y., Bhaumik, A. 2011. Highly ordered acid functionalized SBA-15: A novel organocatalyst for the preparation of xanthenes. *Chemical Communications* 47(23):6677–6679.

Nasr, M.N.A., Said, S.A. 2003. Novel 3,3a,4,5,6,7-hexahydroindazole and arylthiazolylpyrazoline derivatives as anti-inflammatory agents. *Archives of Pharmacology* 336:551–557.

Nasr-Esfahani, M., Montazerozohori, M., Moghadam, M., Akhlaghi, P. 2010. Efficient catalytic synthesis of 2-imidazolines and bis-imidazolines with silica supported tungstosilicic acid. *ARKIVOC* 2010:97–109.

Neef, G., Eder, U., Sauer, G. 1981. One-step conversions of esters to 2-imidazolines, benzimidazoles and benzothiazoles by aluminum organic reagents. *The Journal of Organic Chemistry* 46(13):2824–2826.

Niefang, Y., Aramini, J.M., Germann, M.W., Huang, Z. 2000. Reactions of salicylaldehydes with alkyl cyanoacetates on the surface of solid catalysts: Syntheses of 4H-chromene derivatives. *Tetrahedron Letters* 41:6993–6996.

Nishiguchi, T., Kamio, C. 1989. Dehydration of alcohols catalysed by metallic sulphates supported on silica gel. *Journal of Chemical Society Perkin Transactions* 1:707–715.

Nomoto, Y., Obase, H., Takai, H., Teranishi, M., Nakamura, J., Kubo, K. 1990. Studies on cardiotonic agents. II. Synthesis of novel phthalazine and 1,2,3-benzotriazine derivatives. *Chemical and Pharmaceutical Bulletin* 38:2179–2183.

Ohmura, W., Doi, S., Aoyama, M., Ohara, S. 2000. Antifeedant activity of flavonoids and related compounds against the subterranean termite Coptotermes formosanus Shiraki. *Journal of Wood Science* 46(2):149–153.

Okuhara, T., Mizuno, N., Misono, M. 1996. Catalytic chemistry of heteropoly compounds. *Advances in Catalysis* 41:113–252.

Pathan, M.Y., Paike, V.V., Pachmase, P.R., More, S.P., Ardhapure, S.S., Pawar, R.P. 2006. Microwave-assisted facile synthesis of 2-substituted 2-imidazolines. *ARKIVOC* (xv):205–210.

Pizzio, L., Vázquez, P., Cáceres, C., Blanco, M. 2003. Supported Keggin type heteropolycompounds for ecofriendly reactions. *Applied Catalysis A: General* 256:125–129.

Pope, M.T. 1983. *Heteropoly and Isopoly Oxometalates*. Berlin: Springer.

Potts, K.T., Lovelette, C. 1969. 1,214-Triazoles. XXII. Derivatives of the s-triazolo[3,4-α] phthalazine and related ring systems. *The Journal of Organic Chemistry* 34(11):3221–3230.

Prado, A.G.S., Airoldi, C. 2000. Immobilization of the pesticide 2,4-dichlorophenoxyacetic acid on a silica gel surface. *Pest Management Science* 56(5):419–424.

Price, P.M., Clark, J.H., Macquarrie, D.J. 2000. Modified silicas for clean technology. *Journal of the Chemical Society, Dalton Transactions* (2):101–110.

Puntener, K., Hellman, M.D., Kuester, E., Hegedus, L.S. 2000. Synthesis and complexation properties of poly(ethylene glycol)-linked mono- and bis-dioxocyclams. *Journal of Organic Chemistry* 65(24):8301–8306.

Rafiee, E., Rashidzadeh, S., Azad, A. 2007. Silica-supported heteropoly acids: Highly efficient catalysts for synthesis of alpha-aminonitriles, using trimethylsilyl cyanide or potassium cyanide. *Journal of Molecular Catalysis A: Chemical* 261(1):49–52.

Rafiee, E., Shahbazi, F. 2006. One-pot synthesis of dihydropyrimidones using silica-supported heteropoly acid as an efficient and reusable catalyst: Improved protocol conditions for the Biginelli reaction. *Journal of Molecular Catalysis A: Chemical* 250(1–2):57–61.

Ramalingam, B., Neuburger, M., Pfaltz, A. 2007. Synthesis of chiral C-2-symmetric methylene- and boron-bridged bis(imidazolines). *Synthesis Stuttgart* (4):572–582.

Rapolu, R.K., NabaMukul, B.M., Bommineni, S.R., Potham, R., Mulakayala, N., Oruganti, S. 2013. Silica sulfuric acid: A reusable solid catalyst for the synthesis of N-substituted amides via the Ritter reaction. *RCS Advances* 3:5332–5337.

Rasalkar, M.S., Bhilare, S.V., Deorukhkar, A.R., Darvatkar, N.B., Salunkhe, M.M. 2007. Heteropoly acid in ionic liquid—An efficient and recyclable system for one-pot three-component Mannich reaction. *Canadian Journal of Chemistry Revue Canadienne De Chimie* 85(1):77–80.

Ray, S., Mapolie, S.F., Darkwa, J. 2007. Catalytic hydroxylation of phenol using immobilized late transition metal salicylaldimine complexes. *Journal of Molecular Catalysis A: Chemical* 267(1–2):143–148.

Ren, Y.M., Cai, C. 2008. Convenient and efficient method for synthesis of substituted 2-amino-2-chromenes using catalytic amount of iodine and K(2)CO(3) in aqueous medium. *Catalysis Communications* 9(6):1017–1020.

Ribeiro, S.M., Serra, A.C., Gonsalves, A.M.D. 2011. Silica grafted polyethylenimine as heterogeneous catalyst for condensation reactions. *Applied Catalysis A: General* 399(1–2):126–133.

Rocchiccioli-Deltcheff, C., Amirouche, M., Herve, G., Fournier, M., Che, M., Tatibouet, J.M. 1990. Structure and catalytic properties of silica-supported polyoxomolybdates. 2. Thermal behavior of unsupported and silica-supported 12-molybdosilic acid catalysts from IR and catalytic reactivity studies. *Journal of Catalysis* 126(2):591–599.

Romanelli, G.P., Bennardi, D., Ruiz, D.M., Baronetti, G., Thomas, H.J., Autino, J.C. 2004. A solvent-free synthesis of coumarins using a Wells-Dawson heteropolyacid as catalyst. *Tetrahedron Letters* 45(48):8935–8939.

Rosati, O., Curini, M., Marcotullio, M.C., Macchiarulo, A., Perfumi, M., Mattioli, L., Rismondo, F., Cravotto, G. 2007. Synthesis, docking studies and anti-inflammatory activity of 4,5,6,7-tetrahydro-2*H*-indazole derivatives. *Bioorganic and Medicinal Chemistry* 15:3463–3473.

Rosati, O., Curini, M., Marcotullio, M.C., Oball-Mond, G., Pelucchini, C., Procopio, A. 2010. Synthesis of 5-amino-1,10*b*-dihydro-2*H*-chromeno[3,4-*c*]pyridine-2,4(3*H*)-diones from coumarins and cyanoacetamides under basic conditions. *Synthesis Stuttgart* (2):239–248.

Roy, R.S., Gehring, A.M., Milne, J.C., Belshaw, P.J., Walsh, C.T. 1999. Thiazole and oxazole peptides: Biosynthesis and molecular machinery. *Natural Product Reports* 16(2):249–263.

Sabitha, G., Srinivas, C., Raghavendar, A., Yadav, J.S. 2010. Phosphomolyhdic acid (PMA)-SiO_2 as a heterogeneous solid acid catalyst for the one-pot synthesis of 2*H*-indazolo[1,2-*b*] phthalazine-triones. *Helvetica Chimica Acta* 93(7):1375–1380.

Saheli, P., Dabiri, M., Zolfigol, M.A., Baghbanzadeh, M. 2005. A novel method for the one-pot three-component synthesis of 2,3-dihydroquinazolin-4(1H)-ones. *Synlett* 7:1155–1157.

Salgado-Zamora, H., Campos, E., Jimenez, R., Cervantes, H. 1998. On the reactivity of hydroximoyl chlorides preparation of 2-aryl imidazolines. *Heterocycles* 47(2):1043–1049.

Sarda, S.R., Pathan, M.Y., Paike, V.V., Pachmase, P.R., Jadhav, W.N., Pawar, R.P. 2006. A facile synthesis of flavones using recyclable ionic liquid under microwave irradiation. *ARKIVOC* (xvi):43–48.

Saxena, A.K., Pandey, S.K., Seth, P., Singh, M.P., Dikshit, M., Carpy, A. 2001. Synthesis and QSAR studies in 2-(*N*-aryl-*N*-aroyl)amino-4,5-dihydrothiazole derivatives as potential antithrombotic agents. *Bioorganic and Medicinal Chemistry* 9(8):2025–2034.

Saxena, S., Makrandi, J.K., Grover, S.K. 1985. Synthesis of 5-hydroxyflavones and or 7-hydroxyflavones using a modified phase transfer-catalyzed Baker-Venkataraman transformation. *Synthesis Stuttgart* 6–7:696–697.

Sayyafi, M., Seyyedhamzeh, M., Khavasi, H.R., Bazgir, A. 2008. One-pot, three-component route to 2H-indazolo[2,1-*b*]phthalazine-triones. *Tetrahedron* 64(10):2375–2378.

Schutz, H. 1982. *Benzodiazepines*. Heidelberg: Springer.

Schwegler, M.A., Vinke, P., Vandereijk, M., Van Bekkum, H. 1992. Activated carbon as a support for heteropolyanion catalysts. *Applied Catalysis A: General* 80(1):41–57.

Shaabani, A., Ghadari, R., Ghasemi, S., Pedarpour, M., Rezayan, A.H., Sarvary, A., Ng, S.W. 2009. Novel one-pot three- and pseudo-five-component reactions: Synthesis of functionalized benzo[*g*]- and dihydropyrano[2,3-g]chromene derivatives. *Journal of Combinatorial Chemistry* 11(6):956–959.

Shanthi, G., Perumal, P.T. 2007. An eco-friendly synthesis of 2-aminochromenes and indolyl chromenes catalyzed by $InCl_3$ in aqueous media. *Tetrahedron Letters* 48(38):6785–6789.

Sharma, R.K., Mittal, S., Koel, M. 2003. Analysis of trace amounts of metal ions using silica-based chelating resins: A green analytical method. *Critical Reviews in Analytical Chemistry* 33(3):183–197.

Sharma, R.K., Sharma, C. 2011. Zirconium(IV)-modified silica gel: Preparation, characterization and catalytic activity in the synthesis of some biologically important molecules. *Catalysis Communications* 12(5):327–331.

Shaterian, H.R., Hosseinian, A., Ghashang, M. 2009. Reusable silica supported poly phosphoric acid catalyzed three-component synthesis of 2H-indazolo[2,1-*b*]phthalazine-trione derivatives. *ARKIVOC* (ii):59–67.

Sheldon, R.A., Downing, R.S. 1999. Heterogeneous catalytic transformations for environmentally friendly production. *Applied Catalysis A: General* 189:163–183.

Shikata, S., Nakata, S., Okuhara, T., Misono, M. 1997. Catalysis by heteropoly compounds. 32. Synthesis of methyl tert-butyl ether catalyzed by heteropolyacids supported on silica. *Journal of Catalysis* 166(2):263–271.

Singh, O.V., Kapil, R.S. 1993. A new route to 2-aryl-4-quinolones via thallium(III) *p*-tolylsulphonate mediated oxidation of 2-aryl-1,2,3,4-tetrahydro-4-quinolones. *Synthetic Communications* 23:277–283.

Sirkecioglu, O., Tulinli, N., Akar, A. 1995. Chemical aspects of santalin as a histological stain. *Journal of Chemical Research, Synopses* 502.

Smith, P.W., Sollis, S.L., Howes, P.D., Cherry, P.C., Starkey, I.D., Cobley, K.N., Weston, H., et al. 1998. Dihydropyrancarboxamides related to zanamivir: A new series of inhibitors of influenza virus sialidases. 1. Discovery, synthesis, biological activity, and structure–activity relationships of 4-guanidino- and 4-amino-4*H*-pyran-6-carboxamides. *Journal of Medicinal Chemistry* 41(6):787–797.

Solhy, A., Elmakssoudi, A., Tahir, R., Karkouri, M., Larzek, M., Bousmina, M., Zahouily, M. 2010a. Clean chemical synthesis of 2-amino-chromenes in water catalyzed by nanostructured diphosphate $Na_2CaP_2O_7$. *Green Chemistry* 12:2261–2267.

Solhy, A., Tahir, R., Sebti, S., Skouta, R., Bousmina, M., Zahouily, M., Larzek, M. 2010b. Efficient synthesis of chalcone derivatives catalyzed by re-usable hydroxyapatite. *Applied Catalysis A: General* 374(1–2):189–193.

Soltero-Higgin, M., Carlson, E.E., Phillips, J.H., Kiessling, L.L. 2004. Identification of inhibitors for UDP-galactopyranose mutase. *Journal of the American Chemical Society* 126(34):10532–10533.

Sreekumar, R., Padmakumar, R. 1998. Simple, efficient and convenient synthesis of pyrroles and pyrazoles using zeolites. *Synthetic Communications* 28:1661–1665.

Strappaveccia, G., Lanari, D., Gelman, D., Pizzo, F., Rosati, O., Curini, M., Vaccaro, L. 2013. Efficient synthesis of cyanohydrin trimethylsilyl ethers via 1,2-chemoselective cyanosilylation of carbonyls. *Green Chemistry* 15:199–204.

Street, L.J., Sternfeld, F., Jelley, R.A., Reeve, A.J., Carling, R.W., Moore, K.W., McKernan, R.M., et al. 2004. Synthesis and biological evaluation of 3-heterocyclyl-7,8,9,10-tetrahydro-(7,10-ethano)-1,2,4-triazolo[3,4-*a*]phthalazines and analogues as subtype-selective inverse agonists for the GABA(A)α 5 benzodiazepine binding site. *Journal of Medicinal Chemistry* 47(14):3642–3657.

Torviso, M.D., Alesso, E.N., Moltrasio, G.Y., Vazquez, P.G., Pizzio, L.R., Caceres, C.V., Blanco, M.N. 2006. Effect of the support on a new metanethole synthesis heterogeneously catalyzed by Keggin heteropolyacids. *Applied Catalysis A: General* 301(1):25–31.

Turek, W., Haber, J., Krowiak, A. 2005. Dehydration of isopropyl alcohol used as an indicator of the type and strength of catalyst acid centres. *Applied Surface Science* 252:823–827.

Turk, C., Svete, J., Stanovnik, B., Golic, L., Golic-Grdadolnik, S., Golobic, A., Selic, L. 2001. Regioselective 1,3-dipolar cycloadditions of (1Z)-1-(arylmethylidene)-5,5-dimethyl-3-oxopyrazolidin-1-ium-2-azomethine imines to acetylenic dipolarophiles. *Helvetica Chimica Acta* 84(1):146–156.

Ueno, M., Imaizumi, K., Sugita, T., Takata, I., Takeshita, M. 1995. Effect of a novel anti-rheumatic drug, TA-383, on type-II collagen-induced arthritis. *International Journal of Immunopharmacology* 17(7):597–603.

Varma, R.S., Saini, R.K., Kumar, D. 1998. An expeditious synthesis of flavones on montmorillonite K 10 clay with microwaves. *Journal of Chemical Research, Synopses* 348–349.

Vàzquez, P.G., Blanco, M.N., Caceres, C.V. 1999. Catalysts based on supported 12-molybdophosphoric acid. *Catalysis Letters* 60(4):205–215.

Vázquez, P., Pizzio, L., Romanelli, G., Autino, J., Cáceres, C., Blanco, M. 2002. Mo and W heteropolyacid based catalysts applied to the preparation of flavones and substituted chromones by cyclocondensation of o-hydroxyphenyl aryl 1,3-propanediones. *Applied Catalysis A: General* 235:233–240.

Vizi, E.S. 1986. Compounds acting on α-1-adrenoceptors and α-2-adrenoceptors-agonists and antagonists. *Medicinal Research Reviews* 6(4):431–449.

Vorbruggen, H., Krolikiewicz, K. 1993. A simple synthesis of δ-2-oxazolines, δ-2-oxazines, δ-2-thiazolines and 2-substituted benzoxazoles. *Tetrahedron* 49(41):9353–9372.

Wang, H.K., Morris-Natschke, S.L., Lee, K.H. 1997. Recent advances in the discovery and development of topoisomerase inhibitors as antitumor agents. *Medical Research and Review* 17:367.

Wang, R., Huang, T., Shi, L., Li, B., Lu, X.X. 2007. Heteropoly acids catalyzed direct mannich reactions: Three-component synthesis of N-protected β-amino ketones. *Synlett* 2007(14):2197–2200.

Watanabe, N., Kabasawa, Y., Takase, Y., Matsukura, M., Miyazaki, K., Ishihara, H., Kodoma, K., Adachi, H. 1998. 4-Benzylamino-1-chloro-6-substituted phthalazines: Synthesis and inhibitory activity toward phosphodiesterase 5. *Journal of Medicinal Chemistry* 41:3367-3372.

Wheeler, T.S. 1952. Flavone. *Organic Syntheses* 32:72.

Wu, J.H., Wang, X.H., Yi, Y.H., Lee, K.H. 2003. Anti-AIDS agents 54. A potent anti-HIV chalcone and flavonoids from genus Desmos. *Bioorganic and Medicinal Chemistry Letters* 13(10):1813–1815.

Yadav, G.D., Kirthivasan, N. 1995. Single-pot synthesis of methyl tert-butyl ether from tert-butyl alcohol and methanol: Dodecatungstophosphoric acid supported on clay as an efficient catalyst. *Journal of the Chemical Society, Chemical Communications* (2):203–204.

Yadav, J.S., Reddy, B.V.S., Eshwaraiah, B., Anuradha, K. 2002. Amberlyst-15: A novel and recyclable reagent for the synthesis of 1,5-benzodiazepines in ionic liquids. *Green Chemistry* 6:592–594.

Yang, D., Yip, Y.C., Wang, X.C. 1997. Oxidative cleavage of aryl oxazolines using methyl(trifluoromethyl)dioxirane generated in situ. *Tetrahedron Letters* 38(40): 7083–7086.

Yano, S., Tachibana, H., Yamada, K. 2005. Flavones suppress the expression of the high-affinity IgE receptor Fc epsilon RI in human basophilic KU812 cells. *Journal of Agricultural and Food Chemistry* 53(5):1812–1817.

Zhang, A.Q., Zhang, M., Chen, H.H., Chen, J., Chen, H.Y. 2007. Convenient method for synthesis of substituted 2-amino-2-chromenes. *Synthetic Communications* 37(1–3):231–235.

Zhang, F.M., Wang, J., Yuan, C.S., Ren, X.Q. 2005. A pronounced catalytic activity of heteropoly compounds supported on dealuminated USY for liquid-phase esterification of acetic acid with n-butanol. *Catalysis Letters* 102(3–4):171–174.

Zolfigol, M.A. 2001. Silica sulfuric acid/$NaNO_2$ as a novel heterogeneous system for production of thionitrites and disulfides under mild conditions. *Tetrahedron* 57:9509–9511.

Zolfigol, M.A., Veisi, H., Mohanazadeh, F., Sedrpoushan, A. 2011. Synthesis and application of modified silica sulfuric acid as a solid acid heterogeneous catalyst in Michael addition reactions. *Journal of Heterocyclic Chemistry* 48:977–986.

2 Eco-Benign Synthesis of Indole Derivatives Employing Diverse Heterogeneous Catalysts

Renuka Jain, Kanti Sharma, and Deepak Kumar

CONTENTS

2.1 INTRODUCTION

Heterocycles are important chemical entities finding application as medicines, agrochemicals, and cosmetics. The five-membered *N*-heterocycle indole ring system is the most widely distributed, occurring in nature as alkaloids. Since it was first isolated by treatment of indigo dye with oleum, the name indole is a combination of the words indigo and oleum. Among the numerous structurally diverse derivatives, even the simple indole derivatives such as tryptamine, tryptophan, and serotonin show significant biological activity. Hence, it is not surprising that this structural motif is also an important component in many of today's pharmaceuticals (Figure 2.1).

Indole itself is found to possess psychotropic (Sim 1961), fungicidal (Oimomi et al. 1974), tuberculostatic (Nakhi et al. 2011), and antibacterial (Dodd et al. 2012) activities. Various substituted indole derivatives act as ovulation inhibitors (Braga et al. 1974) and

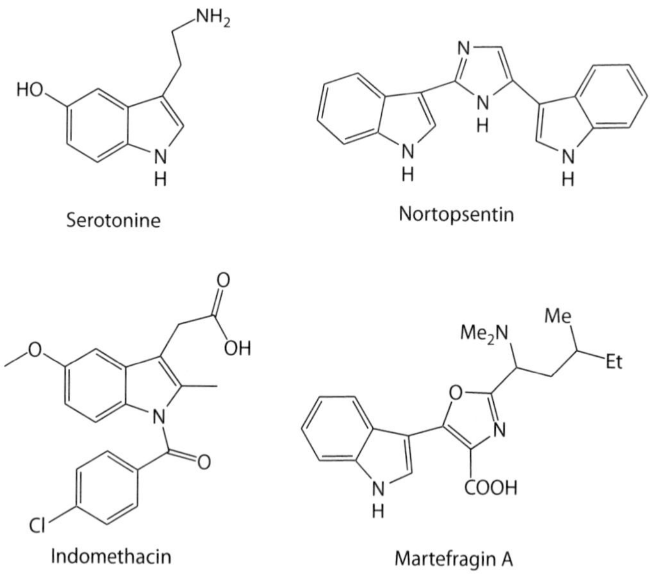

FIGURE 2.1 Examples of some biologically active indoles.

are potent female fertility controlling agents (Kong and Cheng 1989). Methyl derivatives of indole possess combined bacteriostatic and respiration-inhibiting effects and demonstrate insecticidal activity against the body louse (Eddy and Carson 1946; Zetterberg 1947). The phenyl derivatives of indole exhibit virus-inhibiting effects (Avolio et al. 2004), antihistaminic activity (Xu and Xu 2001), and hypocholesterolemic activity (Yoshikuni et al. 1992). Some azo-indoles are well known for antiviral (Boyle et al. 1970) and bactericidal (Chikvaidze et al. 2009) activities. Monge et al. (1984) reported the condensed derivatives of indole, pyridazino[4,5-*b*]indole and pyridazino[4,5-*a*]indole, as inhibitors of human blood platelet aggregation and thromboxane synthetase. Two indoles have received attention as antidepressants are iprindole and tandamine. These drugs appear to function by blocking uptake by andrenergic neurons of catecholamines and serotonin.

This basic skeleton is present in the neurotransmitter serotonin (5-hydroxytryptamine [5-HT]), which plays an important role in a variety of processes through the activation of 5-HT receptors (Kikuchi et al. 1999). The nucleus exhibits a remarkable range of activities as demonstrated by the nonsteroidal anti-inflammatory agent indomethacin (Reinicke 1977), the peptidal mimetic somatostatin agonist (Rohrer et al. 1998), selective dopamine D_4 receptor agonist cyanoindole derivatives (Hubner et al. 2000), and potent and selective factor Xa inhibitors (Heinelt et al. 2001). A variety of natural products containing the indole ring have been identified, such as the antitumoral nortopsentins (Gu et al. 1999), the potent inhibitors of lipid peroxidation martefragin A (Takahashi et al. 1998), other 5-(3-indolyl)oxazoles (Nishida et al. 2000), and the protein kinase C activator indolactam V (Meseguer et al. 1999), as well as fumitremorgin C (van Loevezijn et al. 2001), a recently identified specific reversal agent for the breast cancer resistance protein transporter.

The discovery of new environmentally benign processes for synthesizing organic compounds is an important goal in modern chemistry (Izumi and Onaka 1992; Sheldon and Downing 1999). Without any doubt, there is an increasing tendency to apply catalytic methods for the synthesis and production of multifunctional, complex, fine, and specialty chemicals useful as agrochemicals and pharmaceuticals. In this, the use of heterogeneous catalysts has attracted much attention in a variety of organic chemical fields from sustainable and industrial standpoints because of its advantages such as the ease of preparation (impregnation and calcination) at low cost, high thermal and chemical stabilities, lack of contamination of the products by metallic species, and excellent recyclability of the catalysts (Izumi and Onaka 1992).

The rich structural diversity encountered in indole derivatives, in addition to their biological and pharmaceutical relevance, has motivated more than 10 decades of research aiming at developing economical, efficient, and selective synthesis strategies for these compounds (Sundberg 1984; Tois et al. 2003; Joucla and Djakovitch 2009; Robinson 1982). These methodologies include the Fischer synthesis from aryl hydrazones (Hughes 1993; Robinson 1982), the Leimgruber–Batcho synthesis from *o*-nitrotoluenes and dimethylformamide acetals (Clark and Repke 1984), the Gassman synthesis from *N*-haloanilines (Johnson and Aristoff 1990; Wright et al. 1996), and the Madelung cyclization of *N*-acyl-*o*-toluidines (Wacker and Kasireddy 2002).

These stoichiometric strategies, while successful and commonly applied in the fine chemical industry, suffer from a low structural diversity and are thus not suitable for the selective synthesis of highly functionalized compounds. As a solution, several transition-metal-catalyzed procedures, dedicated either to the construction or the transformation of such heterocycles, have been reported. For instance, the synthesis of indoles has been achieved by palladium-induced cycloadditions of 2-haloanilines with terminal or internal alkynes (Zeni and Larock 2006), and intra- or intermolecular reactions of 2-alkynyl anilides with aryl- or alkylhalides (Cacchi et al. 2003).

However, whatever the target nucleus, these procedures have several drawbacks. The main one is unacceptable metal contamination of the product, generally over the accepted limits as expressed in the medicinal regulations, which often prevents further industrial development of these methods. To resolve this situation, several research groups have developed alternative methodologies based on the use of heterogeneous catalysts, which are advantageous with from environmental and economic points of view, offering viable procedures at both the laboratory and the industrial scales.

Recent reports of the synthesis of indoles and their reaction using diverse heterogeneous catalysts have been classified into five major groups, as follows:

2.1.1 Metal-Based Catalysts

2.1.1.1 Palladium (Pd)

Batail et al. (2009) constructed some indoles in good to high isolated yields via a new ligand- and salt-free heterogeneous palladium-catalyzed Larock reaction (Scheme 2.1). Recycling studies have shown that the palladium catalysts can be readily recovered and reused. Reactions and recovery of the palladium catalysts can be carried out in the presence of air, without any particular precaution.

2.1.1.2 Solid-Supported Palladium (Pd)

Sakai et al. (2008) carried out the one-pot four-step synthesis of indoles using both solid-supported heterogeneous and homogeneous palladium catalysts and reagents (Scheme 2.2). This combination of catalysts and reagents results in a dramatic increase in yield.

2.1.1.3 Zinc Oxide (ZnO)

The synthesis of pentacyclic indole derivatives, 9-phenyl-9,13*c*-dihydro-6*H*-chromeno[4′,3′:4,5]-thiopyrano[2,3-*b*]indoles (Scheme 2.3), has been achieved via domino Knoevenagel-hetero-Diels–Alder reactions of indolin-2-thiones and O-propargylated salicylaldehyde derivatives in the presence of 10 mol% of ZnO as a heterogeneous catalyst in CH_3CN (Kiamehr and Moghaddam 2009). When the

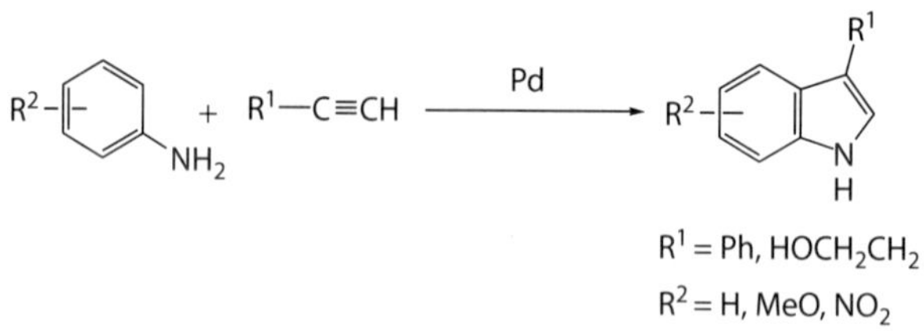

SCHEME 2.1

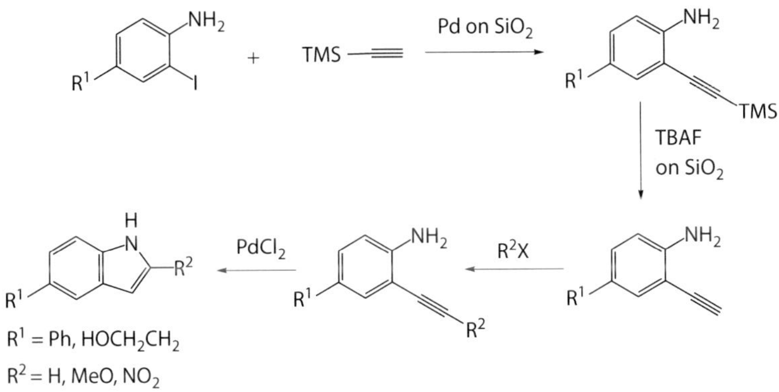

SCHEME 2.2

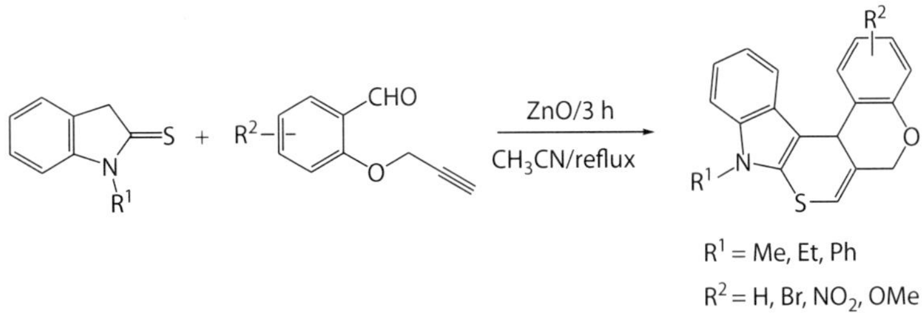

SCHEME 2.3

reaction was carried out at room temperature in toluene or CH_3CN, the reaction rate was slow and the product was obtained in a yield of only 33%–35% after 48 h. Running the reaction in water did not provide the desired product. When the reaction was carried out under refluxing conditions using ZnO (100 mol%) in CH_3CN, the reaction time was reduced to 3 h with 90% yield. Decreasing the ratio of ZnO to 10 mol% afforded the same result, but with 5 mol% of ZnO, the reaction took longer. The use of ZnO as a nontoxic, noncorrosive, commercially available, and inexpensive heterogeneous catalyst makes it a useful and attractive strategy for the synthesis of pentacyclic indole derivatives.

2.1.1.4 Ferric Hydrogen Sulfate ($Fe(HSO_4)_3$)

New bis(indolyl)methanes were synthesized in excellent yields by the reaction of indole derivatives with aromatic and aliphatic aldehydes in the presence of ferric hydrogen sulfate as an efficient, inexpensive, heterogeneous, reusable, and nontoxic catalyst (Scheme 2.4) (Rahimizadeh et al. 2009). This method is effective for aldehydes bearing both electron-withdrawing and electron-donating substituents on the aromatic ring. Aliphatic aldehydes also react satisfactorily under these conditions. Comparing the catalytic efficiency of $Fe(HSO_4)_3$ with Lewis acid–like $FeCl_3$ and Brønsted acid–like HCl showed that ferric hydrogen sulfate was acting as a bifunctional catalyst.

2.1.1.5 Ferric Fluoride (FeF_3)

An efficient method for the construction of bis(indolyl)methanes and bis(indolyl) glycol conjugates from carbonyl compounds with indoles in the presence of a catalytic amount of iron(III) fluoride under solvent-free conditions has been reported by Kamble et al. (2006) (Scheme 2.4). The desired products were obtained in high yields with a simple and environmentally benign procedure. The use of iron(III) fluoride is feasible because of its stability, easy handling, easy recovery, reusability, and good activity.

2.1.1.6 Alumina-Methane Sulfonic Acid (Al_2O_3-$MeSO_3H$)

Indole and its derivatives undergo smooth thiocyanation with ammonium thiocyanate in the presence of a mixture of Al_2O_3-$MeSO_3H$ (AMA) under mild conditions without the use of any organic solvents to afford aryl thiocyanates in excellent yields and with high selectivity at room temperature (Hosseini-Sarvari and Tavakolian 2008) (Scheme 2.5).

2.1.1.7 Hetero-Bimetallic (Pd-Cu)

Chouzier et al. (2004) have described the one-pot indole synthesis catalyzed by new heterogeneous bimetallic (Pd-Cu) catalysts via the Sonogashira C–C coupling reaction with up to 100% yield in short reaction times for the first time (Scheme 2.6). In all cases, a complete selectivity toward the indole compound is observed. The heterogeneous catalysts, while being deactivated during the extraction procedure in air, remain highly active and selective when used in a continuous manner.

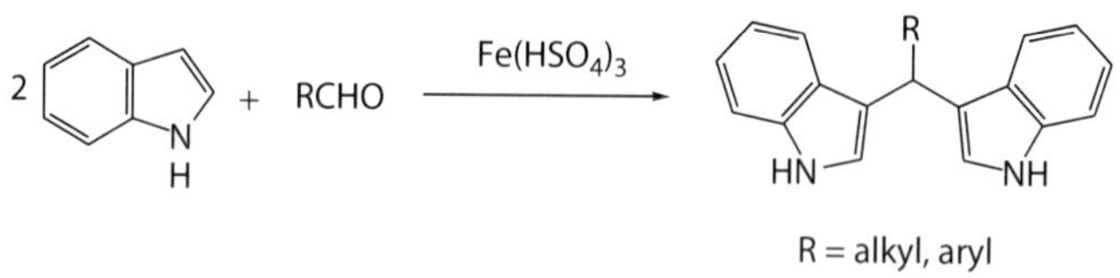

SCHEME 2.4

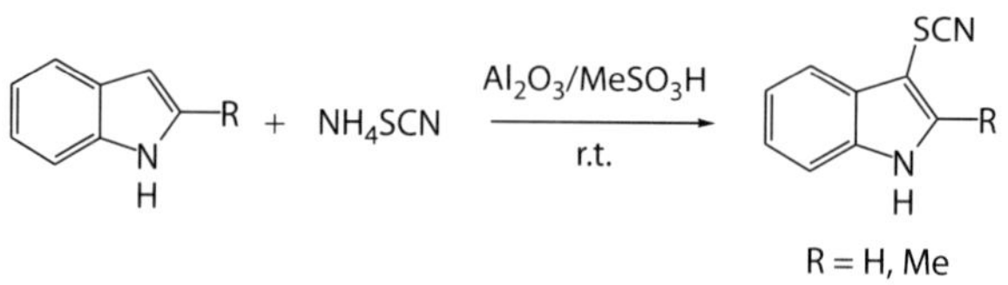

SCHEME 2.5

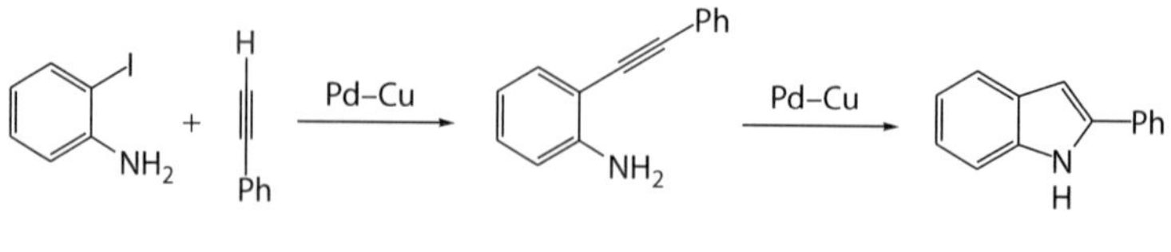

SCHEME 2.6

2.1.1.8 Ceria/Zirconia-Supported Ruthenium (Ru/CeO_2; Ru/ZrO_2)

Simple heterogeneous Ru/CeO_2 catalysts as well as Ru/ZrO_2 catalysts were found to be quite effective for the selective direct synthesis of indole via intramolecular dehydrogenative N-heterocyclization of 2-(2-aminophenyl)ethanol (Scheme 2.7). Ru/CeO_2 catalysts that were calcined at a relatively low temperature (200°C) showed excellent activity and produced indole in a yield of over 99% by the reaction at 140°C for 24 h (Shimura et al. 2011). $Ru^{IV}{=}O$ species interacting with supports are considered to be formed on ceria and zirconia, which act as good precursors for the catalytically active reduced-surface ruthenium species. Hot filtration tests and an ICP-AES analysis indicated that these Ru/CeO_2 catalysts act heterogeneously and leaching of ruthenium species into the solution is negligible. The catalysts could be recycled without a significant loss of activity, which suggests that the present oxide-supported catalysts are promising alternatives to conventional homogeneous catalysts.

2.1.1.9 Combined Phosphated Zirconia (P-Zr) and Bismuth Nitrate ($Bi(NO_3)_3{\cdot}5H_2O$)

Nadkarni and Nagarkar (2011) have synthesized 3-(1*H*-indol-3-yl)-1,1,3-trimethyl-1,2,3,4-tettrahydro-cyclopenta[*b*]indole by using a combined catalytic system of phosphated zirconia (P-Zr) and $Bi(NO_3)_3{\cdot}5H_2O$ (Scheme 2.8). However, the use of $Bi(NO_3)_3{\cdot}5H_2O$ or P-Zr separated as heterogeneous acid catalysts generated 2,2′-diindolylpropanes (DIPs) as a major product. Reactions of indoles and substituted indoles with ketones were carried out and it was observed that the yield was increased for methyl-substituted indoles at the 2-position and *N*-position which could be attributed to the activating effect of the ring due to the +I effect of the methyl group. When reacted with acetone, 3-methylindole did not give any product. The 3-position is the most reactive in the case of indole; blocking this position stops the addition of an acetone molecule, thereby inhibiting the reaction. Reactions with halo-indoles show a slight decrease in the yield of the product. This could be due to the −I effect of the halo group, which decreases the electron density on the ring, thereby deactivating the 3-position of indole.

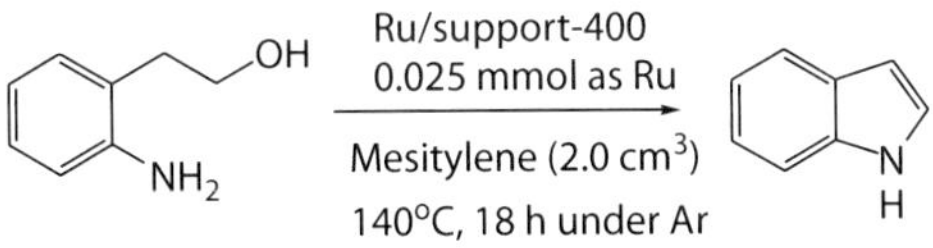

SCHEME 2.7

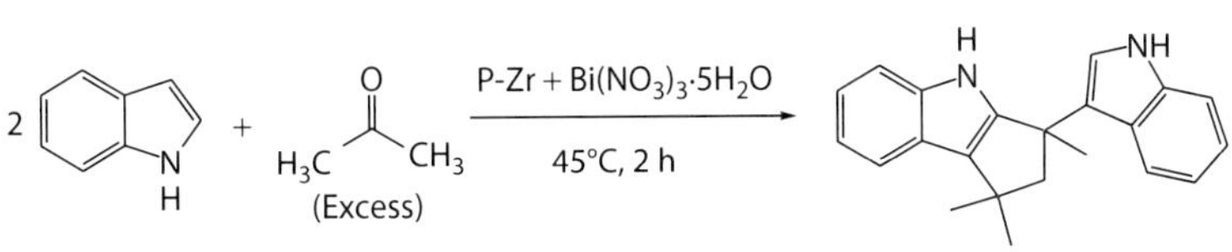

SCHEME 2.8

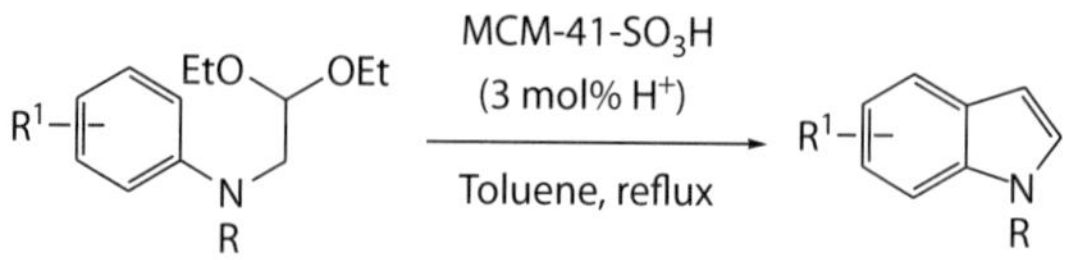

SCHEME 2.9

2.1.2 Polymer/Resin-Based Catalysts

2.1.2.1 Acidic Mesoporous Molecular Sieve (MCM-41-SO_3H)

Sun et al. (2013) reported a general and efficient method for the synthesis of 2,3-unsubstituted indoles by the intramolecular cyclization of *N*-benzyl-2-anilinoacetals using an acidic mesoporous molecular sieve (MCM-41-SO_3H) as a heterogeneous catalyst (Scheme 2.9). 2,3-Unsubstituted indoles bearing 7-substituent or strong electron-withdrawing substituents could also be achieved using this protocol, solving the problem of cyclization of 2-anilinoacetals bearing orthosubstituent or strong electron-withdrawing substituents. The heterogeneous catalyst could be conveniently recovered and reused without obvious loss of the catalytic activity. Therefore, this work provides an economic and environmentally benign method for the construction of various indole derivatives.

2.1.2.2 Indion Ina 225H Resin

Indion Ina 225H resin (a macroreticular sulfonic acid–based cation-exchange resin) was used as a heterogeneous, selective, recyclable, and eco-benign catalyst for liquid-phase electrophilic substitution reactions of substituted indoles with aldehydes to afford the corresponding bis(indolyl)methanes in excellent yields in short reaction times (Scheme 2.4). The versatility of this method has been proved with a wide range of aliphatic and aromatic aldehydes with various stereo-electronic factors. This method shows much better selectivity between aldehydes and ketones (Surasani et al. 2013). The methodology is well demonstrated toward the synthesis of biologically active bis(indolyl)methanes, namely vibrioindole, 4-(di(1*H*-indol-3-yl)methyl) benzene-1,2-diol, and streptindole. The catalyst can be recovered and reused for a minimum of five times without loss of activity, making its use more economical and environmentally friendly.

2.1.2.3 Crosslinked Poly(2-Acrylamido-2-Methyl Propane Sulfonic Acid)

Zohuri et al. (2013) synthesized novel 2,9-dihydro-2-oxo-4-aryl-1*H*-pyrido[2,3-*b*] indole-3-carbonitrile derivatives by condensation of substituted (triethoxymethyl) arene, 1-methyl-1*H*-indol-2-ol, and cyanoacetamide using catalytic amounts of crosslinked poly(2-acrylamido-2-methyl propane sulfonic acid) (AMPS) as an efficient and heterogeneous catalyst (Scheme 2.10). This polymeric solid acid catalyst is stable and can be easily recovered and reused without significant change in its

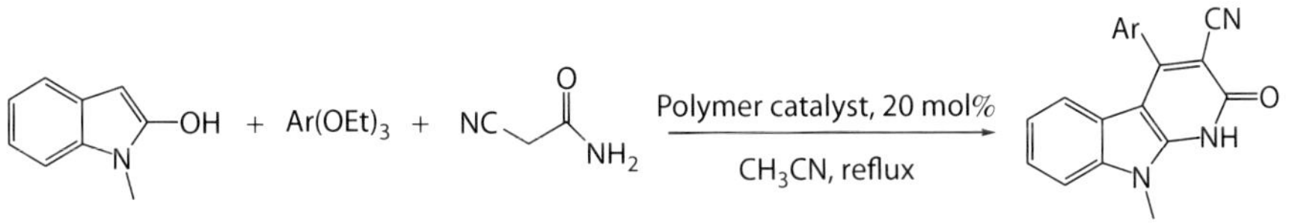

SCHEME 2.10

activity. The substituted functional groups on the aromatic ring of the (triethoxymethyl)arene affected the yield. It was found that the presence of electron-donating groups on the aryl rings of the (triethoxymethyl)arene decreased the yield of the products. However, all the employed (triethoxymethyl)benzene derivatives afforded the expected products in good to high yields.

2.1.2.4 Amberlyst-15

The heterogenous catalyst Amberlyst-15 was used for electrophilic substitution reaction of indoles with isatin derivatives to afford 3,3-di(indolyl)oxindoles in water (Scheme 2.11) (Sarrafi et al. 2012). The presence of electron-withdrawing groups on the aromatic ring of isatin or electron-donating groups on the isatin nitrogen did not affect the reaction time and yields. The notable features of this method are mild reaction conditions, simple operation, cleaner reaction profiles, low cost, and reusability of the catalyst, which make it an attractive and very useful process for the synthesis of important biological oxindoles.

2.1.2.5 Zeokarb-225

Zeokarb-225 is a heterogenous, recyclable, eco-benign catalyst for the liquid-phase electrophilic substitution reactions of indoles with aldehydes or indole aldehydes to afford the corresponding bis(indolyl)methanes or tris(indolyl)methanes in good yields (Scheme 2.4). The catalyst is also effective for the synthesis of new diindolylcarbazolylmethanes in good yields (Magesh et al. 2004).

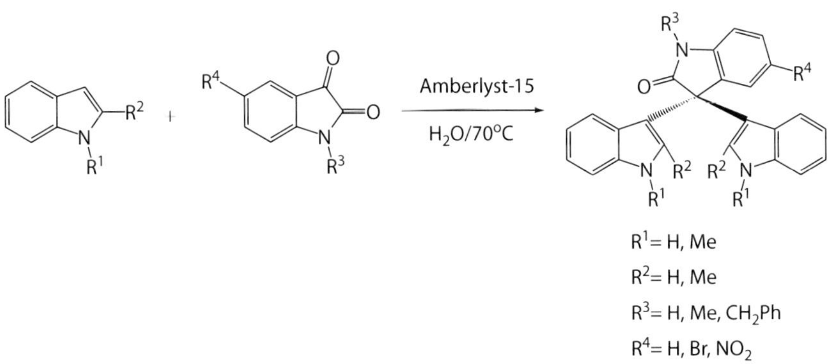

SCHEME 2.11

2.1.2.6 Zeolites

Kunkeler et al. (1997) reported that zeolite HNaY and zeolite-β are active and recyclable heterogeneous catalysts in the Fischer indole synthesis (Scheme 2.12). In certain cases, zeolite-β is capable of producing the linear indole isomer in excess.

2.1.2.7 Ru(III)-Exchanged FAU-Y Zeolite

Khorshidi and Tabatabaeian (2010) used Ru(III)-exchanged FAU-Y zeolite as an efficient reusable heterogeneous catalyst for preparation of oxindoles from condensation reaction of indoles with isatins under very mild reaction conditions (Scheme 2.11). In evaluating the reusability of the solid catalyst, only 8% loss of efficiency in terms of the product yield was observed after five runs, which promises minimization of waste.

2.1.3 Silica-Based Catalysts

2.1.3.1 Silica Sulfuric Acid

Silica sulfuric acid efficiently catalyzed the electrophilic substitution reaction of indoles with various isatins in dichloromethane to afford the corresponding oxindole derivatives in high yields at room temperature. The catalyst exhibited remarkable reusable activity (Scheme 2.11) (Azizian et al. 2006).

2.1.3.2 Modified Silica Sulfuric Acid

Modified silica sulfuric acid (MSSA), as a new type of silica sulfuric acid, was effectively used in the conjugate addition of indoles with Michael acceptors under mild conditions at room temperature (Scheme 2.13) (Zolfigol et al. 2011).

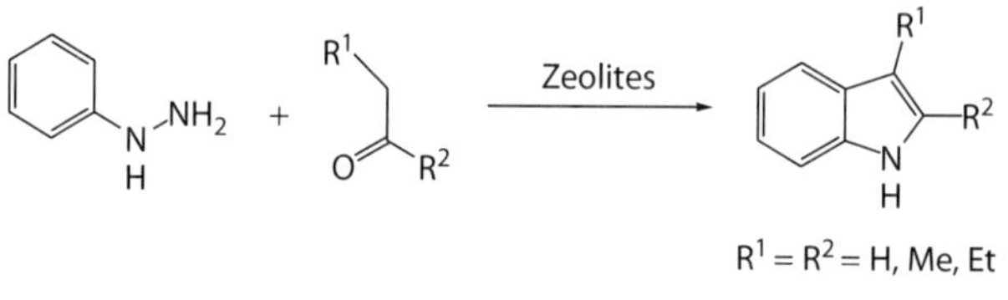

SCHEME 2.12

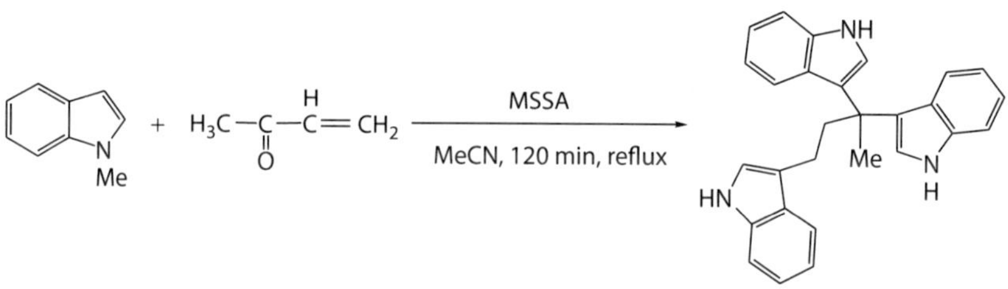

SCHEME 2.13

2.1.3.3 Silica-Supported $LiHSO_4$ (SiO_2-$LiHSO_4$)

Hasaninejad et al. (2009) used silica-supported $LiHSO_4$ (SiO_2-$LiHSO_4$) as a green, cheap, and efficient catalytic system for the synthesis of bis(indolyl)methanes via the condensation of indoles with carbonyl compounds under solvent-free conditions (Scheme 2.4). Indoles were efficiently condensed with a structurally diverse variety of aldehydes, including aromatic aldehydes, possessing electron-withdrawing and electron-releasing substituents as well as halogens on their aromatic rings and aliphatic aldehydes. The reaction of indoles with ketones produced the corresponding bis(indolyl)methanes with lower yields and longer reaction times. This catalyst was also successfully applied in the condensation of indole with terephthaldehyde. One of the most interesting properties of SiO_2-$LiHSO_4$ is its ease of recycling. It can be reused after simple washing with Et_2O, thus rendering the process more economical. The yields of the condensation of indole with benzaldehyde in the second, third, and fourth uses of the catalyst were almost as high as in the first use.

2.1.3.4 Silica Gel–Supported Aluminum Chloride (SiO_2-$AlCl_3$)

Boroujeni and Parvanak (2011) studied stable and nonhygroscopic silica gel–supported aluminum chloride (SiO_2-$AlCl_3$), an environmentally friendly heterogeneous catalyst for the condensation of indole with aldehydes and ketones to afford bis(indolyl)methanes at room temperature under solvent-free conditions (Scheme 2.14). In order to examine the chemoselectivity of the present method, a series of competition experiments with a variety of aldehydes and ketones were performed, which found that SiO_2-$AlCl_3$ is able to discriminate between aldehydes and ketones. SiO_2-$AlCl_3$ showed a higher level of chemoselectivity than $AlCl_3$ in the reaction of indole with carbonyl compounds. Also, the use of $AlCl_3$ resulted in the formation of several unidentified products due to the strong Lewis acidity of $AlCl_3$. In addition, $AlCl_3$ may be required for use in reagent quantities due to its complexation of product molecules. Unlike $AlCl_3$, SiO_2-$AlCl_3$ is a milder catalyst that forms no stable complex with starting materials or products. The efficiency of SiO_2-$AlCl_3$ may also be attributed to its large surface area and its remarkable ability to act as a water scavenger.

2.1.3.5 Kaolin/KOH

The heterogeneous catalyst kaolin, preloaded with KOH, was used for the synthesis of 3-hydroxy-3-indolyl oxindoles from isatins and indoles in good to excellent yields by Srihari and Murthy (2011) (Scheme 2.15). The feasibility of the reaction was studied in various organic solvents such as CH_2Cl_2, CH_3CN, DMF, MeOH, and DMSO and found that the reaction did not proceed in CH_2Cl_2, DMF, and DMSO, whereas the

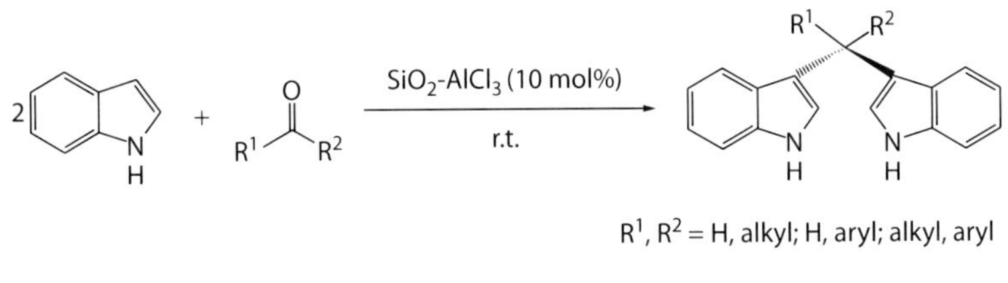

SCHEME 2.14

SCHEME 2.15

product was formed in MeOH and CH_3CN, with better yields in the case of MeOH. Electron-withdrawing and electron-donating groups on isatin did not have a significant effect on the product formation. In the case of substituted indoles, the reaction time of electron-donating groups was shorter. In all these reactions, electrophilic activation occurred only at the 3-position of the carbonyl of isatins. In terms of convenience, cost, and ease of workup, this method is attractive for large-scale operations.

2.1.4 Nanoparticle-Based Catalysts

2.1.4.1 CuO Nanoparticle

Suramwar et al. (2012) used predominant (111) facet CuO nanoparticles as heterogenous catalysts for the N-arylation reactions of indoles using indoacyl derivatives (Scheme 2.16). The nanoparticle catalysts were recycled and reused for further catalytic reactions with minimal loss in activity.

2.1.4.2 Nano-n-propylsulfonated γ-Fe_2O_3 (NPS-γ-Fe_2O_3)

Sobhani and Jahanshahi (2013) synthesized some indole derivatives, such as 2-indolyl-1-nitroalkanes and bis(2-nitrovinyl)benzene, by the reaction of indoles with different β-nitrostyrenes or carbonyl compounds (aldehydes and ketones) using nano-n-propylsulfonated γ-Fe_2O_3 (NPS-γ-Fe_2O_3) as a new magnetically recyclable heterogeneous catalyst (Scheme 2.17). The reaction of indole with a mild electron-donating

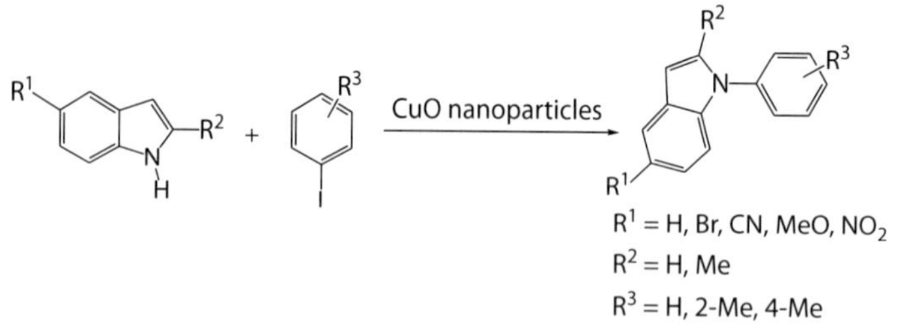

SCHEME 2.16

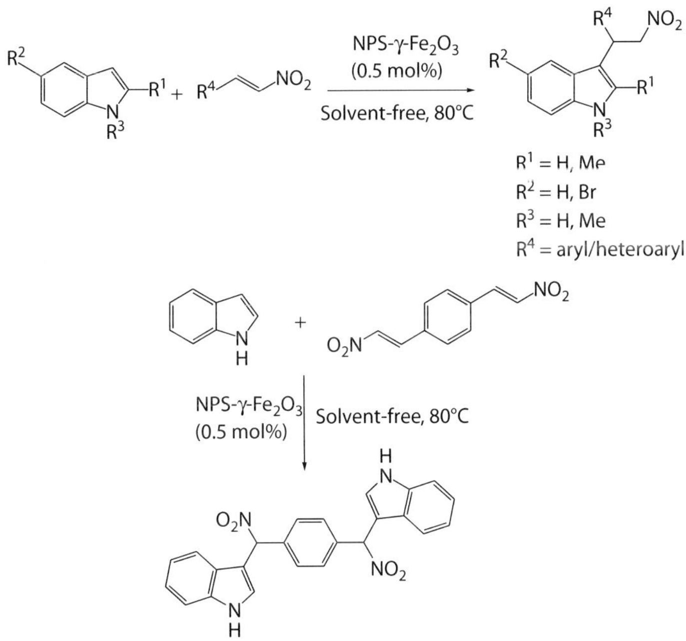

SCHEME 2.17

group such as methyl at the 2-position afforded the desired product in a shorter reaction time than a methyl group at the 1-position. NPS-γ-Fe_2O_3 was easily separated from the reaction mixture by magnetic decantation using a permanent magnet and reused four times without any significant decrease in activity.

2.1.4.3 Boron Trifluoride Supported on Nano-SiO_2

Saffar-Teluri (2013) used boron trifluoride supported on nano-SiO_2 as an efficient and heterogeneous catalyst for the electrophilic substitution reaction of indole with various aromatic aldehydes and isatins in CH_3OH to afford the corresponding bis(indolyl)methanes and oxindole derivatives in high yields at room temperature and under reflux conditions, respectively (Scheme 2.4). The catalyst can be reused several times without loss of catalytic activity. When the reaction was carried out in various solvents, such as CH_3OH, CH_3CH_2OH, H_2O, CH_3CN, CH_2Cl_2, $CHCl_3$, and CCl_4, it was found that the highest product yield was obtained in CH_3OH. The presence of an electron-donating substituent increased the reaction time in comparison with molecules having no substituent, while a substrate with an electron-withdrawing group on isatins decreased the reaction time.

2.1.5 Heteropoly Acid–Based Catalysts

2.1.5.1 Phosphomolybdic Acid ($H_3Mo_{12}O_{40}P$)

Chaskar et al. (2010) developed a simple, efficient, and environmentally friendly method for the synthesis of substituted indole derivatives using aryl hydrazines and

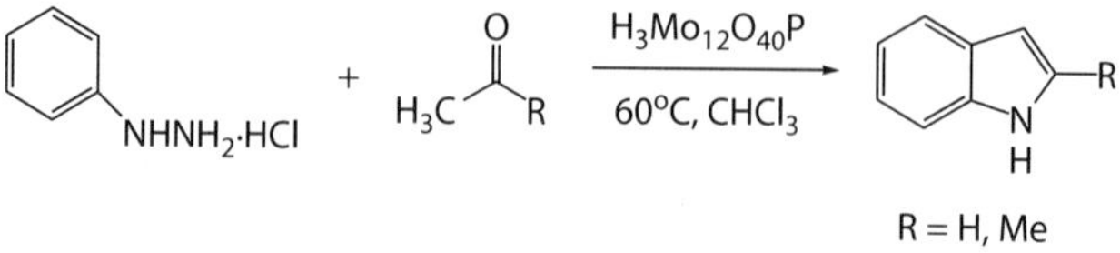

SCHEME 2.18

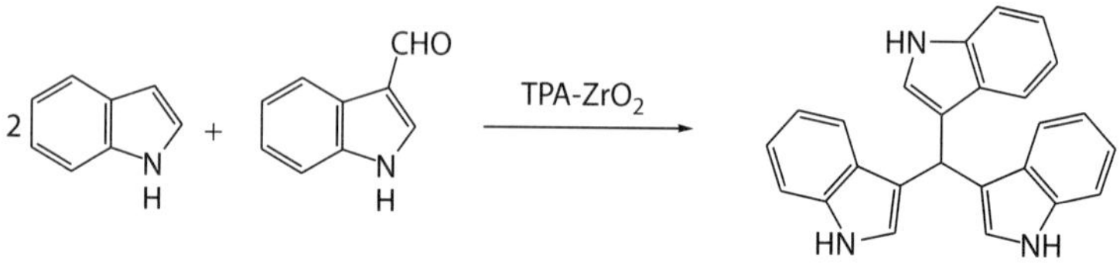

SCHEME 2.19

aldehydes or ketones as reactants, and phosphomolybdic acid, a heteropoly acid (HPA), as a heterogeneous catalyst (Scheme 2.18). In this, an electron-withdrawing group on aryl hydrazine lowers the rate of reaction. The catalytic activities of the heteropoly acid are much higher than those of conventional acid catalysts. It gives a high yield in a short reaction time.

2.1.5.2 Copper Molybdenum Oxide Phosphate ($Cu_{1.5}PMo_{12}O_{40}$)

Seyedi et al. (2009) synthesized some bis(indolyl)methanes in excellent yields in the presence of a catalytic amount of $Cu_{1.5}PMo_{12}O_{40}$ in molten tetraethylammonium chloride as an ionic liquid (Scheme 2.4). In this, as expected, aldehydes reacted more rapidly than ketones and afforded higher yields. The influence of electron-withdrawing and electron-donating substituents on the aromatic ring of aldehydes had no significant effect on the reaction times and the yields. An aliphatic aldehyde such as *n*-butanal afforded the product in a 73% yield in 20 min.

2.1.5.3 12-Tungstophosphoric Acid Supported on Zirconia ($H_3PW_{12}O_{40}$-ZrO_2)

12-Tungstophosphoric acid (TPA) supported on zirconia was employed as an efficient, heterogeneous catalyst for the liquid-phase electrophilic substitution reactions of indoles with aldehydes or indolecarboxaldehyde to afford bis(indolyl)methanes or tris(indolyl)methanes in high yields. The catalytic efficiency of TPA was increased by supporting on zirconia, which was expressed in terms of turnover number (TON) and turnover frequency (TOF; per hour) (Scheme 2.19) (Satam et al. 2008).

2.2 CONCLUSIONS

This chapter deals with recent developments in the discovery of various heterogeneous catalytic procedures for the synthesis or functionalization of the *N*-containing heterocycle, indole. These well-established approaches appear to lead to the development of effective, rapid, and, most of all, environmentally benign synthetic methods, as well as highlight original and practical improvements resulting in reliable tools, competitive with existing methodologies and of great interest for synthetic chemists, whether academic or industrial.

We consider that the operationally simple procedures described in this article represent attractive protocols, most of them contributing to sensible approaches toward greener chemistry. Obviously, the scope of this research topic is not limited to the few examples reported herein and we hope that this chapter will motivate further work in these directions.

REFERENCES

Avolio, S., Di Filippo, M., Harper, S., Narjes, F., Pacini, B., Pompei, M., Rowley, M. and Stansfield, I. 2004. Preparation of indole acetamides as inhibitors of the hepatitis-C virus NS5B polymerase. *PCT Int. Appl.* WO 2004087714 A1 20041014.

Azizian, J., Mohammadi, A. A., Karimi, N., Mohammadizadeh, M. R. and Karimi, A. R. 2006. Silica sulfuric acid a novel and heterogeneous catalyst for the synthesis of some new oxindole derivatives. *Catal. Commun.* 7(10):752–755.

Batail, N., Bendjeriou, A., Lomberget, T., Barret, R., Dufaud, V. and Djakovitch, L. 2009. First heterogeneous ligand- and salt-free Larock indole synthesis. *Adv. Synth. Catal.* 351(13):2055–2062.

Boroujeni, K. P. and Parvanak, K. 2011. Efficient and solvent-free synthesis of bis-indolylmethanes using silica gel-supported aluminium chloride as a reusable catalyst. *Chin. Chem. Lett.* 22(8):939–942.

Boyle, J. J., Raupp, W. G., Stanfield, F. J., Haff, R. F., Dick, E. C., D'Alessio, D. and Dick, C. R. 1970. Progress in rhinovirus chemotherapy. *Ann. N. Y. Acad. Sci.* 173:477–491.

Braga, P., Carraro, A., Corciulo, D., Martini, M. and Genovese, E. 1974. Inhibitory effect caused by some indole compounds of pineal origin on ovulation in rats. *Atti. Accad. Med. Lomb.* 29(1–4):171–178.

Cacchi, S., Fabrizi, G., Lamba, D., Marinelli, F. and Parisi, L. M. 2003. 2-Substituted 3-aryl- and 3-heteroarylindoles by the palladium-catalyzed reaction of *o*-trifluoroacetanilides with aryl bromides and triflates. *Synthesis.* (5):728–734.

Chaskar, A., Deokar, H., Padalkar, V., Phatangare, K. and Patil, S. K. 2010. Highly efficient and facile green approach for one-pot Fischer indole synthesis. *J. Korean Chem. Soc.* 54(4):411–413.

Chikvaidze, I. Sh., Samsoniya, S. A., Targamadze, N. L., Samsonia, N. Sh. and Kadzrishvili, D. O. 2009. Synthesis of some *N*-methyl-2-phenylindole arylazo derivatives. *Saqart. Mecniere. Akad. Macne Kimi. Seria.* 35(2):167–170.

Chouzier, S., Gruber, M. and Djakovitch, L. 2004. New hetero-bimetallic Pd-Cu catalysts for the one-pot indole synthesis via the Sonogashira reaction. *J. Mol. Catal.* 212(1–2):43–52.

Clark, R. D. and Repke, D. B. 1984. The Leimgruber–Batcho indole synthesis. *Heterocycles.* 22(1):195–221.

Dodd, R., Beaumard, F., Kiefer, L., Pages, J.-M. and Chevalier, J. 2012. New indolic molecules with antibacterial activity against a variety of gram-negative and -positive bacteria as well as multidrug-resistant bacteria. *PCT Int. Appl.* WO 2012084971 A1 20120628.

Eddy, G. W. and Carson, N. B. 1946. Further tests of the more promising materials (used as insecticides) against the body louse. *J. Econ. Entomol.* 39:763–767.

Gu, X. H., Wan, X. Z. and Jiang, B. 1999. Syntheses and biological activities of bis(3-indolyl)thiazoles, analogs of marine bis(indole)alkaloid nortopsentins. *Bioorg. Med. Chem. Lett.* 9(4):569–572.

Hasaninejad, A., Zare, A., Sharghi, H., Khalifeh, R. and Shekouhy, M. 2009. Silica-supported $LiHSO_4$ as a highly efficient, mild, heterogeneous, and reusable catalytic system for the solvent-free synthesis of bis(indolyl)methanes. *Phosphorus Sulfur Silicon Relat. Elem.* 184(10):2508–2515.

Heinelt, U., Herok, S., Matter, H. and Wildgoose, P. 2001. Solid-phase optimization of achiral amidinobenzyl indoles as potent and selective factor Xa inhibitors. *Bioorg. Med. Chem. Lett.* 11(2):227–230.

Hosseini-Sarvari, M. and Tavakolian, M. 2008. Synthesis of aryl thiocyanates using Al_2O_3/$MeSO_3H$ (AMA) as a novel heterogeneous system. *J. Chem. Res.* (6):318–321.

Hubner, H., Kraxner, J. and Gmeiner, P. 2000. Cyanoindole derivatives as highly selective dopamine D_4 receptor partial agonists: Solid-phase synthesis, binding assays, and functional experiments. *J. Med. Chem.* 43(23):4563–4569.

Hughes, D. L. 1993. Progress in the Fischer indole reaction: A review. *Org. Prep. Proced. Int.* 25(6):607–632.

Izumi, Y. and Onaka, M. 1992. Organic synthesis using aluminosilicates. *Adv. Catal.* 38:245–282.

Johnson, P. D. and Aristoff, P. A. 1990. General procedure for the synthesis of *o*-aminophenylacetates by a modification of the Gassman reaction. *J. Org. Chem.* 55(4):1374–1375.

Joucla, L. and Djakovitch, L. 2009. Transition metal-catalysed, direct and site-selective N1-, C2- or C3-arylation of the indole nucleus: 20 years of improvements. *Adv. Synth. Catal.* 351(5):673–714.

Kamble, V. T., Bandgar, B. P., Suryawanshi, S. B. and Bavikar, S. N. 2006. Green protocol for synthesis of bis-indolylmethanes and bis-indolylglycoconjugates in the presence of iron(III) fluoride as a heterogeneous, reusable, and eco-friendly catalyst. *Aust. J. Chem.* 59(11):837–840.

Khorshidi, A. and Tabatabaeian, K. 2010. Ru(III)-exchanged FAU-Y zeolite as an efficient heterogeneous catalyst for preparation of oxindoles. *Orient. J. Chem.* 26(3):837–841.

Kiamehr, M. and Moghaddam, F. M. 2009. An efficient ZnO-catalyzed synthesis of novel indole-annulated thiopyrano-chromene derivatives via Domino Knoevenagel-hetero-Diels-Alder reaction. *Tetrahedron Lett.* 50(48):6723–6727.

Kikuchi, C., Nagaso, H., Hiranuma, T. and Koyama, M. 1999. Tetrahydrobenzindoles: Selective antagonists of the 5-HT_7 receptor. *J. Med. Chem.* 42(4):533–535.

Kong, Y. C. and Cheng, K. F. 1989. Yuehchukene analogs and derivatives and related indoles, useful as female fertility-regulating agents, their synthesis, and pharmaceutical preparations. *Brit. U. K. Pat. Appl.* GB 2207670 A 19890208.

Kunkeler, P. J., Rigutto, M. S., Downing, R. S., De Vries, H. J. A. and Van Bekkum, H. 1997. Zeolite catalyzed regioselective synthesis of indoles. *Stud. Surf. Sci. Catal.* 105B:1269–1276.

Magesh, C. J., Nagarajan, R., Karthik, M. and Perumal, P. T. 2004. Synthesis and characterization of bis(indolyl)methanes, tris(indolyl)methanes and new diindolylcarbazolylmethanes mediated by Zeokarb-225, a novel, recyclable, eco-benign heterogeneous catalyst. *Appl. Catal.* 266(1):1–10.

Meseguer, B., Alonso-Diaz, D., Griebenow, N., Herget, T. and Waldmann, H. 1999. Natural product synthesis on polymeric supports-synthesis and biological evaluation of an indolactam library. *Angew. Chem. Int. Ed.* 38(19):2902–2906.

Monge, A., Aldana, I., Erro, A., Parrado, P., Font, M., Rocha, E., Prieto, I., Fremont-Smith, M. and Fernandez-Alvarez, E. 1984. New synthetic thromboxane A2 inhibitors with the pyridazine[4,5-b]indole and pyridazine[4,5-a]indole structures as platelet antiaggregants. *Anal. R. Acad. Farm.* 50(3):365–377.

Nadkarni, S. V. and Nagarkar, J. M. 2011. Synthesis of highly substituted indoles in presence of solid acid catalysts. *Green Chem. Lett. Rev.* 4(2):121–126.

Nakhi, A., Prasad, B., Reddy, U., Rao, R. M., Sandra, S., Kapavarapu, R., Rambabu, D., et al. 2011. A new route to indoles via in situ desilylation–Sonogashira strategy: Identification of novel small molecules as potential anti-tuberculosis agents. *Med. Chem. Commun.* 2(10):1006–1010.

Nishida, A., Fuwa, M., Naruto, S., Sugano, Y., Saito, H. and Nakagawa, M. 2000. Solid-phase synthesis of 5-(3-indolyl)oxazoles that inhibit lipid peroxidation. *Tetrahedron Lett.* 41(24):4791–4794.

Oimomi, M., Hamada, M. and Hara, T. 1974. Antimicrobial activities of indole. *J. Antibiot.* 27(12):987–988.

Rahimizadeh, M., Eshghi, H., Bakhtiarpoor, Z. and Pordel, M. 2009. Ferric hydrogen sulfate as a recyclable catalyst for the synthesis of some new bis(indolyl)methane derivatives. *J. Chem. Res.* (5):269–270.

Reinicke, C. Z. 1977. Nonsteroid antirheumatic agents. *Z. Gesamte Inn. Med.* 32(14):333–337.

Robinson, B. 1982. *The Fischer Indole Synthesis.* Wiley: Chichester, UK.

Rohrer, S. P., Birzin, E. T., Mosley, R. T., Berck, S. C., Hutchins, S. M., Shen, D. M., Xiong, Y., et al. 1998. Rapid identification of subtype-selective agonists of the somatostatin receptor through combinatorial chemistry. *Science.* 282(5389):737–740.

Saffar-Teluri, A. 2013. Boron trifluoride supported on nano-SiO_2: An efficient and reusable heterogeneous catalyst for the synthesis of bis(indolyl)methanes and oxindole derivatives. *Res. Chem. Intermediat.* 40(3):1061–1067.

Sakai, H., Tsutsumi, K., Morimoto, T. and Kakiuchi, K. 2008. One-pot/four-step/palladium-catalyzed synthesis of indole derivatives: The combination of heterogeneous and homogeneous systems. *Adv. Synth. Catal.* 350(16):2498–2502.

Sarrafi, Y., Alimohammadi, K., Sadatshahabi, M. and Norozipoor, N. 2012. An improved catalytic method for the synthesis of 3,3-di(indolyl)oxindoles using Amberlyst 15 as a heterogeneous and reusable catalyst in water. *Monatsh. Chem.* 143(11):1519–1522.

Satam, J. R., Parghi, K. D. and Jayaram, R. V. 2008. 12-Tungstophosphoric acid supported on zirconia as an efficient and heterogeneous catalyst for the synthesis of bis(indolyl) methanes and tris(indolyl)methanes. *Catal. Commun.* 9(6):1071–1078.

Seyedi, N., Khabazzadeh, H. and Saidi, K. 2009. $Cu_{1.5}PMo_{12}O_{40}$ as an efficient, mild and heterogeneous catalyst for the condensation of indole with carbonyl compounds. *Mol. Divers.* 13(3):337–342.

Sheldon, R. A. and Downing, R. S. 1999. Heterogeneous catalytic transformations for environmentally friendly production. *Appl. Catal. A.* 189(2):163–183.

Shimura, S., Miura, H., Wada, K., Hosokawa, S., Yamazoe, S. and Inoue, M. 2011. Ceria-supported ruthenium catalysts for the synthesis of indole via dehydrogenative N-heterocyclizatio. *Catal. Sci. Technol.* 1(8):1340–1346.

Sim, V. M. 1961. The indole nucleus, a common denominator of psychotropism. *J. Forensic Sci.* 6:39–47.

Sobhani, S. and Jahanshahi, R. 2013. Nano n-propylsulfonated γ-Fe_2O_3 (NPS-γ-Fe_2O_3) as a magnetically recyclable heterogeneous catalyst for the efficient synthesis of 2-indolyl-1-nitroalkanes and bis(indolyl)methanes. *New J. Chem.* 37(4):1009–1015.

Srihari, G. and Murthy, M. M. 2011. Kaolin/KOH is an efficient heterogeneous catalyst for the synthesis of 3-hydroxy-3-indolyl oxindoles. *Synth. Commun.* 41(18):2684–2692.

Sun, N., Hong, L., Huang, F., Ren, H., Mo, W., Hu, B., Shen, Z. and Hu, X. 2013. General and efficient synthesis of 2,3-unsubstituted indoles catalyzed by acidic mesoporous molecular sieves. *Tetrahedron.* 69(19):3927–3933.

Sundberg, R. J. 1984. *Pyrroles and their Benzo Derivatives: Synthesis and Applications*, vol. 4. Pergamon: Oxford, UK, pp. 313–376.

Suramwar, N. V., Thakare, S. R., Karade, N. N. and Khaty, N. T. 2012. Green synthesis of predominant (111) facet CuO nanoparticles: Heterogeneous and recyclable catalyst for N-arylation of indoles. *J. Mol. Catal. A Chem.* 359:28–34.

Surasani, R., Kalita, D. and Chandrasekhar, K. B. 2013. Indion Ina 225H resin as a novel, selective, recyclable, eco-benign heterogeneous catalyst for the synthesis of bis(indolyl) methanes. *Green Chem. Lett. Rev.* 6(2):113–122.

Takahashi, S., Matsunaga, T., Hasegawa, C., Saito, H., Fujita, D., Kiuchi, F. and Tsuda, Y. 1998. Martefragin A, a novel indole alkaloid isolated from red alga, inhibits lipid peroxidation. *Chem. Pharm. Bull.* 46(10):1527–1529.

Tois, J., Franzén, R. and Koskinen, A. 2003. Synthetic approaches towards indoles on solid phase recent advances and future directions. *Tetrahedron.* 59(29):5395–5405.

van Loevezijn, A., Allen, J. D., Schinkel, A. H. and Koomen, G. J. 2001. Inhibition of BCRP-mediated drug efflux by fumitremorgin-type indolyl diketopiperazines. *Bioorg. Med. Chem. Lett.* 11(1):29–32.

Wacker, D. A. and Kasireddy, P. 2002. Efficient solid-phase synthesis of 2,3-substituted indoles. *Tetrahedron Lett.* 43(29):5189–5191.

Wright, S. W., McClure, L. D. and Hageman, D. L. 1996. A convenient modification of the Gassman oxindole synthesis. *Tetrahedron Lett.* 37(27):4631–4634.

Xu, L. and Xu, S. 2001. Synthesis and biological activity of 2-[{substituted phenyl}vinyl] indole derivatives. *Yao Xue Xue Bao.* 36(2):100–104.

Yoshikuni, I., Kazuhiko, O. and Hirokazu, T. 1992. Preparation of *N*-benzopyranylmethyl-N′-phenylureas and analogs as cholesterol acyltransferase inhibitors. *Eur. Pat. Appl.* EP 512570 A1 19921111.

Zeni, G. and Larock, R. C. 2006. Synthesis of heterocycles via palladium-catalyzed oxidative addition. *Chem. Rev.* 106(11):4644–4680.

Zetterberg, B. 1947. New substances exerting a combined bacteriostatic and respiration-inhibiting effect on tubercle bacilli. *Nature.* 159(4033):235.

Zohuri, G. H., Damavandi, S. and Sandaroos, R. 2013. Polymeric catalyst for the synthesis of new pyrido[2,3-*b*]indoles. *Res. Chem. Intermediat.* 39(5):2115–2121.

Zolfigol, M. A., Veisi, H., Mohanazadeh, F. and Sedrpoushan, A. 2011. Synthesis and application of modified silica sulfuric acid as a solid acid heterogeneous catalyst in Michael addition reactions. *J. Heterocycl. Chem.* 48(4):977–986.

3 Solid Heterogeneous Catalysts Based on Sulfuric Acid and Transition-Metal Salts

Synthesis of Bioactive Heterocycles

Bahador Karami

CONTENTS

3.1 INTRODUCTION

In recent years, preparation of new catalysts that will improve efficiency or produce a greater yield has been a subject of interest. Catalytic technology plays a major role in many chemical reactions; therefore, preparing and employing green catalysts are extremely valuable (Abrantes et al., 2011; Khalafi-Nezhad et al., 2011). For example, 90%, over a trillion dollars' worth, of manufactured items are produced with the help of catalysts every year. Besides, pressure from environmentalists has led to a search for more eco-friendly forms of catalysis. Recently, synthesis of supported catalysts as clean materials has attracted considerable attention (Kantam et al., 2008; Murkute et al., 2011; Boudart, 1969). Although supported catalysts are available on different supports including charcoal, alumina, silica, and polymer, silica has many other advantages such as no swelling, good mechanical and thermal stabilities, and ease of scalability. In fact, these heterogeneous catalysts decrease reactor and plant corrosion problems and are environmentally safer.

Considerable attention has recently been paid to performance of reactions on the surfaces of solids (Niknam et al., 2007; Metzger, 1998; Tanaka and Tada, 2000; Sharghi and Hosseini-Sarvari, 2002a,b). These reactions are not only of interest from an environmental point of view, but in many cases also offer considerable synthetic advantages with the following features: (i) it is often easy to isolate the products and to separate the catalyst; (ii) comparing the reaction conditions with those of related homogeneous reactions, they are so mild that a high yield of specific products and suppression of by-product formation are expected; (iii) selectivity and activity of the catalysts are often comparable to those of enzymes (Pagni et al., 1998). Several classes of solids have commonly been used for surface organic chemistry, including alumina, silica gel, and clays (Posner, 1978; McKillop and Young, 1979; Cornelis and Laszlo, 1985, 1986; Laszlo, 1986).

The need to implement green chemistry principles (e.g., safer solvents, less hazardous chemical synthesis, atom economy, and catalysis) is a driving force toward the avoidance of the use of toxic organic solvents. A solvent-free or solid-state reaction obviously reduces pollution and reduces handling costs due to simplification of experimental procedure and workup technique and savings on labor. However, interest in the environmental control of chemical processes has increased remarkably in the last three decades as a response to public concern about the use of hazardous chemicals. Therefore, to improve the effectiveness of this method in preventing chemical waste, it is important to investigate its optimal conditions.

The challenge in chemistry to develop practical methods, reaction media, and optimal conditions and materials based on the idea of green chemistry is an important issue in the scientific community. However, the concept of *green chemistry* has emerged as one of the guiding principles of environmentally friendly organic synthesis. An important method in green chemistry is the use of heterogeneous catalysis in organic synthesis. Accordingly, silica sulfuric acid (SSA, **1**) has been synthesized (Scheme 3.1) (Zolfigol, 2001). This heterogeneous solid acid (Zolfigol et al., 2002) and Nafion-H® (Olah et al., 1978) are used for a wide variety of reactions.

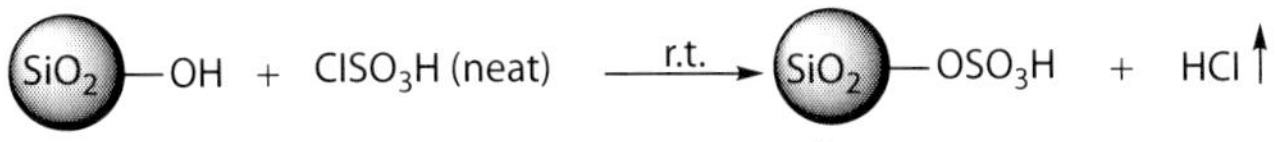

1

SCHEME 3.1

3.2 TUNGSTATE SULFURIC ACID AS A NEW HETEROGENEOUS SOLID ACID CATALYST

Accordingly, Karami et al. (2005a,c) found that anhydrous sodium tungstate reacts with chlorosulfonic acid (1:2 mole ratios) to give tungstate sulfuric acid (TSA, **2**) (Scheme 3.2).

3.2.1 Preparation of Tungstate Sulfuric Acid (TSA, 2)

Anhydrous sodium tungstate (59 g; 0.2 mol) was added to 1 L of dry n-hexane in a 2 L round-bottom flask, equipped with an overhead stirrer and cooled in an ice bath. Then, 27 mL (0.4 mol) of chlorosulfonic acid (CAUTION) was added dropwise over 30 min. This mixture was stirred for 1.5 h. Afterward, the reaction mixture was gradually poured into 1 L of chilled distilled water with stirring. The yellowish catalyst that separated out was collected and washed with distilled water (100 mL) five times until the filtrate showed negative in a test for chloride ion. Drying at 120°C for 5 h gave 80 g (98%) of the yellowish catalyst, mp 285°C (dec).

3.2.2 Characterization of Catalyst

Figure 3.1 shows the x-ray diffraction analysis (XRD) patterns of TSA. It was reported that a high degree of mixing of W–S in chlorosulfonic acid often led to the absence of an XRD pattern for anhydrous sodium tungstate. The broad peak around 25.7° (2θ) (θ is the Bragg's angle) from the smaller inset could be attributed to the linking of WO_3 into the chlorosulfonic acid.

The Fourier transform infrared (FTIR) spectra of anhydrous sodium tungstate and TSA are shown in Figure 3.2. The spectrum of TSA shows the characteristic bonds of anhydrous sodium tungstate and chlorosulfonic acid. The wave numbers in 3406, 1820, 1725, 1702, 1620, 1290, 1060, 1005, and 860 cm^{-1} in catalyst spectra reveal both bonds in anhydrous sodium tungstate and the $-OSO_3H$ group. Firstly, 1 mmol of catalyst was dissolved in 100 mL of water. It was then titrated with NaOH (0.1 N) in the presence of phenolphthalein as an indicator. At the endpoint of titration,

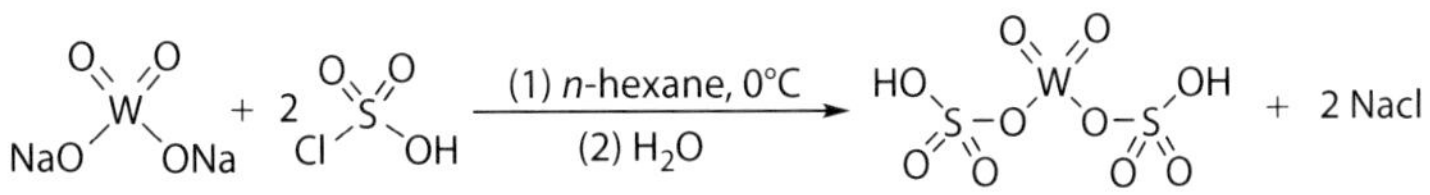

SCHEME 3.2

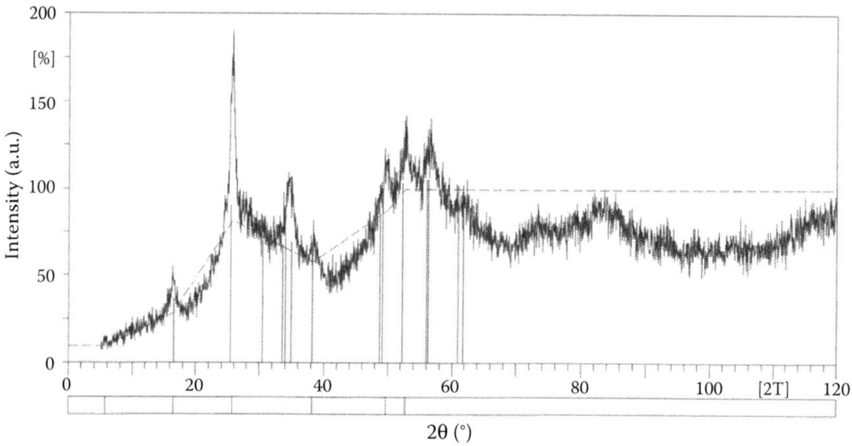

FIGURE 3.1 The powder XRD pattern of TSA.

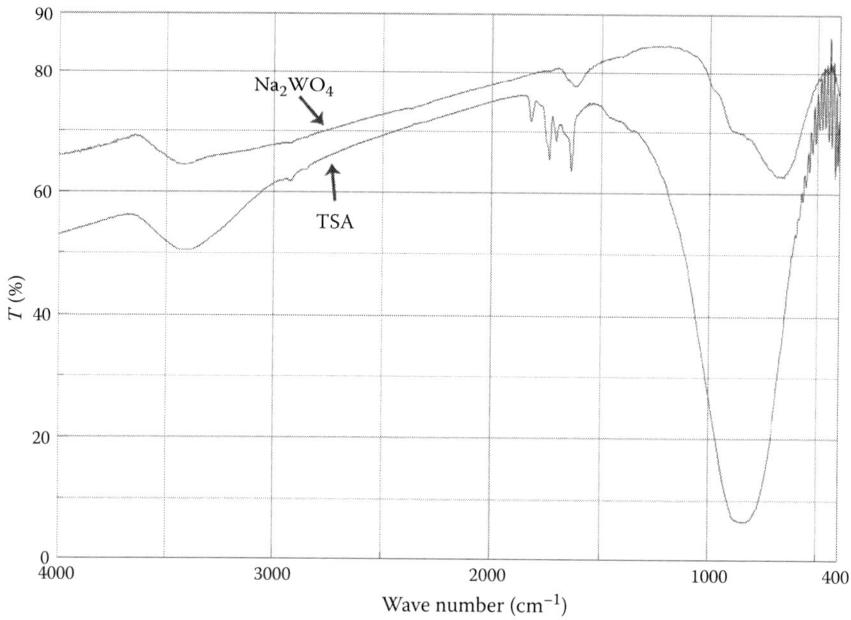

FIGURE 3.2 FTIR spectra of TSA and sodium tungstate.

4 mmol of titrant was consumed. Also, based on potentiometric data, the pK_{a1} value for TSA is 3.17 and pK_{a2} is 8.78. The result of catalyst titration is shown as four acidic valences, because of hydrolysis in the aqueous solvent.

TSA has a number of structural features, for example, an acidic proton and efficiency in solventless conditions, which make it an attractive catalyst for organic transformations. This reagent has good thermal and mechanical stability. It can be filtered from the reaction mixture after use and, therefore, subsequently reused.

3.3 USE OF TSA IN SYNTHESIS OF HETEROCYCLES

Nowadays, catalyst technology is often used as a powerful tool for chemical transformation in both academia and industry because it plays a fundamental role in chemical reactions, allowing faster conversion of a wide variety of starting materials to high-value products at lower cost, with minimum generation of by-products (Karami et al., 2012c; Shafiee and Moloudi, 2011). However, preparation and use of green catalysts that will improve efficiency or produce greater yields has been a subject of interest for some time, especially in the synthesis of heterocyclic compounds (Atienza and Scheidt, 2011; Habibi et al., 2012; Karami and Kiani, 2011). More recently, pressure from environmentalists has led to a search for more environmentally friendly forms of catalysis. In connection with studies on new catalyzed organic reactions, in the past few years, Karami et al. (2011, 2012d,e,g,) have been involved in a program directed toward developing simple, novel, and facial methods for the preparation of organic compounds using various catalysts and readily available starting materials (Damavandi et al., 2002; Mallakpour et al., 1998; Heydari et al., 2000). Solid acids play a significant role in green chemistry, especially in chemical manufacturing processes (Movassaghi and Jacobsen, 2002; Qi et al., 2010; Wedge and Hawthorne, 2003; Stone and Anderson, 2007). Solid acids generally have high turnover numbers and can be easily separated from organic components (Clark, 2002).

Karami et al. (2005a) reported the first application of TSA/$NaNO_2$ as a novel heterogeneous system for the N-nitrosation of secondary amines under mild conditions (Scheme 3.3).

TSA/$NaNO_2$ or TSA/$KMnO_4$ have been reported as novel heterogeneous systems for rapid deoximation (Scheme 3.4) (Karami et al., 2005c; Karami and Montazertozohori, 2006).

TSA/$NaNO_2$ was also reported as a new and efficient heterogeneous system for rapid aromatization of Hantzsch 1,4-dihydropyridine under mild conditions (Scheme 3.5) (Karami et al., 2005b).

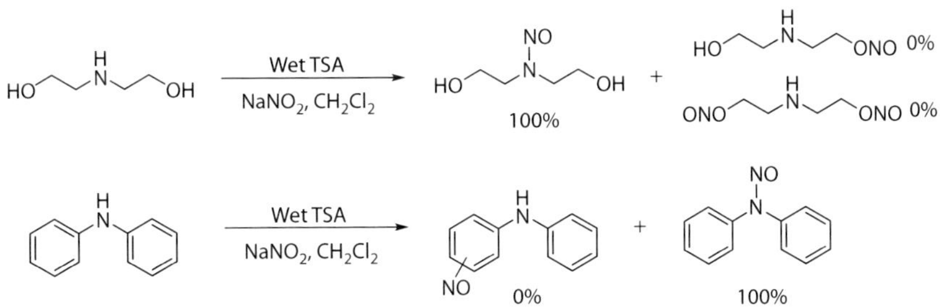

SCHEME 3.3

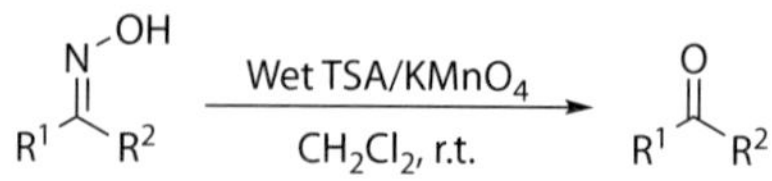

SCHEME 3.4

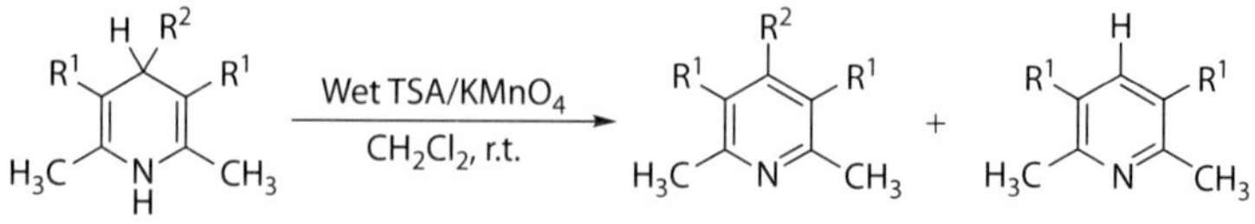

SCHEME 3.5

Furthermore, TSA was reported as a novel and efficient solid acidic reagent for the oxidation of thiols to disulfide and oxidative demasking of 1,3-dithianes (Scheme 3.6) (Karami et al., 2006).

Also, synthesis of aza-polycyclic aromatic compounds such as novel phenazines and quinoxalines using TSA was reported by Karami et al. (2011) (Scheme 3.7). This stimulated our interest in further studies on the chemistry of catalysts.

Calix[4]resorcinarenes, a subclass of calixarenes, are large cyclic tetramers that have found applications as macrocyclic receptors, host molecules, host–guest complexes, and so on (Botta et al., 2005; Martinez et al., 2000; Leyton et al., 2005; Yonetake et al., 2001). It should be noted that the synthesis of calix[4]resorcinarenes was first reported in the late nineteenth century by Bayer based on the concentrated sulfuric acid–catalyzed cyclocondensation of benzaldehyde and resorcinol (Bayer, 1872). These cyclic compounds were synthesized by TSA (Scheme 3.8) (Karami et al., 2012f).

In this process, as indicated in Figure 3.3, the recycled catalyst was used for four cycles during which a little appreciable loss was observed in the catalytic activities.

Synthetically, the pyrrole ring, as a class of important heterocycles, has attracted considerable attention (Bharadwaj and Scheidt, 2004), because the pyrroles are key

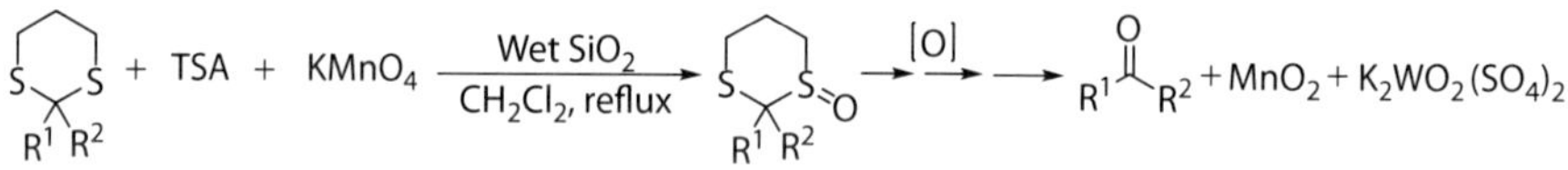

SCHEME 3.6

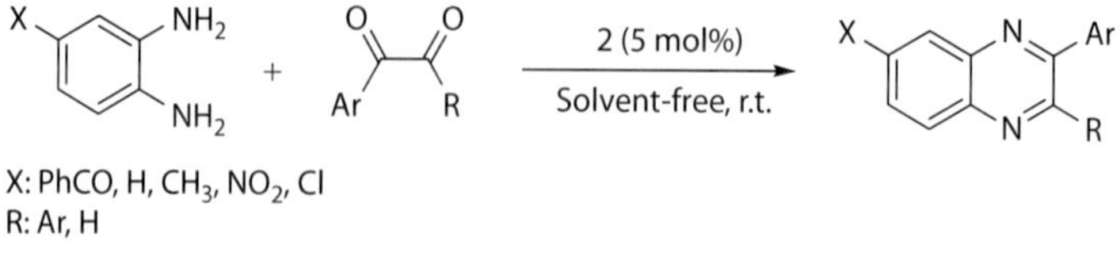

SCHEME 3.7

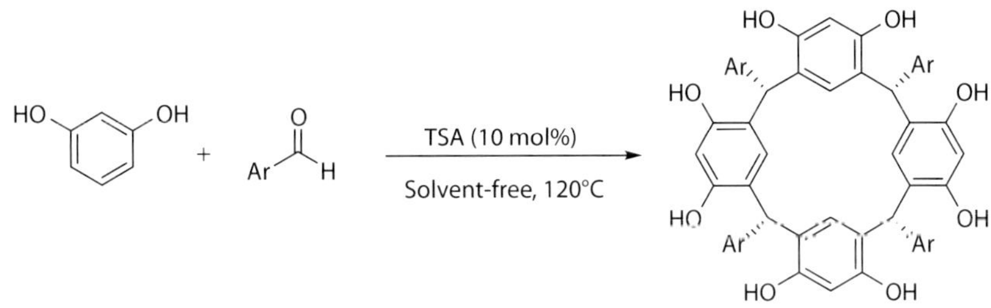

SCHEME 3.8

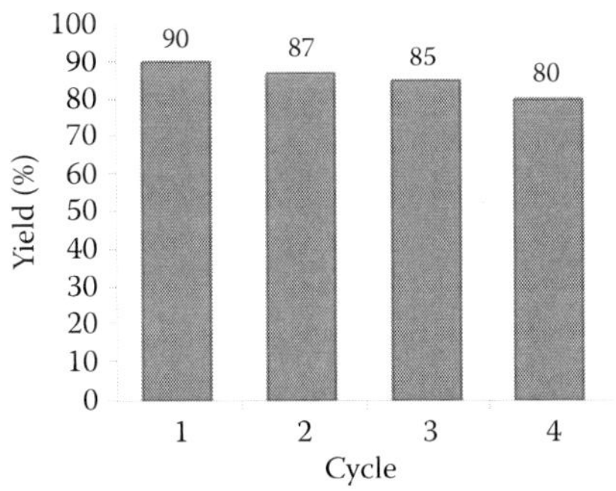

FIGURE 3.3 Recyclability of TSA (**2**) as catalyst for model reaction.

core elements of various natural and biologically active molecules (Bandyopadhyay et al., 2010) including anticancer (Shinohara et al., 2010), antimycobacterial (Ragno et al., 2000), and antiviral agents (Dabur et al., 2004). Besides, pyrroles serve as building blocks for porphyrin synthesis (Nonn, 1995; Weidner and Sigurdsson, 1990; Woo and Sigurdsson, 1993).

TSA (**2**) can act as a safe and recyclable catalyst for the reaction of aromatic primary amines and 2,5-hexadione under thermal and solvent-free conditions to afford some known pyrrole derivatives in good to excellent yields (Scheme 3.9). This also afforded some novel pyrrole derivatives (Scheme 3.10) (Karami et al., 2013a).

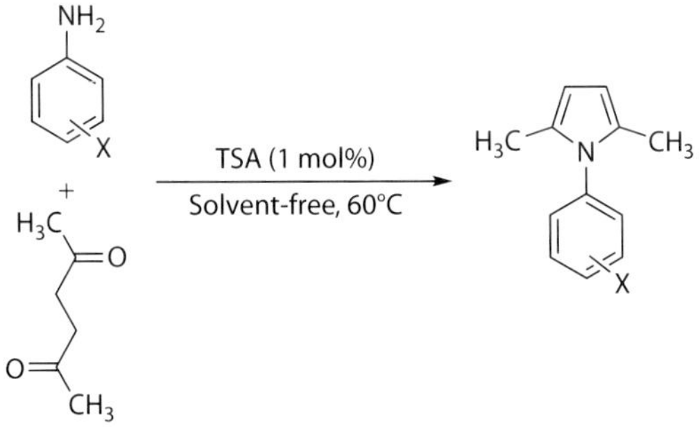

SCHEME 3.9

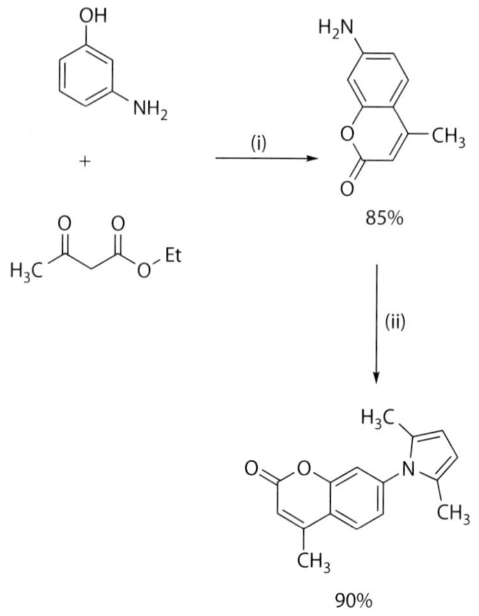

(i) $ZrO_2Cl–SiO_2$ (10 mol%), 90°C, solvent-free
(ii) TSA (1 mol%), 120°C, solvent-free

SCHEME 3.10

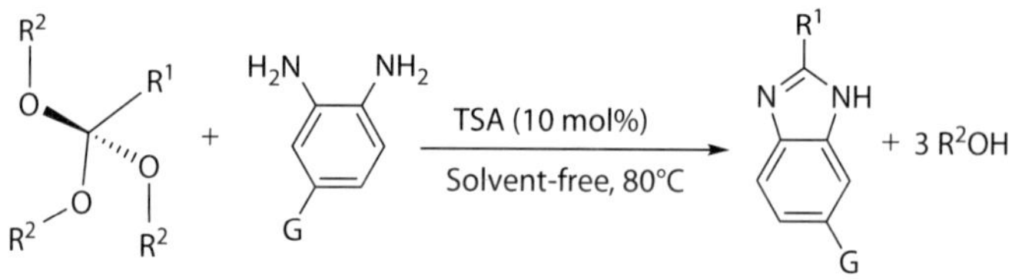

G: H, Me, NO_2, benzoyl R[1] and R[2]: H, Me, Et

SCHEME 3.11

Many reports on the application of solid acids as efficient catalysts in organic transformations (Martin and Narayana, 2010; Romanelli et al., 2010; Marziano et al., 2005) and recent studies on the synthesis of organic compounds (Karami and Khodabakhshi, 2011; Karami et al., 2006; Mallakpour et al., 1998; Heydari et al., 2000; Damavandi et al., 2002) found that TSA can be used as an efficient and safe catalyst for the condensation of ortho esters and *o*-phenylenediamines under thermal and solvent-free conditions to afford benzimidazole derivatives in good to excellent yields (Scheme 3.11).

As shown in Scheme 3.12, a plausible mechanism for the synthesis of benzimidazoles and recyclability of TSA was proposed. TSA acts as a Brønsted acid so that it can release a proton.

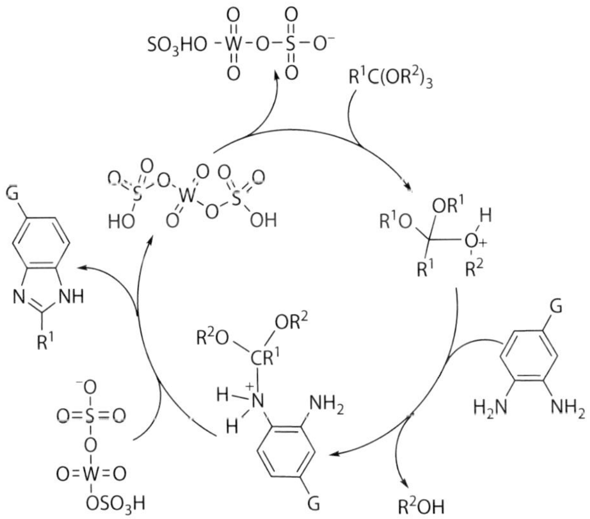

SCHEME 3.12

Numerous methods are available in the literature for preparation of 3,4-dihydropyrimidine-2-(1*H*)-one/thione derivatives. Generally, these compounds can be prepared by the Biginelli condensation of aldehydes with β-diketones and urea or thiourea derivatives in refluxing ethanolic HCl (Biginelli, 1983; Yu et al., 2007; Phucho et al., 2010; Ranu et al., 2000; Hu et al., 1998; Yadav et al., 2001a; Ma et al., 2000; Lu et al., 2000; Horacio et al., 2009; Liu and Wang, 2009). In connection with previous programs on developing TSA in organic transformations (Karami et al., 2005a,c, 2006; Montazerozohori et al., 2007), a simple and convenient route for the condensation of aryl aldehydes with β-dicarbonyls and thiourea using TSA as catalyst under solvent-free conditions to afford the 3,4-dihydropyrimidine-2-(1H) thione derivatives (Scheme 3.13) was reported (Karami et al., 2012d).

Until now, several reagents for the synthesis of dithiolane and dithiane derivatives (Burczyk and Kortylewticz, 1982; Kamitori et al., 1986; Srinivasulu et al., 2007) in connection with previous programs on developing TSA in organic transformations (Karami et al., 2006, 2011; Asgarian-Damavandi et al., 2001; Karami and Khodabakhshi, 2011) have been reported. The thioacetalization of carbonyl compounds was reported as a simple and convenient route for using catalytic amounts of

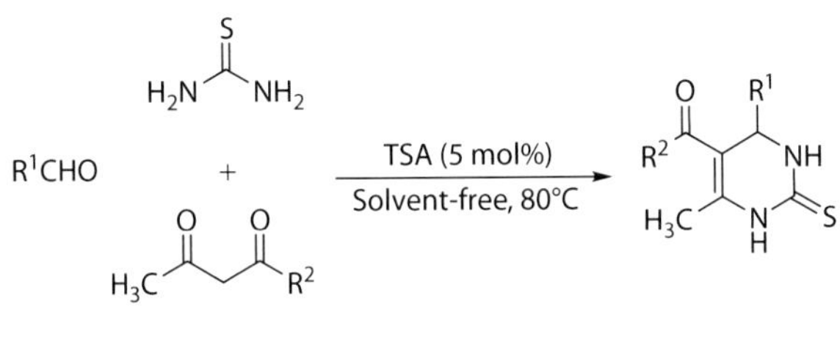

SCHEME 3.13

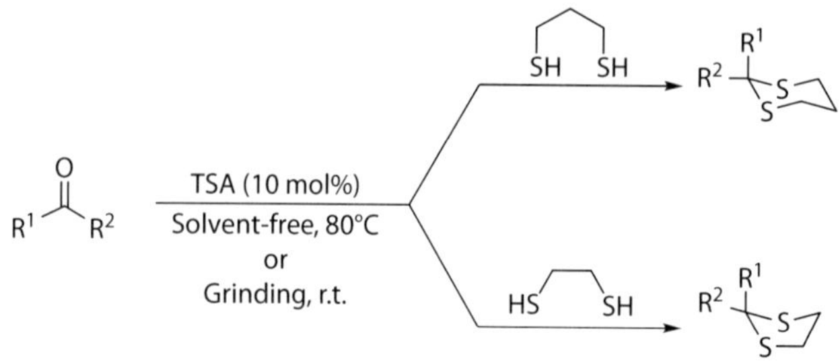

SCHEME 3.14

TSA under solvent-free conditions to afford 1,3-dithiane and 1,3-dithiolane derivatives (Scheme 3.14). TSA is a strong solid acid that can activate the carbonyl group to decrease the energy of the transition state of the nucleophilic attack step (Karami et al., 2012i).

In another variation, the chemoselectivity of the method was investigated. It is noteworthy that a few of the reported methods have demonstrated chemoselective thioacetalization of aldehydes in the presence of ketones. Therefore, this prompted us to explore the chemoselective protection of aldehydes in the presence of ketones. As can be seen in Scheme 3.15, when an equimolar mixture of methoxybenzaldehyde and acetophenone was allowed to react with 1,2-ethanedithiol in the presence of a catalytic amount of TSA, benzaldehyde was exclusively protected, whereas acetophenone was intact under the same experimental conditions.

There has been considerable interest in benzoxazole and benzothiazole derivatives, not least because of their value for a variety of industrial (Belhouchet et al., 2012; Eshghi et al., 2012), biological (Fukuzawa et al., 2009), and medicinal chemistry uses (Praveen et al., 2012; Wen et al., 2012). In continuation of the interest in the use of heterogeneous solid catalysts (Karami et al., 2008, 2013b), a very simple procedure for the synthesis of benzoxazole and benzothiazole derivatives using TSA as a superior solid acid catalyst was reported (Scheme 3.16) (Farahi et al., 2013).

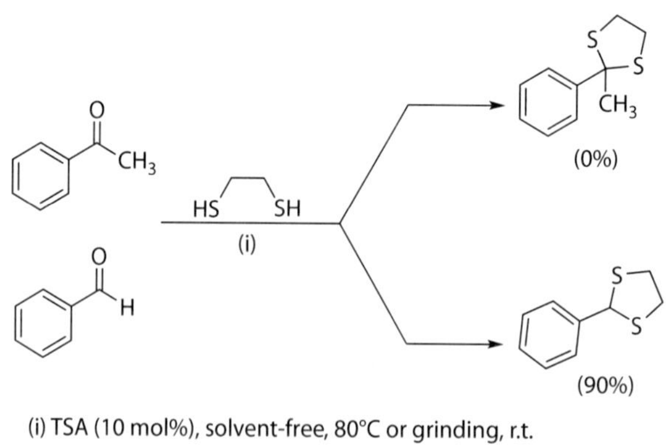

SCHEME 3.15

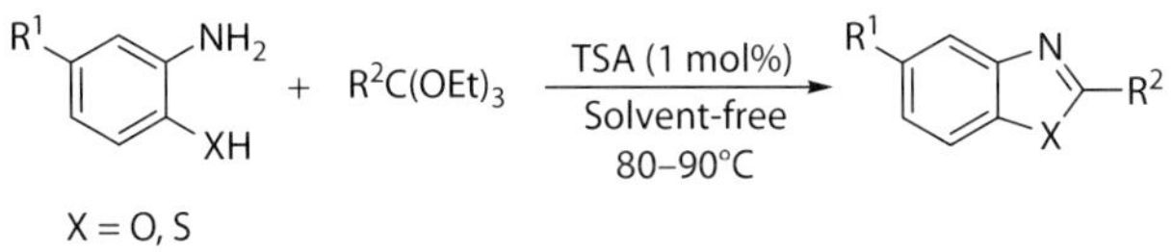

SCHEME 3.16

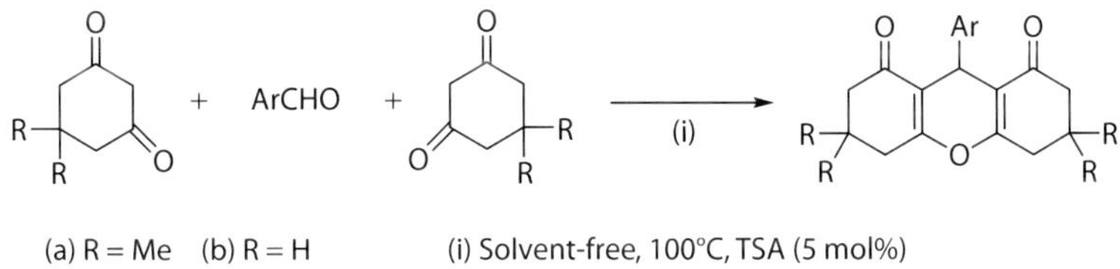

SCHEME 3.17

Xanthenes and their derivatives have received significant attention in recent years due to their wide range of biological and therapeutic properties (El-Brashy et al., 2004; Chibale et al., 2003; Kinjo et al., 1995) and they have been prepared by various methods (Casiraghi et al., 1973; Bekaert et al., 1992; Knight and Little, 2001; Seyyedhamzeh et al., 2008). TSA was reported as a new inorganic solid acid. Karami et al. (2013a) have described the use of TSA for the synthesis of 1,8-dioxooctahydroxanthenes by condensation of 1,3-cyclohexanediones and aromatic aldehydes under solvent-free conditions (Scheme 3.17).

3.4 MOLYBDATE SULFURIC ACID AS A NEW HETEROGENEOUS SOLID ACID CATALYST

According to reported studies on the applications of solid acids in organic synthesis, such as introducing TSA, the preparation and use of molybdate sulfuric acid (MSA, **3**) was reported for the first time in organic synthesis (Scheme 3.18) (Montazerozohori et al., 2007; Montazerozohori and Karami, 2006).

3.4.1 Preparation of Molybdate Sulfuric Acid (MSA)

Anhydrous sodium molybdate (20.58 g, 0.1 mol) was added gradually to chlorosulfonic acid (23.304 g, 0.2 mol) in a 250 mL round-bottom flask immersed in an ice bath. After the completion of the addition, the mixture was shaken for 1 h, which gave rise to crude MSA as a bluish-white solid, which was filtered off and washed

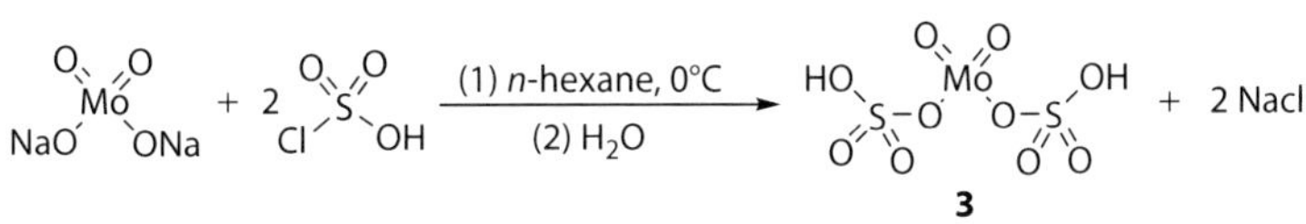

SCHEME 3.18

with cold H_2O. Yield: 28 g (87.5%; mp 356°C dec). The compound was dissolved a little in water but not in the organic solvents. Characteristic IR bands (KBr, cm^{-1}); 3600–2200 (OH, bs), 1230–1150 (S=O, bs), 1050 (S–O, m), 1010 (S–O, m), 880–840 (Mo=O, m), 450 (Mo–O, m).

3.4.2 Characterization of Catalyst

Figure 3.4 shows the XRD patterns of MSA. It was reported that a high degree of mixing of Mo–S in chlorosulfonic acid often led to the absence of an XRD pattern for anhydrous sodium molybdate.

The broad peaks around 23°, 29°, and 34° (2θ) (θ is the Bragg's angle) from the smaller inset could be attributed to linking of Mo to the sulfonic acid. The FTIR spectra of anhydrous sodium molybdate and MSA are shown in Figure 3.5. The spectrum of MSA shows the characteristic bonds of anhydrous sodium molybdate and chlorosulfonic acid. The absorptions in 3459, 2110, 1635, 1129, 909, 771, 637, 616, and 451 cm^{-1} in the catalyst spectrum reveal both bonds in anhydrous sodium molybdate and the $-OSO_3H$ group.

In addition, titration of catalyst with NaOH (0.1 N) was done for characterization of the catalyst. Firstly, 1 mmol of catalyst was dissolved in 100 mL of water then titrated with NaOH (0.1 N) in the presence of phenolphthalein as an indicator. It was observed that for 1 mmol of catalyst, 2 mmol of NaOH was utilized. The result of catalyst titration is shown as two acidic valences that coincide with the proposed structure of the catalyst. MSA has a number of structural features, for example, an acidic proton and efficiency in solvent-free conditions, which make it an attractive catalyst for organic transformations. This reagent has good thermal and mechanical stability. Because it can be filtered from the reaction mixture after use, it can be subsequently reused.

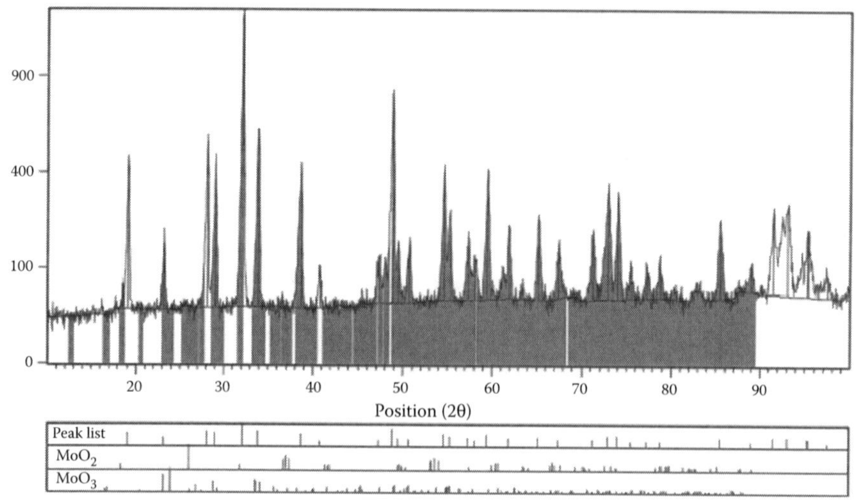

FIGURE 3.4 The powder XRD pattern of MSA.

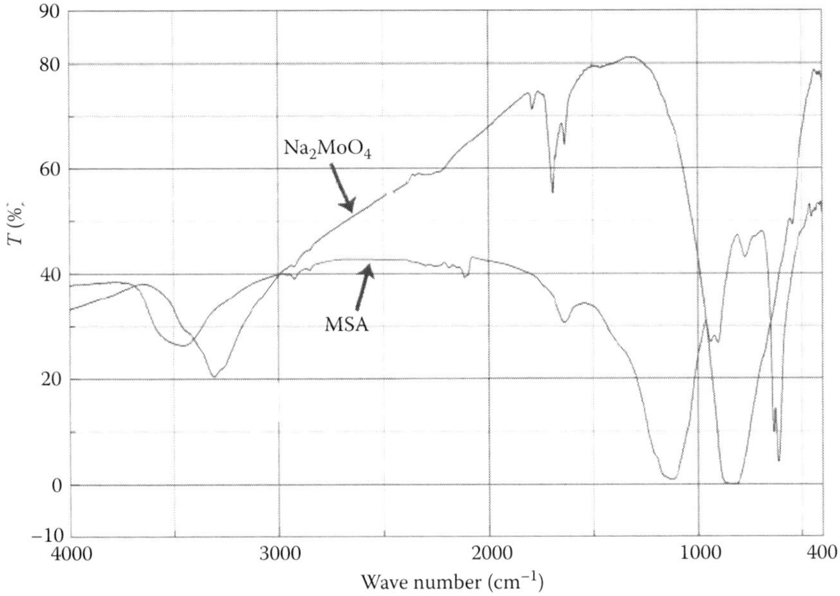

FIGURE 3.5 FTIR spectra of MSA.

3.5 USE OF MSA IN SYNTHESIS OF HETEROCYCLES

The first application of MSA was in nitrosation of secondary amines under mild conditions (Scheme 3.19) (Montazerozohori and Karami, 2006).

Another use for MSA in organic synthesis was for the purpose of the oxidation of thiols to symmetrical disulfides (Scheme 3.20) (Montazerozohori et al., 2007).

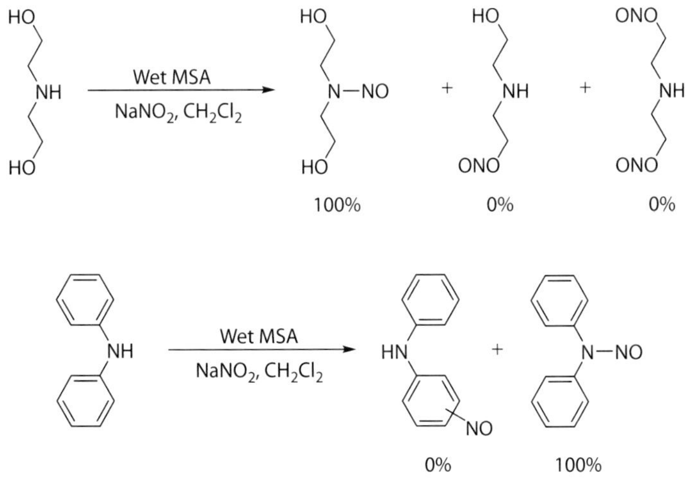

SCHEME 3.19

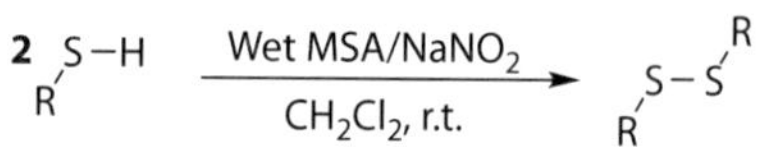

SCHEME 3.20

Until now, several types of polycyclic aromatic nitrogen heterocycles have been known. In fact, heterocyclic synthesis plays an important role in modern life. Among the nitrogen heterocycles, quinoxaline and phenazine derivatives have attracted great interest because of their wide application in organic and medicinal chemistry. Many phenazine compounds are found in nature and are produced by bacteria. Quinoxalines are the fundamental components of several pharmacological active compounds and some antibiotics such as levomycin, echinomycin, and actinomycin (Bailly et al., 1999; Dell et al., 1975).

In this part, the synthesis of some polycyclic aromatic nitrogen heterocyclic compounds through a convenient and economical method is reported by the use of MSA (Scheme 3.21) (Karami et al., 2011). MSA, as an inorganic solid acid, has been proved to be an efficient and heterogeneous catalyst for rapid synthesis of quinoxaline and phenazine derivatives from the condensation of *o*-phenylenediamines and 1,2-diketones in ethanol as solvent at room temperature in excellent yields (Scheme 3.21) (Karami et al., 2011).

For example, as illustrated in Scheme 3.22, dibenzo[f,h]pyrido[2,3-b]quinoxaline was prepared in refluxing ethanol as a novel aza-heterocyclic compound. This approach shows an interesting synthesis of the new polycyclic aromatic heterocycles, which was the main purpose of this work (Scheme 3.22).

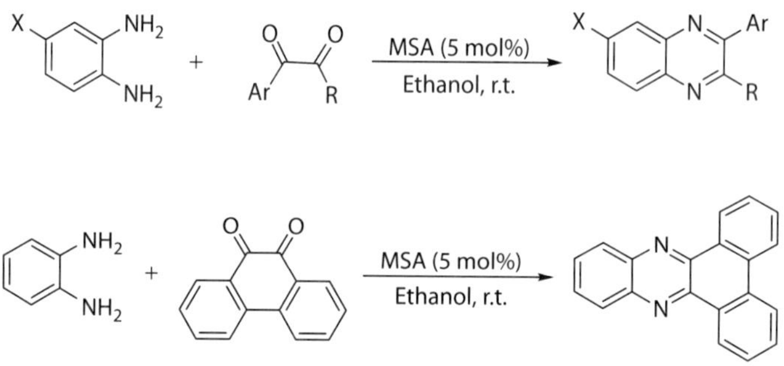

SCHEME 3.21

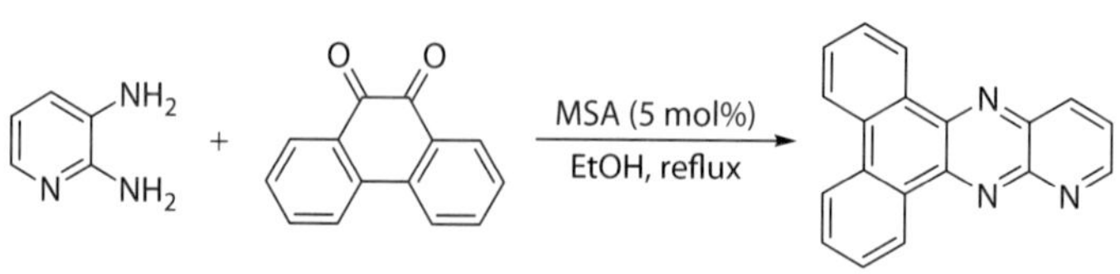

SCHEME 3.22

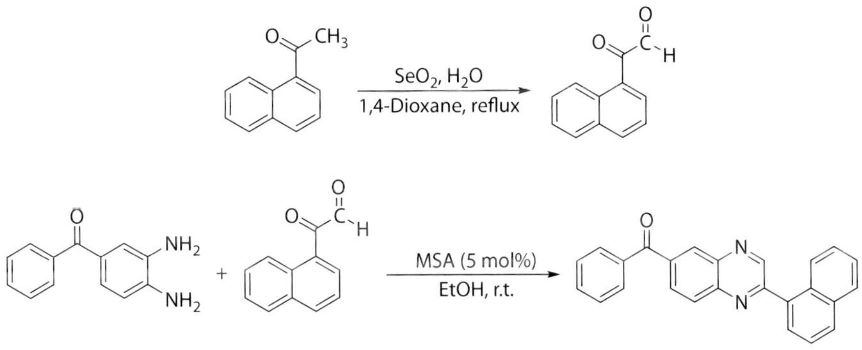

SCHEME 3.23

Scheme 3.23 shows another variation: 1′-acetonaphthone oxidized by selenium dioxide to afford the corresponding 1,2-diketone, which is called 1-naphthyl glyoxal. Subsequently, the produced compound was used as a precursor for preparing the new quinoxaline.

The possibility of recycling the catalyst was also examined. For the model reaction, the recycled catalyst was used for four cycles during which no appreciable loss was observed in the catalytic activities (Figure 3.6). Although the reaction mechanism is not totally clear, a mechanistic study in this area, which is outlined in Scheme 3.24, shows that the coordination of carbonyl to the catalyst in the first step makes it susceptible to nucleophilic attack by *o*-diamine nitrogen. The next step in this process is dehydration, in which MSA again plays a significant catalytic role, and finally, product is afforded (Karami et al., 2012b).

Among organic compounds, xanthenes and their derivatives have received significant attention in recent years due to their wide range of biological and therapeutic properties (El-Brashy et al., 2004; Chibale et al., 2003). In view of the importance of xanthene derivatives, many methods for their synthesis have been reported including condensation of β-naphthol and aldehydes or acetals catalyzed

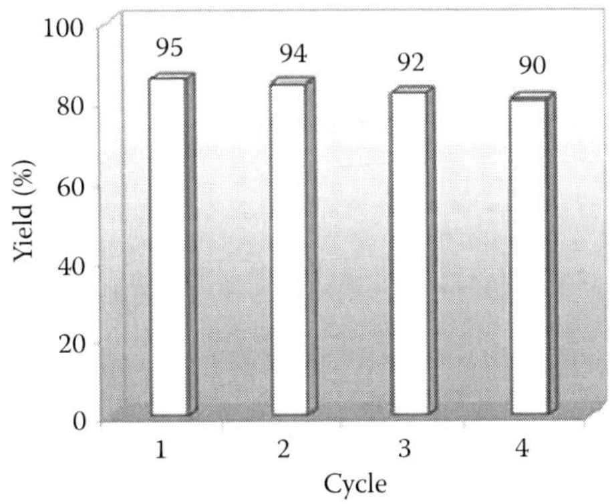

FIGURE 3.6 Recyclability of MSA (**3**) as catalyst for model reaction.

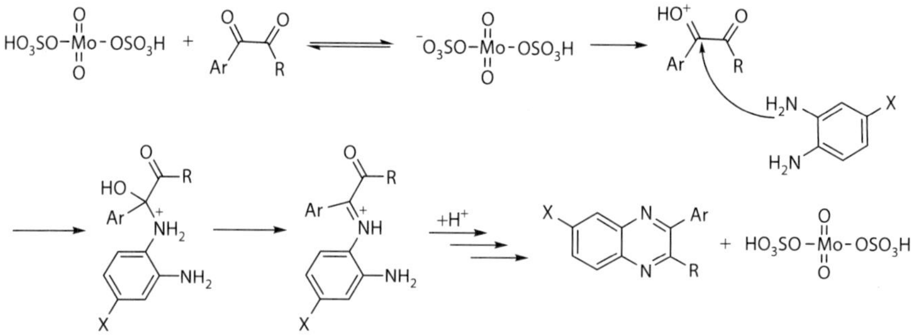

SCHEME 3.24

by silica sulfuric acid, HCl/CH_3COOH, or H_3PO_4 (Seyyedhamzeh et al., 2008). However, some of these methods involve long reaction times, harsh reaction conditions, and unsatisfactory yields. Therefore, improvements in these syntheses have been sought continuously. Karami et al. (2013c) found a new route to preparation of xanthene derivatives by employing MSA as an eco-friendly, highly efficient, and reusable catalyst. This method not only affords the products in excellent yields but also avoids the problems associated with catalyst cost, handling, safety, and pollution (Karami et al., 2013c).

MSA can be used as a powerful, safe, and recyclable catalyst for the condensation reaction of cyclic 1,3-diketones with arylaldehydes (Scheme 3.25). Firstly, synthesis of 2,2′-(arylmethylene)bis(3-hydroxycyclohex-2-enone) from reaction of cyclic 1,3-diketones with several aromatic aldehydes was expected, but, as can be seen from Scheme 3.25, under the given conditions, 2,2′-(arylmethylene)bis(3-hydroxycyclohex-2-enone) was not formed and cyclic 1,3-diketones with aldehydes were effectively cyclized to give 9-aryl-substituted 1,8-dioxooctahydroxanthenes.

In another variation, when using aromatic dialdehyde substrate, benzaldehyde derivatives led to condensation with 1,3-cyclic diketones (1:4 ratio) to afford bisxanthene products. In this case, four 1,3-diketones with dialdehyde were

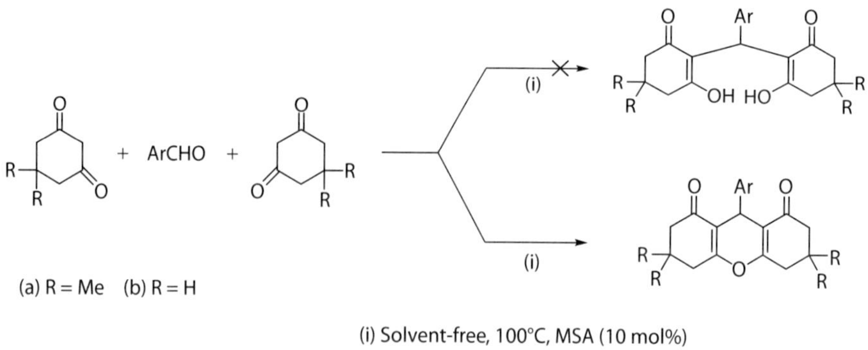

SCHEME 3.25

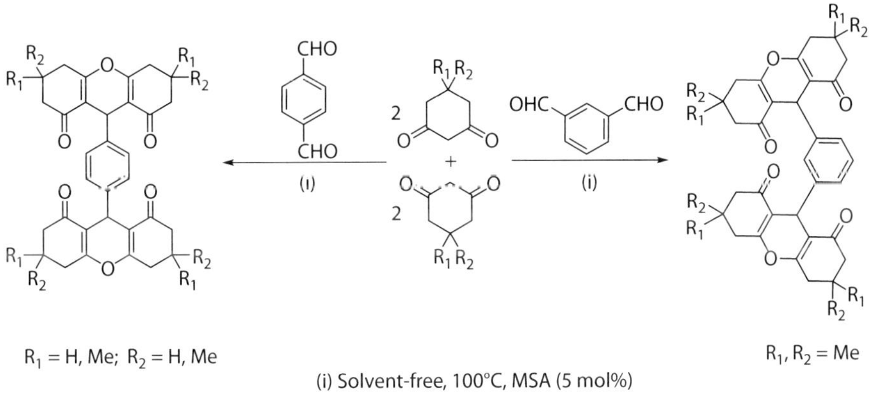

SCHEME 3.26

effectively cyclized to obtain bis(9-aryl-substituted 1,8-dioxooctahydroxanthenes) (Scheme 3.26).

We found that MSA showed high catalytic activity with very short reaction times.

3.6 SILICA TUNGSTIC ACID (STA, 4) AS A NEW HETEROGENEOUS SOLID ACID CATALYST

Silica tungstic acid (STA) has been used as a new silica-supported catalyst in some organic transformations (Karami et al., 2012c). This inexpensive and reusable catalyst can be readily handled and easily separated from the reaction mixture, making the reaction cleaner and faster with higher yields (Scheme 3.27).

3.6.1 Preparation of Silica Tungstic Acid (STA)

A mixture of silica chloride (6.00 g) and sodium tungstate (7.03 g) was added to *n*-hexane (10 mL) and the resulting mixture was stirred under refluxing conditions (70°C) for 4 h. After completion of the reaction, the reaction mixture was filtered,

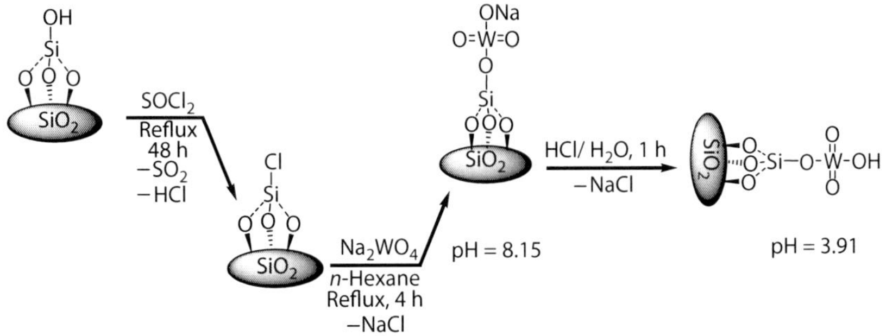

SCHEME 3.27

washed with distilled water, dried, and then stirred in the presence of 0.1 N HCl (40 mL) for 1 h. Finally, the mixture was filtered, washed with distilled water, and dried to afford STA.

3.6.2 Characterization of Catalyst

STA was characterized by x-ray fluorescence (XRF), XRD, and FTIR spectra. Figure 3.7 shows the XRD patterns for STA, which shows the presence of a tungstic acid crystalline phase supported on amorphous silica as a broad peak around 22° (2θ) (θ is the Bragg's angle). The three peaks in the 23°–25° region of the XRD spectrum could be attributed to the presence and linking of WO_3 to the silica gel (Karami et al., 2012a; Santato et al., 2001).

The FTIR spectra for the anhydrous sodium tungstate, silica chloride, and STA are shown in Figure 3.8. This spectrum shows the characteristic bonds of anhydrous sodium tungstate and silica chloride.

The adsorption in 3459, 1636, 1096, 970, 1620, and 799 cm^{-1} in the catalyst spectrum reveal both bonds in SiO_2–Cl and the WO_4 group. We evaluated the amounts of tungstic acid supported on SiO_2 using two methods: (a) titration with 0.1 N NaOH (neutralization reaction) and (b) calculating the weight difference between primary SiO_2–Cl and produced STA. After these experiments, we found that 1 g of catalyst

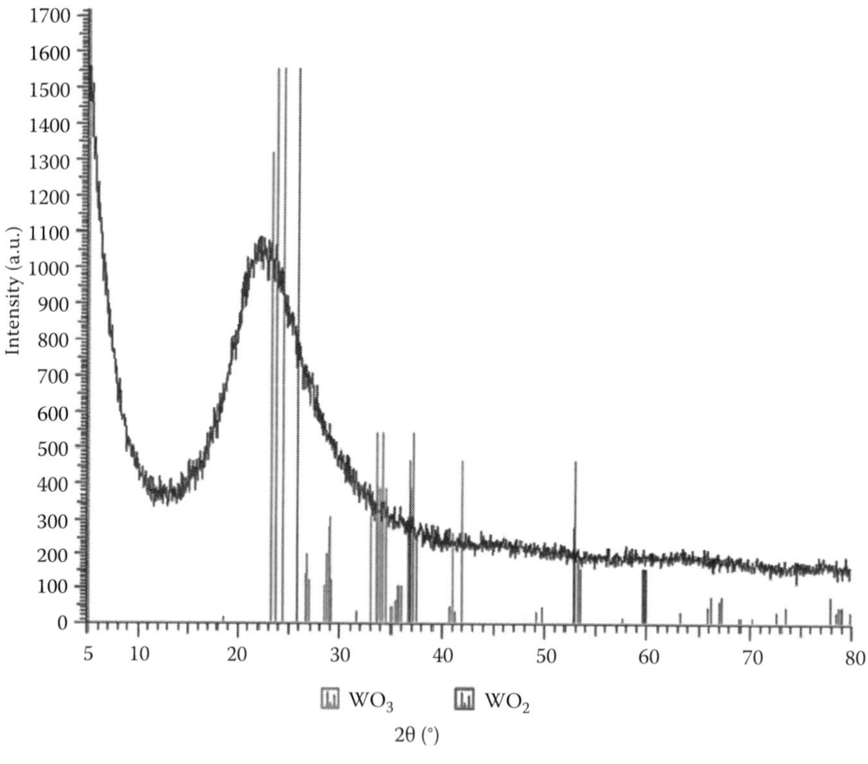

FIGURE 3.7 XRD patterns for STA.

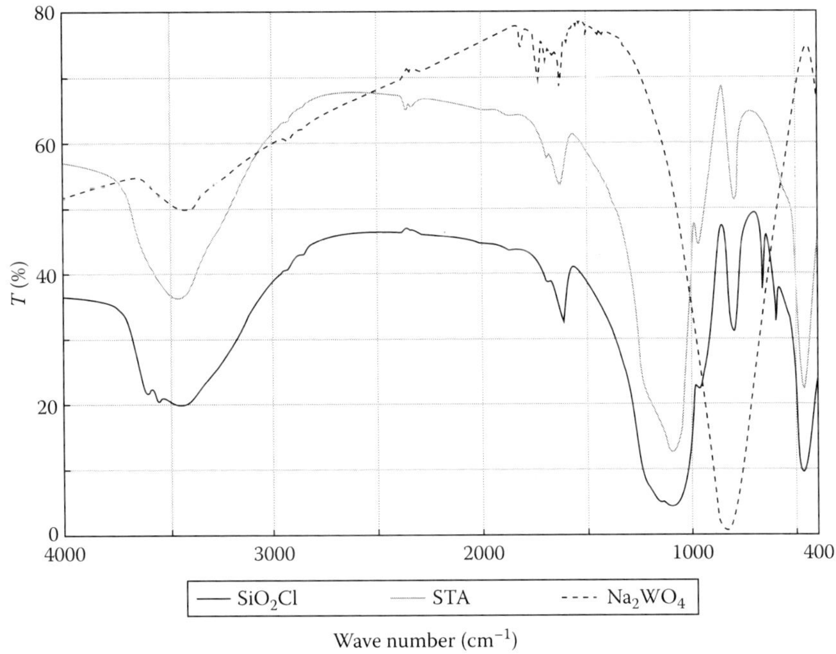

FIGURE 3.8 The FTIR spectra for the anhydrous sodium tungstate, silica chloride, and STA.

includes 0.05 g $-OWO_3H$. Regarding the molecular weight of WO_4H (249 g), therefore, 1 g catalyst is equal to 0.2 mmol. Generally, in this protocol, STA has valuable and special features. It has high thermal stability and because a small amount of catalyst was used, it makes the reaction economical and eco-friendly.

3.7 USE OF STA IN SYNTHESIS OF HETEROCYCLES

Among the wide variety of nitrogen heterocycles that have been explored as pharmaceutically important compounds, benzimidazoles exhibit relatively high biological activities. They have exhibited activity against cancer (Huang et al., 2006), HIV (Porcari et al., 1998; Roth et al., 1997), herpes (HSV-1) (Migawa et al., 1998), influenza (Tamm, 1957), certain fungi (Kus, 2003), and raf kinase (Buchstaller et al., 2011). To determine the simple and suitable conditions for the synthesis of benzimidazole derivatives using STA as a solid acid catalyst, the treatment of triethyl orthoformate with *o*-phenylenediamine was chosen as a model reaction (Scheme 3.28) (Karami et al., 2012b,c). In this process, as indicated in Figure 3.9, the recycled catalyst was used for four cycles during which a little appreciable loss was observed in the catalytic activities.

Chromene derivatives are an important class of heterocyclic compounds having significant biological activities (Lazarenkow et al., 2012). Due to the advantages of multicomponent reactions and heterogeneous catalysts and in continuation of our interest in the use of supported catalysts (Karami et al., 2008, 2012b; Karami and Kiani, 2011), a new methodology was reported for the synthesis of 2-amino-4*H*-chromenes using STA as an efficient heterogeneous catalyst (Scheme 3.29).

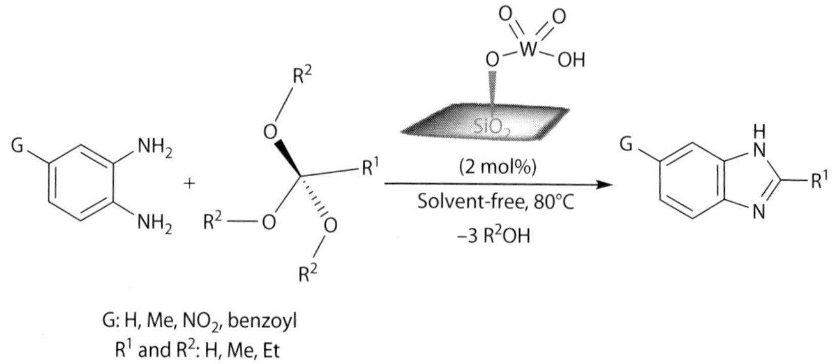

SCHEME 3.28

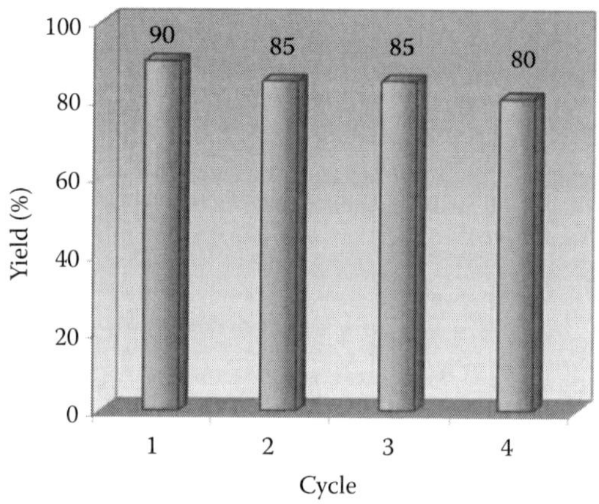

FIGURE 3.9 Recyclability of STA as catalyst for model reaction.

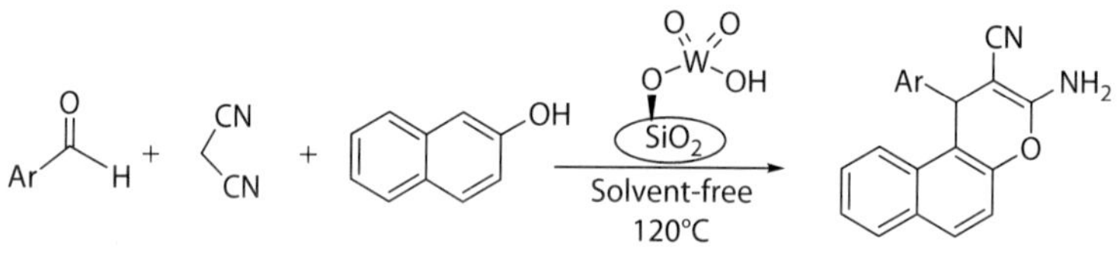

SCHEME 3.29

The probable mechanism for the synthesis of 2-amino-4*H*-chromene in the presence of a catalytic amount of STA is outlined in Scheme 3.30. It seems that STA acts as a Brønsted acid so that it can release a proton to activate aromatic aldehyde. It can also prepare malononitrile for attack by the activated aromatic aldehyde. After condensation of aldehyde and malononitrile, β-naphthol reacts with the produced intermediate to give 2-amino-4*H*-chromene as a product.

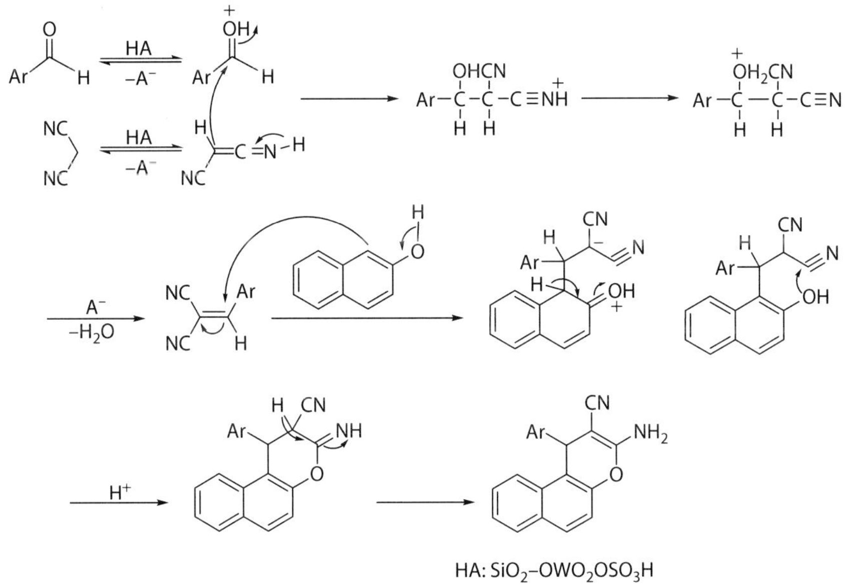

SCHEME 3.30

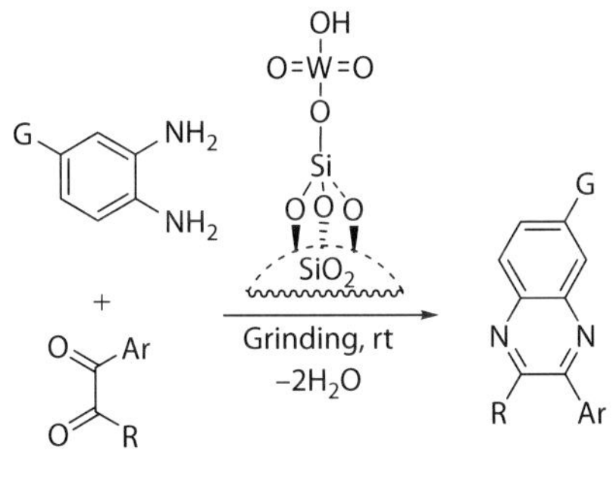

SCHEME 3.31

STA is employed as a green catalyst for solvent-free synthesis of novel benzopyrazines. Catalyst loadings as low as 2 mol% can be used to afford high yields of pure products. Scheme 3.31 shows a rapid and eco-friendly synthesis of novel and known benzopyrazines using STA as a new and recyclable catalyst (Khodabakhshi and Karami, 2012).

In continuation of previous studies on the development of various catalysts in the synthesis of organic compounds (Karami et al., 2012f,g,h), STA was used as a powerful and reusable catalyst for the new and solvent-free synthesis of 5,7-dihydroxy-4-substituted coumarin via the Pechmann condensation of phloroglucinol with β-ketoester (Scheme 3.32).

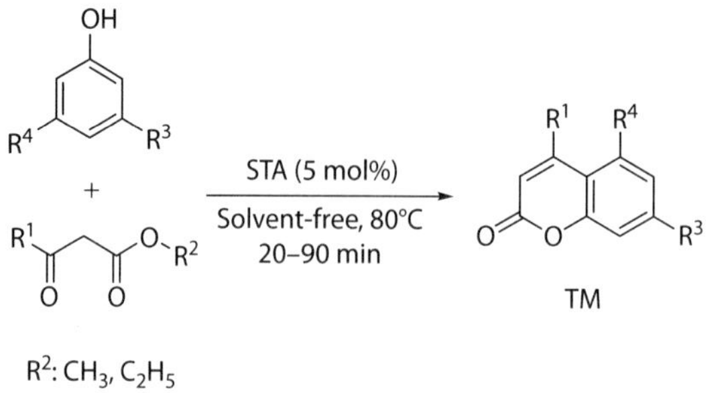

TM	R^1	R^3	R^4
a	CH_3	OH	OH
b	CH_2Cl	OH	OH
c	Ph	OH	OH
d	CH_3	NH_2	H
e	Ph	OH	H
f	Ph	OH	H

SCHEME 3.32

3.8 $ZrOCl_2{\cdot}8H_2O/SiO_2$ (5) AS A NEW HETEROGENEOUS SOLID ACID CATALYST

The importance of zirconium as a homogeneous catalyst has already been cited in the literature in a number of biologically and pharmaceutically significant organic transformations (Firouzabadi et al., 2006; Eftekhari-Sis et al., 2006; Wu et al., 2009; Chakraborti and Kondaskar, 2003; Bhagat and Chakraborti, 2007; Chakraborti and Gulhane, 2004). To improve catalyst selectivity and activity, ease of separation, and catalyst reusability, immobilization of the zirconium complex onto the common solid support was reported (Karami et al., 2011). The support material plays a critical role in the performance of the resulting supported reagent catalyst. Silica gel support was chosen due to its high surface area, excellent stability (chemical and thermal), good accessibility, and ease of functionalization of the surface groups (Price et al., 2000; Sharma et al., 2003). The catalyst does not need activation and is recycled many times under the same conditions with fresh reactants to yield similar results without significant loss of activity (Yakaiah et al., 2005).

3.8.1 Preparation of Catalyst $ZrOCl_2{\cdot}8H_2O/SiO_2$

The grafted silica gel (1 g) was stirred with zirconium(IV) oxychloride (0.161 g, 0.5 mmol) in chloroform (10 mL), which was heated under reflux conditions for 2 h. The mixture was then filtered and washed thoroughly with chloroform (3 × 10 mL), and the obtained catalyst was dried.

3.8.2 Characterization of Catalyst

The XRD pattern of $ZrOCl_2{\cdot}8H_2O/SiO_2$ is shown in Figure 3.10. It was reported that a high degree of mixing of Zr–Si in oxide form often led to the absence of an XRD pattern for $ZrOCl_2{\cdot}8H_2O$. The broad peak around 22° (2θ) from the smaller inset could be attributed to insertion of Zr into the framework of amorphous SiO_2. The XRF data of $ZrOCl_2{\cdot}8H_2O/SiO_2$ is shown in Table 3.1 and indicates the presence of Cl and ZrO in the supported $ZrOCl_2{\cdot}8H_2O/SiO_2$ matrix.

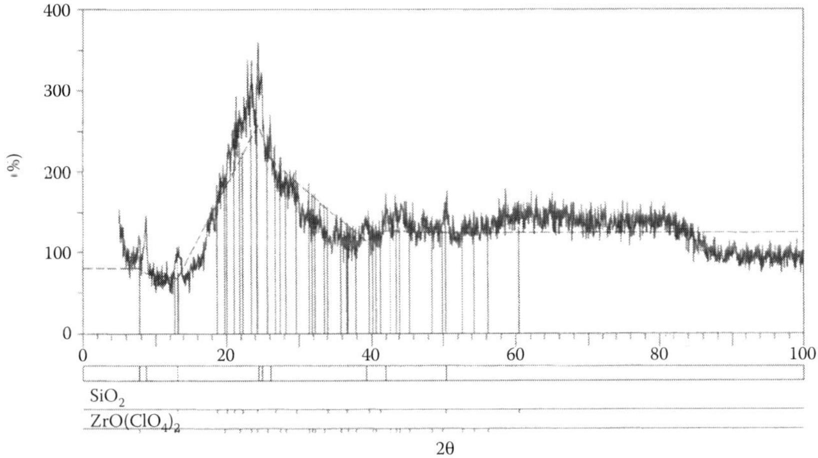

FIGURE 3.10 XRD pattern of $ZrOCl_2{\cdot}8H_2O/SiO_2$.

TABLE 3.1
XRF Data of $ZrOCl_2{\cdot}8H_2O/SiO_2$

Compound	Concentration (% w/w)
SiO_2	79.999
Cl	2.07
ZrO_2	2.00
SO_3	0.140
CaO	0.108
Na_2O	0.084
HfO_2	0.081
Y_2O_3	0.039
Fe_2O_3	0.027
CuO	0.012
LOI	15.85
Total	**100.40**

LOI, loss on ignition.

The FTIR spectra of $ZrOCl_2{\cdot}8H_2O$, SiO_2, and $ZrOCl_2{\cdot}8H_2O/SiO_2$ are shown in Figure 3.11. The spectrum of $ZrOCl_2{\cdot}8H_2O/SiO_2$ shows the characteristic bonds of $ZrOCl_2{\cdot}8H_2O$ and SiO_2. The wave numbers in 800, 1020, 1150, 1610, 2040, and 3400 cm^{-1} of $ZrOCl_2$ were observed in catalyst spectra.

3.9 USE OF $ZrOCl_2{\cdot}8H_2O/SiO_2$ IN SYNTHESIS OF HETEROCYCLES

Coumarins and their derivatives are very important organic compounds. They are the structural unit of several natural products (Murray et al., 1982). Their applications

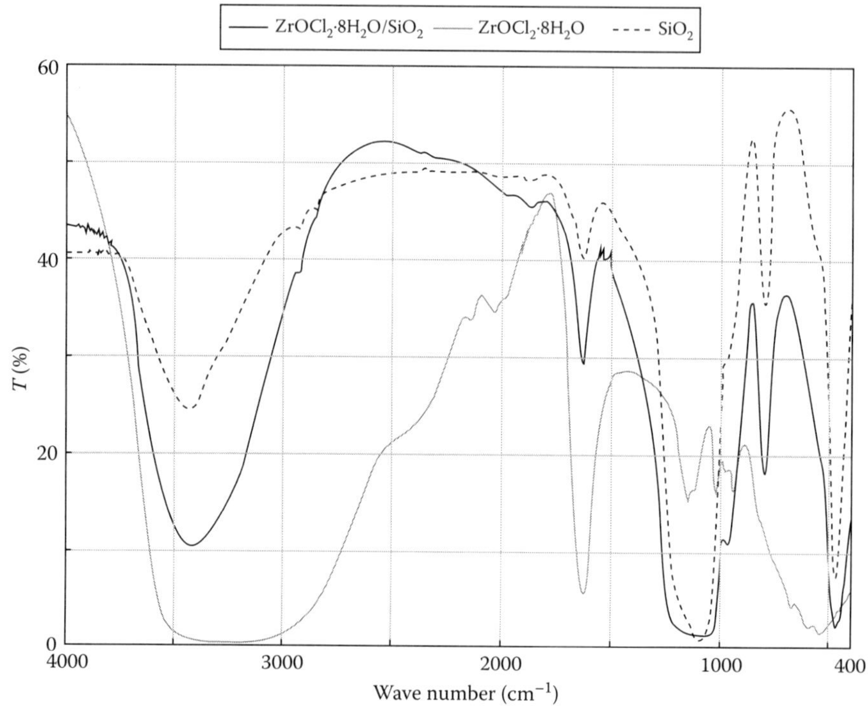

FIGURE 3.11 FTIR spectra of $ZrOCl_2{\cdot}8H_2O$, SiO_2 and $ZrOCl_2{\cdot}8H_2O/SiO_2$.

range from pharmaceuticals (Yu et al., 2003) to optical brighteners (Zabradink, 1992) and laser dyes (Raju and Varadarajan, 1995). Silica gel–supported zirconyl chloride octahydrate was found to be an efficient and recyclable catalyst for the synthesis of a series of biologically important molecules at high turnover numbers and rates. Several substituted coumarins can be prepared in high yield and purity by direct reaction of β-keto esters and phenol derivatives in the presence of a catalytic amount of $ZrOCl_2{\cdot}8H_2O/SiO_2$ as Lewis acid and at ambient temperature under solvent-free conditions. This method, which is called Pechmann condensation, is a very easy and rapid reaction for the synthesis of coumarin derivatives (Scheme 3.33) (Karami et al., 2011).

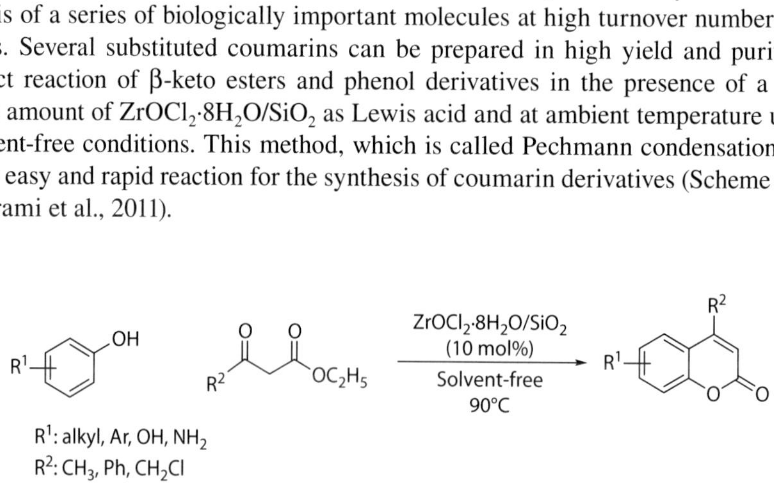

SCHEME 3.33

TABLE 3.2
Reusability of $ZrOCl_2{\cdot}8H_2O/SiO_2$

Product	Recovery	Time (min)	Yields (%)
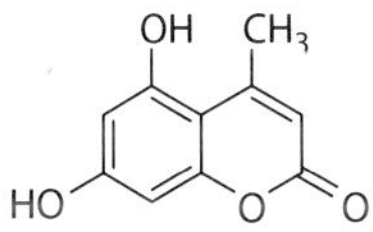	—	10	98
	1	15	83
	2	18	82
	3	20	78

In order to check the stability of the catalyst, the catalyst was recycled (fresh + 3 cycles) using $ZrOCl_2{\cdot}8H_2O/SiO_2$ in Pechmann condensation. The results are presented in Table 3.2. After workup of the reaction mixture, the catalyst was separated by filtration, washed with chloroform, and dried. The recovered catalyst after each reaction was reused in another reaction.

The suggested mechanism for the Pechmann condensation of phenols with β-ketoesters in the presence of $ZrOCl_2{\cdot}8H_2O/SiO_2$ catalyst (Scheme 3.34) has been described using condensation as the probe reaction. The $ZrOCl_2{\cdot}8H_2O/SiO_2$ catalyst would cause dehydration and produce an olefinic bond; at the same time ethyl alcohol would be eliminated with the formation of a coumarin ring.

3.10 CONCLUSIONS

To conclude, in this study, we presented a simple and clean method for the synthesis of several heterogeneous solid acid catalysts that are capable of efficiently catalyzing synthesis of various heterocyclic biological compounds. We believe that these eco-friendly, inexpensive, recyclable catalysts can be used for simple experimental procedures with solvent-free reaction conditions, low catalyst loading, short reaction

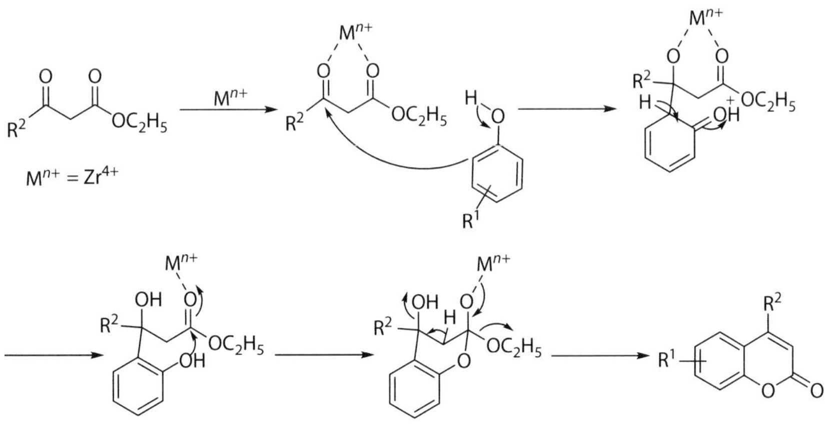

SCHEME 3.34

periods, and good to excellent yields. This is a valid contribution to the existing methodologies for organic reactions.

REFERENCES

Abrantes, M., Bruno, S. M., Tome, C., Pillinger, M., Gonçalves, I. S. and Valente, A. 2011. Epoxidation of DL-limonene using an indenyl molybdenum(II) tricarbonyl complex as catalyst precursor. *Catal. Commun.* 15(1):64–67.

Asgarian-Damavandi, J., Zolfigol, M. A. and Karami, B. 2001. Oxidation of 1,2-dihydroquinoline under mild and heterogeneous conditions. *Synth. Commun.* 31(20):3183–3187.

Atienza, R. L. and Scheidt, K. A. 2011. *N*-Heterocyclic carbene-promoted Rauhut–Currier reactions between vinyl sulfones and α,β-unsaturated aldehydes. *Aust. J. Chem.* 64(8):1158–1164.

Bailly, C., Echepare, S., Gago, F. and Waring, M. 1999. Recognition elements that determine affinity and sequence-specific binding to DNA of 2QN, a biosynthetic bis-quinoline analogue of echinomycin. *J. Anti-Cancer Drug Des.* 14(3):291–301.

Bandyopadhyay, D., Rivera, G., Salinas, I., Aguilar, H. and Banik, B. K. 2010. Remarkable iodine-catalyzed synthesis of novel pyrrole-bearing *N*-polyaromatic β-lactams. *Molecules.* 15(2):1082–1088.

Bayer, A. 1872. Ueber die Verbingdungen der Aldehyde mit den Phenolen. *Ber. Dtsch. Chem. Ges.* 5:25.

Bekaert, A., Andrieux, J. and Plat, M. 1992. New total synthesis of bikaverin. *Tetrahedron Lett.* 33(20):2805–2806.

Belhouchet, M., Youssef, C., Ammar, H. B., Salem, R. B. and Mhiri, T. 2012. Synthesis and crystal structure of a benzoxazole derivative: (*E*)-2-(6-methylbenzoxazol-2-yl)-3-phenylacrylonitrile. *X-Ray Struct. Anal.* 28(1):3–4.

Bhagat, S. and Chakraborti, A. K. 2007. An extremely efficient three-component reaction of aldehydes/ketones, amines, and phosphites (Kabachnik–Fields reaction) for the synthesis of α-aminophosphonates catalyzed by magnesium perchlorate. *J. Org. Chem.* 72(4):1263–1270.

Bharadwaj, A. R. and Scheidt, K. A. 2004. Catalytic multicomponent synthesis of highly substituted pyrroles utilizing a one-pot Sila-Stetter/Paal–Knorr strategy. *Org. Lett.* 6:2465.

Biginelli, P. 1983. Synthesis of 3,4-dihydropyrimidin-2(1*H*)-ones. *Gazz. Chem. Ital.* 23:360–416.

Botta, B., Cassani, M. D. I., Misiti, D., Subissati, D. and Monache, G. D. 2005. Resorcarenes: Emerging class of macrocyclic receptors. *Curr. Org. Chem.* 9(4):337–355.

Boudart, M. 1969. Catalysis by supported metals. *Adv. Catal.* 20:153–177.

Buchstaller, P., Burgdorf, L., Finsinger, D., Stieber, F., Sirrenberg, C., Amendt, C., Grell, M., Zenke, F. and Krier, M. 2011. Design and synthesis of isoquinolines and benzimidazoles as RAF kinase inhibitors. *Bioorg. Med. Chem. Lett.* 21(8):2264–2269.

Burczyk, B. and Kortylewticz, Z. 1982. Organic sulfur compounds; II1. Sulfur dioxide as catalyst in the synthesis of thioacetals from aldehydes or ketones and alkanethiols, alkanedithiols, or hydroxyalkanethiols. *Synthesis.* 10:831–833.

Casiraghi, G., Casnati, G. and Cornia, M. 1973. Regiospecific reactions of phenol salts: Reaction-pathways of alkylphenoxy-magnesiumhalides with triethyl orthoformate. *Tetrahedron Lett.* 14(9):679–682.

Chakraborti, A. K. and Gulhane, R. 2004. Zirconium(IV) chloride as a new, highly efficient, and reusable catalyst for acetylation of phenols, thiols, amines, and alcohols under solvent-free conditions. *Synlett.* 4:627–630.

Chakraborti, A. K. and Kondaskar, A. 2003. $ZrCl_4$ as a new and efficient catalyst for the opening of epoxide rings by amines. *Tetrahedron Lett.* 44(45):8315–8319.

Chibale, K., Visser, M., Schalkwyk, D. V., Smith, P. J., Saravanamuthu, A. and Fairlamb, A. H. 2003. Exploring the potential of xanthene derivatives as trypanothione reductase inhibitors and chloroquine potentiating agents. *Tetrahedron.* 59(13):2289–2296.

Clark, J. H. 2002. Solid acids for green chemistry. *Acc. Chem. Res.* 35:791.

Cornelis, A. and Laszlo, P. 1985. Clay-supported copper(II) and iron(III) nitrates: Novel multipurpose reagents for organic synthesis. *Synthesis.* 1985(10):909–918.

Cornelis, A. and Laszlo, P. 1986. *Chemical Reactions in Organic and Inorganic Constrained Systems*, R. Setton (ed.). Reider: Dordrecht, p. 212.

Dabur, R., Ali, M., Singh, H., Gupta, J. and Sharma, G. L. 2004. A novel antifungal pyrrole derivative from *Datura metel* leaves. *Pharmazie.* 59:568–570.

Damavandi, J. A., Karami, B. and Zolfigol, M. A. 2002. Selective oxidation of *N*-alkyl imines to oxaziridines using UHP/maleic anhydride system. *Synlett.* (6):933–934.

Dell, A., William, D. H., Morris, H. R., Smith, G. A., Feeney, J. and Roberts, G. C. K. 1975. Structure revision of the antibiotic echinomycin. *J. Am. Chem. Soc.* 97(9):2497–2502.

Eftekhari–Sis, B., Abdollahifar, A., Hashemi, M.M. and Zirak, M. 2006. Stereoselective synthesis of β-amino ketones via direct Mannich-type reactions, catalyzed with ZrOCl2·8H2O under solvent-free conditions. *Eur. J. Org. Chem.* 5152.

El-Brashy, A. M., Metwally, M. E. and El-Sepai, F. A. 2004. Spectrophotometric determination of some fluoroquinolone antibacterials by binary complex formation with xanthene dyes. *Il Farmaco.* 59:809–817.

Eshghi, H., Rahimizadeh, M., Shiri, A. and Sedaghat, P. 2012. One-pot synthesis of benzimidazoles and benzothiazoles in the presence of $Fe(HSO_4)_3$ as a new and efficient oxidant. *Bull. Korean Chem. Soc.* 33(2):515–518.

Farahi, M., Karami, B. and Azari, M. 2013. Tungstate sulfuric acid as an efficient catalyst for the synthesis of benzoxazoles and benzothiazoles under solvent-free conditions. *C. R. Chimie.* 16(11):1029–1034.

Firouzabadi, H., Iranpoor, N., Jafarpour, M. and Ghaderi, A. 2006a. $ZrOCl_2{\cdot}8H_2O$ as a highly efficient and the moisture tolerant Lewis acid catalyst for Michael addition of amines and indoles to α,β-unsaturated ketones under solvent-free conditions. *J. Mol. Catal. A Chem.* 252:150–155.

Fukuzawa, S.-I., Shimizu, E., Atsuumi, Y., Haga, M. and Ogata, K. 2009. Copper-catalyzed direct thiolation of benzoxazole with diaryl disulfides and aryl thiols. *Tetrahedron Lett.* 50(20):2374–2376.

Habibi, A., Tarameshloo, Z., Rostamizadeh, S. and Amani, A. M. 2012. Efficient synthesis of 3-aminoimidazo[1,2-*a*] pyridines using silica-supported perchloric acid ($HClO_4$-SiO_2) as a novel heterogenous catalyst. *Lett. Org. Chem.* 9(3):155–159.

Heydari, A., Larijani, H., Emami, J. and Karami, B. 2000. Lithium perchlorate/diethylether-catalyzed three-component coupling reactions of aldehydes, hydroxylamines and trimethylsilyl cyanide leading to α-cyanohydroxylamines. *Tetrahedron Lett.* 41(14):2471–2473.

Horacio, C., David-Alexandre, B., Romain, N., Frank, K., Bernard, R. and Roman, L. 2009. Synthesis of 5-acyl-3,4-dihydropyrimidine-2-thiones via solvent-free, solution-phase and solid-phase Biginelli procedures. *Synlett.* 2009(11):1737–1740.

Hu, E. H., Sidler, D. R. and Dolling, U. H. 1998. Unprecedented catalytic three component one-pot condensation reaction: An efficient synthesis of 5-alkoxycarbonyl-4-aryl-3,4-dihydropyrimidin-2(1*H*)-ones. *J. Org. Chem.* 63(10):3454–3457.

Huang, S. T., Hsei, I. J. and Chen, C. 2006. Synthesis and anticancer evaluation of bis(benzimidazoles), bis(benzoxazoles), and benzothiazoles. *Bioorg. Med. Chem.* 14(17):6106–6119.

Kamitori, Y., Hojo, M., Masuda, R., Kimura, T. and Yoshida, T. 1986. Selective protection of carbonyl compounds. Silica gel treated with thionyl chloride as an effective catalyst for thioacetalization. *J. Org. Chem.* 51(9):1427–1431.

Kantam, M. L., Roy, M., Roya, S., Sreedhara, B. and De, R. L. 2008. Polyaniline supported CuI: An efficient catalyst for C–N bond formation by N-arylation of *N*(*H*)-heterocycles and benzyl amines with aryl halides and arylboronic acids, and aza-Michael reactions of amines with activated alkenes. *Catal. Commun.* 9(13):2226–2230.

Karami, B., Eskandari, K., Farahi, M. and Barmas, A. 2012a. An effective and new method for the synthesis of polysubstituted imidazoles by the use of $CrCl_3·6H_2O$ as a green and reusable catalyst: Synthasis of some novel imidazole derivatives. *J. Chin. Chem. Soc.* 59(4):473–479.

Karami, B., Eskandari, K., Gholipour, S. and Jamshidi, M. 2013a. Facile and rapid synthesis of 9-aryl 1,8-dioxoöctahydroxanthenes derivatives using tungstate sulfuric acid. *Org. Prep. Proced. Int.* 45(3):220–226.

Karami, B., Ghashghayee, V. and Khodabakhshi, S. 2012b. A rapid access to novel and known benzimidazole derivatives using silica chloride as a reusable catalyst. *Chin. J. Chem.* 30(4):959–964.

Karami, B., Ghashghaee, V. and Khodabakhshi, S. 2012c. Novel silica tungstic acid (STA): Preparation, characterization and its first catalytic application in synthesis of new benzimidazoles. *Catal. Commun.* 20:71–75.

Karami, B., Haghighijou, Z., Farahi, M. and Khodabakhshi, S. 2012d. One-pot synthesis of dihydropyrimidine-thione derivatives using tungstate sulfuric acid (TSA) as recyclable catalyst. *Phosphorus Sulfur Silicon Relat. Elem.* 187(6):754–761.

Karami, B., Hoseini, S. J., Eskandari, K., Ghasemi, A. and Nasrabadi, H. 2012e. Synthesis of xanthene derivatives by employing Fe_3O_4 nanoparticles as an effective and magnetically recoverable catalyst in water. *Catal. Sci. Technol.* 2:331–338.

Karami, B., Jamshidi, M. and Khodabakhshi, S. 2013b. Modified Paal–Knorr synthesis of novel and known pyrroles using tungstate sulfuric acid as a recyclable catalyst. *Lett. Org. Chem.* 10(1):12–16.

Karami, B. and Khodabakhshi, S. 2011. A facile synthesis of phenazine and quinoxaline derivatives using magnesium sulfate heptahydrate as a catalyst. *J. Serb. Chem. Soc.* 76(9):1191–1198.

Karami, B., Khodabakhshi, S., Safikhani, N. and Arami, A. 2012f. A green and highly efficient solvent-free synthesis of novel calicx[4]resorcinarene derivatives using tungstate sulfuric acid. *Bull. Korean Chem. Soc.* 33(1):123–127.

Karami, B. and Kiani, M. 2011. An efficient and recyclable catalyst for the preparation of coumarin derivatives by Pechmann condensation reaction. *Catal. Commun.* 14(1):62–67.

Karami, B., Khodabakhshi, S. and Haghighijou, Z. 2012g. Tungstate sulfuric acid: Preparation, characterization, and application in catalytic synthesis of novel benzimidazoles. *Chem. Pap.* 66(7):684–690.

Karami, B., Khodabakhshi, S. and Nikrooz, M. 2011. Synthesis of aza-polycyclic compounds: Novel phenazines and quinoxalines using molybdate sulfuric acid (MSA). *Polycycl. Aromat. Compd.* 31(2):97–109.

Karami, B., Mallakpour, S. and Farahi, M. 2008. Silica supported ICl as a novel heterogeneous system for the rapid oxidation of urazoles to triazolinediones. *Heteroatom Chem.* 19(14):389–393.

Karami, B. and Montazertozohori, M. 2006. Tungstate sulfuric acid (TSA)/$KMnO_4$ as a novel heterogeneous system for rapid deoximation. *Molecules.* 11(9):720–725.

Karami, B., Montazertozohori, M. and Habibi, M. H. 2005a. Tungstate sulfuric acid (TSA)/$NaNO_2$ as a novel heterogeneous system for the N-nitrosation of secondary amines under mild conditions. *Bull. Kor. Chem. Soc.* 26(7):1125–1128.

Karami, B., Montazerozohori, M. and Habibi, M. H. 2006. Tungstate sulfuric acid (TSA): A novel and efficient heterogeneous system for rapid deoximation solid acidic reagent for the oxidation of thiols to disulfides and oxidative demasking of 1,3-dithianes. *Phosphorous Sulfur Silicon Relat. Elem.* 181(12):2825–2831.

Karami, B., Montazerozohori, M., Habibi, M. H. and Zolfigol, M. A. 2005b. Tungstate sulfuric acid/$KMnO_4$ as a novel heterogeneous system for the rapid aromatization of Hantzsch1, 4-dihydropyridines under mild conditions. *Heterocycl. Commun.* 11(6):513–516.

Karami, B., Montazerozohori, M., Karimipour, G. R. and Habibi, M. H. 2005c. Tungstate sulfuric acid (TSA)/$NaNO_2$ as a novel heterogeneous system for rapid deoximation. *Bull. Kor. Chem. Soc.* 26(9):1431–1433.

Karami, B., Nikrooz, M. and Khodabakhshi, S. 2012h. A modified synthesis of some novel polycyclic aromatic phenazines and quinoxalines by using the tungstate sulfuric acid (TSA) as a reusable catalyst under solvent-free conditions. *J. Chin. Chem. Soc.* 59(2):187–192.

Karami, B., Taei, M., Khodabakhshi, S. and Jamshidi, M. 2012i. Synthesis of 1,3-dithiane and 1,3-dithiolane derivatives by tungstate sulfuric acid (TSA): Recyclable and green catalyst. *J. Sulfur Chem.* 33(1):65–74.

Karami, B., Zare, Z. and Eskandari, K. 2013c. Molybdate sulfonic acid: Preparation, characterization, and application as an effective and reusable catalyst for octahydroxanthene-1,8-dione synthesis. *Chem. Pap.* 67(2):145–154.

Khalafi-Nezhad, A., Foroughi, H., Doroodmand, M. M. and Panahi, F. 2011. Silica boron–sulfuric acid nanoparticles (SBSANs): Preparation, characterization and their catalytic application in the Ritter reaction for the synthesis of amide derivatives. *J. Mater. Chem.* 21:12842–12851.

Khodabakhshi, S. and Karami, B. 2012. A rapid and eco-friendly synthesis of novel and known benzopyrazines using silica tungstic acid (STA) as a new and recyclable catalyst. *Catal. Sci. Technol.* 2(9):1940–1944.

Kinjo, J., Uemura, H., Nohara, T., Yamashita, M., Marubayashi, N. and Yoshihira, K. 1995. Novel yellow pigment from Pterocarpus santalinus: Biogenetic hypothesis for santalin analogs. *Tetrahedron Lett.* 36(31):5599–5602.

Knight, D. W. and Little, P. B. 2001. The first efficient method for the intramolecular trapping of benzynes by phenols: A new approach to xanthenes. *J. Chem. Soc. Perkin Trans. 1.* 14(15):1771–1777.

Kus, C. 2003. Synthesis of some new benzimidazole carbamate derivatives for evaluation of antifungal activity. *Turk. J. Chem.* 27(1):35–39.

Laszlo, P. 1986. Catalysis of organic reactions by inorganic solids. *Acc. Chem. Res.* 19(4):121–127.

Lazarenkow, A., Nawrot-Modranka, J., Brzezinska, E., Krajewska, U. and Rozalski, M. 2012. Synthesis, preliminary cytotoxicity evaluation of new 3-formylchromone hydrazones and phosphorohydrazone derivatives of coumarin and chromone. *Med. Chem. Res.* 21(8):1861–1868.

Leyton, P., Sanchez-Cortes, S., Campos-Vallette, M., Domingo, C., Garcia-Ramos, J. V. and Saitz, C. 2005. Surface-enhanced Micro-Raman detection and characterization of calix[4]arene–polycyclic aromatic hydrocarbon Host–Guest complexes. *Appl. Spectrosc.* 59(8):1009–1015.

Liu, C. J. and Wang, J. D. 2009. Copper(II) sulfamate: An efficient catalyst for the one-pot synthesis of 3,4-dihydropyrimidine-2(1*H*)-ones and thiones. *Molecules.* 14(2):763–770.

Lu, J., Bai, Y., Wang, Z., Yang, B. and Ma, H. 2000. One-pot synthesis of 3,4-dihydropyrimidin-2(1H)-ones using lanthanum chloride as a catalyst. *Tetrahedron Lett.* 41(47):9075–9078.

Ma, Y., Qian, C., Wang, L. and Yang, M. 2000. Lanthanide triflate catalyzed Biginelli Reaction. One-pot synthesis of dihydropyrimidinones under solvent-free conditions. *J. Org. Chem.* 65(12):3864–3868.

Mallakpour, S. E., Karami, B. and Sheikholeslami, B. 1998. Polymerization of 1-methyl-2,5-bis[1-(4-phenylurazolyl)] pyrrole dianion with alkyldihalides. *Polym. Int.* 45(1):98–102.

Martin, A. and Narayana, V. 2010. Heterogeneously catalyzed ammoxidation: A valuable tool for one-step synthesis of nitriles. *ChemCatChem.* 2(12):1504–1522.

Martinez, G. M., Teran, C. R., Tlapanco, O. A., Toscano, A. and Cruz-Almanza, R. 2000. Supramolecular complexes between fullerene [60] and [70], and resorcinarene. *Fullerene Sci. Technol.* 8(6):475.

Marziano, N. C., Ronchin, L., Tortato, C., Ronchin, S. and Vavasori, A. 2005. Selective oxidations by nitrosating agents: Part 2: Oxidations of alcohols and ketones over solid acid catalysts. *J. Mol. Catal. A Chem.* 235(1):26–34.

McDonald, M. and Mavrodi, D. V. 2001. Phenazine biosynthesis in *Pseudomonas fluorescens*: Branchpoint from the primary shikimate biosynthetic pathway and role of phenazine-1,6-dicarboxylic acid. *J. Am. Chem. Soc.* 123(38):9459.

McKillop, A. and Young, D. W. 1979. Organic synthesis using supported reagents—Part I. *Synthesis.* (6):401–422.

Metzger, J. D. 1998. Solvent-free organic syntheses. *Angew. Chem. Int. Ed. Engl.* 37(21):2975–2978.

Migawa, M. T., Girardet, J. L., Walker, J. A., Koszalka, G. W., Chamberlain, S. D., Drach, J. C. and Townsend, L. B. 1998. Design, synthesis, and antiviral activity of α-nucleosides: D- and L-isomers of lyxofuranosyl- and (5-deoxylyxofuranosyl)benzimidazoles. *J. Med. Chem.* 41(8):1242–1251.

Montazerozohori, M. and Karami, B. 2006. Molybdate sulfuric acid (MSA)/$NaNO_2$ as a novel heterogeneous system for the N-nitrosation of secondary amines under mild conditions. *Helv. Chim. Acta.* 89(12):2922–2926.

Montazerozohori, M., Karami, B. and Azizi, M. 2007. Molybdate sulfuric acid (MSA): A novel and efficient solid acid reagent for the oxidation of thiols to symmetrical disulfides. *Arkivoc.* (i):99–104.

Movassaghi, M. and Jacobsen, E. N. 2002. A direct method for the conversion of terminal epoxides into γ-butanolides. *J. Am. Chem. Soc.* 124(11):2456.

Murkute, A. D., Jackson, J. E. and Miller, D. J. 2011. Supported mesoporous solid base catalysts for condensation of carboxylic acids. *J. Catal.* 278(2):189–199.

Murray, R. D. H., Mendezand, J. and Brown, S. A. 1982. *The Natural Coumarins: Occurrence, Chemistry and Biochemistry.* Wiley: New York.

Niknam, K., Kiasat, A. R., Karami, B. and Heydari, N. 2007. Basic Al_2O_3 as a recyclable reagent for the protection of carbonyl groups with phenylhydrazine derivatives and semicarbazides. *Turk J. Chem.* 31(2):135–139.

Nonn, A. and Burchard, F. 1995. Novel porphyrinoids for chemistry and medicine by biomimetic syntheses. *Angew. Chem. Int. Ed.* 34(17):1795–1811.

Olah, G. A., Molhotra, R. and Narang, S. C. 1978. Aromatic substitution. 43. Perfluorinated resinsulfonic acid catalyzed nitration of aromatics. *J. Org. Chem.* 43(24):4628–4630.

Pagni, R. M., Kobalka, G. W., Boothe, R., Gaetano, K., Stewart, L. J. and Conawaya, R. 1998. Reactions of unsaturated compounds with iodine and bromine on gamma alumina. *J. Org. Chem.* 53(19):4477–4482.

Phucho, I. T., Nongpiur, A., Nongrum, R. and Nongkhlaw, R. L. 2010. Synthesis of 3,4-dihydropyrimidin-2(1*H*)ones and 3,4,5,6,7,8-hexahydroquinazolin-2(1*H*)-ones via three component cyclocondensation. *Ind. J. Chem.* 49(3):346–350.

Porcari, A. R., Devivar, R. V., Kucera, L. S., Drach, J. C. and Townsend, L. B. 1998. Design, synthesis, and antiviral evaluations of 1-(substituted benzyl)-2-substituted-5,6-dichlorobenzimidazoles as nonnucleoside analogues of 2,5,6-trichloro-1-(β-D-ribofuranosyl)benzimidazole. *J. Med. Chem.* 41(8):1252–1262.

Posner, G. H. 1978. Organic reactions at alumina surfaces. *Angew. Chem. Int. Ed. Engl.* 17(7):487–496.

Praveen, C., Namdakumar, A., Dheenkumar, P., Muralidharan, D. and Perumal, P. T. 2012. Microwave-assisted one-pot synthesis of benzothiazole and benzoxazole libraries as analgesic agents. *J. Chem. Sci.* 124(3):609–624.

Price, P. M., Clark, J. H. and Macquarrie, D. J. 2000. Modified silicas for clean technology. *J. Chem. Soc. Dalton Trans.* 101–110.

Qi, X., Rice, G. T., Lall, M. S., Plummer, M. S. and White, M. C. 2010. Diversification of a β-lactam pharmacophore via allylic C–H amination: Accelerating effect of Lewis acid co-catalyst. *Tetrahedron.* 66(26):4816–4826.

Ragno, R., Marshall, G. R., Di-Santo, R., Costi, R., Massa, S., Rompei, R. and Artico, M. 2000. Antimycobacterial pyrroles: Synthesis, anti-*Mycobacterium tuberculosis* activity and QSAR studies. *Bioorg. Med. Chem.* 8(6):1423–1432.

Raju, B. B. and Varadarajan, T. S. 1995. Photophysical properties and energy transfer dye laser characteristics of 7-diethylamino-3-heteroaryl coumarin in solution. *Laser Chem.* 16(2):109–120.

Ranu, B. C., Hajra, A. and Jana, U. 2000. Indium (III) chloride-catalyzed one-pot synthesis of dihydropyrimidinones by a three-component coupling of 1,3-dicarbonyl compounds, aldehydes, and urea: An improved procedure for the Biginelli reaction. *J. Org. Chem.* 65(19):6270–6272.

Romanelli, G. P., Ruiz, D. M., Autino, J. C. and Giaccio, H. E. 2010. A suitable preparation of *N*-sulfonyl-1,2,3,4-tetrahydroisoquinolines and their ring homologs with a reusable Preyssler heteropolyacid as catalyst. *Mol. Divers.* 14(4):803–807.

Roth, T., Morningstar, M. L., Boyer, P. L., Hughes, S. H., Buckheitjr, R. W. and Michejda, C. J. 1997. Synthesis and biological activity of novel nonnucleoside inhibitors of HIV-1 reverse transcriptase. 2-Aryl-substituted benzimidazoles. *J. Med. Chem.* 40(26):4199–4207.

Santato, C., Odziemkowski, M., Ulmann, M. and Augustynski, J. 2001. Crystallographically oriented mesoporous WO_3 films: Synthesis, characterization, and applications. *J. Am. Chem. Soc.* 123(43):10639–10649.

Seyyedhamzeh, M., Mirzaei, P. and Bazgir, A. 2008. Solvent-free synthesis of aryl-14*H*-dibenzo[*a,j*]xanthenes and 1,8-dioxo-octahydro-xanthenes using silica sulfuric acid as catalyst. *Dyes Pigm.* 76:836–839.

Shafiee, M. R. M. and Moloudi, R. 2011. Solvent-free preparation of 2,4,6-triaryl pyridines using silver(I) nitrate adsorbed on silica gel nanoparticles ($AgNO_3$–nano-SiO_2) as an efficient catalyst. *Lett. Org. Chem.* 8(10):717.

Sharghi, H. and Hosseini-Sarvari, M. 2002a. A direct synthesis of nitriles and amides from aldehydes using dry or wet alumina in solvent free conditions. *Tetrahedron.* 58(52):10323–10328.

Sharghi, H. and Hosseini-Sarvari, M. 2002b. Solvent-free and one-step beckmann rearrangement of ketones and aldehydes by zinc oxide. *Synthesis.* (8):1057–1060.

Sharma, R. K., Mittal, S. and Koel, M. 2003. Analysis of trace amounts of metal ions using silica-based chelating resins: A green analytical method. *Crit. Rev. Anal. Chem.* 33:183–197.

Shinohara, K., Bando, T. and Sugiyama, H. 2010. Anticancer activities of alkylating pyrrole–imidazole polyamides with specific sequence recognition. *Anticancer Drugs.* 21(3):228–242.

Srinivasulu, M., Rajesh, K., Suryakiran, N., Selvam, J. J. P. and Venkateswarlu, Y. 2007. Lanthanum(III) nitrate hexahydrate catalyzed chemoselective thioacetalization of aldehydes. *J. Sulfur Chem.* 28(3):245–249.

Stone, M. T. and Anderson, H. L. 2007. A cyclodextrin-insulated anthracene rotaxane with enhanced fluorescence and photostability. *Chem. Commun.* 126(23):2387–2389.

Tamm, I. 1957. Ribonucleic acid synthesis and infulenza virus multiplication. *Science.* 126:1235–1236.

Tanaka, K. and Tada, F. 2000. Solvent-free organic synthesis. *Chem. Rev.* 100(3):1025–1074.

Turner, J. M. and Messenger, A. J. 1986. Occurrence, biochemistry and physiology of phenazine pigment production. *Adv. Microb. Physiol.* 27(C):211–275.

Wedge, T. J. and Hawthorne, M. F. 2003. Multidentate carborane-containing Lewis acids and their chemistry: Mercuracarborands. *Coord. Chem. Rev.* 240:111–128.

Weidner, M. F., Sigurdsson, S. T. and Hopkins, P. B. 1990. Sequence preferences of DNA interstrand cross-linking agents: dG-to-dG cross-linking at 5′-CG by structurally simplified analogs of mitomycin C. *Biochemistry.* 29(39):9225–9233.

Wen, X., Bakali, J. E., Deprez-Poulain, R. and Deprez, B. 2012. Efficient propylphosphonic anhydride (®T3P) mediated synthesis of benzothiazoles, benzoxazoles and benzimidazoles. *Tetrahedron Lett.* 53(19):2440–2443.

Woo, J. and Sigurdsson, S.T. 1993. DNA interstrand cross-linking reactions of pyrrole-derived, bifunctional electrophiles: Evidence for a common target site in DNA. *J. Am. Chem. Soc.* 115(9):3407–3415.

Wu, Y., He, L.-N., Du, Y., Wang, J.-Q., Miao, C.-X. and Li, W. 2009. Zirconyl chloride: An efficient recyclable catalyst for synthesis of 5-aryl-2-oxazolidinones from aziridines and CO_2 under solvent-free conditions. *Tetrahedron.* 65(31):6204–6210.

Yadav, J. S., Reddy, B. S. and Pandey, S. K. 2001a. $LiBF_4$ catalyzed chemoselective conversion of aldehydes to 1,3-oxathiolanes and 1,3-dithianes. *Synlett.* (2):238–239.

Yakaiah, T., Venkat Reddy, G., Lingaiah, B. P. V., Shanthan Rao, P. and Narsaiah, B. 2005. $ZrOCl_2{\cdot}8H_2O$ as a new solid phase and recyclable catalyst for an efficient Knoevenagel condensation under solvent-free microwave irradiation conditions. *Indian J. Chem.* 44B(6):1301–1303.

Yonetake, K., Nakayama, T. and Ueda, M. 2001. New liquid crystals based on calixarenes. *J. Mater. Chem.* 11:761–767.

Yu, Y., Liu, D., Liu, C. and Luo, G. 2007. One-pot synthesis of 3,4-dihydropyrimidin-2(1*H*)-ones using chloroacetic acid as catalyst. *Bioorg. Med. Chem. Lett.* 17(12):3508–3510.

Yu, D., Xie, M. L., Morris-Natschke, S. L. and Lee, K. H. 2003. Recent progress in the development of coumarin derivatives as potent anti-HIV agents. *Med. Res. Rev.* 23(3):322–345.

Zabradink, M. 1992. *The Production and Application of Fluorescent Brightening Agent.* Wiley: New York.

Zolfigol, M. A. 2001. Silica sulfuric acid/$NaNO_2$ as a novel heterogeneous system for production of thionitrites and disulfides under mild conditions. *Tetrahedron.* 57(46):9509–9511.

Zolfigol, M. A., Shirin, F., Ghorbani-Choghamarani, A. and Mohammadpoor-Baltork, I. 2002. Silica modified sulfuric acid/$NaNO_2$ as a novel heterogeneous system for the oxidation of 1,4-dihydropyridines under mild conditions. *Green Chem.* 4(6):562–564.

4 Heterogeneous Copper-Catalyzed Synthesis of Bioactive Heterocycles

Xin Lv and Guodong Yuan

CONTENTS

4.1 INTRODUCTION

Developing novel catalytic systems that meet the requirements of green chemistry has emerged as one of the most attractive and challenging subjects in chemistry.

In the last two decades, copper catalysis has been intensively utilized for numerous useful transformations including oxidation (Diaz-Requejo et al., 2000; Kantam et al., 2009a), cycloaddition (Reymond and Cossy, 2008; Hein and Fokin, 2010), cross-coupling (Ma and Cai, 2008), C–H activation (Daugulis et al., 2009; Wendlandt et al., 2011), domino reactions (Liu and Wan, 2011, 2012), and so on. These Cu-mediated reactions are also involved in many valuable synthetic applications such as the assembly of biologically and medicinally useful molecules.

Despite its efficiency, homogeneous catalysis may suffer from inherent limitations such as tedious purification procedures and the difficulties of removing and recycling the catalysts. Notably, it is essential to remove the trace amount of catalyst from the final product because metal contamination is strictly regulated, particularly by the medicinal and biological industries. To overcome the shortcomings of homogeneous catalysis, heterogeneous catalysis has been developed as an appropriate solution. Compared with the homogeneous version, heterogeneous catalysis possesses more attractive properties since the products can be simply isolated and the catalysts can be efficiently separated and recovered (Xia et al., 2005).

In recent years, heterogeneous Cu catalysis has been successfully applied for the assembly of various bioactive molecules including many linear and cyclic compounds (Chassaing et al., 2010; Climent et al., 2011). This chapter will concentrate on the recent advances in heterogeneous copper-catalyzed reactions for the synthesis of bioactive heterocycles. Since the subject is vast, it is beyond the scope of this chapter to include all the relevant examples. Thus, we have placed more emphasis on important and recent developments. The chapters are organized according to type of synthetic protocol.

4.1.1 Synthesis of Heterocycles via Heterogeneous Cu-Catalyzed Cycloadditions

Up to now, a variety of heterogeneous Cu-catalyzed cycloaddition reactions for the assembly of bioactive heterocycles have been reported. Most examples are the Huisgen-type 1,3-dipolar cycloadditions, but there are rare reports on the heterogeneous Cu-catalyzed synthesis of heterocycles through other types of cycloaddition reactions.

4.1.1.1 Synthesis of 1,2,3-Triazoles via Huisgen Cycloaddition

1,2,3-Triazole derivatives are known as useful molecules with interesting biological activities (Buckle et al., 1986; Alvarez et al., 1994; Genin et al., 2000). One of the most popular approaches to 1,2,3-triazoles is the Huisgen cycloaddition of terminal alkynes with azides (Huisgen, 1963, 1984, 1989). The application of the original method was mainly limited by the requirement for high reaction temperatures and low regioselectivity. To overcome these limitations, Meldal et al. and Sharpless et al. have independently developed the copper(I)-catalyzed regioselective Huisgen cycloaddition (Tornøe et al., 2002; Rostovtsev et al., 2002). However, the traditional Cu-mediated cycloaddition may suffer from potential contamination of the products by the Cu complexes. In recent years, heterogeneous Cu-mediated azide–alkyne click chemistry has appeared as an attractive synthetic protocol since it possesses unique advantages in comparison with the homogeneous version (Dervaux and Du Prez, 2012). Generally, the heterogeneous catalysts could be removed completely and well recovered after simple treatments.

4.1.1.1.1 Synthesis of 1,2,3-Triazoles Using Silica Gel–Supported Cu Catalysts

Silica-based supports are easily available and environmentally benign. They have been widely utilized for heterogeneous Cu catalysis, which opens up new opportunities for the synthesis of greener clicked compounds. Wang's group immobilized a

Cu(I) catalyst on primary amine-modified silica gel and utilized this Cu(I) complex as the promoter for the three-component reactions between alkylhalides, sodium azide, and terminal acids (Scheme 4.1) (Miao and Wang, 2008). The azides were delivered *in situ* and the handling of potentially explosive organic azides was avoided, indicating that the method was more convenient and safer than the traditional synthesis. However, this catalyst does not seem very efficient for the reactions.

They also employed *N*-heterocyclic carbene (NHC)-modified silica particles as efficient and recyclable ligands for the Huisgen cycloaddition (Scheme 4.2) (Li et al., 2008). A variety of 1,2,3-triazoles were generated in high yields from various organic azides and alkynes.

Shamim and Paul (2010) found that the silica modified by a Schiff base–type ligand promoted the Cu-catalyzed, three-component synthesis of 1,2,3-triazoles in water at room temperature (Scheme 4.3). It also exhibited higher efficiency than Wang's promoter, probably because the Schiff bases possess much better coordinating ability than the primary amines.

Several silica-supported chelating absorbents containing multidentate nitrogenated ligands were developed by Santoyo-Gonzalez et al., and their Cu(I) complexes could be utilized for the click synthesis of the triazole products (Scheme 4.4) (Megia-Fernandez et al., 2010). These solid supports acted as heterogeneous catalysts when complexed with Cu(I) (Si-Lm·Cu⁺), and also participated as efficient Cu scavengers

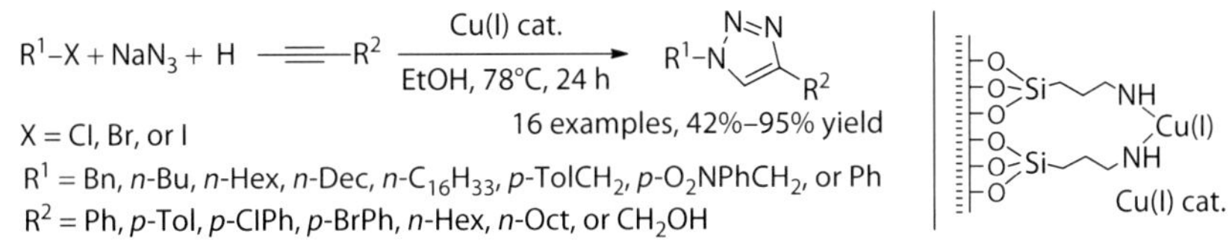

SCHEME 4.1

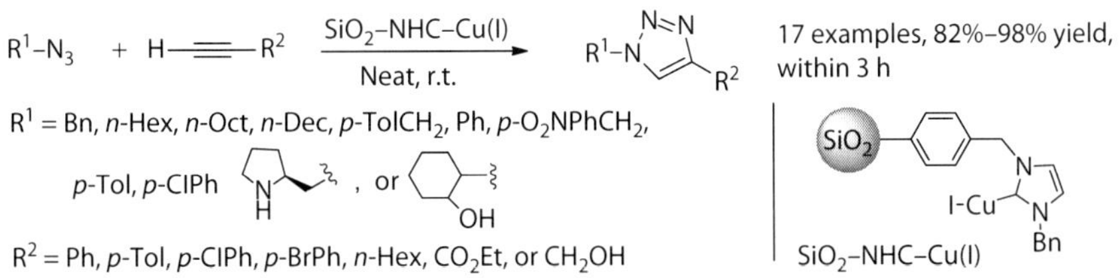

SCHEME 4.2

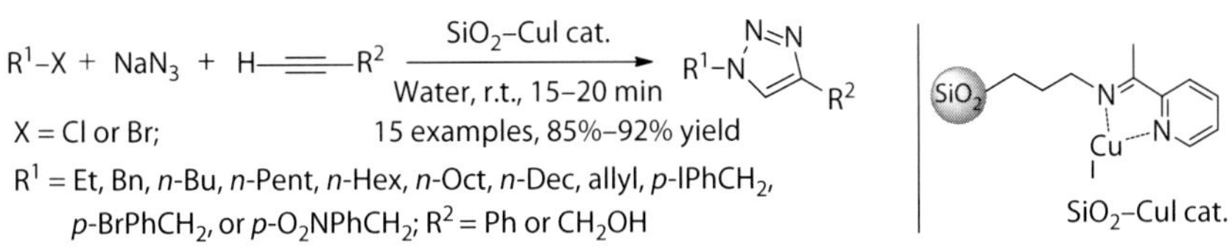

SCHEME 4.3

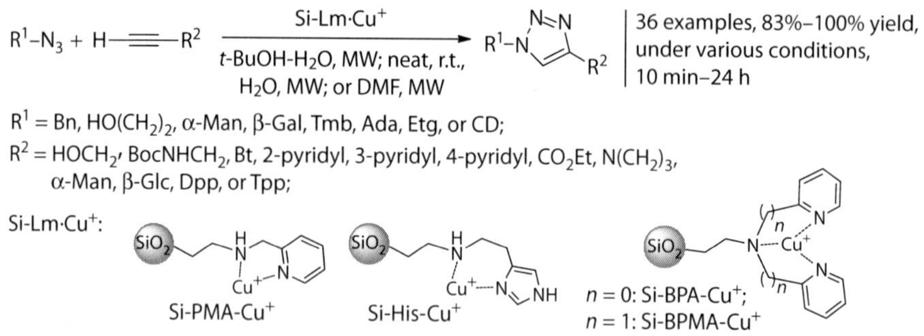

SCHEME 4.4

to remove traces of Cu in their uncomplexed form. Thus, the catalysts can be rapidly and easily removed by simple filtration and they have excellent recyclability properties. Recently, Santoyo-Gonzalez et al. also successfully applied Si-Lm·Cu+ to the three-component click reactions (Scheme 4.5) (Megia-Fernandez et al., 2012).

4.1.1.1.2 *Synthesis of 1,2,3-Triazoles Using Magnetic Catalysts*

The azide–alkyne cycloaddition reactions could be well promoted by magnetic catalysts including bimetallic catalysts (Cu/Fe) and three-component catalysts (Cu/Fe/SiO_2). Compared with homogeneous Cu catalysis, these methods provide 1,2,3-triazole derivatives containing much lower amounts of Cu contaminants. It is worth noting that magnetic iron plays a dual role as an appropriate support for Cu and also as a scavenger that efficiently removes catalyst from the product. Moreover, the unique properties of the magnetic catalysts make them efficient to separate and recycle several times without any loss in activities, using an external magnetic field (Kovacs et al., 2012). Therefore, magnetic Cu catalysis has gained extensive attention.

Hur and coworkers conducted tentative investigations on the click reactions promoted by Cu_3N nanoparticles, which were immobilized on a superparamagnetic mesoporous silica microsphere ($Cu_3N/Fe_3N@SiO_2$) (Scheme 4.6) (Lee et al., 2010). Due to their nanometer size range, these particles have much broader surface areas that promote the catalysis process. It is noteworthy that the catalyst supported on a magnetic microsphere could be efficiently and simply removed by a magnet and

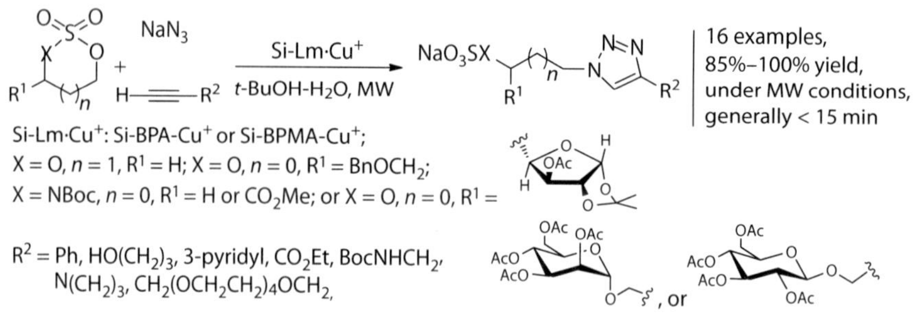

SCHEME 4.5

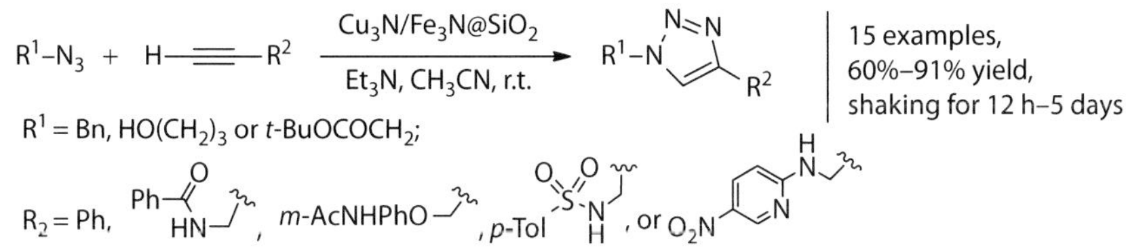

SCHEME 4.6

recycled several times without any loss of activity; thus, it is expected to possess the potential for applications in living and environmentally benign systems.

Santoyo-Gonzalez et al. prepared a ligand immobilized on silica-coated Fe_3O_4 particles (Megia-Fernandez et al., 2010). The ligand was complexed with Cu(I) to generate a magnetic heterogeneous catalyst (Fe_3O_4@Si-BPA·Cu^+) (Scheme 4.7). This catalyst was used for a set of click reactions, and the results showed that it possessed a good catalytic profile similar to that of the Si-Lm·Cu^+ promoters (Scheme 4.8). Compared with Si-Lm·Cu^+ catalysts, this magnetic silica-coated catalyst could be conveniently and rapidly recovered by magnetic decantation, and no discernible loss of catalytic activity was observed after three cycles. Moreover, the use of this promoter allows the reaction to successfully proceed in aqueous media or even without any solvent.

$CuFe_2O_4$ nanoparticles have also been used as magnetically separable catalysts for the multicomponent assembly of 1,2,3-triazoles in water (Scheme 4.9) (Kumar et al., 2012).

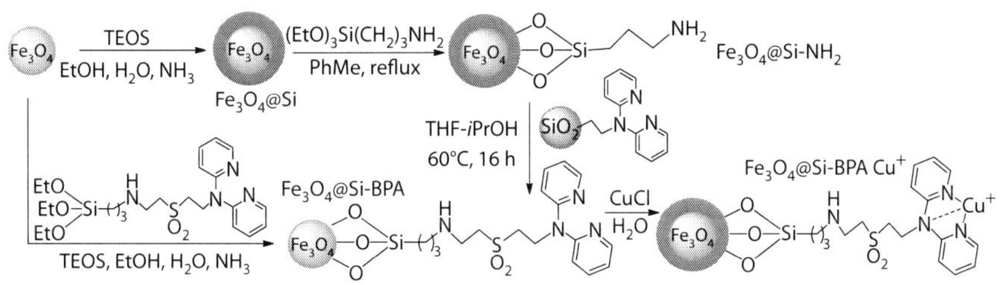

SCHEME 4.7

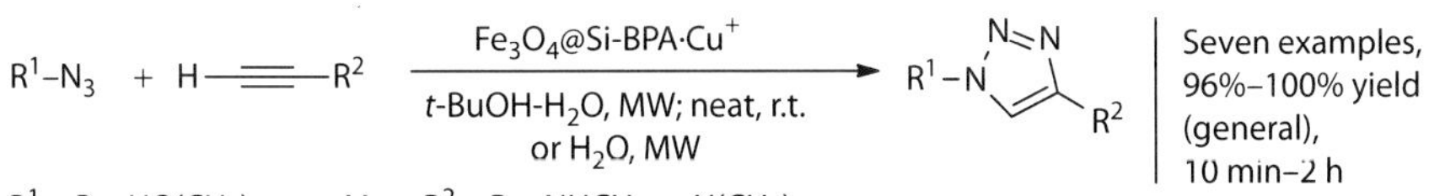

SCHEME 4.8

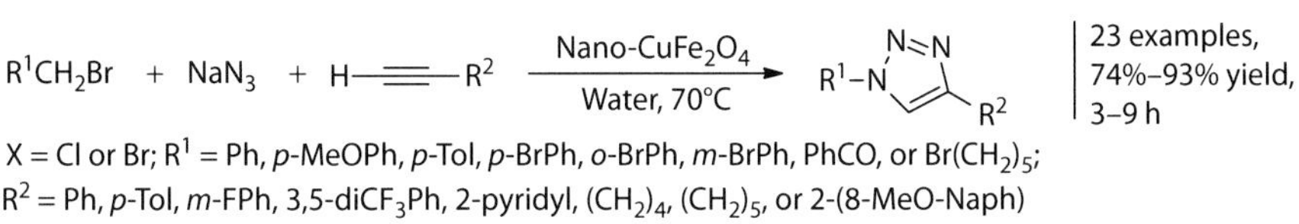

SCHEME 4.9

The magnetic bimetallic catalyst Cu@Fe nanoparticles facilitated the azide–alkyne click reactions in aqueous media (Scheme 4.10) (Hudson et al., 2012). However, this catalyst lost its catalytic activity in the air atmosphere after each run. Concurrently, Nadit et al. found that nano-ferrite-glutathione-copper (nano-FGT-Cu) could be utilized as a magnetic heterogeneous catalyst for click reactions (Nasir Baig and Varma, 2012). Under microwave irradiation conditions, the one-pot reactions of primary alkyl bromides, sodium azide, and alkynes in aqueous media gave the desired 1,2,3-triazole (Scheme 4.11). The magnetically separable catalyst could be recycled at least three times without any loss of activity. However, microwave irradiation and high temperature were required to achieve good conversion.

4.1.1.1.3 Synthesis of 1,2,3-Triazoles Using Charcoal-Based Cu Catalysts

Huisgen cycloaddition can also be promoted by charcoal-based Cu catalysts (Lipshutz and Taft, 2006; Sharghi et al., 2009b; Alonso et al., 2010). Despite their low cost, most of these catalytic systems suffer from high reaction temperatures and the difficulty to reuse.

Lipshutz et al. (2008) developed a novel bimetallic heterogeneous catalyst composed of Cu and Ni oxide particles supported within charcoal (Cu–Ni/C), and they successfully applied it to the cycloaddition reactions between several organic azides and terminal alkynes (Scheme 4.12). However, the reaction mixtures needed to be heated in order to obtain good yields.

Khalifeh and coworkers prepared a novel heterogeneous Cu catalyst [T(*o*-Cl)PPCu-AMWCNT] based on carbon by impregnating Cu(I) onto hydroxyl-modified multiwalled carbon nanotubes (MWCNTs). It efficiently promoted the

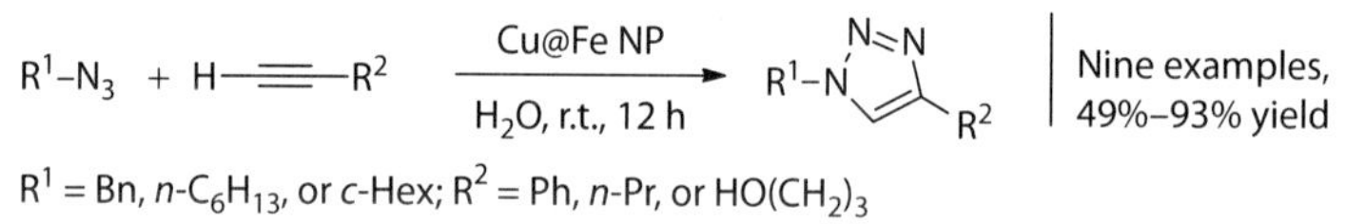

SCHEME 4.10

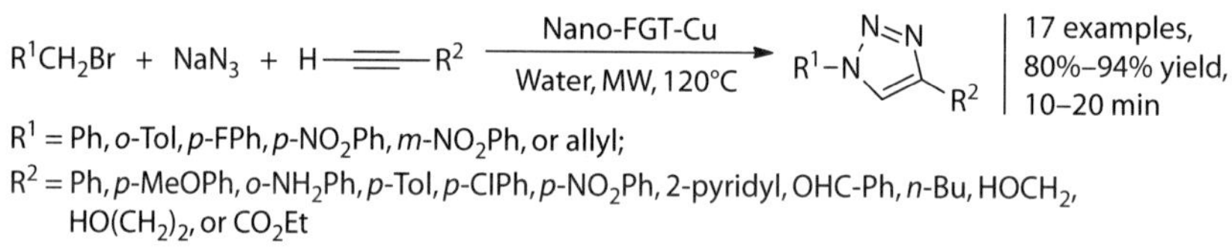

SCHEME 4.11

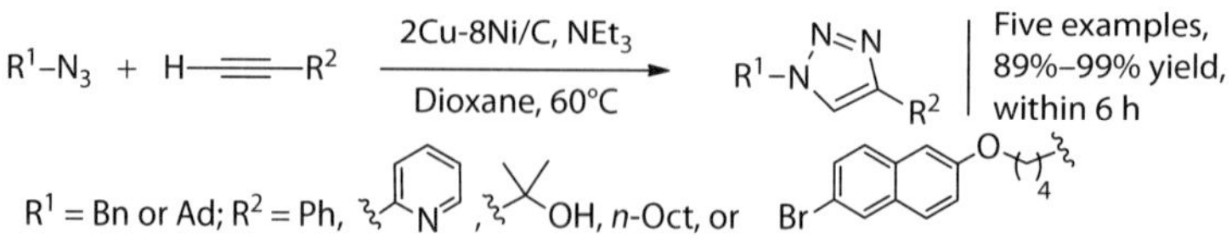

SCHEME 4.12

three-component synthesis of hydroxymethylated 1,2,3-triazoles in water (Sharghi et al., 2009a). Compared with the homogeneous catalyst T(*o*-Cl)PPCu, this heterogeneous catalyst exhibits higher efficiency and can be reused for more than 10 runs.

Alonso et al. (2010) have demonstrated the multicomponent synthesis of 1,2,3-triazoles in water promoted by Cu nanoparticles on activated carbon (CuNPs/C) (Scheme 4.13).

Recently, Cu_2O on charcoal was also utilized as a heterogeneous catalyst for azide–alkyne reactions (Lopez-Ruiz et al., 2013). A variety of 1,2,3-triazoles were generated in moderate to excellent yields (35 examples, at 80°C, 69%–94% yield). However, heating was required to achieve good conversions and the catalyst lost its catalytic activity after three runs.

4.1.1.1.4 Synthesis of 1,2,3-Triazoles Using Other Heterogeneous Catalysts Based on Inorganic Supports

Park et al. (2008) utilized a recyclable catalyst Cu/AlO(OH), which was composed of Cu nanoparticles in AlO(OH) nanofiber, to efficiently facilitate the Huisgen cycloaddition of a wide range of azides and alkynes at room temperature (Scheme 4.14).

Cu(II)–hydrotalcite (Cu(II)–HT) was reported to be a heterogeneous catalyst for Huisgen [3+2] cycloaddition (15 examples, at room temperature for 6–12 h, 75%–89% yield) (Namitharan et al., 2009). However, this catalyst is not very efficient compared with other heterogeneous promoters.

Mizuno and coworkers found that an alumina-supported copper hydroxide $[Cu(OH)_x/Al_2O_3]$ could be utilized for the 1,3-dipolar cycloaddition (Katayama et al., 2009). However, the method was limited due to its low efficiency. Another supported copper hydroxide $[Cu(OH)_x/TiO_2]$ was also used as a heterogeneous catalyst for click reactions by the same group (Yamaguchi et al., 2009). Compared with the Al_2O_3-supported catalyst, this promoter was more efficient.

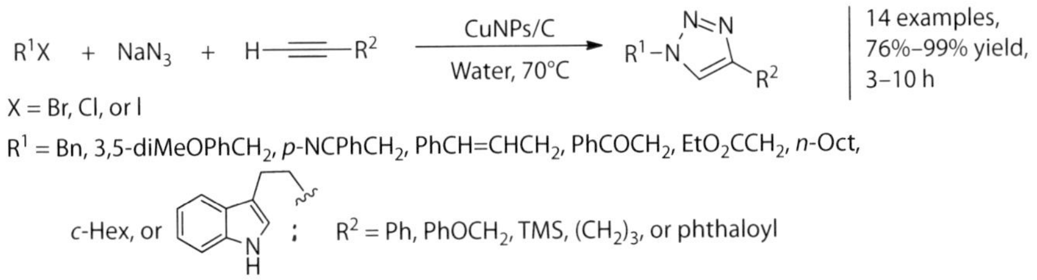

SCHEME 4.13

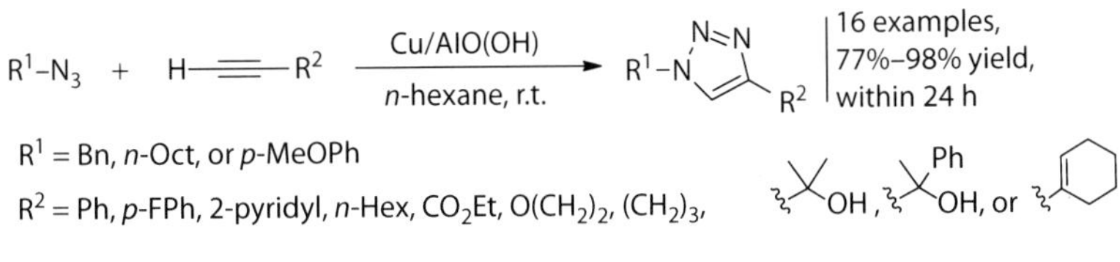

SCHEME 4.14

Pale et al. developed the one-pot synthesis of 1,2,3-triazoles from alkylhalides, NaN_3, and terminal alkynes mediated by a Cu(I)-modified zeolite (Scheme 4.15) (Beneteau et al., 2010). Also, Cu(I)-zeolites could smoothly promote the multicomponent synthesis of hydroxymethylated 1,2,3-trizoles from epoxides, NaN_3, and terminal alkynes (Scheme 4.16) (Boningari et al., 2010). However, a drawback of the method is that long reaction times are required.

A bimetallic catalyst based on Cu–Mn spinel oxide was used for azide–alkyne reactions. Microwave irradiation conditions were required due to its low intrinsic catalytic activity. It was observed that Mn significantly influenced the efficiency and selectivity of the bimetallic catalyst (Yousuf et al., 2010). Nanoporous Cu catalysts (CuNPore) can be employed for the efficient assembly of 1,2,3-triazoles (Jin et al., 2011). Additionally, their tunable ligament sizes have a significant impact on the catalytic activity. Bharate et al. reported a recyclable clay-supported, Cu(II)-catalyzed, one-pot synthesis of 1,2,3-triazoles (Mohammed et al., 2012). In this case, montmorillonite KSF clay-supported CuO nanoparticles efficiently promoted the one-pot aromatic azidonation of the arylboronic acids/azide–alkyne cycloaddition process at room temperature.

4.1.1.1.5 Synthesis of 1,2,3-Triazoles Using Polymer-Based Cu Catalysts

Cu(I) on Amberlyst A-21 was shown to be a highly efficient catalyst for the click reactions (Girard et al., 2006; Jlalia et al., 2010). Supported by a dimethylaminomethyl-grafted poly(styrene)/divinylbenzene (PS/DVB) resin, the Cu(I) catalyst facilitated the click reactions to afford the desired triazoles in good to excellent yields at room temperature (Scheme 4.17). The support contained a tertiary amino group that acted as a chelating agent and also as a base.

Chi's group developed several Merrifield resins with quaternary ammonium salts and used their Cu(I) complexes as heterogeneous promoters for the synthesis of 1,2,3-triazoles (Sirion et al., 2008). The ionic polymers could be simply prepared and

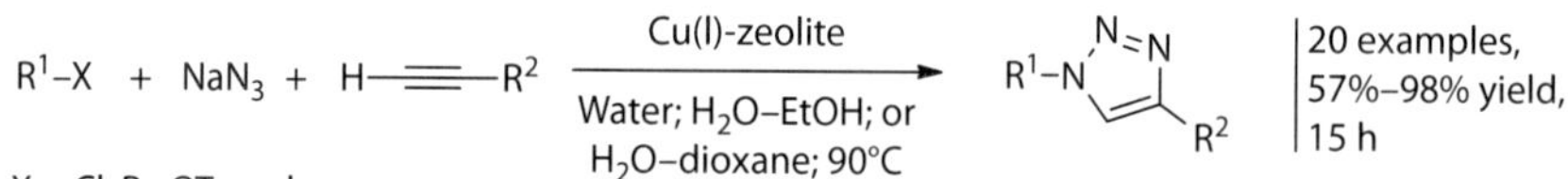

X = Cl, Br, OTs, or I;
R^1 = Bn, Et, *n*-Pr, 2,6-diClPhCH_2, $CH_3CH{=}CHCH_2$, MeO_2CCH_2, Ph$COCH_2$, or *s*-Bu;
R^2 = Ph, *p*-Tol, *p*-CF_3Ph, *n*-Pr, *n*-Hex, $HO(CH_2)_3$, or $CH_3(CH_2)_4CHOH$

SCHEME 4.15

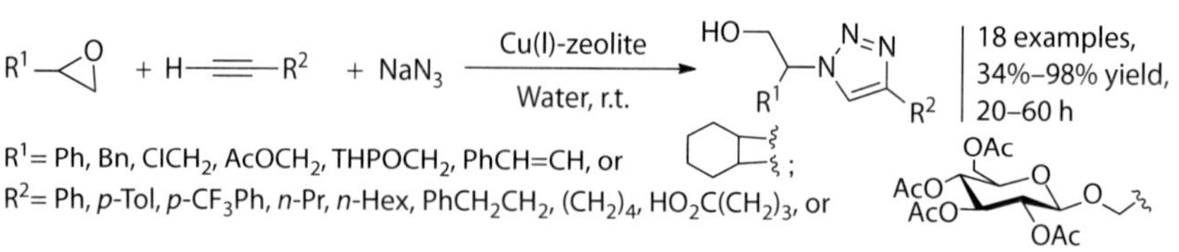

SCHEME 4.16

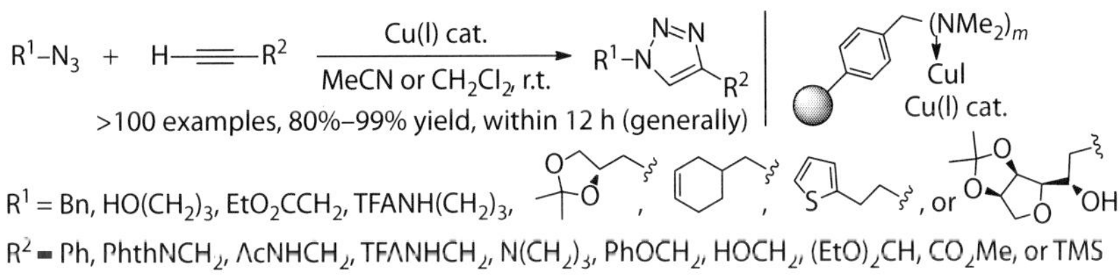

SCHEME 4.17

represented negligible levels of Cu(I) leaching. The support can be recycled 10 times without any loss of activity.

Wang et al. (2011) found that a Merrifield resin containing imidazolium could also facilitate the Cu(II)-catalyzed azide–alkyne cycloaddition, but it seemed quite inefficient.

PS-supported 1,5,7-triazabicyclo[4.4.0]dec-5-ene (TBD) was used as a polyvalent ligand for the Cu-catalyzed Huisgen reactions (Coelho et al., 2010) (Scheme 4.18). The three-component synthesis of 1,2,3-triazoles was successfully achieved under the same conditions. However, the catalytic system was less active, maybe due to its low catalyst loading.

Prez and coworkers found that poly(*N*-methylethylene imine) (PMEI) was an efficient solid support for the Cu(I)-catalyzed Huisgen cycloaddition. Despite its low catalyst loading, it afforded the desired products in good yields at room temperature (Bonami et al., 2009).

In several cases, modified TentaGel resins were chosen as the supports because they could swell in a variety of solvents and efficiently promote the click reactions while simultaneously stabilizing the Cu(I) species against oxidation (Chan and Fokin, 2007; Suzuka et al., 2010). In addition, hardly any Cu leaching was observed and the resin could be efficiently reused. However, the catalytic activity was unsatisfactory due to the low Cu loading on the supports. Thus, the scope of their application is limited to a certain degree (Dervaux and Du Prez, 2012).

Recently, Yamada et al. (2012) reported that the self-assembly of $CuSO_4$ and a poly(imidazole-acrylamide) amphiphile afforded a highly efficient and recyclable solid-phase catalyst for click reactions. Even 4.5–45 mol ppm of the amphiphilic

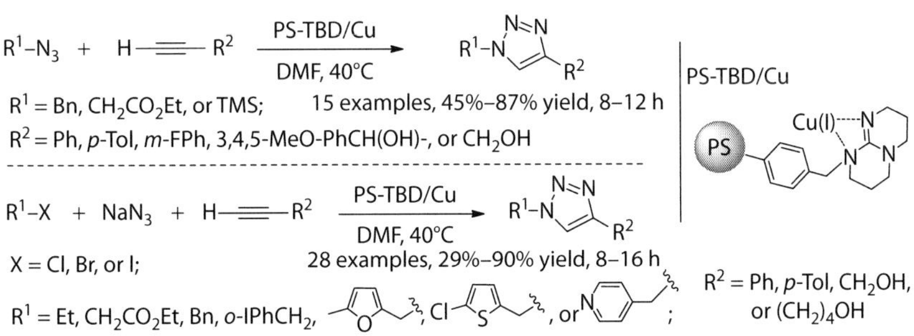

SCHEME 4.18

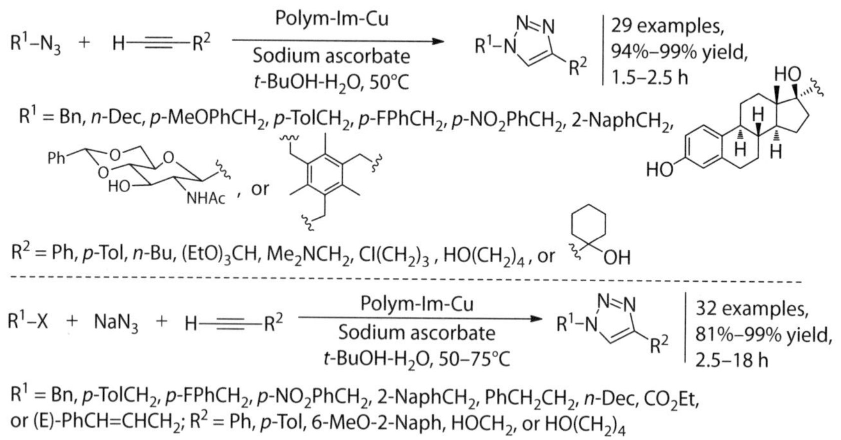

SCHEME 4.19

polymeric promoter can efficiently facilitate the Huisgen cycloaddition and the three-component reactions of NaN_3, alkynes, and alkylhalides (Scheme 4.19).

Other types of heterogeneous copper catalysts have also been developed for the azide–alkyne cycloaddition. Zhang et al. (2013) utilized porous copper(0) to facilitate the three-component reactions in water, and the desired 1,2,3-triazoles could be assembled in good to excellent yields, though long reaction times were usually required (at 55°C for 20–54 h).

4.1.1.2 Synthesis of Heterocycles via Other [3 + 2] Cycloaddition Reactions

Sommer et al. found that Cu(I)-exchanged zeolites (Cu(I)-USY) could be utilized as heterogeneous catalysts for [3 + 2] cycloaddition of azomethine imines with terminal alkynes (Scheme 4.20). This method provides an efficient, versatile, and highly regioselective approach to *N*,*N*-bicyclic pyrazolidinone derivatives, which might exhibit useful bioactivities. The catalysts were readily available, convenient to remove, and reusable (Keller et al., 2009).

The 1,3-dipolar cycloaddition of azomethine imines with alkynes could also be well facilitated by an Al_2O_3-supported copper hydroxide [$Cu(OH)_x/Al_2O_3$] (Scheme 4.21). Generally, the desired pyrazolidinones were obtained in good to excellent yields (Yoshimura et al., 2011).

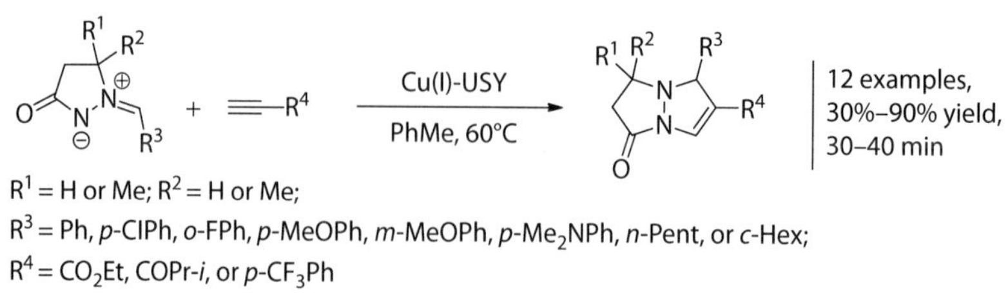

SCHEME 4.20

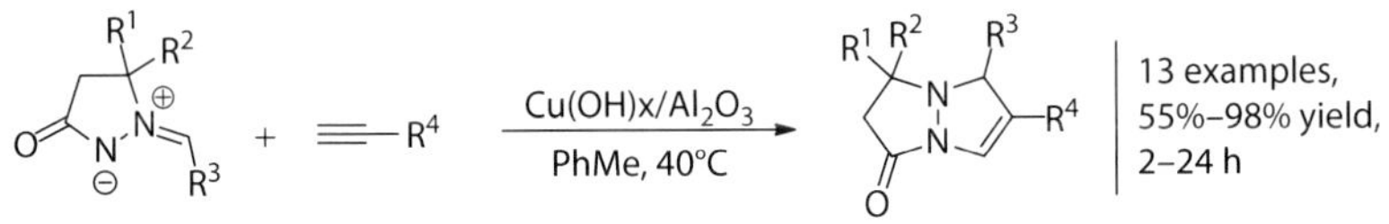

R^1 = H or Me; R^2 = H or Me; R^3 = Ph, *p*-ClPh, *p*-MeOPh, *n*-Pent, or *c*-Hex;
R^4 = CO_2Me, CO_2Et, COMe, COPh, or *p*-Ts

SCHEME 4.21

Recently, montmorillonite clay Cu(II)/NaN_3 was found to be a suitable heterogeneous catalytic system for the cycloaddition of nitrile oxide and phenylacetylene in water (Bharate et al., 2013) (Scheme 4.22).

4.1.1.3 Heterogeneous Cu-Catalyzed Hetero [4 + 2] Reactions

Gebbink et al. reported that the immobilized Cu(II)-chiral bis(oxazoline) complexes were employed as new heterogeneous catalysts for the hetero-Diels–Alder reactions (O'Leary et al., 2004). The hetero-Diels–Alder cycloaddition between ethyl (*E*)-2-oxopent-3-enoate and ethyl vinyl ether could be facilitated by these immobilized catalysts, but only up to 42% ee, with a low conversion of 25% being obtained (Scheme 4.23).

4.1.2 Synthesis of Heterocycles via Heterogeneous Cu-Catalyzed Coupling Reactions

Many aromatic compounds such as *N*-aryl/*N*-heteroaryl azoles and heteroaryl thioethers are known as important molecules in the biological and medicinal fields (Craig, 1991; Negwer, 1994). One of the most efficient approaches to these heterocycles is transition-metal-catalyzed cross-coupling (Jiang and Buchwald, 2004). Although Pd- and Ni-promoted coupling reactions have achieved remarkable progress, modern Cu-mediated sp^2 C–C and sp^2 C–Y (Y=N, O, S, etc.) formations have attracted great attention due to their high efficiency, low cost, and green properties (Ma and Cai, 2008; Monnier and Taillefer, 2009; Surry and Buchwald, 2010). Recently, Cu-mediated coupling has become an important protocol for the assembly

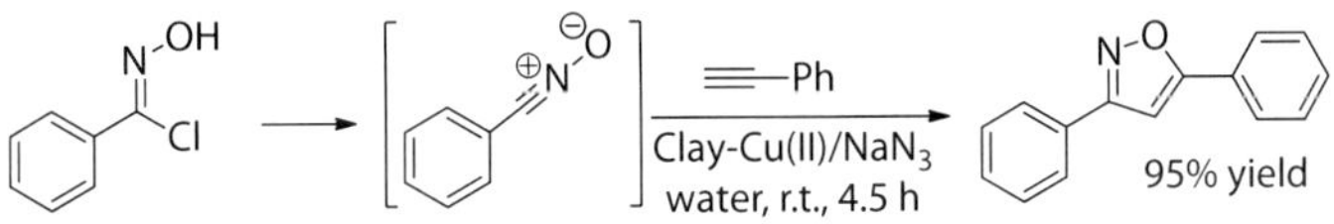

SCHEME 4.22

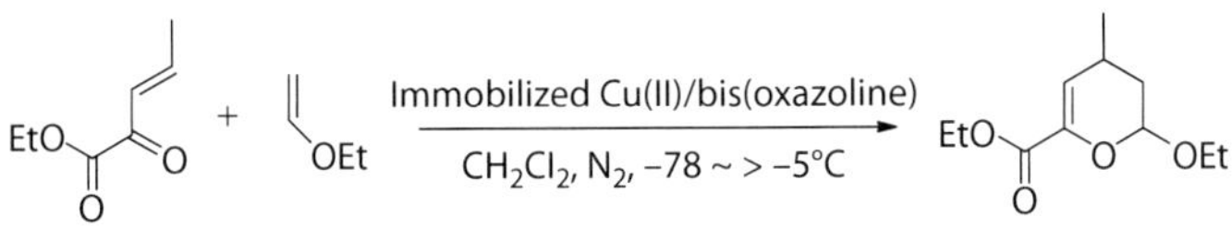

SCHEME 4.23

of aromatic products, which are useful in the biological and pharmaceutical sciences. However, the classic homogeneous Cu-mediated coupling may suffer from contamination of its products by Cu complexes, tedious isolation procedures, and the infeasibility of recycling catalysts. These shortcomings stimulated considerable efforts to develop greener and simpler Cu-catalyzed coupling methods in recent years. One of the most successful solutions is the use of heterogeneous Cu-catalyzed coupling. Generally, these protocols not only have high synthetic efficiency, but also provide more convenient and environmentally benign approaches to the coupled products (Benyahya et al., 2008; Zhang et al., 2008; Swapna et al., 2011; Saha et al., 2012; Bhong et al., 2013).

4.1.2.1 Synthesis of Heterocycles via Intermolecular Coupling Reactions

In recent years, heterogeneous copper-catalyzed intermolecular cross-coupling has been successfully applied for the synthesis of a variety of heterocyclic compounds.

4.1.2.1.1 Heterogeneous Cu-Catalyzed N-Arylation Using Aryl/Heteroaryl Halides

As early as 2004, Hyeon et al. reported the application of Cu_2O-coated Cu nanoparticles (Cu_2O/nano-Cu) to Ullmann-type coupling of aryl chlorides with *N*-heterocycles for the synthesis of *N*-aryl imidazoles. Benzimidazole, pyrazole, and pyrrole also smoothly coupled with electron-poor chloroarenes (Scheme 4.24) (Son et al., 2004).

Kantam et al. (2007) studied the synthesis of *N*-arylheterocycles via the nanocrystalline CuO-catalyzed N-arylation of heterocycles with activated chloro- and fluoroarenes (Scheme 4.25). Various *N*-aryl/*N*-heteroaryl imidazoles, benzimidazoles, indoles, pyrroles, and pyrazoles were synthesized in moderate to excellent yields (76%–94% yield in 23 h). Much higher reactivity was observed when fluoroarenes were used (81%–94% yield in 3 h). The Cu catalyst could be recovered by centrifugation and reused for five runs without obvious loss of activity. Transmission electron microscopy (TEM) studies of both fresh and used catalysts revealed that the shape and size of the particles remained unchanged, supporting the assumption that the morphology of the Cu catalyst remained the same after recycling. However, one of the major shortcomings of the above methods developed by Hyeon and Kantam

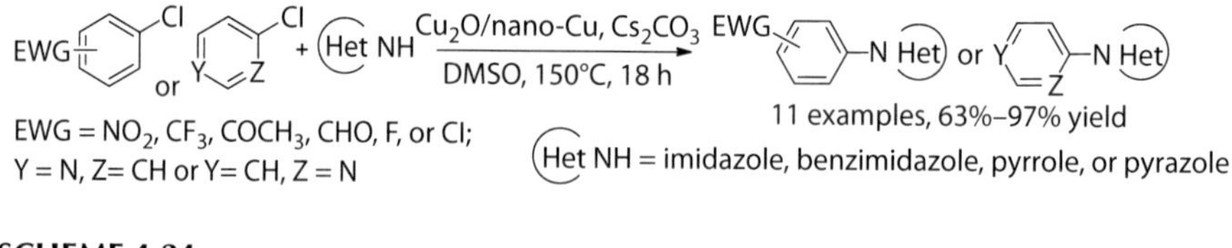

SCHEME 4.24

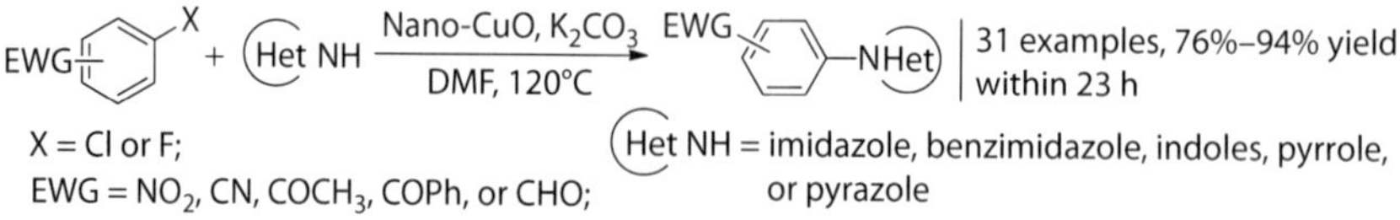

SCHEME 4.25

is that the aryl halides should be preactivated by modification of the phenyl rings with strong electron-withdrawing groups. Additionally, harsh conditions were usually required (at 120°C–150°C).

Cu-exchanged fluorapatite (CuFAP) and *tert*-buoxyapatite (CuTBAP) were first used as recyclable heterogeneous catalysts by Choudary et al. (2005). Under the promotion of CuFAP, the C–N coupling of imidazoles, benzimidazole, pyrrole, pyrazole, and piperidine with chloroarenes and fluoroarenes afforded the desired *N*-aryl heterocyclic products in good to excellent yields (Scheme 4.26). Although the fluoroarenes should have been premodified with electron-withdrawing groups, the reactions of the chloroarenes bearing electron-donating groups proceeded smoothly. During their optimization procedures, they observed that the decreasing order of activity was: CuFAP (+*t*BuOK) > CuTBAP (+*t*BuOK) > CuFAP > CuTBAP, and other Cu catalysts such as CuHAP, $Cu(OAc)_2$, Cu(I), and Cu showed inactivity.

Chouhan et al. (2007) found that Cu(I)-catalyzed C–N coupling reactions between aryl/heteroaryl bromides and azoles could be efficiently facilitated by the proline-functionalized magnetic nanoparticles (MNPs). Generally, the desired *N*-aryl/*N*-heteroaryl azoles (including pyrazoles, indoles, and benzimidazoles) were synthesized in good to excellent yields (Scheme 4.27).

Lipshutz et al. (2008) also found that several types of cross-coupling reactions could be smoothly achieved by using a new heterogeneous catalyst composed of Cu and Ni oxide particles supported within charcoal (Cu–Ni/C). Several heterocyclic products were smoothly obtained from the coupling of aryl/heterarylchlorides with heterocyclic secondary amines. However, these reactions needed to be conducted in a microwave reactor at high temperature (200°C) to achieve complete conversions.

Li et al. developed the solvent-free C–N coupling between *N*-heterocycles and aryl/heteroaryl halides promoted by the nano-Cu_2O/1,10-phen/tetra-*n*-butylammonium fluoride (TBAF) catalytic system. Different types of Cu_2O were evaluated, and the studies showed that Cu_2O nanoparticles (especially the cubic form) were the most efficient for the C–N coupling reactions (Scheme 4.28) (Tang et al., 2008).

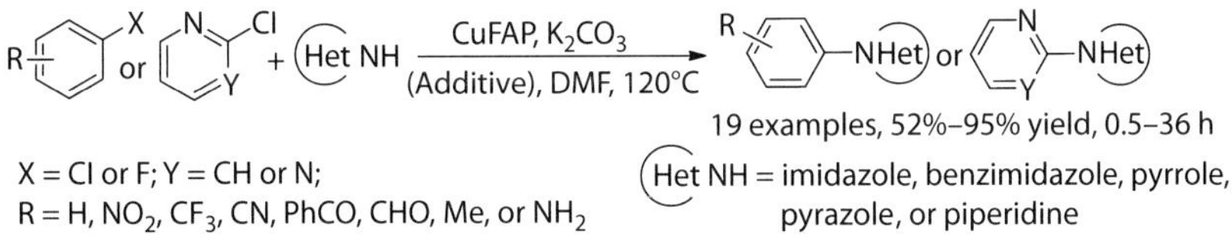

SCHEME 4.26

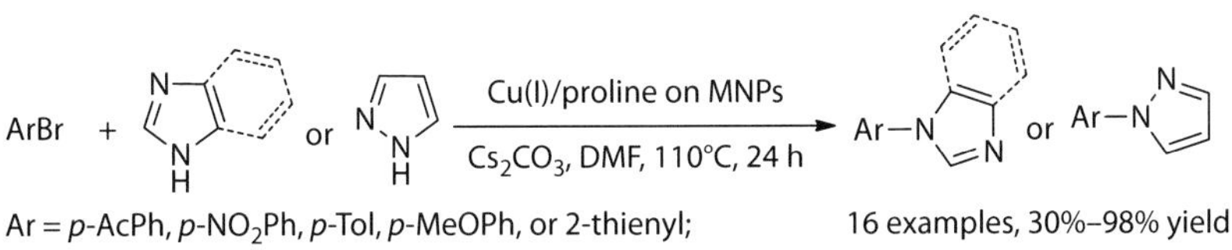

SCHEME 4.27

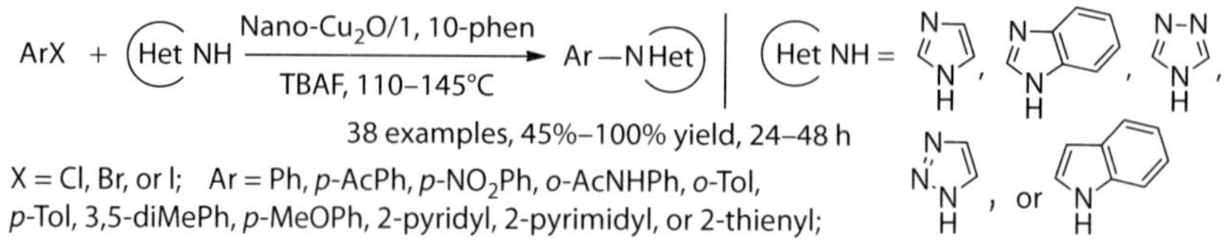

SCHEME 4.28

Acetylene black–immobilized CuO hollow nanospheres (CuO/AB) can be used for the N-arylation of heterocycles (Kim et al., 2009). This novel catalytic system was suitable for a variety of substrates, including the electron-rich aryl halides and *N*-heterocycles. Heteroarylamines such as pyridine-2-amine, pyridine-3-amine, and pyrimidin-2-amine also afforded the desired *N*-aryl heterocyclic amines. However, a high temperature (180°C) was required to achieve good conversions.

Kaushik and coworkers developed a ligand-free heterogeneous Cu-catalyzed N-arylation of amines with diheteroaryl halides (Scheme 4.29) (Verma et al., 2011). In this case, benzyltributyl ammonium bromide was used as the phase transfer catalyst (PTC), and potassium hydroxide (KOH) was employed as the base. Under optimized conditions (Cu(I)/PTC/KOH, in H_2O–$CHCl_3$ or H_2O–EtOH), a variety of heteroaromatic amines were obtained in high yields with good chemoselectivity at 35°C–40°C.

Cu nanoparticle-doped silica cuprous sulfate (CN-DSCS) can be utilized as an efficient and reusable heterogeneous catalyst for the C–N coupling of nucleobases with aryl halides (Scheme 4.30) (Rad et al., 2011). Various bioactive *N*-aryl nucleobases and their analogs were delivered in moderate to good yields under the catalysis of CN-DSCS, which could be easily recycled and reused for five runs without a decrease in the catalytic activity.

Nageswar et al. used nano-CuO for the efficient assembly of N-substituted indoles from indolines via the aromatization/C–N coupling process. No external ligand was

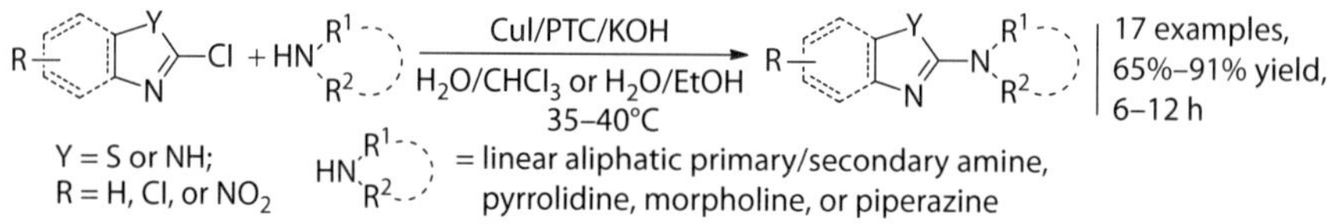

SCHEME 4.29

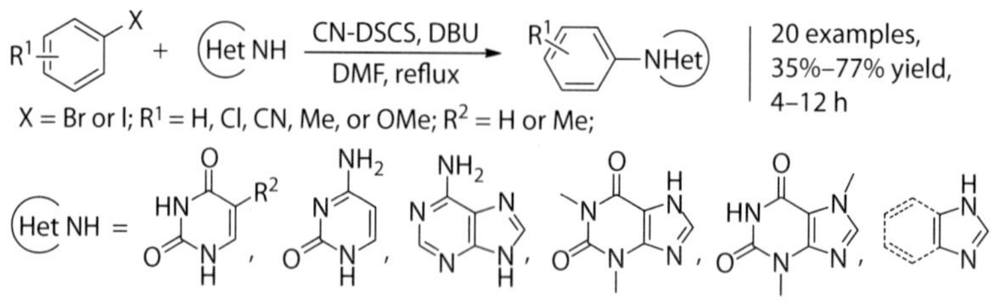

SCHEME 4.30

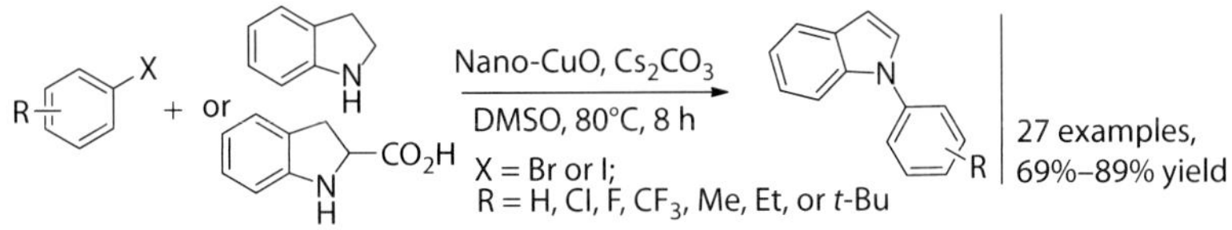

SCHEME 4.31

added and the catalyst could be easily reused for up to four runs without loss of activity (Scheme 4.31) (Reddy et al., 2012).

Very recently, Cai and coworkers also found that the N-arylation of indoles could be successfully achieved by employing the MCM-41-immobilized bidentate nitrogen Cu(I) complex (MCM-41-2*N*-Cu(I)) (Xiao et al., 2013). Moderate to excellent yields of the desired *N*-arylindoles/*N*-heteroarylindoles were assembled from the coupling of indoles and (hetero)aryl halides in toluene at 110°C (Scheme 4.32). The catalyst was recovered by a simple filtration and reused for 10 cycles with almost consistent activity.

A stable and reusable CuO/carbon nanotube catalyst (CuO/MWCNT) for N-arylation of imdazole was developed by Karvembu et al. TEM images of the nanocatalyst showed good adhesion of CuO nanoparticles to anchoring sites of acid-treated MWCNTs. Even a small amount of the catalyst (0.98 mol%) was sufficient for the coupling reactions of imidazole and aryl halides. Additionally, a variety of the desired *N*-arylimidazoles were smoothly generated (11 examples, at 120°C for 24 h, 44%–96% yields) (Gopiraman et al., 2013). However, the application scope of the method is limited because only electron-poor aryl halides have been investigated.

4.1.2.1.2 Heterogeneous Cu-Catalyzed N-Arylation Using Arylboronic Acids

An efficient base-free N-arylation of imidazoles with arylboronic acids using CuFAP was also developed by Kantam et al. (2006) (Scheme 4.33). The advantages of this method were (1) mild conditions (base free and room temperature); (2) broad application scope (a variety of substituents were tolerated) with high efficiency; and (3) reusability of the catalyst.

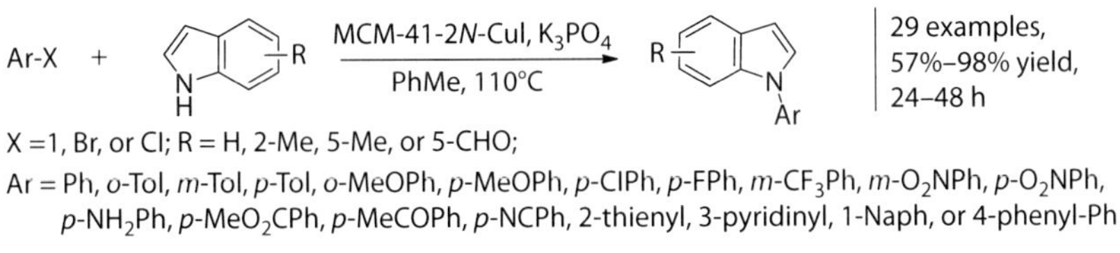

SCHEME 4.32

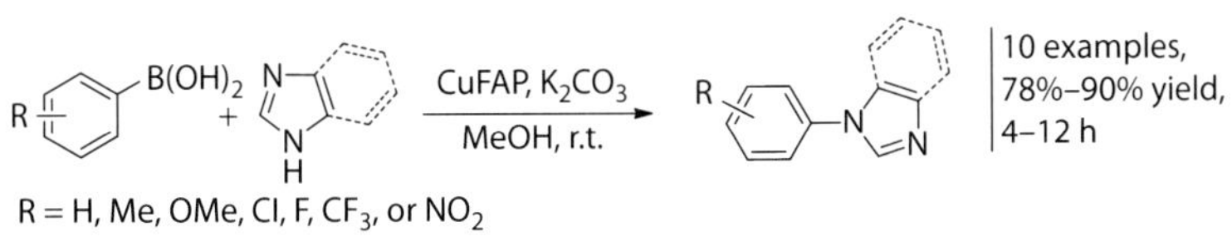

SCHEME 4.33

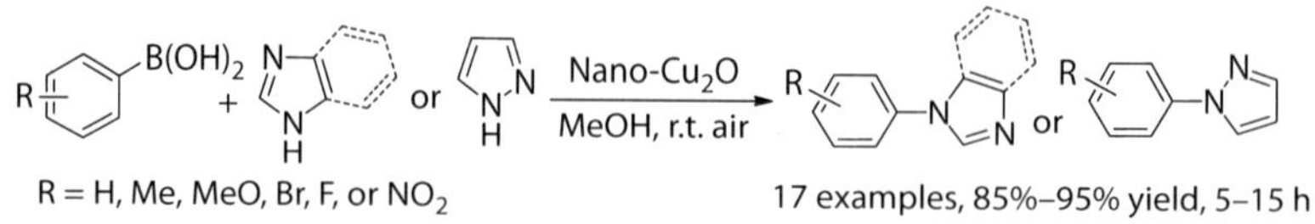

SCHEME 4.34

During the same year, nano-Cu_2O was employed to facilitate the N-arylation of azoles and amines with arylboronic acids (Scheme 4.34). Several *N*-aryl azole derivatives could be efficiently assembled at room temperature under base-free conditions (Sreedhar et al., 2008).

Recently, Islam et al. (2012) reported that a polymer-supported Cu(II) complex could promote the C–N and C–O coupling of arylboronic acids with *N*-/*O*-nucleophiles. Three polymer-supported Cu complexes were prepared from the polymer-anchored ligand and the corresponding Cu salts; Cu(II) complex [CuL$(OAc)_2$] from Cu$(OAc)_2$ was proven to be the best catalyst for N-arylation. A variety of *N*-aryl imidazoles/benzimidazoles were synthesized in good to excellent yields in methanol (MeOH) at 40°C (Scheme 4.35). The system is air and moisture stable and the catalyst can be reused for several cycles with consistent catalytic activity.

The N-arylation of imidazole was also successfully accomplished by using a two-dimensional $[Cu(ima)_2]_n$ metal–organic framework (Cu-MOF) in MeOH at room temperature (Li et al., 2013). Several *N*-arylimidazoles were generated in good to excellent yields from the coupling of imidazole and arylboronic acids (six examples, 76%–92% yields). The catalyst is highly stable and can be easily reused.

4.1.2.1.3 Heterogeneous Cu-Catalyzed S-Arylation Using Diaryliodonium Salts

Recently, nano-CuO-mediated ligand-free C–S coupling and concomitant oxidative aromatization of 4-aryl-3,4-dihydropyrimidin-2(1*H*)-thione (DHPM) with diaryliodonium salts was reported by Karade and coworkers (Scheme 4.36) (Bhong et al., 2013). This is the first example of a nano-CuO-catalyzed C–S coupling of

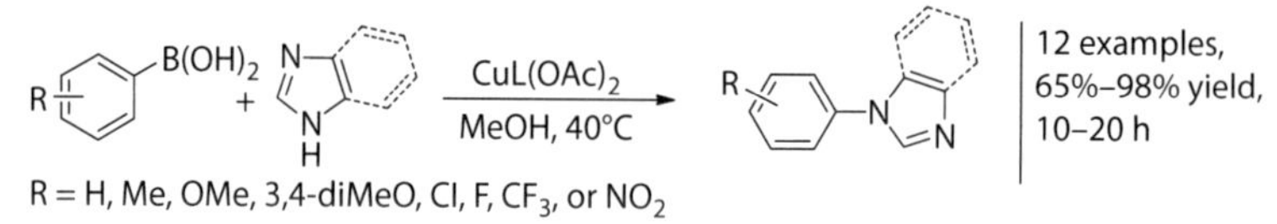

SCHEME 4.35

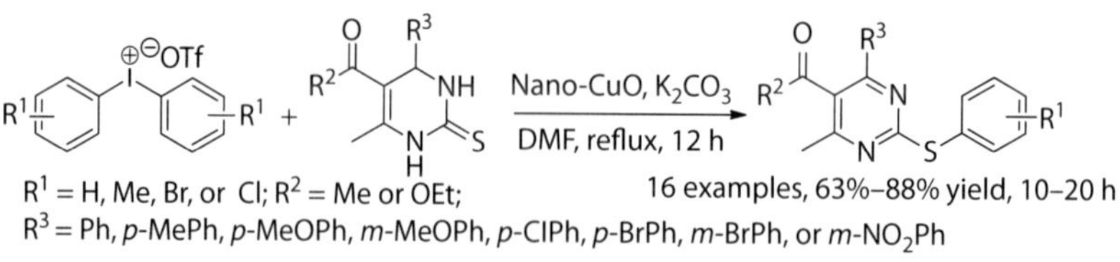

SCHEME 4.36

thiocarbonyl functional groups of DHPMs, and the obtained 2-(phenylthio)pyrimidines are biologically and synthetically important. The catalyst can be well reused at least three times.

4.1.2.2 Synthesis of Heterocycles via Intramolecular Coupling Reactions

Using appropriate functionalized substrates, the Cu-mediated coupling reactions proceed intramolecularly, affording the corresponding cyclized products with high efficiency (Evindar and Batey, 2003; Joyce et al., 2004; Hu and Li, 2005; Jiang et al., 2008; Zhao and Li, 2008). Generally, these substrates include both electrophilic moiety (sp^2 C–X) and a nucleophilic center (NH, OH, SH, etc.) in the same molecule. During recent years, heterogeneous Cu-catalyzed intramolecular coupling has also been successfully applied for the efficient construction of various bioactive heterocycles.

Punniyamurthy and coworkers reported a ligand-free nano-CuO-catalyzed synthesis of substituted benzo-azole derivatives via intramolecular cross-coupling (Saha et al., 2009). Various benzimidazoles, 2-aminobenzimidazoles, 2-aminobenzothiazoles, and benzoxazoles were efficiently assembled from the corresponding *o*-bromoaryl substrates using heterogeneous CuO nanoparticles as the catalyst in dimethyl sulfoxide (DMSO) under air (Scheme 4.37). The procedure is quite simple, versatile, and efficient. The nano-CuO could be recycled without loss of its activity, and no external chelating ligands were required.

In the same year, a series of benzoxazoles were synthesized through the intramolecular C–O coupling of *o*-halobenzanilides using CuFAP as the catalyst (Scheme 4.38) (Kantam et al., 2009b). A variety of 2-substituted benzoxazoles were generated in good to excellent yields at 110°C. The catalyst can be readily recycled three times without loss of its activity.

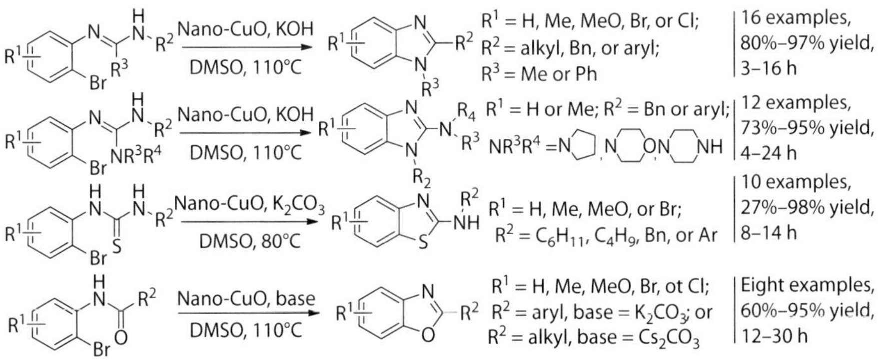

SCHEME 4.37

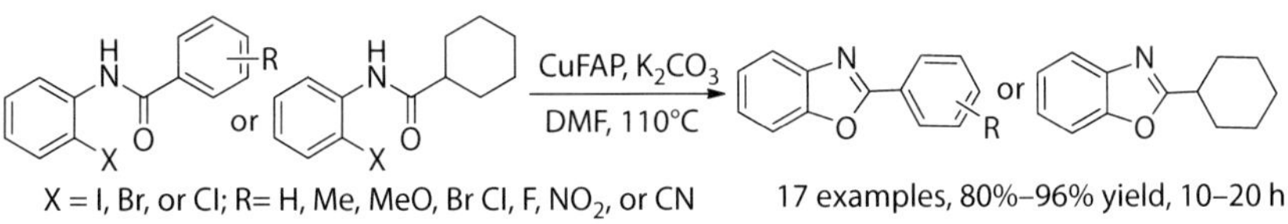

SCHEME 4.38

Recently, polystyrene-supported pyrrole-2-carbohydrazide (PSP) was combined with Cu(I) to make up a recyclable catalytic system (Cu(I)/PSP) for the N-(hetero)arylation of amines and imidazole (Huang et al., 2013). This heterogeneous catalyst could also be successfully applied to the assembly of imidazo[1,2-*a*]quinoxaline through the intramolecular cyclization of *N*-(2-iodophenyl)-1*H*-imidazole-2-carboxamide (Scheme 4.39).

4.1.3 Synthesis of Heterocycles via Heterogeneous Cu-Catalyzed Domino Reactions

The domino reaction allows the formation of two or more bonds in one process and meets the requirements of the modern synthesis (Tietze et al., 2006). In contrast with the traditional stepwise synthesis, such a process provides a wide range of possibilities for the efficient assembly of complex molecules from simple starting materials in a single step, and the tedious workup and isolation procedures of intermediates are avoided. Moreover, additional solvents and reagents can be saved by using such a method. Recently, heterogeneous Cu-catalyzed domino reactions have been applied for the efficient and convenient elaboration of heterocyclic molecules.

4.1.3.1 Two-Component Domino Synthesis of Heterocycles

4.1.3.1.1 Synthesis of Furan Derivatives

Nano-Cu_2O can be used as a heterogeneous and recyclable catalyst for the domino synthesis of α-carbonyl furan derivatives (Cao et al., 2010, 2011). A variety of α-carbonyl furans can be efficiently and regioselectively obtained in moderate to good yields from electron-deficient alkynes and 2-yn-1-ols under air atmosphere (Scheme 4.40). Notably, a novel 2,4,5-trisubstituted 3-ynyl-furan was simply generated without tedious procedures.

Ramon et al. found that impregnated copper on magnetite (CuO–Fe_3O_4) could facilitate the domino Sonogashira-cyclization processes between 2-iodophenol and various alkynes to afford the 2-substituted benzo[*b*]furan derivatives (Scheme 4.41) (Cano et al., 2012).

4.1.3.1.2 Synthesis of Quinoline/Pyrimidine Derivatives

Recently, Phan et al. (2013) prepared an open metal site MOF Cu(BDC) and successfully applied it to the modified Friedländer reaction. In the presence of 3–5 mol% Cu(BDC), several 2-aryl quinolines were efficiently synthesized from 2-aminobenzyl alcohol and methyl ketones (Scheme 4.42). Different from the previous reports, the reaction of *p*-methoxyacetophenone or *p*-nitroacetophenone did not occur in this catalytic system.

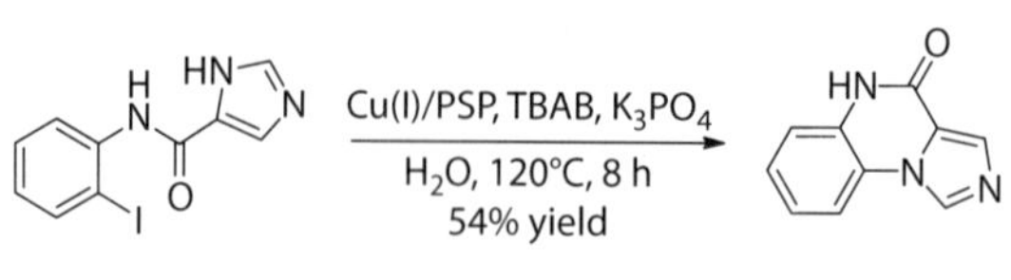

SCHEME 4.39

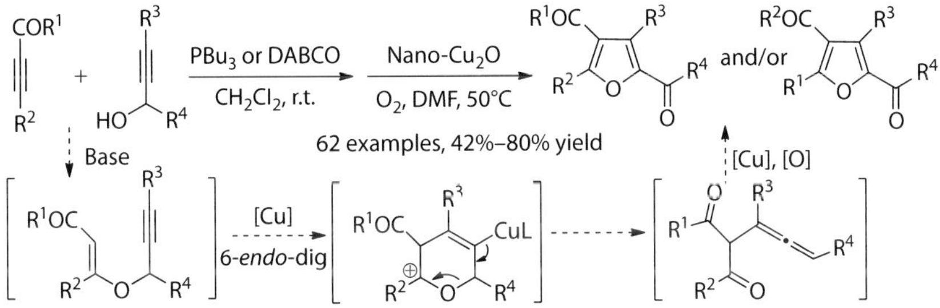

R^1 = OEt, Ph, *p*-Tol or 2-thienyl;
R^2 = Me, Et, Ph, *p*-Tol, *o*-Tol, *p*-EtPh, *o*-MeOPh, *o*-FPh, *p*-MeO_2CPh or Co_2Et;
R^3 = H, Me, Ph, *p*-MeOPh, *m*-Tol, *p*-NO_2Ph, 2-pyridyl, 2-thienyl, Ph—≡—, or *p*-Tol—≡—;
R^4 = H, Et or *n*-Pr

SCHEME 4.40

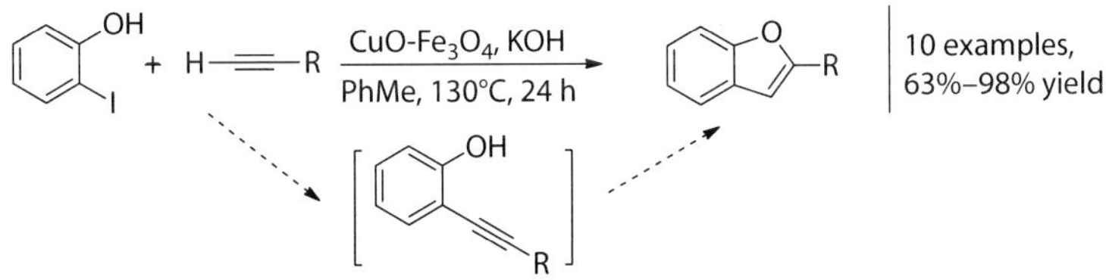

R = Ph, *p*-CF_3Ph, 3-(HC≡C)Ph, *p*-MeOPh, *p*-(Me_2N)Ph, 2-pyridyl, *n*-hex,

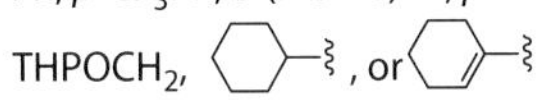

SCHEME 4.41

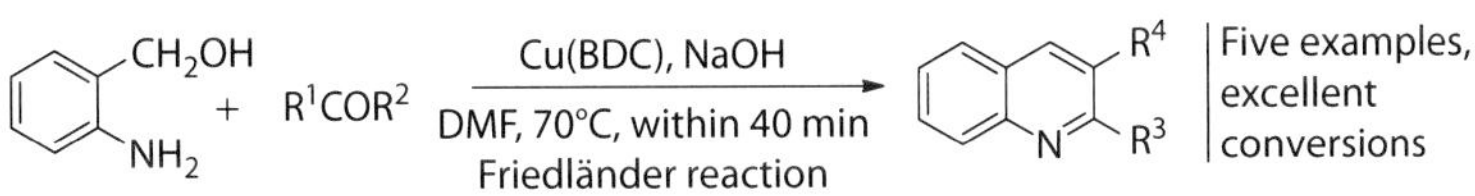

R^1COR^2 = $PhCOCH_3$, PhCOEt, *p*-$MePhCOCH_3$, *p*-$ClPhCOCH_3$, or cyclohexanone;
R^3 = Ph, *p*-Tol, or *p*-ClPh, R^4 = H; R^3 = Ph, R^4 = Me; or R^3, R^4 = $(CH_2)_4$

SCHEME 4.42

4.1.3.1.3 Synthesis of Azole Derivatives

Cu(BHPPDAH)·H_2O-catalyzed one-pot synthesis of benzimidazoles was first reported by Sharghi et al. (2008). The catalyst was prepared from $Cu(OAc)_2$ and *N*,*N*-bis(2-hydroxyphenyl)-pyridine-2,6-dicarboxamide (BHPPDAH) (Scheme 4.43). Facilitated by the recyclable Cu(II) complex, various *o*-phenylenediamines smoothly cyclized with aromatic aldehydes to give benzimidazoles at room temperature under air atmosphere (Scheme 4.44).

Recently, Sharghi et al. (2011) also investigated the assembly of 2-aryl/2-heteroaryl benzimidazoles from *o*-phenylenediamines and aromatic aldehydes using Cu nanoparticles on activated carbon (CuNPs/C) as the reusable heterogeneous catalyst (Scheme 4.45). The substrates can directly condensate at room temperature to

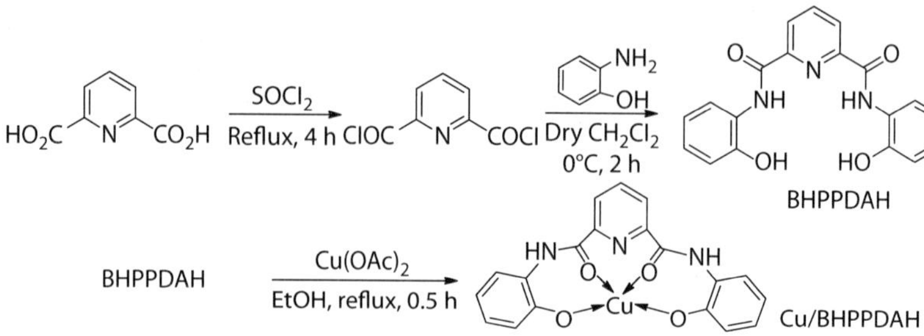

SCHEME 4.43

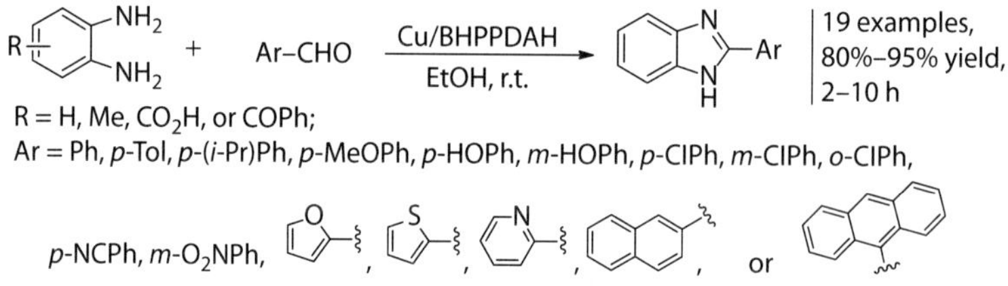

SCHEME 4.44

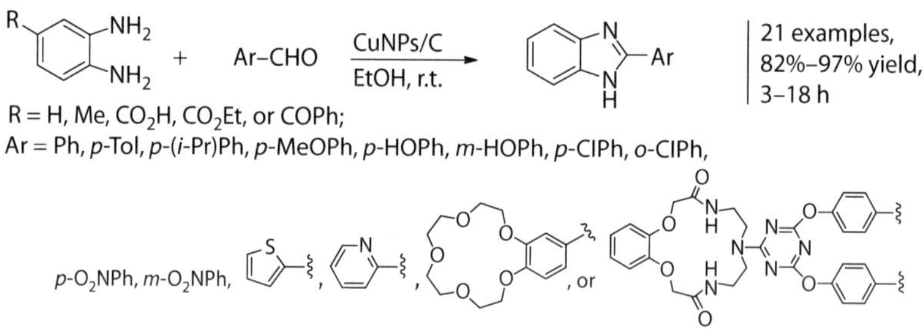

SCHEME 4.45

give the desired products in 82%–97% yields. Interestingly, several benzimidazoles containing macrocyclic moieties (such as crown ethers and azacrown ethers), which might possess special biological activities/coordination properties, have been successfully assembled. The CuNPs/C can be recycled by simple filtration and reused for at least eight runs without a significant decrease in activity.

Very recently, Mandal and coworkers reported that CuO nanoparticles supported on silica (CuO-NPs/SiO_2) could be utilized as a heterogeneous and recyclable catalyst for the one-pot synthesis of benzimidazoles, benzothiazoles, and benzoxazoles (Scheme 4.46) (Inamdar et al., 2013). The Cu-NPs/SiO_2-catalyzed cyclization of *o*-phenylenediamines, *o*-aminothiophenols, and *o*-aminophenols with various aldehydes efficiently delivered the corresponding heterocycles.

Cai et al. developed a synthesis of 2-aminobenzothiazoles using MCM-41-2*N*-$CuSO_4$-catalyzed domino reactions of 2-haloanilines with isothiocyanates (Xiao et al., 2012). The heterogeneous Cu catalyst was easily prepared from commercially

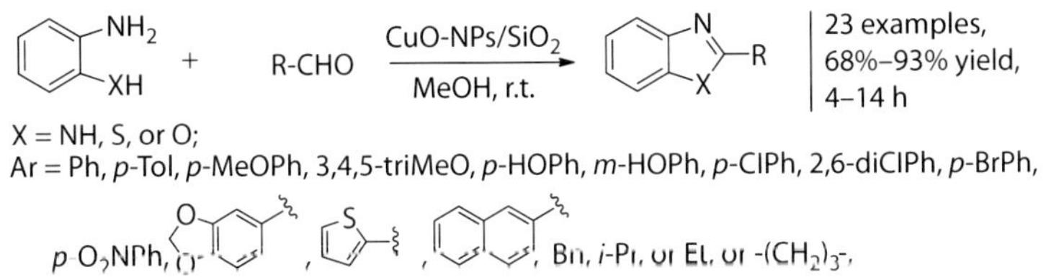

SCHEME 4.46

available and inexpensive reagents (Scheme 4.47). Promoted by only 0.25 mol% of the catalyst, which can be reused at least 10 times without any decrease in its activity, the domino reactions smoothly proceeded under air atmosphere and afforded the desired benzothiazoles in good to excellent yields (Scheme 4.48).

Hekmatshoar and coworkers synthesized benzo-*N*-heterocycles using nano-CuO as the heterogeneous catalyst (Sadjadi et al., 2010). Several benzoxazinone, benzothiazinone, and quinoxalinone derivatives were rapidly assembled in good to excellent yields from the aromatic amines and acetylenedicarboxylic acid esters (Scheme 4.49). The water-resistant catalysts were environmentally benign, and could be easily synthesized, stored, and recycled.

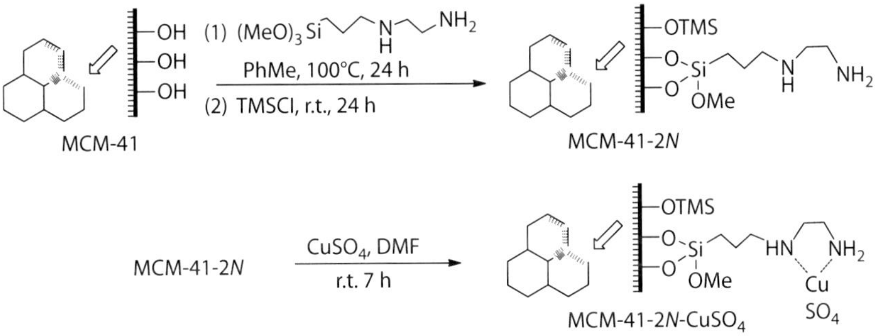

SCHEME 4.47

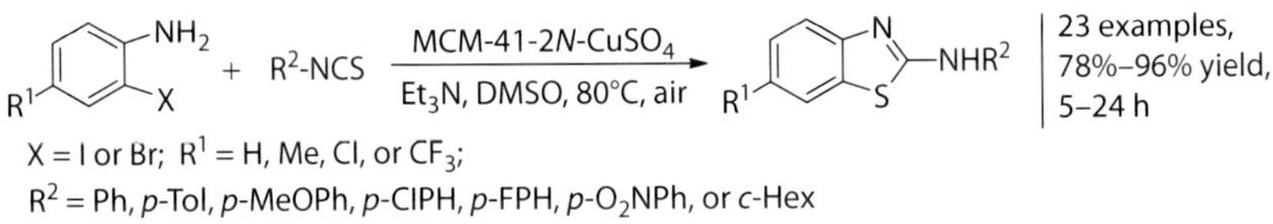

SCHEME 4.48

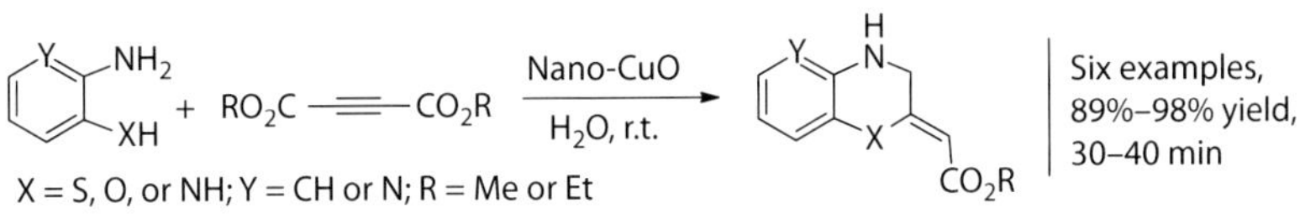

SCHEME 4.49

The Fe_3O_4 nanoparticle-supported Cu(I) complex (Fe_3O_4@SiO_2-2*N*-Cu(I); Scheme 4.50) was utilized as a magnetically recoverable and reusable catalyst for the preparation of quinazolinones and bicyclic pyrimidinones (Yu et al., 2012). A variety of the quinazolinone and bicyclic pyrimidinone derivatives were smoothly generated from the corresponding 2-halobenzoic acids/2-bromocyclohex-1-enecarboxylic acid and amidines via the C–N coupling/condensation process (Scheme 4.51). The catalyst could be easily and efficiently recycled.

Recently, Wang et al. found that kaolin-supported nanoparticulate CuO (SCONP-3) is an efficient catalyst for the domino synthesis of quinazolines. Using 2-aminobenzophenones and benzylic amines as the starting materials, a series of substituted quinazolines were delivered in moderate to excellent yields (Scheme 4.52) (Gao et al., 2013).

Maiti and coworkers utilized Cu(0) nanoparticles (Cu(0)-NPs) as a heterogeneous catalyst for the preparation of 4-hydroxy-2-pyrroline-5-ones in a water medium (Gayen et al., 2012). Through a domino reductive cleavage/cyclization process, a variety of isoxazolines afforded the corresponding 4-hydroxy-2-pyrroline-5-one derivatives in good to excellent yields (Scheme 4.53).

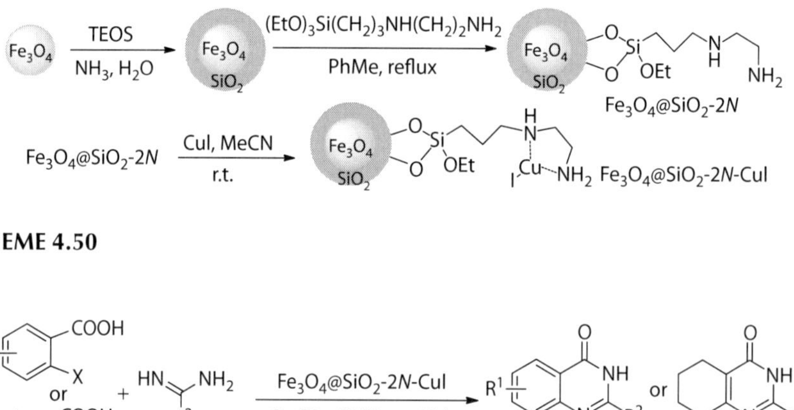

SCHEME 4.50

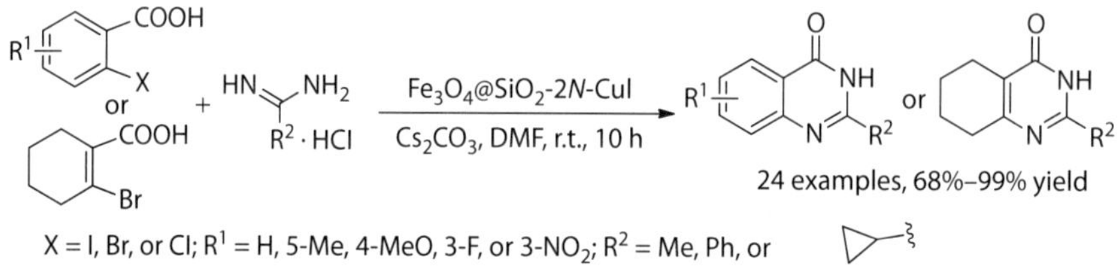

SCHEME 4.51

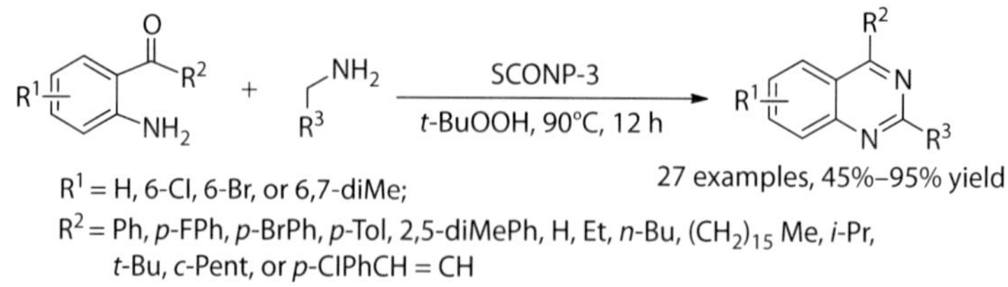

SCHEME 4.52

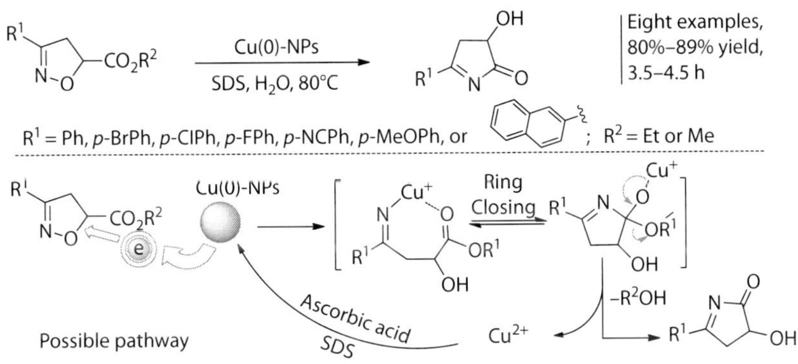

SCHEME 4.53

4.1.3.2 Multicomponent Domino Synthesis of Heterocycles

A multicomponent reaction (MCR) utilizes three or more reactants to generate products in one pot. An MCR provides an atom-economic, environmentally benign, and easy approach to the complex compounds (Zhu and Bienayme, 2005).

Namitharan and Pitchumani (2011) proposed a novel Cu-catalyzed one-pot sulfonyl azide–alkyne/ketenimine-nitrone cycloaddition sequence for the synthesis of imidazolidin-4-ones, which might be utilized as antimalarial and antiproliferative drugs. The heterogeneous version employing Cu(I)-zeolites as recyclable catalysts is attractive for its improved efficiency and diastereoselectivity (Scheme 4.54). In this case, a three-component heterogeneous domino process was utilized and a variety of imidazolidin-4-ones were synthesized in moderate to excellent yields. The products could be easily isolated by simple filtration.

The pyrido[1,2-*a*]pyrimidin-4-imine skeleton exists in some medicinally useful molecules and possesses diverse biological activities (Hermecz and Vasvari-Debreczy, 2007). Recently, the same group developed a heterogeneous Cu-catalyzed

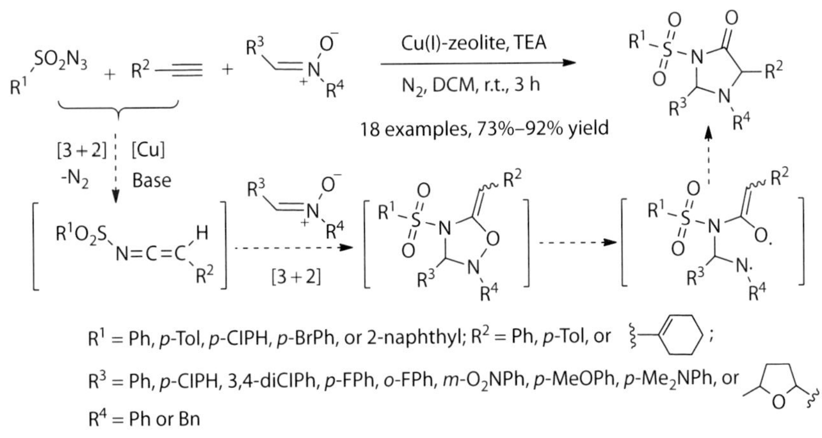

SCHEME 4.54

[3 + 2] cycloaddition/ring-opening rearrangement/[4 + 2] cycloaddition/aromatization cascade. Under the catalysis of Cu(I)-zeolites, three-component domino cyclization of sulfonyl azide, alkynes, and *N*-arylidenepyridin-2-amines afforded the desired pyrido[1,2-*a*]pyrimidin-4-imines in moderate to excellent yield (Scheme 4.55) (Namitharan and Pitchumani, 2013).

Mishra and Ghosh (2011) found that Cu(I)–$NaHSO_4{\cdot}SiO_2$ could be utilized as a novel and efficient heterogeneous catalyst for the synthesis of imidazo[1,2-*a*]pyridines using three-component reactions of 2-aminopyridines, aldehydes, and terminal alkynes in one pot (Scheme 4.56). Under this catalytic system, diverse imidazo[1,2-*a*]pyridine derivatives were generated in moderate to excellent yields.

Sadjadi et al. reported a nano-CuO-catalyzed one-pot three-component synthesis of imidazo[1,2-*a*]quinolines and quinolino[1,2-*a*]quinazolines in water (Scheme 4.57) (Ahmadi et al., 2011). Using aromatic aldehyde, malononitrile, and enaminones as the substrates, the corresponding tricyclic and tetracyclic products, which might be of biological and medicinal importance, were assembled in good to excellent yields within 1.5 h. The isolation of the catalyst is simple, and it can be reused in several runs without loss of catalytic activity.

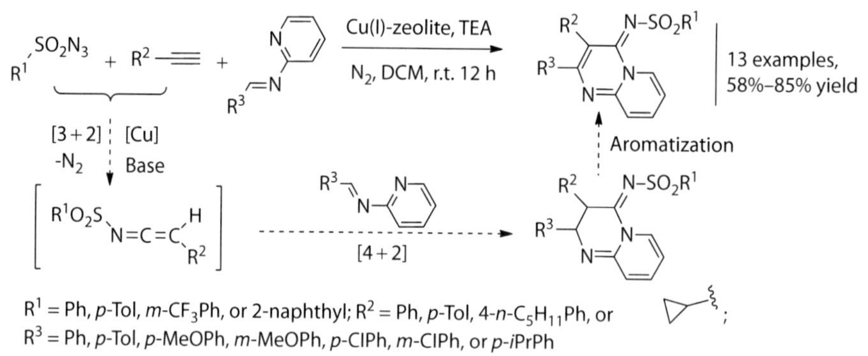

SCHEME 4.55

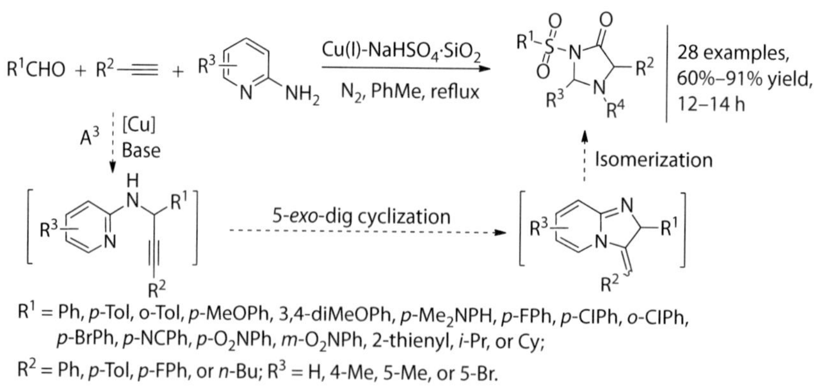

SCHEME 4.56

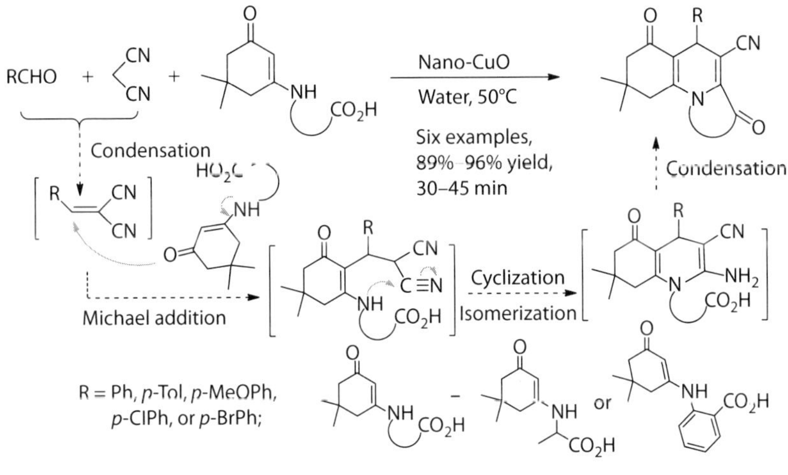

SCHEME 4.57

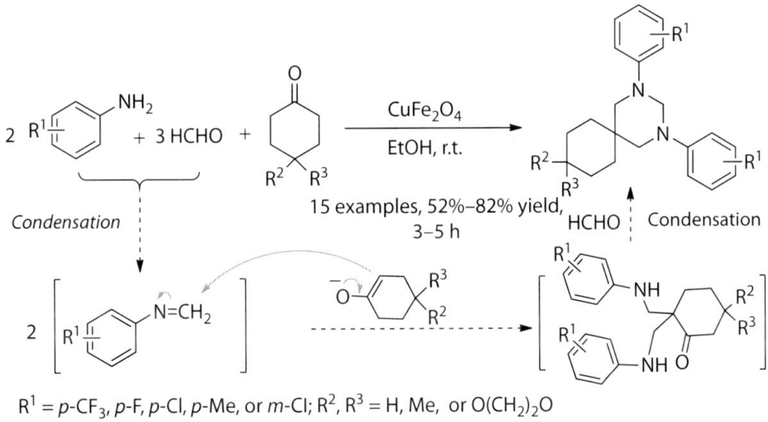

SCHEME 4.58

Recently, Dandia et al. (2013) employed $CuFe_2O_4$ nanoparticles as a highly efficient and magnetically recoverable catalyst for the assembly of medicinally privileged spiropyrimidine scaffolds. The method provides a green, rapid, and highly efficient approach to fluorine containing spirohexahydropyrimidines (Scheme 4.58), which might be of biological and medicinal value. The magnetic nature of the catalyst allows its easy removal from the reaction mixture and it can be recycled well.

Latterly, Reddy et al. reported a novel three-component assembly of 2*H*-indazoles catalyzed by Cu(II)-HT (Prasad et al., 2013). Through consecutive condensation, and C–N and N–N bond formation processes, the domino reactions of 2-bromobenzaldehydes, primary amines, and sodium azide smoothly gave the desired 2*H*-indazole derivatives (Scheme 4.59). The heterogeneous catalyst could be readily recovered and reused several times without significant loss of activity.

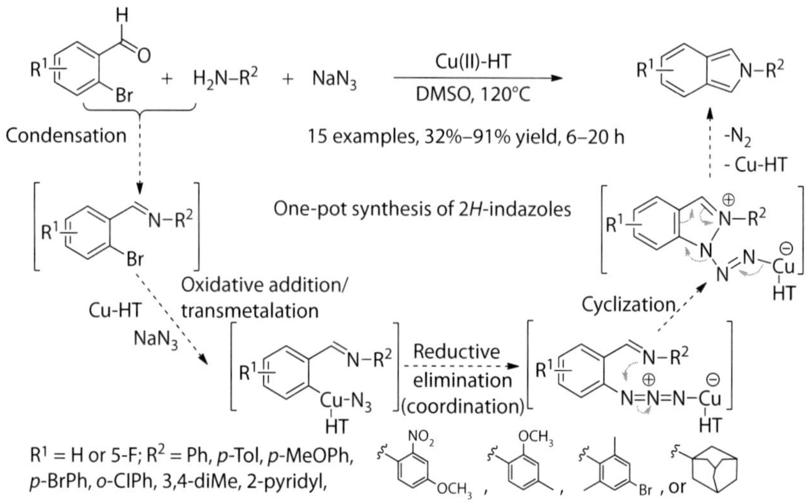

SCHEME 4.59

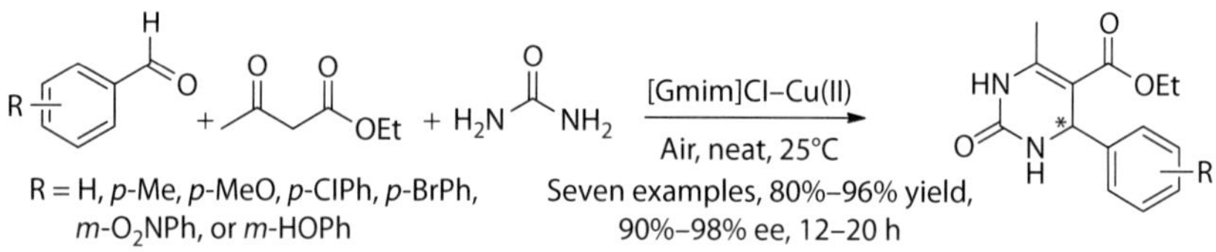

SCHEME 4.60

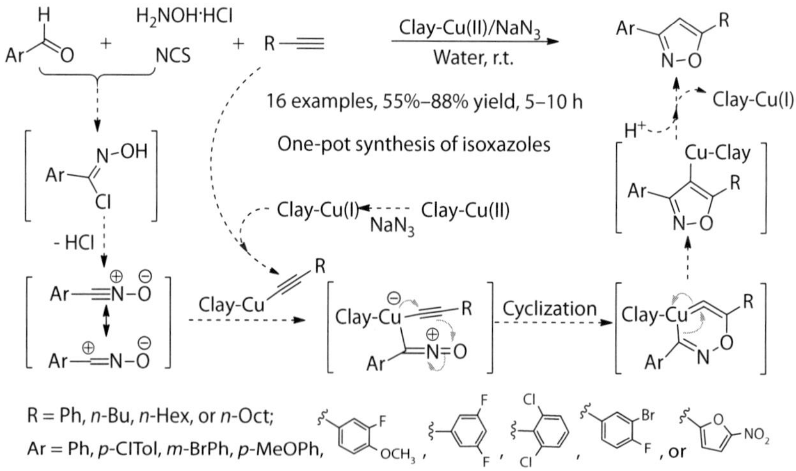

SCHEME 4.61

A novel, efficient, and recyclable asymmetric 1-glycyl-3-methyl imidazolium chloride–copper(II) complex [Gmim]Cl–Cu(II) was studied as a heterogeneous catalyst for the enantioselective Biginelli reaction (Karthikeyan et al., 2013). The reaction was conducted under air atmosphere and solvent-free conditions, and the desired cyclized products were conveniently obtained in high yields with excellent enantioselectivities (up to 98% ee) (Scheme 4.60).

Bharate et al. (2013) applied montmorillonite clay Cu(II)/NaN_3 as a heterogeneous catalytic system for the four-component synthesis of isoxazole derivatives in water. The desired 3,5-disubstituted isoxazoles were obtained in moderate to good yields using the domino reactions of aldehydes, hydroxylamine, *N*-chlorosuccinimide (NCS), and acetylenes (Scheme 4.61); the catalytic system could be efficiently recycled.

4.2 CONCLUSIONS AND OUTLOOK

There has been great progress in heterogeneous copper-catalyzed synthesis during the last few years. A variety of heterocyclic molecules, which might be biologically active and medicinally useful, can be successfully assembled.

By using the heterogeneous copper catalysts in their immobilized or mobilized forms, it is feasible to recover the catalysts in a convenient and efficient way. Additionally, in most cases, the purification procedures are very simple and rapid.

These pioneering achievements in heterogeneous Cu-catalyzed synthesis of heterocycles are just the tip of the iceberg, and much more remains to be discovered and explored.

REFERENCES

Ahmadi, S. J., Hosseinpour, M., and Sadjadi, S. 2011. Nanocrystalline copper(II) oxide-catalyzed one-pot synthesis of imidazo[1,2-*a*]quinoline and quinolino[1,2-*a*]quinazoline derivatives via a three-component condensation. *Synth. Commun.* 41(3):426–435.

Alonso, F., Moglie, Y., Radivoy, G., and Yusa, M. 2010. Multicomponent synthesis of 1,2,3-triazoles in water catalyzed by copper nanoparticles on activated carbon. *Adv. Synth. Catal.* 352(18):3208–3214.

Alvarez, R., Velazquez, S., Felix, A. S., Aquaro, S., De Clercq, E., Perno, C. F., Karlsson, A., Balzariniand, J., and Camarasa, M. J. 1994. 1,2,3-Triazole-[2,5-bis-*O*-(*tert*-butyldimethyl-silyl)-β-D-ribofuranosyl]-3′-spiro-5″-(4″-amino-1″,2″-oxathiole 2″,2″-dioxide) (TSAO) analogs: Synthesis and anti-HIV-1 activity. *J. Med. Chem.* 37(24):4185–4194.

Beneteau, V., Olmos, A., Boningari, T., Sommer, J., and Pale, P. 2010. Zeo-click synthesis: Cu^I-zeolite-catalyzed one-pot two-step synthesis of triazoles from halides and related compounds. *Tetrahedron Lett.* 51(28):3673–3677.

Benyahya, S., Monnier, F., Taillefer, M., Man, M. W. C., Bied, C., and Ouazzani, F. 2008. Efficient and versatile sol-gel immobilized copper catalyst for Ullmann arylation of phenols. *Adv. Synth. Catal.* 350(14–15):2205–2208.

Bharate, S. B., Padala, A. K., Dar, B. A., Yadav, R. R., Singh, B., and Vishwakarma, R. A. 2013. Montmorillonite clay Cu(II) catalyzed domino one-pot multicomponent synthesis of 3,5-disubstituted isoxazoles. *Tetrahedron Lett.* 54(27):3558–3561.

Bhong, B. Y., Shelke, A. V., and Karade, N. N. 2013. Nano copper oxide mediated ligand-free C–S cross-coupling and concomitant oxidative aromatization of 4-aryl-3,4-dihydropyrimidin-2(1*H*)-thione with diaryliodonium salts. *Tetrahedron Lett.* 54(8):739–743.

Bonami, L., Van Camp, W., Van Rijckegem, D., and Du Prez, F. E. 2009. Facile access to an efficient solid-supported click catalyst system based on poly(ethyleneimine). *Macromol. Rapid Commun.* 30(1):34–38.

Boningari, T., Olmos, A., Reddy, B. M., Sommer, J., and Pale, P. 2010. Zeo-click chemistry: Copper(I)-zeolite-catalyzed cascade reaction; one-pot epoxide ring-opening and cycloaddition. *Eur. J. Org. Chem.* 2010(33):6338–6347.

Buckle, D. R., Rockell, C. J. M., Smith, H., and Spicer, B. A. 1986. Studies on 1,2,3-triazoles. 13. (Piperazinylalkoxy)-[1]benzopyrano[2,3-*d*]-1,2,3-triazol-9(1*H*)-ones with combined H1-antihistamine and mast cell stabilizing properties. *J. Med. Chem.* 29(11):2262–2267.

Cano, R., Yus, M., and Ramon, D. J. 2012. Impregnated copper or palladium-copper on magnetite as catalysts for the domino and stepwise Sonogashira-cyclization processes: A straightforward synthesis of benzo[*b*]furans and indoles. *Tetrahedron.* 68(5):1393–1400.

Cao, H., Jiang, H. F., and Huang, H. W. 2011. Transition-metal-catalyzed domino reactions: Efficient one-pot regiospecific synthesis of highly functionalized polysubstituted furans from electron-deficient alkynes and 2-yn-1-ols. *Synthesis.* (7):1019–1036.

Cao, H., Jiang, H. F., Yuan, G. Q., Chen, Z. W., Qi, C. R., and Huang, H. W. 2010. Nano-Cu_2O-catalyzed formation of C–C and C–O bonds: One-pot domino process for regioselective synthesis of α-carbonyl furans from electron-deficient alkynes and 2-yn-1-ols. *Chem. Eur. J.* 16(34):10553–10559.

Chan, T. R. and Fokin, V. V. 2007. Polymer-supported copper(I) catalysts for the experimentally simplified azide–alkyne cycloaddition. *QSAR Comb. Sci.* 26(11–12):1274–1279.

Chassaing, S., Alix, A., Boningari, T., Sido, K. S. S., Keller, M., Kuhn, P., Louis, B., Sommer, J., and Pale, P. 2010. Copper(I)-zeolites as new heterogeneous and green catalysts for organic synthesis. *Synthesis.* (9):1557–1567.

Choudary, B. M., Sridhar, C., Kantam, M. L., Venkanna, G. T., and Sreedhar, B. 2005. Design and evolution of copper apatite catalysts for N-arylation of heterocycles with chloro- and fluoroarenes. *J. Am. Chem. Soc.* 127(28):9948–9949.

Chouhan, G., Wang, D., and Alper, H. 2007. Magnetic nanoparticle-supported proline as a recyclable and recoverable ligand for the CuI catalyzed arylation of nitrogen nucleophiles. *Chem. Commun.* (45):4809–4811.

Climent, M. J., Corma, A., and Iborra, S. 2011. Heterogeneous catalysts for the one-pot synthesis of chemicals and fine chemicals. *Chem. Rev.* 111(2):1072–1033.

Coelho, A., Diz, P., Caamano, O., and Sotelo, E. 2010. Polymer-supported 1,5,7-triazabicyclo[4.4.0]dec-5-ene as polyvalent ligands in the copper-catalyzed Huisgen 1,3-dipolar cycloaddition. *Adv. Synth. Catal.* 352(7):1179–1192.

Craig, P. N. 1991. *Comprehensive Medicinal Chemistry*, Vol. 8, Drayton, C. J. (ed.), New York: Pergamon Press.

Dandia, A., Jain, A. K., and Sharma, S. 2013. $CuFe_2O_4$ nanoparticles as a highly efficient and magnetically recoverable catalyst for the assembly of medicinally privileged spiropyrimidine scaffolds. *RSC Adv.* 3(9):2924–2934.

Daugulis, O., Do, H. Q., and Shabashov, D. 2009. Palladium- and copper-catalyzed arylation of carbon-hydrogen bonds. *Acc. Chem. Res.* 42(8):1074–1086.

Dervaux, B. and Du Prez, F. E. 2012. Heterogeneous azide–alkyne click chemistry: Towards metal-free end products. *Chem. Sci.* 3(4):959–966.

Diaz-Requejo, M. M., Belderrain, T. R., and Perez, P. J. 2000. Polypyrazolylborate copper(I) complexes as catalysts of the homogeneous and heterogeneous styrene epoxidation reaction. *Chem. Commun.* (19):1853–1854.

Evindar, G. and Batey, R. A. 2003. Copper- and palladium-catalyzed intramolecular aryl guanidinylation: An efficient method for the synthesis of 2-aminobenzimidazoles. *Org. Lett.* 5(2):133–136.
Gao, L. F., Xiong, S. S., Wan, C. F., and Wang, Z. Y. 2013. An expedition through the last decade of heterocycle construction by using palladium, iron, copper, or iodine/*tert*-butyl hydroperoxide. *Synlett.* 24(11):1322–1339.
Gayen, K. S., Sengupta, T., Saima, Y., Das, A., Maiti, D. K., and Mitra, A. 2012. Cu(0) nanoparticle catalyzed efficient reductive cleavage of isoxazoline, carbonyl azide and domino cyclization in water medium. *Green Chem.* 14(6):1589–1592.
Genin, M. J., Allwine, D. A., Anderson, D. J., Barbachyn, M. R., Emmert, D. E., Garmon, S. A., Graber, D. R., et al. 2000. Substituent effects on the antibacterial activity of nitrogen-carbon-linked (azolyl-phenyl)oxazolidinones with expanded activity against the fastidious Gram-negative organisms *Haemophilus influenzae* and *Moraxella catarrhalis*. *J. Med. Chem.* 43(5):953–970.
Girard, C., Onen, E., Aufort, M., Beauviere, S., Samson, E., and Herscovici, J. 2006. Reusable polymer-supported catalyst for the [3+2] Huisgen cycloaddition in automation protocols. *Org. Lett.* 8(8):1689–1692.
Gopiraman, M., Badu, S. G., Khatri, Z., Kai, W., Kim, Y. A., Endo, M., Karvembu, R., and Kim, I. S. 2013. An efficient, reusable copper-oxide/carbon-nanotube catalyst for N-arylation of imidazole. *Carbon.* 62:135–148.
Hein, J. E. and Fokin, V. V. 2010. Copper-catalyzed azide–alkyne cycloaddition (CuAAC) and beyond: New reactivity of copper(I) acetylides. *Chem. Soc. Rev.* 39(4):1302–1315.
Hermecz, I. and Vasvari-Debreczy, L. 2007. Bicyclic 6-6 systems with one bridgehead (ring junction) nitrogen atom: One extra heteroatom 1:0. In *Comprehensive Heterocyclic Chemistry III*, Vol. 11, Jones, K. and Katritzky, A. R. (eds), pp. 77–79. Oxford: Elsevier Press.
Hu, T. S. and Li, C. Z. 2005. Synthesis of lactams via copper-catalyzed intramolecular vinylation of amides. *Org. Lett.* 7(10):2035–2038.
Huang, L. Y., Yu, R. N., Zhu, X. H., and Wan, Y. Q. 2013. A recyclable Cu-catalyzed C–N coupling reaction in water and its application to synthesis of imidazo[1,2-*a*]quinoxaline. *Tetrahedron.* 69:8974–8977.
Hudson, R., Li, C. J., and Moores, A. 2012. Magnetic copper–iron nanoparticles as simple heterogeneous catalysts for the azide–alkyne click reaction in water. *Green Chem.* 14(3):622–624.
Huisgen, R. 1963. 1,3-Dipolar cycloadditions. Past and future. *Angew. Chem. Int. Ed.* 2(10):565–598.
Huisgen, R. 1984. 1,3-Dipolar cycloadditions: Introduction, survey, mechanism. In *1,3-Dipolar Cycloaddition Chemistry*, Vol. 1, Padwa, A. (ed.), pp. 1–176. New York: Wiley.
Huisgen, R. 1989. Kinetics and reaction mechanisms: Selected examples from the experience of forty years. *Pure Appl. Chem.* 61(4):613–628.
Inamdar, S. M., More, V. K., and Mandal, S. K. 2013. CuO nano-particles supported on silica, a new catalyst for facile synthesis of benzimidazoles, benzothiazoles and benzoxazoles. *Tetrahedron Lett.* 54(6):579–583.
Islam, S. M., Mondal, S., Mondal, P., Roy, A. S., Tuhina, K., Salam, N., and Mobarak, M. 2012. A reusable polymer supported copper catalyst for the C–N and C--O bond cross-coupling reaction of aryl halides as well as arylboronic acids. *J. Organometal. Chem.* 696(26):4264–4274.
Jiang, L. and Buchwald, S. L. 2004. Palladium-catalyzed aromatic carbon–nitrogen bond formation. In *Metal-Catalyzed Cross-Coupling Reactions*, 2nd edn., de Meijere, A. and Diederich, F. (eds), pp. 699–760. Weinheim: Wiley-VCH.
Jiang, B., Tian, H., Huang, Z. G., and Xu, M. 2008. Successive copper(I)-catalyzed cross-couplings in one pot: A novel and efficient starting point for synthesis of carbapenems. *Org. Lett.* 10(13):2737–2740.

Jin, T. N., Yan, M., Menggenbateer, Minato, T., Bao, M., and Yamamoto, Y. 2011. Nanoporous copper metal catalyst in click chemistry: Nanoporosity-dependent activity without supports and bases. *Adv. Synth. Catal.* 353(17):3095–3100.

Jlalia, I., Beauvineau, C., Beauviere, S., Onen, E., Aufort, M., Beauvineau, A., Khaba, E., Herscovici, J., Meganem, F., and Girard, C. 2010. Automated synthesis of a 96 product-sized library of triazole derivatives using a solid phase supported copper catalyst. *Molecules.* 15(5):3087–3120.

Joyce, L. L., Evindar, G., and Batey, R. A. 2004. Copper- and palladium-catalyzed intramolecular C–S bond formation: A convenient synthesis of 2-aminobenzothiazoles. *Chem. Commun.* (4):446–447.

Kantam, M. L., Arundhathi, R., Likhar, P. R., and Damodara, D. 2009a. Reusable copper-aluminum hydrotalcite/*rac*-BINOL system for room temperature selective aerobic oxidation of alcohols. *Adv. Synth. Catal.* 351(16):2633–2637.

Kantam, M. L., Venkanna, G. T., S. Kumar, K. B., Balasubrahmanyam, V., and Bhargava, S. 2009b. Synthesis of benzoxazoles via intramolecular cyclization of *ortho*-halobenzanilides using copper fluorapatite catalyst. *Synlett.* (11):1753–1756.

Kantam, M. L., Venkanna, G. T., Sridhar, C., Sreedhar, B., and Choudary, B. M. 2006. An efficient base-free N-arylation of imidazoles and amines with arylboronic acids using copper-exchanged fluorapatite. *J. Org. Chem.* 71(25):9522–9524.

Kantam, M. L., Yadav, J., Laha, S., Sreedhar, B., and Jhab, S. 2007. N-arylation of heterocycles with activated chloro- and fluoroarenes using nanocrystalline copper(II) oxide. *Adv. Synth. Catal.* 349(11–12):1938–1942.

Karthikeyan, P., Aswar, S. A., Muskawar, P. N., Bhagat, P. R., and Kumar, S. S. 2013. Development and efficient 1-glycyl-3-methyl imidazolium chloride-copper(II) complex catalyzed highly enantioselective synthesis of 3,4-dihydropyrimidin-2(1H)-ones. *J. Organomet. Chem.* 723:154–162.

Katayama, T., Kamata, K., Yamaguchi, K., and Mizuno, N. 2009. A supported copper hydroxide as an efficient, ligand-free, and heterogeneous precatalyst for 1,3-dipolar cycloadditions of organic azides to terminal alkynes. *ChemSusChem.* 2(1):59–62.

Keller, M., Sido, A. S. S., Pale, P., and Sommer, J. 2009. Copper(I) zeolites as heterogeneous and ligand-free catalysts: [3+2] cycloaddition of azomethine imines. *Chem. Eur. J.* 15(12):2810–2817.

Kim, A. Y., Lee, H. J., Park, J. C., Kang, H., Yang, H., Song, H., and Park, K. H. 2009. Highly efficient and reusable copper-catalyzed N-arylation of nitrogen-containing heterocycles with aryl halides. *Molecules.* 14(12):5169–5178.

Kovacs, S., Zih-Perenyi, K., Revesz, A., and Novak, Z. 2012. Copper on iron: Catalyst and scavenger for azide–alkyne cycloaddition. *Synthesis.* 44(24):3722–3730.

Kumar, B. S. P. A., Reddy, K. H. V., Madhav, B., Ramesh, K., and Nageswar, Y. V. D. 2012. Magnetically separable $CuFe_2O_4$ nano particles catalyzed multi-component synthesis of 1,4-disubstituted 1,2,3-triazoles in tap water using "click chemistry". *Tetrahedron Lett.* 53(34):4595–4599.

Lee, B. S., Yi, M., Chu, S. Y., Lee, J. Y., Kwon, H. R., Lee, K. R., Kang, D., et al. 2010. Copper nitride nanoparticles supported on a superparamagnetic mesoporous microsphere for toxic-free click chemistry. *Chem. Commun.* 46(22):3935–3937.

Li, P. H., Wang, L., and Zhang, Y. C. 2008. SiO_2–NHC–Cu(I): An efficient and reusable catalyst for [3+2] cycloaddition of organic azides and terminal alkynes under solvent-free reaction conditions at room temperature. *Tetrahedron.* 64(48):10825–10830.

Li, Z.-H., Xue, L.-P., Wang, L., Zhang, S.-T., and Zhao, B.-T. 2013. Two-dimensional copper-based metal–organic framework as a robust heterogeneous catalyst for the N-arylation of imidazole with arylboronic acids. *Inorg. Chem. Commun.* 27:119–121.

Lipshutz, B. H., Nihan, D. M., Vinogradova, E., Taft, B. R., and Boskovic, Z. V. 2008. Copper + nickel-in-charcoal (Cu–Ni/C): A bimetallic, heterogeneous catalyst for cross-couplings. *Org. Lett.* 10(19):4279–4282.

Lipshutz, B. H. and Taft, B. R. 2006. Heterogeneous copper-in-charcoal-catalyzed click chemistry. *Angew. Chem. Int. Ed.* 45(48):8235–8238.

Liu, Y. Y. and Wan, J.-P. 2011. Tandem reactions initiated by copper-catalyzed cross-coupling: A new strategy towards heterocycle synthesis. *Org. Biomol. Chem.* 9(20):6873–6894.

Liu, Y. Y. and Wan, J.-P. 2012. Advances in copper-catalyzed C-C coupling reactions and related domino reactions based on active methylene compounds. *Chem. Eur. J.* 7(7):1488–1501.

Lopez-Ruiz, H., de la Cerda-Pedro, J. E., Rojas-Lima, S., Perez-Perez, I., Rodriguez-Sanchez, B. V., Santillan, R., and Coreno, O. 2013. Cuprous oxide on charcoal-catalyzed ligand-free synthesis of 1,4-disubstituted 1,2,3-triazoles via click chemistry. *Arkivoc.* (iii):139–164.

Ma, D. W. and Cai, Q. 2008. Copper/amino acid catalyzed cross-couplings of aryl and vinyl halides with nucleophiles. *Acc. Chem. Res.* 41(11):1450–1460.

Megia-Fernandez, A., Ortega-Munoz, M., Hernandez-Mateo, F., and Santoyo-Gonzalez, F. 2012. One-pot three-component click reaction of cyclic sulfates and cyclic sulfamidates. *Adv. Synth. Catal.* 354(9):1797–1803.

Megia-Fernandez, A., Ortega-Munoz, M., Lopez-Jaramillo, J., Hernandez-Mateo, F., and Santoyo-Gonzalez, F. 2010. Non-magnetic and magnetic supported copper(I) chelating adsorbents as efficient heterogeneous catalysts and copper scavengers for click chemistry. *Adv. Synth. Catal.* 352(18):3306–3320.

Miao, T. and Wang, L. 2008. Regioselective synthesis of 1,2,3-triazoles by use of a silica-supported copper(I) catalyst. *Synthesis.* (3):363–368.

Mishra, S. and Ghosh, R. 2011. Mechanistic studies on a new catalyst system (CuI–$NaHSO_4$·SiO_2) leading to the one-pot synthesis of imidazo[1,2-*a*]-pyridines from reactions of 2-aminopyridines, aldehydes, and terminal alkynes. *Synthesis.* (21):3463–3470.

Mohammed, S., Padala, A. K., Dar, B. A., Singh, B., Sreedhar, B., Vishwakarma, R. A., and Bharate, S. B. 2012. Recyclable clay supported Cu(II)-catalyzed tandem one-pot synthesis of 1-aryl-1,2,3-triazoles. *Tetrahedron.* 68(39):8156–8162.

Monnier, F. and Taillefer, M. 2009. Catalytic C–C, C–N, and C–O Ullmann-type coupling reactions. *Angew. Chem. Int. Ed.* 48(38):6954–6971.

Namitharan, K., Kumarraja, M., and Pitchumani, K. 2009. Cu(II)-Hydrotalcite as an efficient heterogeneous catalyst for Huisgen [3+2] cycloaddition. *Chem. Eur. J.* 15(12):2755–2758.

Namitharan, K. and Pitchumani, K. 2011. Copper(I)-catalyzed three component reaction of sulfonyl azide, alkyne, and nitrone cycloaddition/rearrangement cascades: A novel one-step synthesis of imidazolidin-4-ones. *Org. Lett.* 13(21):5728–5731.

Namitharan, K. and Pitchumani, K. 2013. Copper(I)-catalyzed [3+2] cycloaddition/ring-opening rearrangement/[4+2] cycloaddition/aromatization cascade: An unprecedented chemo- and stereoselective three component coupling of sulfonyl azide, alkyne and *N*-arylidenepyridin-2-amine to pyrido[1,2-*a*]pyrimidin-4-imine. *Adv. Synth. Catal.* 355(1):93–98.

Nasir Baig, R. B. and Varma, R. S. 2012. A highly active magnetically recoverable nano ferrite-glutathione-copper (nano-FGT-Cu) catalyst for Huisgen 1,3-dipolar cycloadditions. *Green Chem.* 14(3):625–632.

Negwer, M. 1994. *Organic-Chemical Drugs and Their Synonyms: An International Survey*, 7th edn., Vol. 4, Verlag, A. (ed.), pp. 2857–3937 Weinheim: Wiley-VCH Press.

O'Leary, P., Krosveld, N. P., De Jong, K. P., van Koten, G., and Klein Gebbink, R. J. M. 2004. Facile and rapid immobilization of copper(II) bis(oxazoline) catalysts on silica: Application to Diels–Alder reactions, recycling, and unexpected effects on enantioselectivity. *Tetrahedron Lett.* 45(16):3177–3180.

Park, I. S., Kwon, M. S., Kim, Y., Lee, J. S., and Park, J. 2008. Heterogeneous copper catalyst for the cycloaddition of azides and alkynes without additives under ambient conditions. *Org. Lett.* 10(3):497–500.

Phan, N. T. S., Nguyen, T. T., Nguyen, K. D., and Vo, A. X. T. 2013. An open metal site metal–organic framework Cu(BDC) as a promising heterogeneous catalyst for the modified Friedländer reaction. *Appl. Catal. A.* 464–465:128–135.

Prasad, A. N., Srinivas, R., and Reddy, B. M. 2013. Cu^{II}–hydrotalcite catalyzed one-pot three component synthesis of 2*H*-indazoles by consecutive condensation, C–N and N–N bond formations. *Catal. Sci. Tech.* 3(3):654–658.

Rad, M. N. S., Behrouz, S., Doroodmand, M. M., and Moghtaderi, N. 2011. Copper nanoparticle-doped silica cuprous sulfate as a highly efficient and reusable heterogeneous catalysis for N-arylation of nucleobases and *N*-heterocyclic compounds. *Synthesis.* (23):3915–3924.

Reddy, K. H. V., Satish, G., Ramesh, K., Karnakar, K., and Nageswar, Y. V. D. 2012. An efficient synthesis of N-substituted indoles from indoline/indoline carboxylic acid via aromatization followed by C–N cross-coupling reaction by using nano copper oxide as a recyclable catalyst. *Tetrahedron Lett.* 53(24):3061–3065.

Reymond, S. and Cossy, J. 2008. Copper-catalyzed Diels–Alder reactions. *Chem. Rev.* 108(12):5359–5406.

Rostovtsev, V. V., Green, L. G., Fokin, V. V., and Sharpless, K. B. 2002. A stepwise Huisgen cycloaddition process: Copper(I)-catalyzed regioselective "ligation" of azides and terminal alkynes. *Angew. Chem. Int. Ed.* 41(14):2596–2599.

Sadjadi, S., Hekmatshoar, R., Ahmadi, S. J., Hosseinpour, M., and Outokesh, M. 2010. On water: A practical and efficient synthesis of benzoheterocycle derivatives catalyzed by nanocrystalline copper(II) oxide. *Synth. Commun.* 40(4):607–614.

Saha, D., Adak, L., Mukherjee, M., and Ranu, B. C. 2012. Hydroxyapatite-supported Cu(I)-catalysed cyanation of styrenyl bromides with $K_4[Fe(CN)_6]$: An easy access to cinnamonitriles. *Org. Biomol. Chem.* 10(5):952–957.

Saha, P., Ramana, T., Purkait, N., Ali, M. A., Paul, R., and Punniyamurthy, T. 2009. Ligand-free copper-catalyzed synthesis of substituted benzimidazoles, 2-aminobenzimidazoles, 2-aminobenzothiazoles, and benzoxazoles. *J. Org. Chem.* 74(22):8719–8725.

Shamim, T. and Paul, S. 2010. Silica functionalized Cu(I) as a green and recyclable heterogeneous catalyst for the Huisgen 1,3-dipolar cycloaddition in water at room temperature. *Catal. Lett.* 136(3–4):260–265.

Sharghi, H., Beyzavi, M. H., Safavi, A., Doroodm, M. M., and Khalifeh, R. 2009a. Immobilization of porphyrinatocopper nanoparticles onto activated multi-walled carbon nanotubes and a study of its catalytic activity as an efficient heterogeneous catalyst for a click approach to the three-component synthesis of 1,2,3-triazoles in water. *Adv. Synth. Catal.* 351(14–15):2391–2410.

Sharghi, H., Hosseini-Sarvari, M., and Moeini, F. 2008. Copper-catalyzed one-pot synthesis of benzimidazole derivatives. *Can. J. Chem.* 86(11):1044–1051.

Sharghi, H., Khalifeh, R., and Doroodmand, M. M. 2009b. Copper nanoparticles on charcoal for multicomponent catalytic synthesis of 1,2,3-triazole derivatives from benzyl halides or alkyl halides, terminal alkynes and sodium azide in water as a "green" solvent. *Adv. Synth. Catal.* 351(1–2):207–218.

Sharghi, H., Khalifeh, R., Mansouri, S. G., Aberi, M., and Eskandari, M. M. 2011. Simple, efficient, and applicable route for synthesis of 2-aryl(heteroaryl)-benzimidazoles at room temperature using copper nanoparticles on activated carbon as a reusable heterogeneous catalyst. *Catal. Lett.* 141(12):1845–1850.

Sirion, U., Bae, Y. J., Lee, B. S., and Chi, D. Y. 2008. Ionic polymer supported copper(I): A reusable catalyst for Huisgen's 1,3-dipolar cycloaddition. *Synlett.* 2008(15):2326–2330.

Son, S. U., Park, I. K., Park, J., and Hyeon, T. 2004. Synthesis of Cu_2O coated Cu nanoparticles and their successful applications to Ullmann-type amination coupling reactions of aryl chlorides. *Chem. Commun.* (7):778–779.

Sreedhar, B., Venkanna, G. T., Kumar, K. B. S., and Balasubrahmanyam, V. 2008. Copper(I) oxide catalyzed N-arylation of azoles and amines with arylboronic acid at room temperature under base-free conditions. *Synthesis.* (5):795–799.

Surry, D. S. and Buchwald, S. L. 2010. Diamine ligands in copper-catalyzed reactions. *Chem. Sci.* 1(1):13–31.

Suzuka, T., Ooshiro, K., and Kina, K. 2010. Reusable polymer-supported terpyridine copper complex for [3+2] Huisgen cycloaddition in water. *Heterocycles.* 81(3):601–610.

Swapna, K., Murthy, S. N., Jyothi, M. T., and Nageswar, Y. V. D. 2011. Nano-$CuFe_2O_4$ as a magnetically separable and reusable catalyst for the synthesis of diaryl/aryl alkyl sulfides via cross-coupling process under ligand-free conditions. *Org. Biomol. Chem.* 9(17):5989–5996.

Tang, B.-X., Guo, S.-M., Zhang, M.-B., and Li, J.-H. 2008. N-arylations of nitrogen-containing heterocycles with aryl and heteroaryl halides using a copper(I) oxide nanoparticle/1,10-phenanthroline catalytic system. *Synthesis.* 2008(11):1707–1716.

Tietze, L. F., Brasche, G., and Gericke, K. 2006. *Domino Reactions in Organic Synthesis.* Weinheim: Wiley-VCH Press.

Tornøe, C. W., Christensen, C., and Meldal, M. 2002. Peptidotriazoles on solid phase: [1,2,3]-Triazoles by regiospecific copper(I)-catalyzed 1,3-dipolar cycloadditions of terminal alkynes to azides. *J. Org. Chem.* 67(9):3057–3064.

Verma, S. K., Acharya, B. N., and Kaushik, M. P. 2011. Chemospecific and ligand free CuI catalysed heterogeneous N-arylation of amines with diheteroaryl halides at room temperature. *Org. Biomol. Chem.* 9(5):1324–1327.

Wang, Y., Liu, J. H., and Xia, C. G. 2011. Insights into supported copper(II)-catalyzed azide–alkyne cycloaddition in water. *Adv. Synth. Catal.* 353(9):1534–1542.

Wendlandt, A. E., Suess, A. M., and Stahl, S. S. 2011. Copper-catalyzed aerobic oxidative C–H functionalizations: Trends and mechanistic insights. *Angew. Chem. Int. Ed.* 50(47):11062–11087.

Xia, Q.-H., Ge, H.-Q., Ye, C.-P., Liu, Z.-M., and Su, K.-X. 2005. Advances in homogeneous and heterogeneous catalytic asymmetric epoxidation. *Chem. Rev.* 105(5):1603–1662.

Xiao, R. A., Hao, W. Y., Ai, J. T., and Cai, M. Z. 2012. A practical synthesis of 2-aminobenzothiazoles via the tandem reactions of 2-haloanilines with isothiocyanates catalyzed by immobilization of copper in MCM-41. *J. Organometal. Chem.* 705:44–50.

Xiao, R. A., Zhao, H., and Cai, M. Z. 2013. MCM-41-immobilized bidentate nitrogen copper(I) complex: A highly efficient and recyclable catalyst for Buchwald N-arylation of indoles. *Tetrahedron.* 69(26):5444–5450.

Yamada, Y. M. A., Sarkar, S. M., and Uozumi, Y. 2012. Amphiphilic self-assembled polymeric copper catalyst to parts per million levels: Click chemistry. *J. Am. Chem. Soc.* 134(22):9285–9290.

Yamaguchi, K., Oishi, T., Katayama, T., and Mizuno, N. 2009. A supported copper hydroxide on titanium oxide as an efficient reusable heterogeneous catalyst for 1,3-dipolar cycloaddition of organic azides to terminal alkynes. *Chem. Eur. J.* 15(40):10464–10472.

Yoshimura, K., Oishi, T., Yamaguchi, K., and Mizuno, N. 2011. An efficient, ligand-free, heterogeneous supported copper hydroxide catalyst for the synthesis of *N,N*-bicyclic pyrazolidinone derivatives. *Chem. Eur. J.* 17(14):3827–3831.

Yousuf, S. K., Mukherjee, D., Singh, B., Maity, S., and Taneja, S. C. 2010. Cu–Mn bimetallic catalyst for Huisgen [3+2]-cycloaddition. *Green Chem.* 12(9):1568–1572.

Yu, L., Wang, M., Li, P. H., and Wang, L. 2012. Fe_3O_4 nanoparticle-supported copper(I): Magnetically recoverable and reusable catalyst for the synthesis of quinazolinones and bicyclic pyrimidinones. *Appl. Organometal. Chem.* 26(11):576–582.

Zhang, C., Huang, B., Chen, Y., and Cui, D.-M. 2013. Porous copper catalyzed click reaction in water. *New J. Chem.* 37:2606–2609.

Zhang, J. T., Zhang, Z. H., Wang, Y., Zeng, X. Q., and Wang, Z. Y. 2008. Nano-CuO-catalyzed Ullmann coupling of phenols with aryl halides under ligand-free conditions. *Eur. J. Org. Chem.* 2008(30):5112–5116.

Zhao, Q. W. and Li, C. Z. 2008. Preference of β-lactam formation in Cu(I)-catalyzed intramolecular coupling of amides with vinyl bromides. *Org. Lett.* 10(18):4037–4040.

Zhu, J. and Bienayme, H. (eds). 2005. *Multicomponent Reactions*. Weinheim: Wiley-VCH Press.

5 Silica Sulfuric Acid

A Simple and Powerful Heterogeneous Catalyst in Organic Synthesis

Sushma S. Kauthale, Sunil U. Tekale, Ambadas B. Rode, Sandeep V. Shinde, K. L. Ameta, and Rajendra P. Pawar

CONTENTS

5.1 INTRODUCTION

Heterogeneous catalysis is a type of catalysis in which the catalyst and reactants are present in different phases. It provides an important avenue for the growth and development of sustainable science and technologies, and also highly efficient and eco-friendly synthetic protocols with minimal environmental impact. It involves the application of solid-supported reagents, preferably under solvent-free conditions. The use of reusable heterogeneous catalysts minimizes the consumption of auxiliary substances. It is useful in terms of saving energy and time and achieving separation, and has significant economic and environmental benefits. It assists in the development of safer, cleaner, and more economically feasible processes. There has been increasing interest in the development of heterogeneous catalysts for clean and environmentally friendly technologies with minimization of toxic and hazardous waste. Heterogeneous catalysis has become a sustainable and powerful tool in organic synthesis. Thus, heterogeneous acid catalysts are gaining importance for the exploration of green technologies (Sheldon et al., 2007).

The development of efficient reaction protocols using reusable heterogeneous catalysts is an important target in synthetic organic chemistry. Typically, an ideal heterogeneous catalyst should meet the following requirements:

- Simple
- Efficient
- Thermally and chemically stable
- Easily separable from the reaction mass after completion of reaction
- Minimum leaching during reaction
- Good recycling ability

Commonly, three types of inorganic supports are available as heterogeneous catalysts:

- Amorphous supports: for example, metal oxide-alumina, silica, and so on
- Layered supports: for example, clays
- Microporous supports: for example, zeolites

In recent years, besides the application of microporous, mesoporous, and polymer-supported materials, inorganic oxides have attracted significant attention from synthetic organic chemists as efficient heterogeneous catalysts for a wide variety of organic transformations (Marmaduke, 2008). Solid acid catalysts are environmentally benign in terms of corrosiveness, ease of separation, safety, recycling ability, and replacement to the corrosive liquid reagents in organic synthesis. Many inorganic oxides such as ZnO, CuO, and Al_2O_3-supported reagents have become popular reusable heterogeneous catalysts. A supported catalyst has the combined advantages of both homogeneous and heterogeneous catalysis. A supported surface in close proximity to the anchored catalytic site not only influences the extent of the reaction but also determines the selectivity.

Silica gel is a granular, vitreous, and porous form of silicon dioxide. It is a tough and hard material with good ability to adsorb compounds on its surface. It has a large surface area (800 m^2/g) that can be effectively used for catalytic applications. The silanol (Si-OH) groups of silica gel provide an attractive center for the synthesis of silica-supported/functionalized novel heterogeneous catalysts. Furthermore, chelating groups can also be covalently bonded to silica gel. When silica is used as a support, its polar surface helps to change the spatial orientation of reactants toward catalysts through hydrogen bonding with functional groups of reactant molecules. This is an advantage over the organic polymeric supports, which usually have little porosity, and hence there is a spatial constriction of catalysts for better orientation.

Silica sulfuric acid (SSA) is a simple, cost-effective, and readily available solid Brønsted acid catalyst that has attracted significant attention as a reusable heterogeneous catalyst during the past decade. SSA is an advantageous catalyst compared to

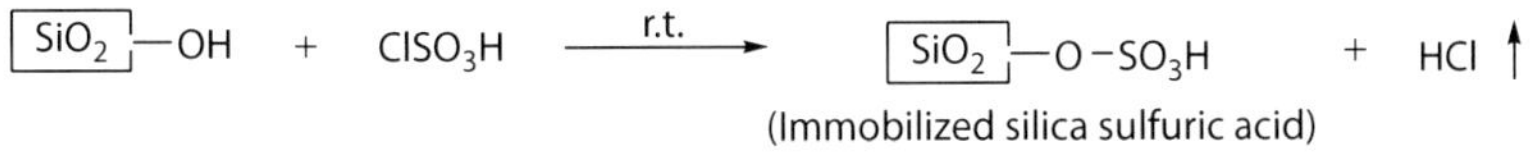

SCHEME 5.1

dangerous concentrated sulfuric acid. SSA is a safe, easy to handle, environmentally benign heterogeneous catalyst with fewer disposal problems and has emerged as an effective heterogeneous catalyst in various organic transformations.

5.2 PREPARATION OF SILICA SULFURIC ACID

SSA catalyst can be easily prepared by simply adding a requisite amount of H_2SO_4 or chlorosulfonic acid to a slurry of silica in a volatile solvent such as *n*-hexane or acetone and then evaporating the solvent under reduced pressure, followed by drying the catalyst at around 100°C–120°C (Scheme 5.1). The preparation method is simple, short, and clean, and avoids tedious workup procedure.

5.3 BENEFITS OF SILICA SULFURIC ACID CATALYST

- Heterogeneous nature
- Low cost
- Ease of preparation
- Nontoxic nature
- Ease of handling
- Recycling ability
- High stability
- High activity and selectivity
- No need for inert atmosphere

Because of these advantages, SSA is extensively used as an efficient heterogeneous catalyst for several organic transformations. It is also used in synthetic carbohydrate chemistry. In many cases, solvent-free reaction protocols have been developed using SSA catalyst; thus, it is useful from a green chemistry point of view. The present chapter provides an insight into organic transformations using an SSA catalyst.

5.4 SILICA SULFURIC ACID–CATALYZED REACTIONS

SSA is used as a heterogeneous catalyst for a variety of organic reactions, which can be summarized under the following subheadings:

1. Synthesis of heterocyclic compounds
2. Protection–deprotection of functional groups
3. Alkylation/acylation/sulfonation reactions
4. Oxidation reactions
5. Miscellaneous reactions

5.4.1 Synthesis of Heterocyclic Compounds

The class of heterocyclic compounds constitutes one the largest areas of research in organic and medicinal chemistry. These structures comprise simple aromatic rings and nonaromatic rings. Heterocyclic compounds usually possess a stable ring structure and offer a high degree of structural diversity as therapeutic agents. The majority of pharmaceuticals and biologically active agrochemicals are heterocyclic in nature. More than 90% of existing drugs contain heterocycles; they are at the interface between chemistry and biology. Several new active heterocycles have been shown to exhibit considerable biological activities such as antibiotic, antifungal, anti-inflammatory, antiviral, anticancer, anticonvulsant, anthelmintic, antidepressant, and so on. Thus, the chemistry of heterocyclic compounds is involved in day-to-day life including in pharmaceuticals, dyes, textiles, biochemistry, and so on. SSA can be used as an effective heterogeneous catalyst for the synthesis of various heterocyclic compounds containing *S*, *N*, and *O* heteroatoms.

5.4.1.1 Synthesis of Nitrogen-Containing Heterocyclic Compounds

The nitrogen-containing heterocyclic compounds include imidazoles, pyrimidinones, indazoles, and so on. The synthesis of these heterocyclic compounds can be accomplished by silica-supported sulfuric acid catalyst. Dihydropyrimidinones and thiopyrimidinones constitute the pharmacores associated with many biological activities such as antibacterial, antiviral, antihypertensive, and so on, and act as calcium channel modulators as well as antagonists. The Biginelli reaction is the simple and straightforward one-pot multicomponent reaction for the synthesis of these compounds. SSA was documented to catalyze the Biginelli reaction of an aldehyde, β-dicarbonyl compound, and urea or thiourea in refluxing ethanol (Salehi et al., 2003). The corresponding dihydropyrimidinones and their thio-analogs were obtained in excellent yields (Scheme 5.2). The catalyst can be reused and showed good reusability results without any decrease in the yield of the reactions. The Biginelli reaction was also studied under solvent-free conditions using SSA/sulfamic acid (Chen et al., 2007). Octahydroquinazolinone derivatives were synthesized using the multicomponent reaction of aromatic aldehydes, dimedone, and urea or thiourea in the presence of SSA via the Biginelli reaction in ethanol under reflux conditions in excellent yields (Scheme 5.3). This constitutes a simple, cheap, and convenient method for the synthesis of octahydroquinazolinones. Astonishingly, no formation of 1,8-dioxo-octahydroxanthenes was observed in the presence of SSA. Hence, the method is applicable for the selective synthesis of octahydroquinazolinone derivatives (Mobinikhaledi et al., 2010).

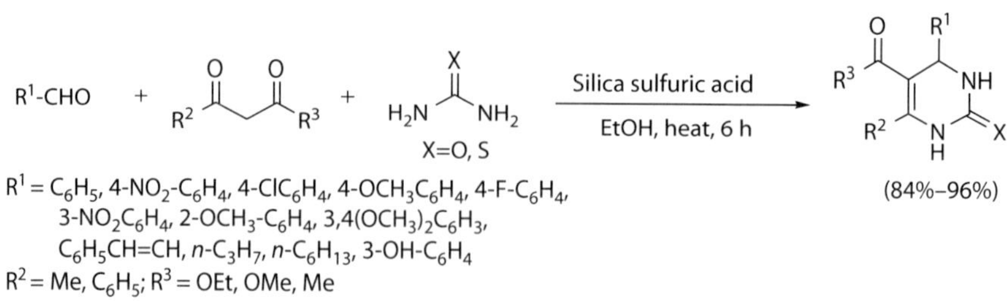

SCHEME 5.2

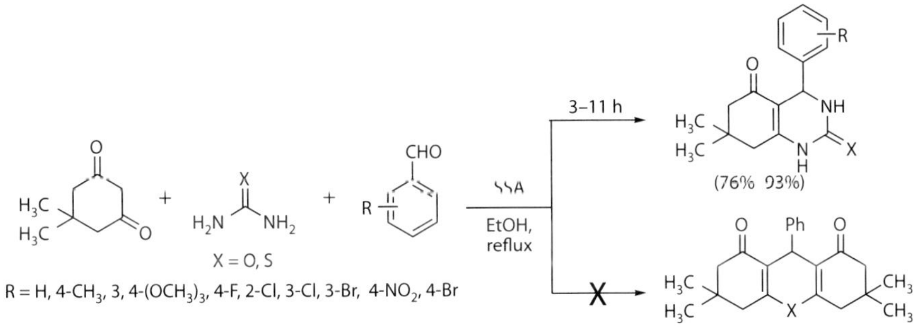

SCHEME 5.3

Hantzsch's dihydropyridine derivatives are the analogs of NADH coenzymes and constitute the skeleton of many drug molecules such as nifedepine, amlodepine, besylate, and so on. A mild protocol for the synthesis of 1,4-dihydropyridine derivatives by the condensation of aldehydes, 1,3-dicarbonyl compounds, and ammonium acetate in the presence of SSA under solvent-free conditions was developed by Datta and Pasha et al. (2011) (Scheme 5.4). This method has several distinct advantages such as an environmentally benign nature, high yield, short reaction time, and so on.

One-pot three-component synthesis of 2*H*-indazolo[2,1-*b*]phthalazine-1,6,11(13*H*)-trione derivatives by the condensation of phthalhydrazide, dimedone, and aromatic aldehydes was carried out using SSA under solvent-free conditions (Shaterian et al., 2008) (Scheme 5.5). Recently, Kiasat and Davarpanah (2013) synthesized iron oxide Fe_3O_4 supported on SSA as a novel catalyst and evaluated its catalytic activity for one-pot

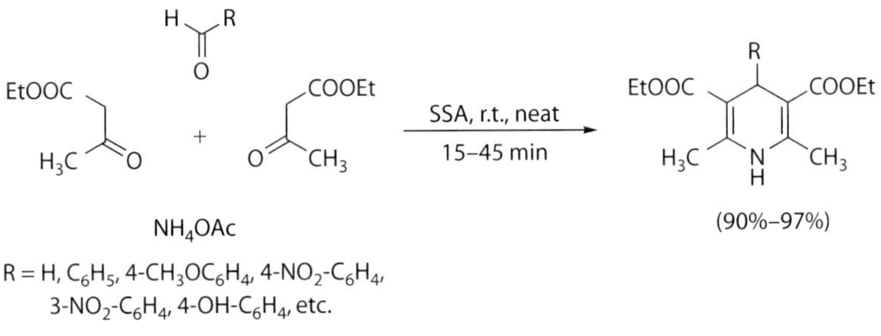

SCHEME 5.4

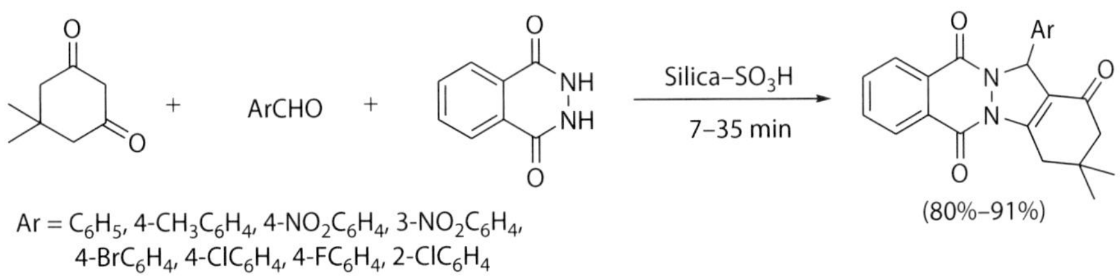

SCHEME 5.5

three-component synthesis of indazolo[2,1-*b*]phthalazine-triones and pyrazolo[1,2-*b*]phthalazine-diones from aldehyde, phthalhydrazide, and 1,3-dicarbonyl compounds—dimedone or acetyl acetone (Scheme 5.6). Mobinikhaledi et al. (2009) reported one-pot four-component synthesis of Hantzsch's polyhydroquinoline derivatives from aldehydes, dimedone, ethyl acetoacetate, and ammonium acetate by two different methods: solvent-free conventional heating and microwave irradiation. Since microwave is a green chemistry tool, the present protocol is an environmentally benign approach for the synthesis of polyhydroquinolines (Scheme 5.7).

The mono- and disubstituted quinazolin-4(3*H*)-one derivatives were firstly synthesized under solvent-free conditions by one-pot three-component reaction of isatoic anhydride, ortho ester, and ammonium acetate or an aliphatic and aromatic primary amine in the presence of SSA (Salehi et al., 2005) (Scheme 5.8). After completion of the reaction, the crude mass was diluted with hot ethanol and filtered. The filtrate was concentrated and products were purified by recrystallization from ethanol. Solvent-free

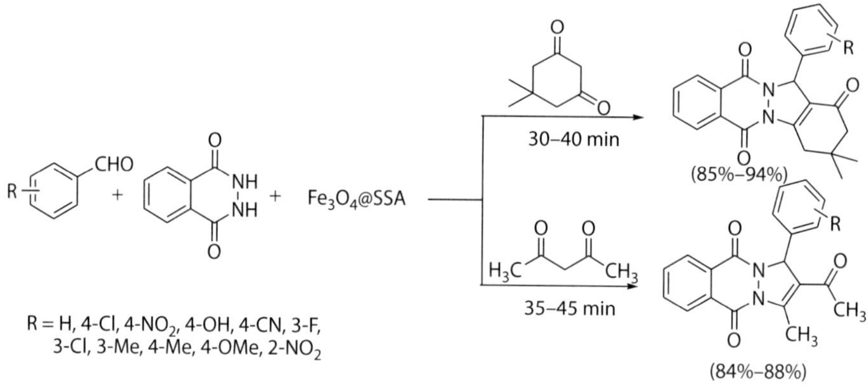

SCHEME 5.6

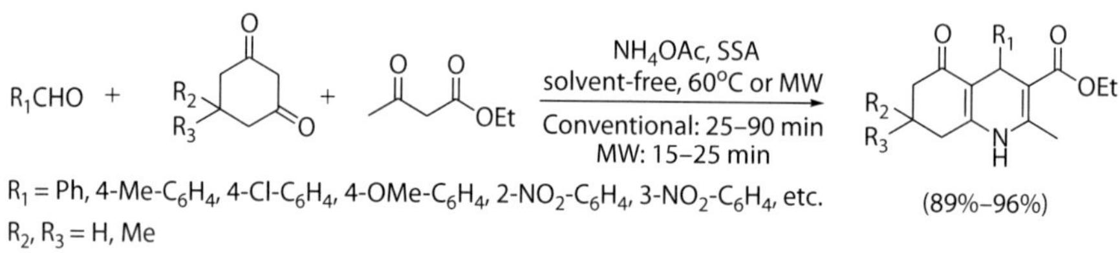

SCHEME 5.7

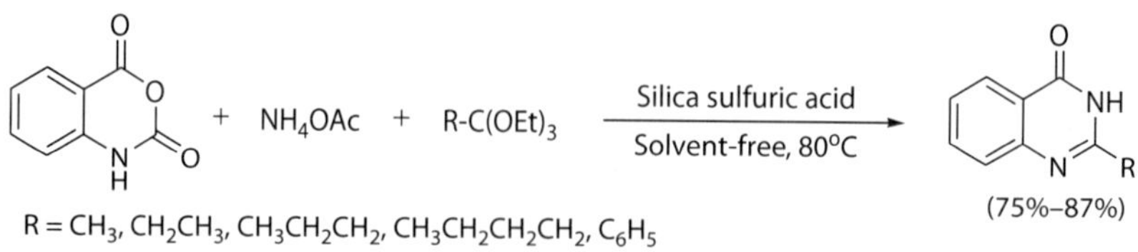

SCHEME 5.8

conditions and reusability of catalyst make it a green approach for the one-pot three-component synthesis of quinazolin-4(3*H*)-ones. SSA was used as a heterogeneous catalyst for the one-pot three-component synthesis of 3,4-dihydroquinoxalin-2-amine derivatives by the condensation of *o*-phenylenediamines, aliphatic/aromatic ketones, and isocyanides at room temperature (Scheme 5.9). The activity of SSA was compared with the studies carried out with *p*-toluenesulfonic acid (PTSA), which showed better results in terms of yield of the corresponding products using SSA catalyst. Furthermore, some of these synthesized compounds exhibited good anti-neuroinflammatory activity (Shobha et al., 2012). Quinoline derivatives are the important compounds in the chemistry of electronic and photonic materials. A Friedländer synthesis of quinolines from 2-aminoaryl ketones with an activated α-CH compound under solvent-free conditions at 100°C using SSA as an inexpensive and recyclable heterogeneous catalyst was reported in the literature (Shaabani et al., 2006) (Scheme 5.10). This method affords products in high yields and avoids the problems associated with the use of solvents and liquid reagents used to promote the Friedländer synthesis of quinolines. The reaction could not proceed in the absence of catalyst even after continuation for 24 h. A similar transformation was reported by Narasimhulu et al. (2007).

Dabiri et al. (2008) demonstrated the green and efficient synthesis of 2,3-dihydroquinazolinones from isatoic anhydride, a primary amine or ammonium acetate, and an aromatic aldehyde using SSA as the catalyst (Scheme 5.11). The catalyst exhibited good efficiency for the synthesis of 2,3-dihydroquinazolin-4(1*H*)-ones in water and also under solvent-free conditions. The present method has several advantages such as environmental compatibility, economic feasibility, operational

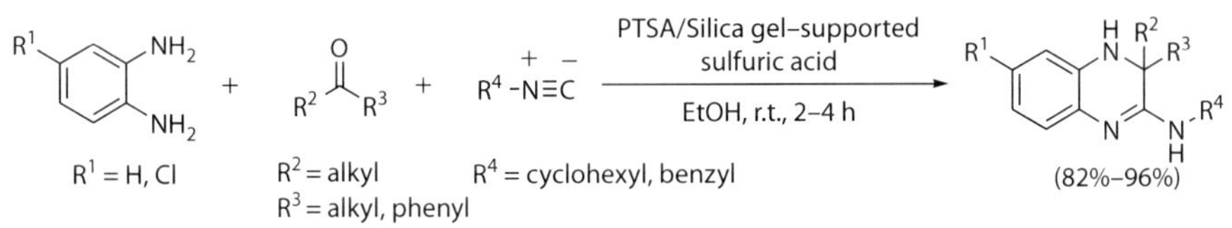

SCHEME 5.9

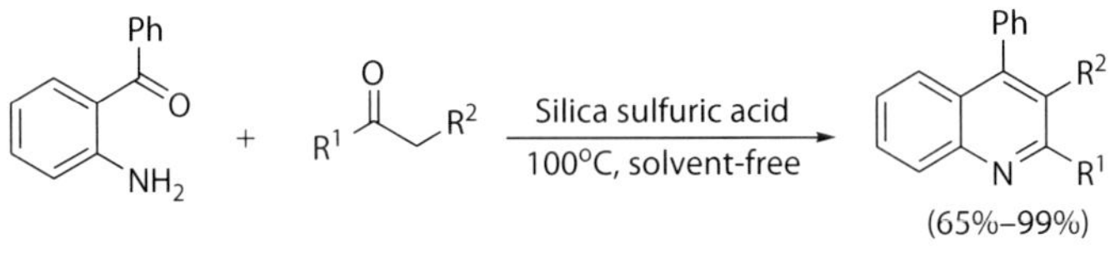

SCHEME 5.10

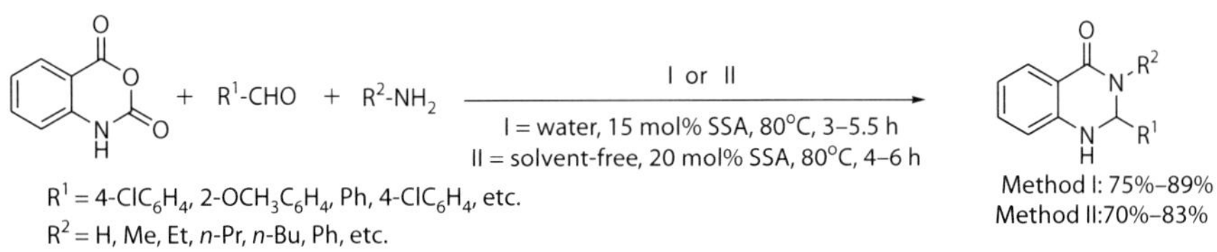

SCHEME 5.11

simplicity, use of low-cost catalyst, and so on. The catalyst recovered after completion of the reaction showed no significant leaching, as confirmed by its UV–vis spectrum, and it could be reused up to five times without much loss of activity.

1,5-Benzodiazepines have anti-inflammatory activity and are of immense importance as dyes for acrylic fibers and valuable synthons in organic chemistry. Hence, the synthesis of these compounds by efficient methods is essential. A mild and efficient synthesis of 1,5-benzodiazepine derivatives was accomplished by the condensation of *o*-phenylenediamines with ketones, both linear and cyclic, to afford the corresponding products in quantitative yields (Scheme 5.12). The reactions were carried out in the presence of SSA catalyst under solvent-free and room temperature conditions (Shaabani et al., 2007). Excellent yields, short reaction time, and high selectivity make this process a valuable addition to the existing protocols for the synthesis of benzodiazepine derivatives. Polyakov et al. (2009) reported a novel approach for the synthesis of imidazo[1,2-*a*]pyridines from the three-component reaction of aromatic aldehydes, trimethyl silyl cyanide, or cyanohydrins and 2-aminopyridines in the presence of SSA in methanol (Scheme 5.13). Recently, Rostamnia et al. (2012) used novel nanomagnetically modified sulfuric acid, that is, γ-Fe_2O_3@SiO_2–OSO_3H catalyst for one-pot three-component synthesis of aminoimidazopyridine derivatives from 2-amino pyridines, aldehydes, and isocyanides via the Ugi-like Groebke–Blackburn–Bienayme reaction under solvent-free conditions without any additives in a short reaction time (45–70 min) (Scheme 5.14). This constitutes a green and rapid synthesis of imidazopyridine derivatives.

2,4,5-Triaryl-substituted imidazoles were synthesized from benzil or benzoin, aldehyde, and ammonium acetate. A literature survey reveals several reports for the synthesis of triarylimidazoles using SSA as a catalyst. Shaabani et al. (2007) reported the synthesis of triarylimidazoles using SSA under conventional heating as well as microwave irradiation (Scheme 5.15). Microwave irradiation afforded the corresponding products in shorter reaction times (5–10 min). Catalytic reusability

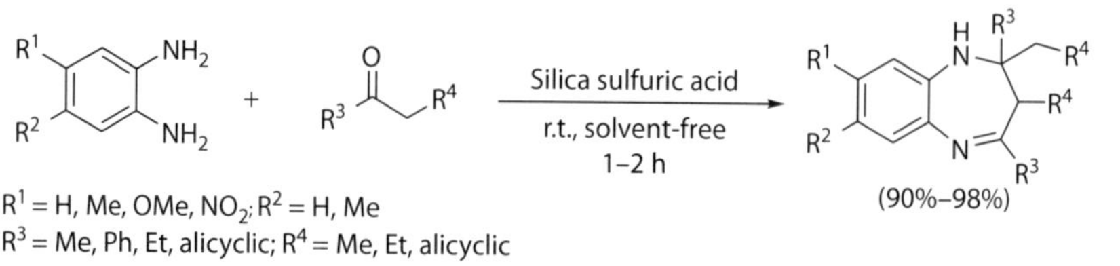

SCHEME 5.12

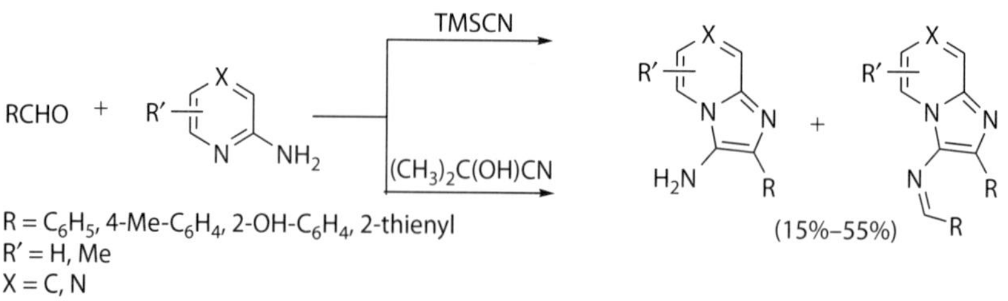

SCHEME 5.13

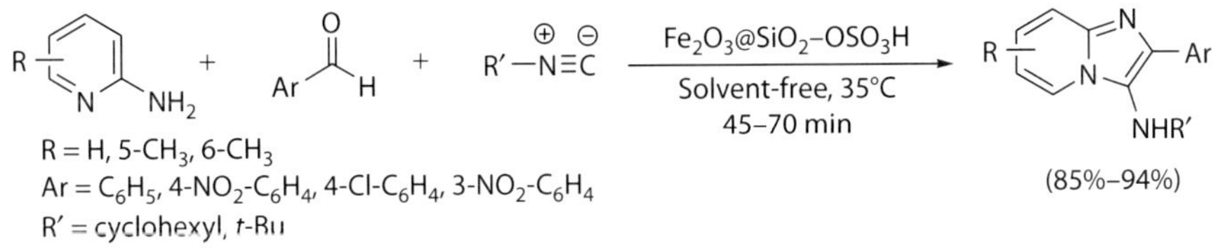

SCHEME 5.14

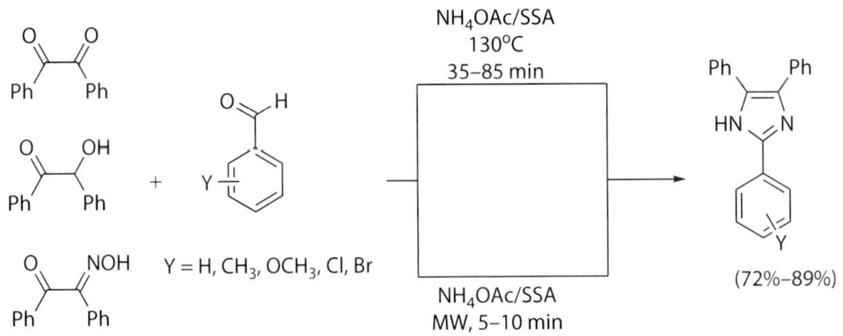

SCHEME 5.15

was also studied, which showed that the catalyst can be reused several times without appreciable loss of catalytic efficiency. The synthesis of these heterocyclic compounds can also be accomplished by refluxing the contents in aqueous medium for 4–8 h (Shaabani and Rahmati, 2006) or by the conventional method under solvent-free conditions (Maleki et al., 2012).

Benzimidazole ring–bearing heterocyclic compounds are important in pharmacology and medicinal chemistry. The synthesis of benzimidazoles from *o*-phenylenediamines and carboxylic acids or nitriles needs harsh conditions. Hence, the synthesis of benzimidazoles can be accomplished with *o*-phenylenediamines and aldehydes. Benzimidazole derivatives were synthesized by the reaction of *o*-phenylenediamine with either aldehydes or nitriles in the presence of SSA as a cheap and readily available heterogeneous catalyst in refluxing ethanol (Sadeghi and Nejad, 2013) (Scheme 5.16). The protocol is beneficial in terms of short reaction time, high yield, and operational simplicity. Salehi et al. (2006) synthesized 2-aryl-1-arylmethyl-1*H*-1,3-benzimidazoles by the SSA-catalyzed reaction between *o*-phenylenediamines and aromatic aldehydes (Scheme 5.17). The reactions were studied in ethanol and aqueous medium and were successfully completed in ethanol (67%–94%) in 1–2 h to afford the corresponding products in good to excellent yield; this was a shorter reaction time than in aqueous medium (71%–90%). The reactions were found to be highly selective and the catalyst could be reused in subsequent runs.

SSA was used as an efficient catalyst for the rapid synthesis of various heterocyclic compounds such as benzoxazoles, benzimidazoles, and oxazolo[4,5-*b*]pyridines from the reaction of ortho esters with *o*-aminophenols, *o*-phenylenediamine, and 2-amino-3-hydroxypyridine under solvent-free conditions (Mohammadpoor-Baltorka et al., 2008) (Scheme 5.18). The recovered catalyst could be reused five times without loss of

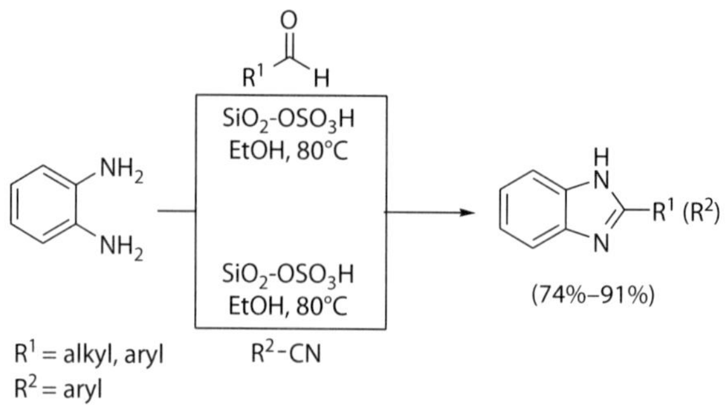

SCHEME 5.16

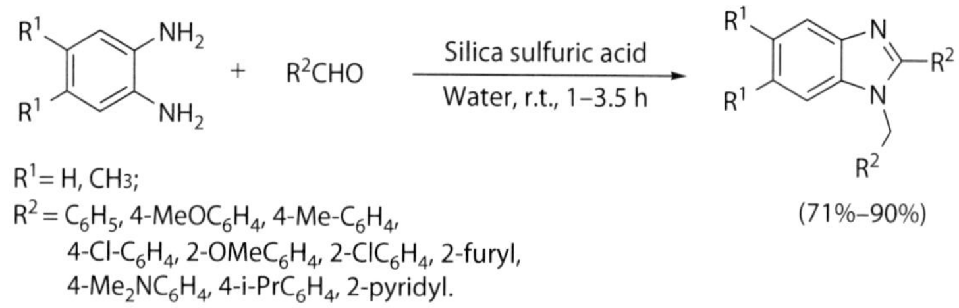

SCHEME 5.17

SCHEME 5.18

activity. Mild reaction conditions, high yield of products, short reaction time, solvent-free conditions, and nontoxic reusable catalyst make the present protocol a viable and economically accepted method for the construction of these heterocyclic compounds.

Synthesis of 2-oxazolines and bis-oxazolines from nitriles and ethanol amine was reported under conventional reflux or ultrasound irradiation using SSA as the catalyst (Schemes 5.19 and 5.20). 2-Imidazolines and bis-imidazolines can also be obtained under similar conditions. Eventually, the synthesis of such heterocyclic compounds was found to be better using ultrasound (Mohammadpoor-Baltork et al., 2008).

A pyrrole ring is the core structure found in the skeletons of many naturally occurring biomolecules such as globins, porphyrins, vitamins, and so on and constitutes a unit of pharmaceuticals. Most commonly, pyrroles are synthesized by multistep synthetic methods. A one-pot three-component synthesis of 2,3,4,5-tetrasubstituted pyrroles from benzoin, 1,3-dicarbonyls, and ammonium acetate was carried out using SSA under solvent-free conditions (Tamaddon and Farahi, 2012) (Scheme 5.21). Veisi (2010) reported a room temperature synthesis of N-substituted

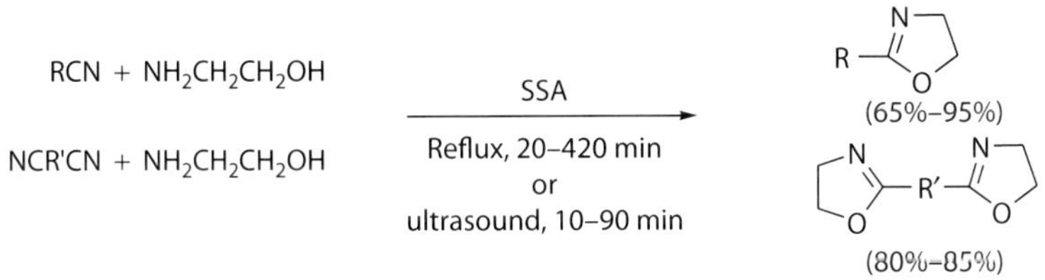

SCHEME 5.19

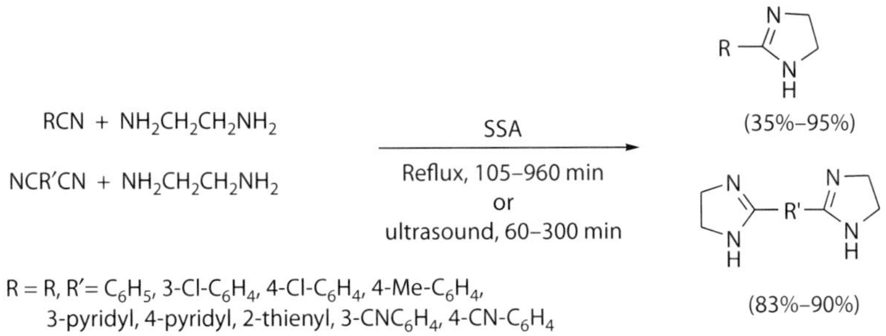

SCHEME 5.20

pyrroles by the reaction of γ-diketones with various aliphatic and aromatic primary amines using SSA catalyst under solvent-free conditions (Scheme 5.22). It requires a low-cost SSA catalyst and has a simple procedure, short reaction time, and mild reaction conditions, which make it a practical approach for the synthesis of pyrroles. Initially, the synthesis of 2,5-dimethyl-*N*-benzylpyrrole was tried as a model reaction using various solvents such as CCl_4, CH_3CN, EtOH, and so on. Acetonitrile afforded better yield of the products as compared to other solvents, but solvent-free conditions are favored from an environmental point of view. Du et al. (2012) documented the first solid-supported acid-catalyzed one-pot synthesis of 5-substituted-1*H*-tetrazoles involving [3 + 2] cycloaddition of nitriles and sodium azide using SSA in DMF under reflux conditions (Scheme 5.23). This is a method with operational simplicity, cost-effectiveness, and high yield of the products.

In continuation of our efforts in the development of new synthetic methods for the synthesis of various heterocyclic compounds, we have developed a green and eco-friendly protocol for the synthesis of indazole derivatives from *o*-hydroxy aromatic

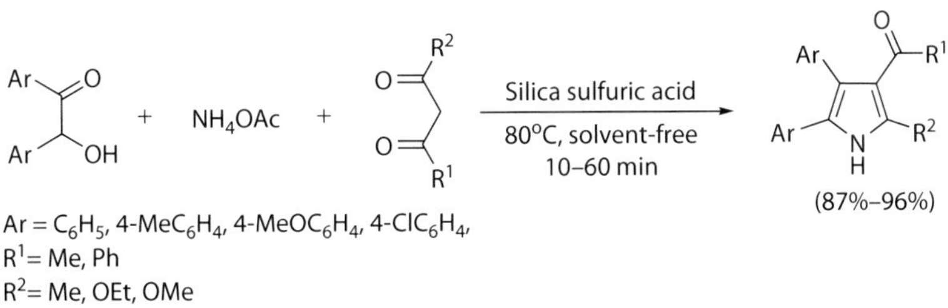

SCHEME 5.21

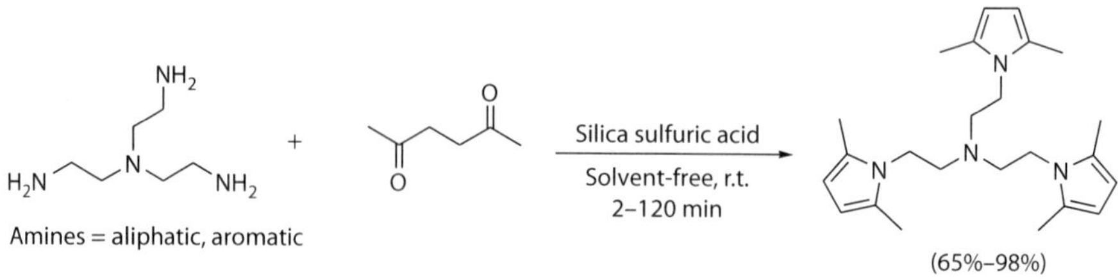

SCHEME 5.22

SCHEME 5.23

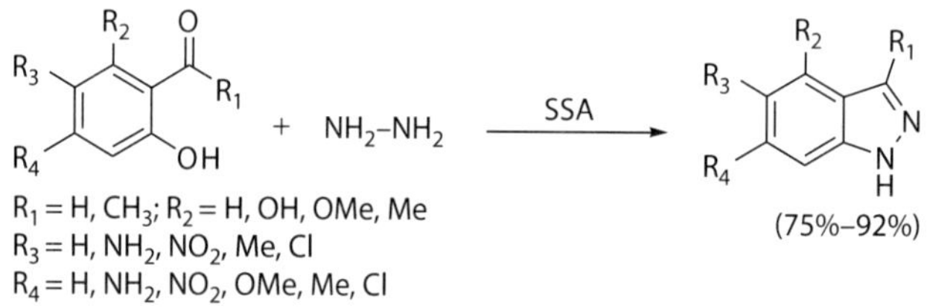

SCHEME 5.24

aldehydes or acetophenone and hydrazine hydrate in DMSO at room temperature using a catalytic amount of SSA (Gaikwad and Pawar, 2010) (Scheme 5.24). We also examined the reusability of SSA catalyst under optimized reaction conditions and the catalyst was found to have good reusability results for several times. This protocol offers a rapid method and good yield of the indazoles as compared to the existing methods for the synthesis of these heterocyclic compounds.

5.4.1.2 Synthesis of Oxygen-Containing Heterocyclic Compounds

These include the SSA-catalyzed synthesis of heterocyclic compounds such as xanthenes, coumarins, oxazoles, and so on. Xanthenes are of great therapeutic and biological interest on account of their many biological activities such as anti-inflammatory, antiviral, antibacterial properties, and so on. Seyyed Hamzeh et al. (2008) carried out the SSA-catalyzed synthesis of aryl-14*H*-dibenzo[*a*,*j*]xanthenes from aldehydes and β-naphthol and 1,8-dioxo-octahydro-xanthenes from aldehydes and 1,3-dicarbonyl compound such as dimedone under solvent-free conditions (Schemes 5.25, 5.26). Reactions were carried out at 80°C. Nazeruddin et al. (2011) documented an effective method for 9,10-dihydro-12-aryl-8*H*-benzo[α]xanthenes-11(12*H*)-one derivatives in excellent yield and short reaction time using SSA under solvent-free conditions (Scheme 5.27).

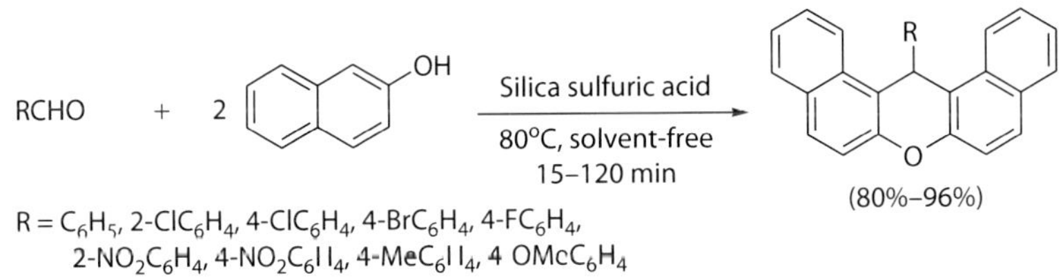

SCHEME 5.25

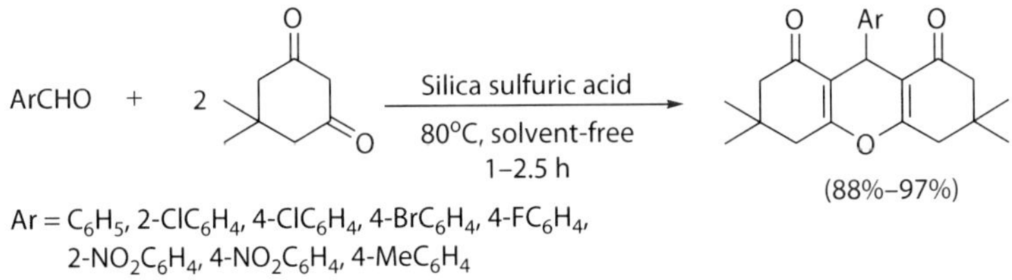

SCHEME 5.26

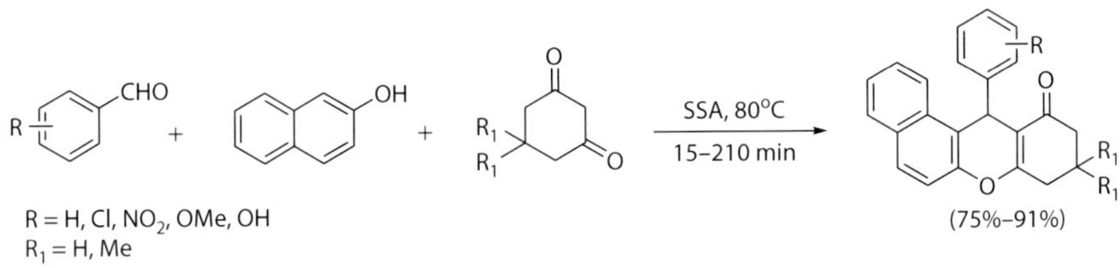

SCHEME 5.27

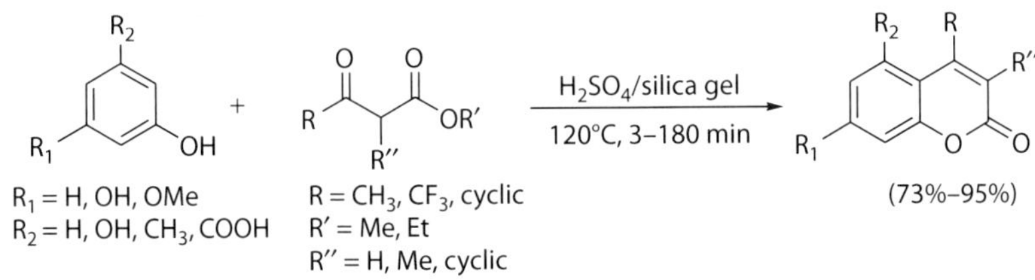

SCHEME 5.28

Reddy et al. (2009) documented a novel protocol for a Pechmann reaction involving the synthesis of substituted coumarins from phenols and β-ketoesters catalyzed by sulfuric acid supported on the silica gel surface (Scheme 5.28). The method has several distinct advantages such as a clean reaction profile, operational simplicity, use of nontoxic catalysts, and higher yields in short reaction time and provides a valuable addition to the existing methods for the synthesis of coumarins.

The oxa-Pictet–Spengler reaction, an intramolecular cyclization, is an important method for the synthesis of phthalans. It requires strong Brønsted acid and the reactants of this reaction are limited in number. 3-Hydroxyphthalans can be synthesized via the condensation of aromatic 3-hydroxybenzyl alcohols and aldehydes under conventional heating as well as microwave irradiation using nano-silica sulfuric acid (NSSA) (Khorsandi et al., 2011) (Scheme 5.29). The method is applicable for the scale up process also.

Benzoxazole derivatives have many biological activities such as anti-inflammatory, antiviral, antibiotic, anticancer, and so on. Anand et al. (2011) reported SSA-catalyzed microwave-assisted rapid synthesis of substituted benzoxazole derivatives and studied their antimicrobial properties on various strains such as *Staphylococcus aureus*, *Escherichia coli*, *Candida glabrata*, and so on (Scheme 5.30).

5.4.1.3 Synthesis of Sulfur-Containing Heterocyclic Compounds

Benzothiazole ring skeletons are present in many natural products and constitute the building blocks of many drugs. Aryl benzothiazoles were effectively synthesized by the condensation reaction between *o*-aminothiophenol and aromatic aldehydes in the presence of SSA in ethanol or methanol medium (Patil et al., 2011; Chen et al., 2013) (Scheme 5.31). The catalyst recovered after completion of reaction could be reused three times without any change in its activity in terms of yield and reaction time. Almost all types of aldehydes possessing either electron-donating or electron-withdrawing groups afforded the corresponding products in excellent yields.

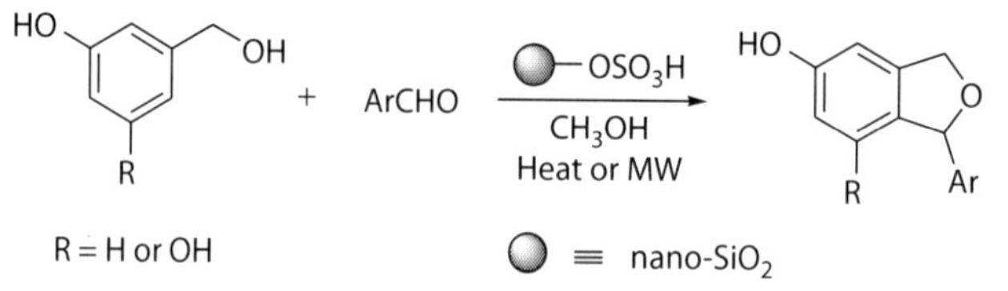

SCHEME 5.29

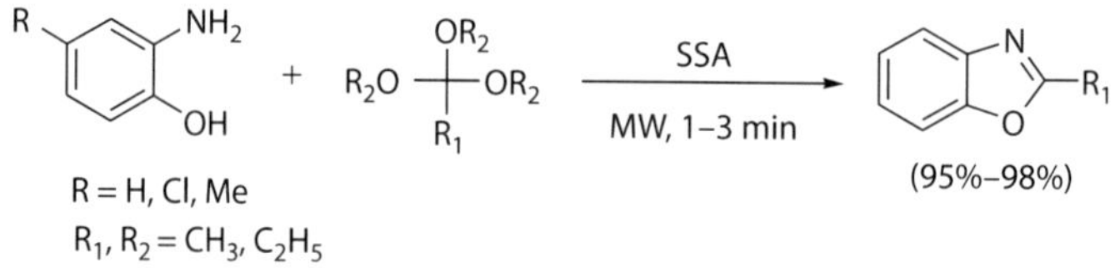

SCHEME 5.30

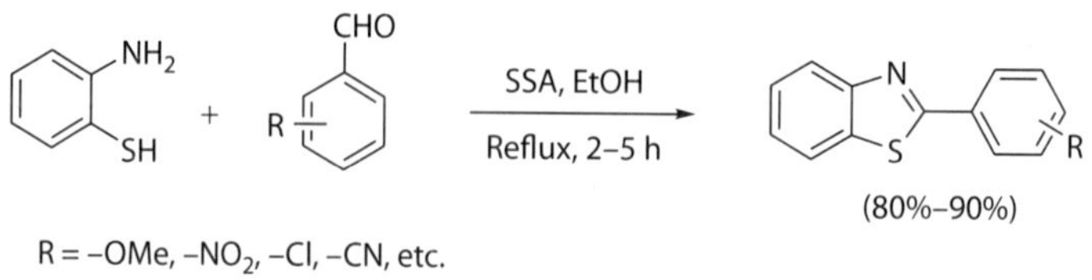

SCHEME 5.31

5.4.2 Protection–Deprotection of Functional Groups

Protection–deprotection of functional groups is an inevitable event in multistep organic synthesis. The carbonyl group is commonly protected as acylals or ketals. Acylals are the important intermediates in organic synthesis, particularly for the synthesis of dienes and for Diels–Alder cycloadditions. Aldehydes, on treatment with acetic anhydride in the presence of a catalytic amount of SSA under solvent-free conditions, afford the corresponding 1,1-diacetates (acylals) in excellent yields (Pourmousavi and Zinati, 2009) (Scheme 5.32). However, ketones are not reacted under these reaction conditions, mainly due to steric hindrance around the carbonyl group. Desai et al. (2006) documented the protection–deprotection of the carbonyl group using reusable SSA solid acid catalyst at room temperature (Scheme 5.33). The carbonyl protection was carried out under solvent-free conditions, whereas the deprotection was reported in methanol.

The combination of SSA and wet SiO_2 is strong enough for the transformation of acetals and ketals into the corresponding parent carbonyl compounds. The combination provides an excellent deacetalizing reagent under thermal conditions. The reactions proceed smoothly in toluene at 60°C–70°C (Mirjalili et al., 2002) (Scheme 5.34). SSA-catalyzed acetalization of carbonyl compounds was successfully accomplished by refluxing a mixture of aldehydes or ketones and diol in *n*-hexane (Mirjalili et al., 2004). Under these conditions, various aldehydes and ketones can be condensed with different diols to form acetals (Scheme 5.35). Oxathioacetalization of carbonyl compounds with 2-mercaptoethanol and deprotection of the obtained 1,3-oxathiolanes is easily achieved in the presence of SSA (Shirini et al., 2009) (Scheme 5.36).

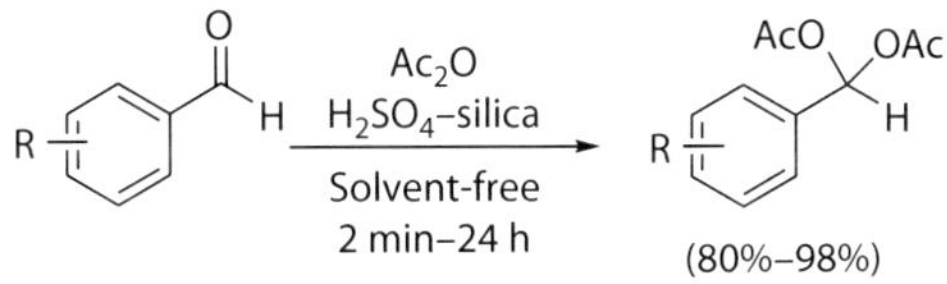

SCHEME 5.32

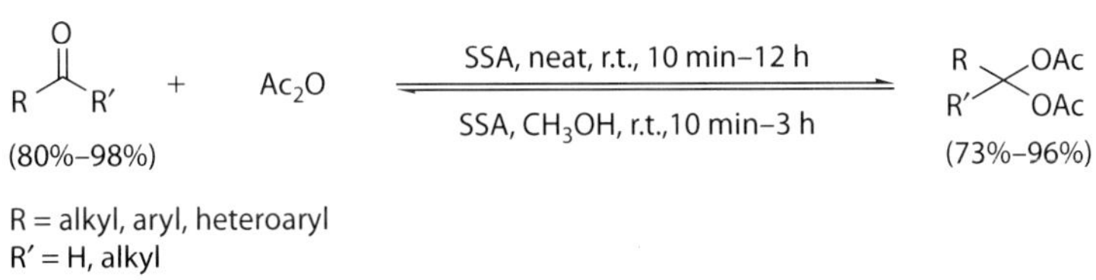

SCHEME 5.33

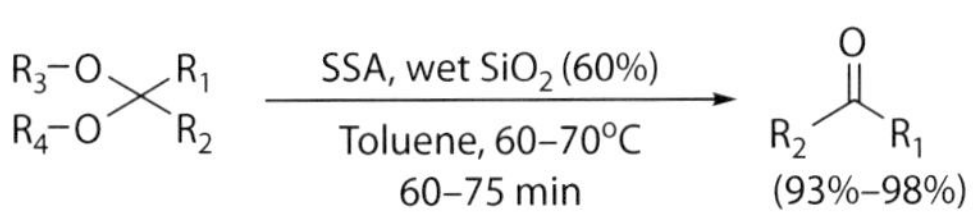

SCHEME 5.34

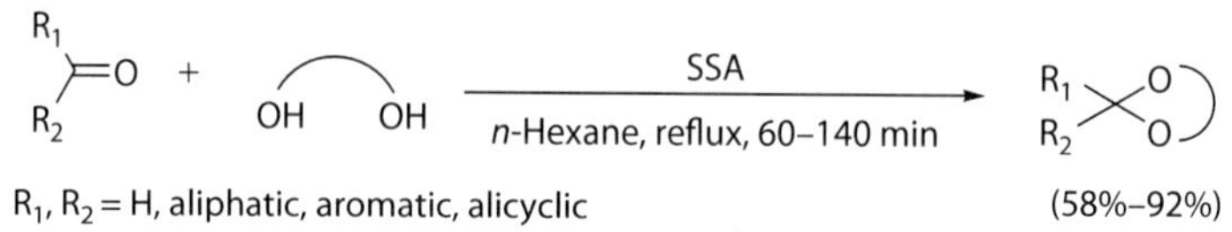

SCHEME 5.35

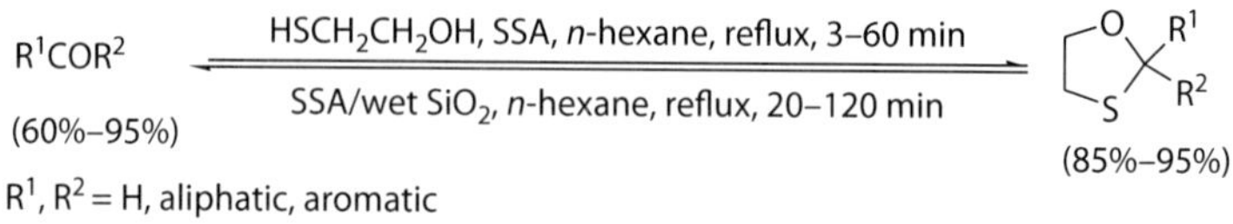

SCHEME 5.36

All reactions were performed under mild and completely heterogeneous conditions, offering good to high yield of the protected compounds.

Sulfuric acid immobilized on silica can be used as a simple, economic, and effective catalyst for the selective hydrolysis of a terminal *O*-isopropylidene group of sugar derivatives (Rajput et al., 2006). Using this protocol, di-*O*-isopropylidene derivatives of D-glucose, D-mannose, D-fructose, and so on resulted in the formation of corresponding mono-*O*-isopropylidene derivatives (Scheme 5.37). The reaction occurs smoothly, even on a large scale (20 g). Moreover, good recycling ability of the catalyst, operational simplicity, and excellent yield of the products make this an attractive process for the deprotection of terminal *O*-isopropylidene groups in carbohydrate chemistry. Gawande et al. (2007) reported a simple and mild protocol for the chemoselective *N*-benzyloxycarbonylation of amines using SSA under solvent-free conditions at room temperature (Scheme 5.38). Several aliphatic, aromatic, and cyclic amines containing both electron-donating as well as electron-withdrawing groups were successfully protected using the present protocol to afford excellent yield of the products (87%–96%). After completion of the reaction, the reaction mass

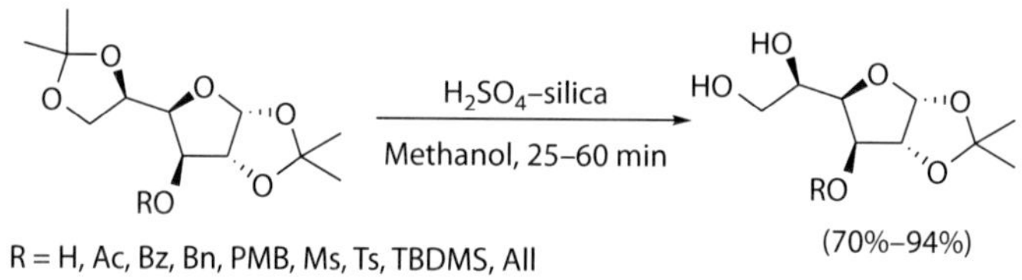

SCHEME 5.37

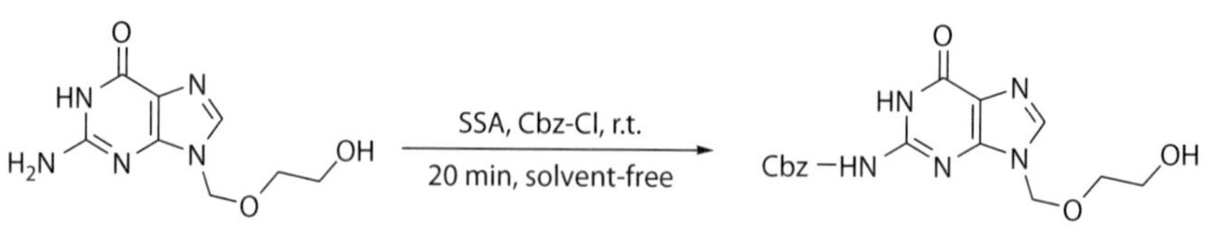

SCHEME 5.38

was diluted with ethyl acetate, filtered off to separate the catalyst, and further subjected to recycle studies. The recovered catalyst can be reused at least five times without significant loss of its activity.

Cleavage of ether linkage usually requires drastic conditions. The ether cleavage is particularly important in deprotection of hydroxyl (OH) groups involved in carbohydrate chemistry. Nucleoside trityl ethers and purine and pyrimidine derivatives can be deprotected under mild conditions within a short reaction time at room temperature using SSA in acetonitrile (Ali et al., 2007). This strategy is compatible for acid-sensitive OH-protecting groups such as *p*-methoxybenzyl (PMB), isopropylidene, cyclohexylidene, di-(*p*-anisyl)methylidene, and so on (Scheme 5.39). When the deprotection was studied using unsupported silica, no reaction took place, even after a prolonged reaction time (72 h). It constitutes a rapid, efficient, mild, and chemoselective method for deprotection from the protected nucleosides.

5.4.3 Alkylation/Acylation/Sulfonation Reactions

The methoxymethylation of alcohols with formaldehyde dimethoxyacetal was carried out using SSA to afford the corresponding methoxymethyl (MOM) ethers under solvent-free conditions at room temperature (Niknam et al., 2006) (Scheme 5.40). The most commonly used reagent for formylation reactions is formic acid, which is corrosive and toxic in nature. Furthermore, with the use of formic acid as the formylating agent, acid-sensitive groups may decompose during the course of reactions. Zolfigol et al. (2008) demonstrated that a wide range of alcohols can be successfully formylated with ethyl formate in the presence of a catalytic amount of SSA or $Al(HSO_4)_3$ to form the corresponding ester derivatives under solvent-free conditions at ambient temperature (Scheme 5.41).

Mukhopadhyay (2006) used sulfuric acid immobilized on silica gel as the promoter for acetylation of sugar glycosides to afford the corresponding products in good to excellent yields (Scheme 5.42). Furthermore, the per-*O*-acetylated acetals/ketals were successfully synthesized by this method.

SCHEME 5.39

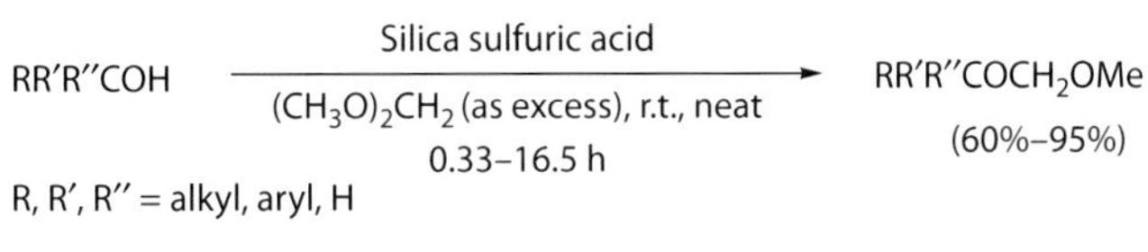

SCHEME 5.40

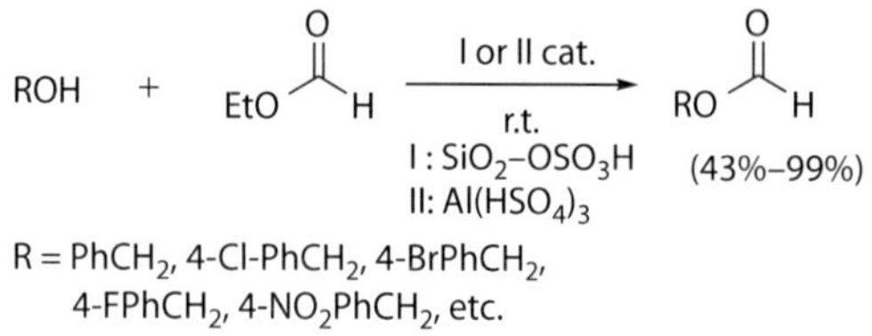

SCHEME 5.41

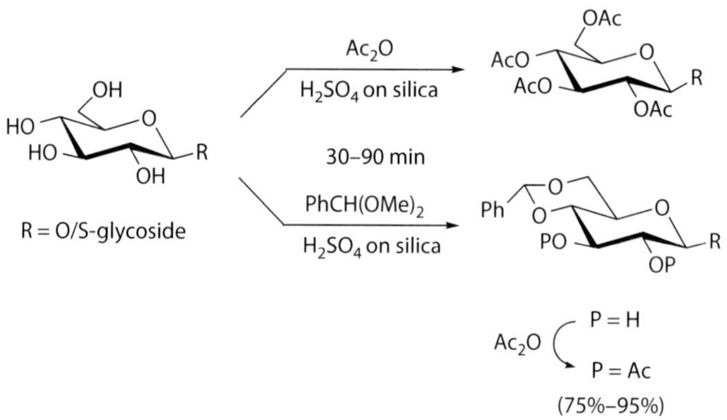

SCHEME 5.42

The acetylation of alcohol is usually carried out under basic conditions. An effective mild method for the acetylation of alcohols with acetic anhydride was developed using SSA at room temperature either under solvent-free conditions or using a nonpolar solvent such as *n*-hexane (Shirini et al., 2004) (Scheme 5.43). The reactions were found to be clean, without the formation of any detectable by-products. The reaction required a short reaction time under solvent-free conditions rather than with the use of a solvent, and slightly higher yields were obtained under solvent-free conditions.

Friedel–Crafts acylation involves the direct introduction of an acyl group on the aromatic rings. This reaction, when carried out using corrosive and liquid catalysts, poses tedious workup and separation problems. An effort to minimize such limitations was made by Alizadeh et al. (2007), who reported a Friedel–Crafts reaction involving the treatment of acetic anhydride with aromatic compounds in the presence of SSA as a reusable, nontoxic, and heterogeneous catalyst (Scheme 5.44).

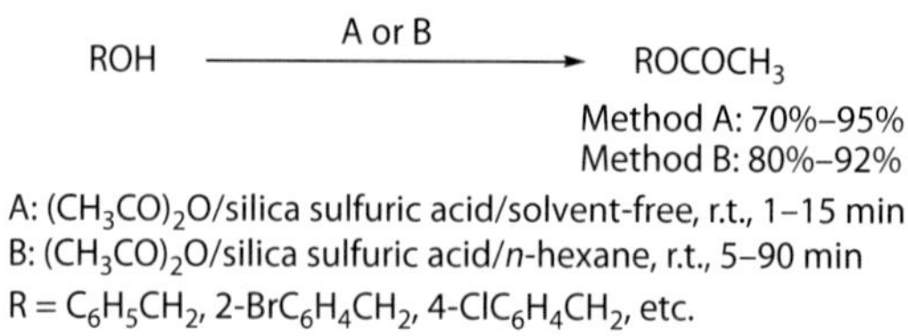

SCHEME 5.43

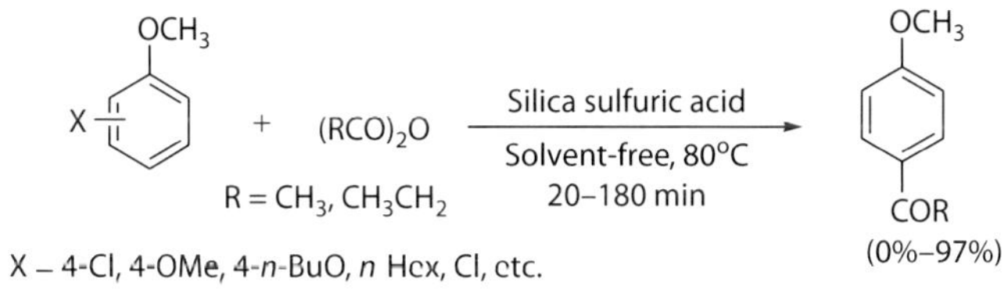

SCHEME 5.44

A highly chemoselective and direct sulfonation of aromatics was accomplished with SSA in 1,2-dichloroethane or under solvent-free conditions (Hajipour et al., 2004) (Scheme 5.45). Aromatic sulfonic acids, which are used for the synthesis of detergents, as bleaching and dying agents, are the industrially important targets. Hence, the synthesis of sulfonic acids is essential. $NaHSO_4$ adsorbed on silica gel support is an excellent proton source and has a higher catalytic activity than many other heterogeneous acid catalysts. Recently, direct sulfonation of aromatic and polyaromatic compounds was reported in the literature using SSA/$NaHSO_4$ and wet SiO_2 as the reusable heterogeneous catalyst (Joshi et al., 2011). The sulfonation was reported under mild conditions (room temperature) (Scheme 5.46).

5.4.4 Oxidation Reactions

In organic chemistry, oxidation and reduction processes are different from ordinary redox reactions because in many cases they do not involve direct electron transfer but may involve a decrease in electron density around a molecule or loss/gain of hydrogen. Oxidation reactions are useful to convert alcohols into carbonyl compounds, nitriles into acids, and amines into imines. SSA along with a suitable reagent such as oxone or sodium nitrite serves as a powerful oxidant. This part of the chapter encompasses the oxidation reactions catalyzed by SSA.

An environmentally friendly, mild, and chemoselective oxidation of sulfides into sulfoxides was achieved using SSA, NH_4NO_3, wet SiO_2, and a catalytic amount of KBr in dichloromethane in a short reaction time with excellent yields (Arash et al., 2009)

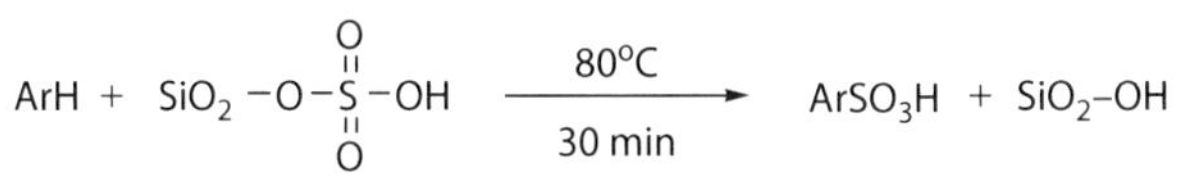

SCHEME 5.45

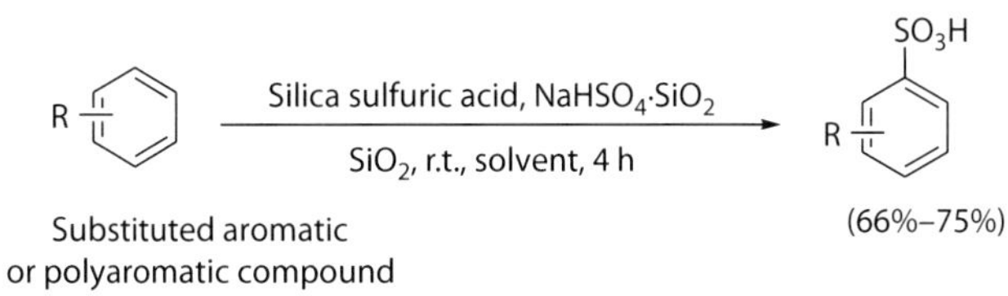

SCHEME 5.46

(Scheme 5.47). The oxidant nitric acid required for the reaction was proposed to be generated *in situ* by the reaction of ammonium nitrate and SSA. The use of cost-effective reagents and catalysts, easy workup, and high selectivity are the salient features of the present method that make it a significant addition to the existing methods for the conversion of sulfides into sulfoxides. (Shaabani and Rezayan, 2007) reported the oxidation of alkyl and aryl sulfides into sulfones or sulfoxides using aqueous hydrogen peroxide in the presence of SSA at 25°C–90°C (Scheme 5.48). The formation of sulfones or sulfoxides is governed by controlling the catalyst concentration and amount of oxidant H_2O_2.

Zolfigol et al. (2002) documented the combination of SSA, potassium bromate, and wet SiO_2 as a powerful oxidizing agent for the conversion of alcohols into aldehydes or ketones in various solvents such as acetonitrile, dichloromethane, and toluene (Scheme 5.49).

A combination of silica-supported sulfuric acid, sodium nitrite, and wet silica was used as an oxidant for the effective oxidation of dihydropyridines to pyridine derivatives (Zolfigol et al., 2002) (Scheme 5.50). Products were obtained in excellent yields within short reaction times. Niknam et al. (2007) reported the same combination for

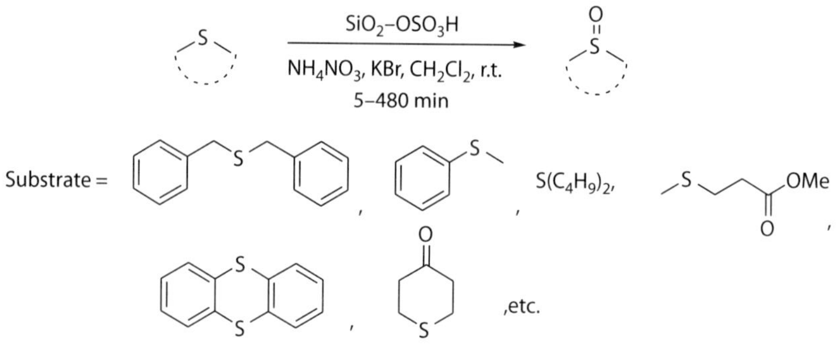

SCHEME 5.47

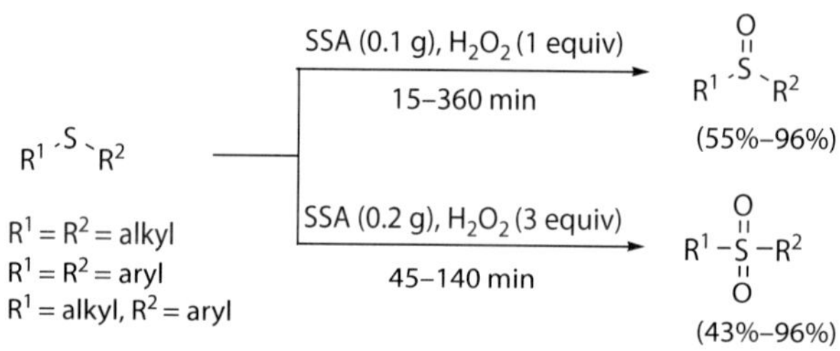

SCHEME 5.48

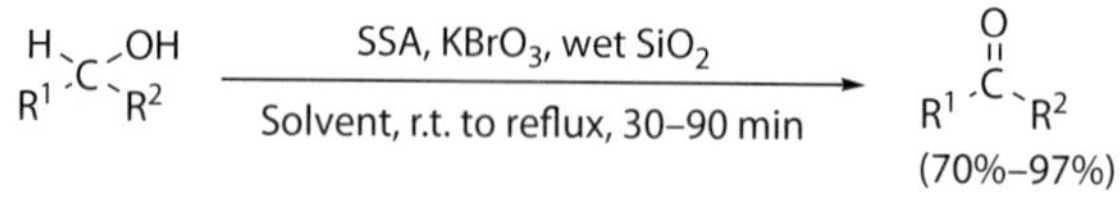

SCHEME 5.49

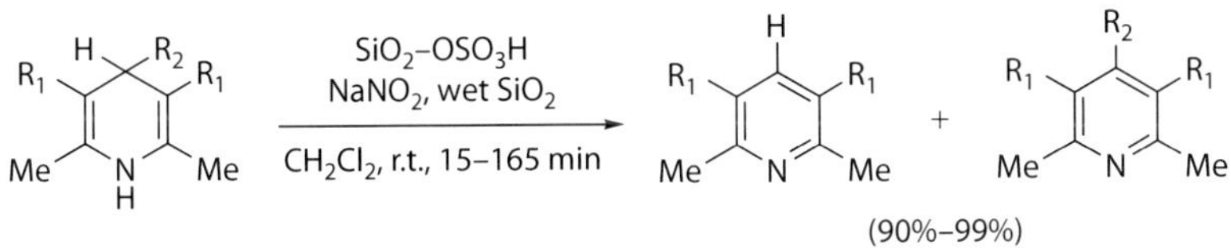

R_1 = COOEt; R_2 = H, Me, Et, Ph, 2-thienyl, 2-NO_2-C_6H_4, 3-NO_2-C_6H_4, 2-OCH_3 C_6H_4, 2,5-$(OCH_3)_2$-C_6H_3, 4-Br-C_6H_4, 2-furyl, 5-methyl-2-furyl

SCHEME 5.50

the oxidation of 1,2-dihydroquinoline derivatives into their corresponding quinolines in dichloromethane as the reaction medium at room temperature (Scheme 5.51). The combined use of SSA/potassium perchlorate or SSA/oxone was used for urazoles and bis-urazoles to afford the corresponding triazolinediones in excellent yields (Zolfigol et al., 2007) (Scheme 5.52).

5.4.5 Miscellaneous Reactions

Various miscellaneous reactions have also been catalyzed by SSA and these are discussed in this last part of the chapter.

Pore et al. (2006) documented the thia-Michael addition involving the conjugate addition of thiols to α,β-unsaturated ketones in the presence of SSA as a heterogeneous catalyst under solvent-free conditions at room temperature (Scheme 5.53). The catalytic activity was comparatively studied with potassium phosphate. Studies revealed that SSA is suitable for electron-deficient α,β-unsaturated compounds, whereas K_3PO_4 catalyzed the reactions of both electron-deficient as well as electron-rich enones. SSA catalyzes a rapid, green, and chemoselective aza-Michael addition of amines and thiols

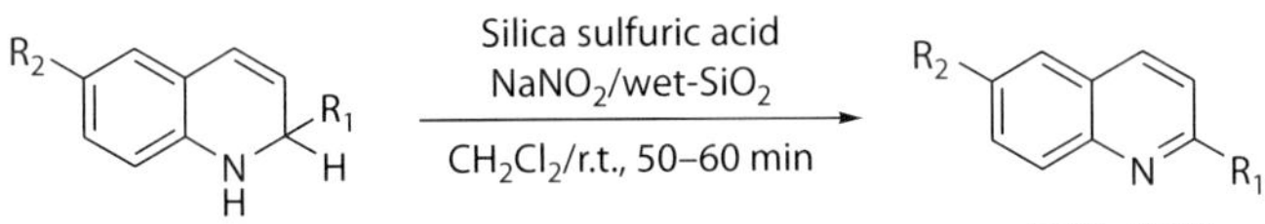

(86%–93%)

R_1 = H, Me, *n*-Bu, Ph
R_2 = H, Me

SCHEME 5.51

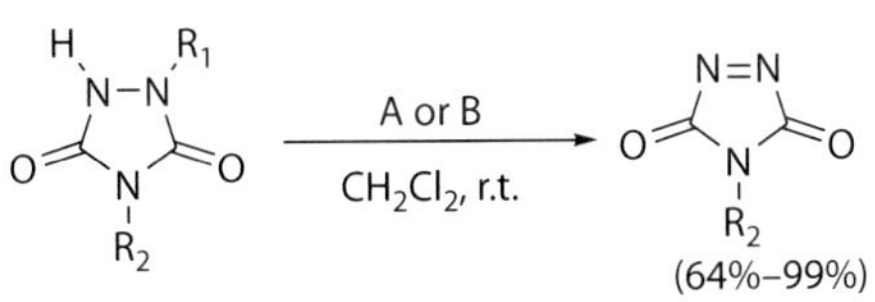

A: SiO_2–OSO_3H/$KClO_3$/wet SiO_2, 30–120 min
B: SiO_2–Cl/oxane/wet SiO_2, 3–32 min
R_1 = H, Na
R_2 = Me, Et, *n*-Pr, *n*-Bu, cyclohexyl, Ph, 4-Cl-Ph, 3, 4-Cl_2-Ph, 4-NO_2-Ph

SCHEME 5.52

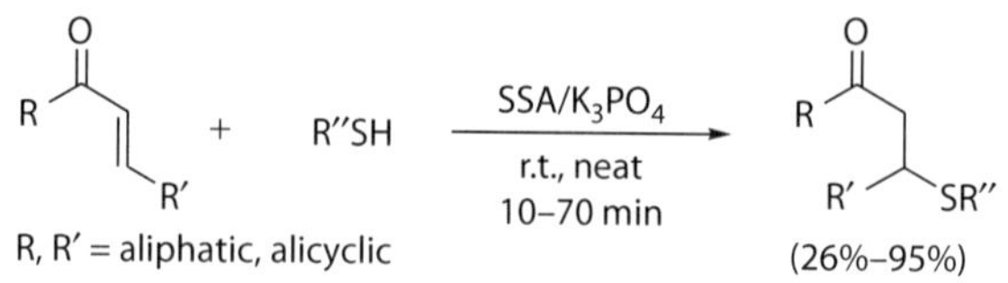

SCHEME 5.53

to α,β-unsaturated carbonyl compounds (Wang et al., 2009). Various amine nucleophiles were allowed to add on the β-carbon of acrylic acid 2-phenylsulfanyl-ethyl ester (PTEA) and PEEA (Scheme 5.54). The Michael addition of various aliphatic amines and aryl amines with electron-donating and electron-withdrawing groups was smoothly carried out to afford the corresponding Michael adducts in high yields. Aliphatic amines reacted rapidly as compared to the aromatic amines and primary amines such as benzylamine afforded only the monoalkylated adducts. Furthermore, no side products were obtained even by using excess amines.

Recently, Rapolu et al. (2013) synthesized primary amides by the Ritter reaction of secondary and tertiary alcohols with nitriles catalyzed by SSA catalyst in toluene at 90°C (Scheme 5.55). As the catalyst is cost-effective, stable to air, and recyclable, the present method constitutes a noteworthy modification to the Ritter reaction for the synthesis of amides. Under these conditions, a variety of amines including sterically hindered amines were successfully reacted to form the corresponding amides in excellent yields. The reaction was observed to be chemoselective in nature. Mirjalilia and Sadeghi (2009) studied the Ritter reaction under solvent-free conditions (Scheme 5.56).

Eshghi and Hassankhani (2007) studied the Beckmann rearrangement using sulfuric acid immobilized on the surface of silica gel (Scheme 5.57). Under these conditions, high conversion and excellent selectivity were found in the Beckmann rearrangement of cyclohexanone oxime. Treatment of carbamates and oxazolidinones with anhydrides in the presence of SSA under solvent-free conditions at room temperature results in the formation of *N*-acyl carbamates and oxazolidinones

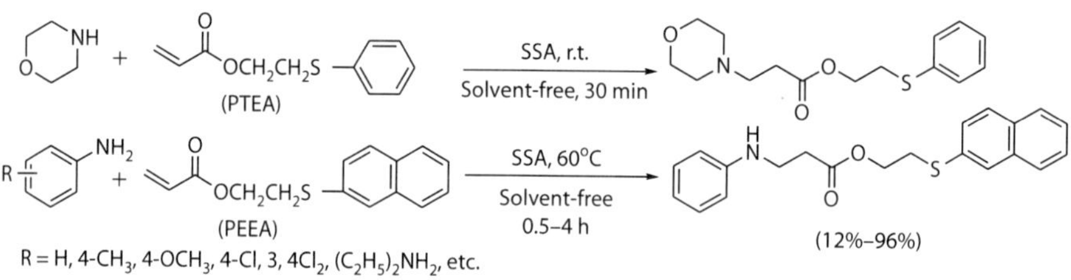

SCHEME 5.54

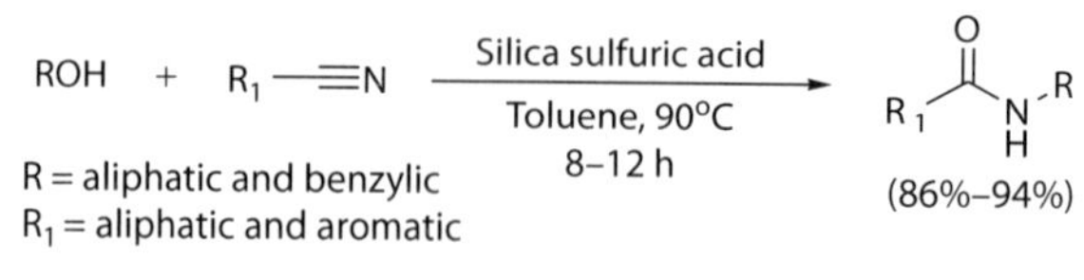

SCHEME 5.55

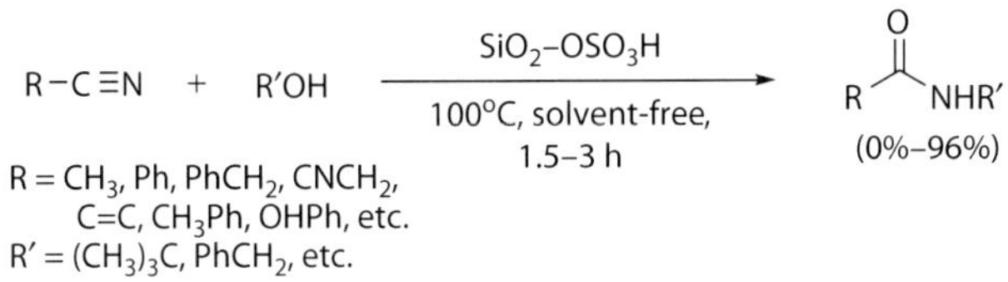

SCHEME 5.56

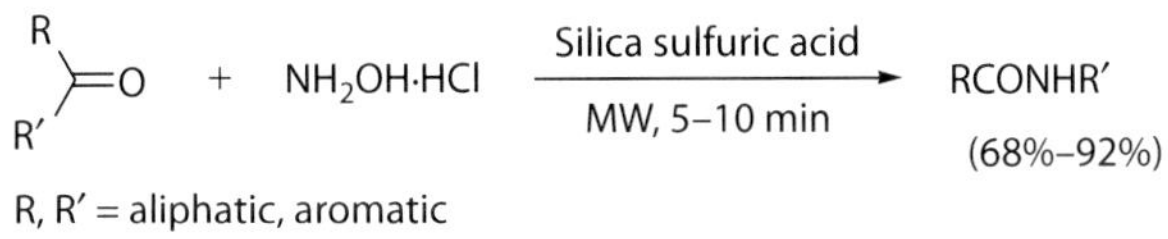

SCHEME 5.57

(Wu et al., 2011) (Scheme 5.58). The hydroxyl group of alcohols and phenols can be converted into primary carbamates under solvent-free conditions using the grinding technique in the presence of SSA as the reusable catalyst (Modarresi-Alam et al., 2007) (Scheme 5.59). No epimerization was observed in the absence of solvent.

β-Acetamido ketones were successfully synthesized by one-pot four-component reaction of an aryl aldehyde, ketone, acetyl chloride, and acetonitrile at 80°C in the presence of SSA (Khodaei et al., 2005) (Scheme 5.60). The reaction does not require

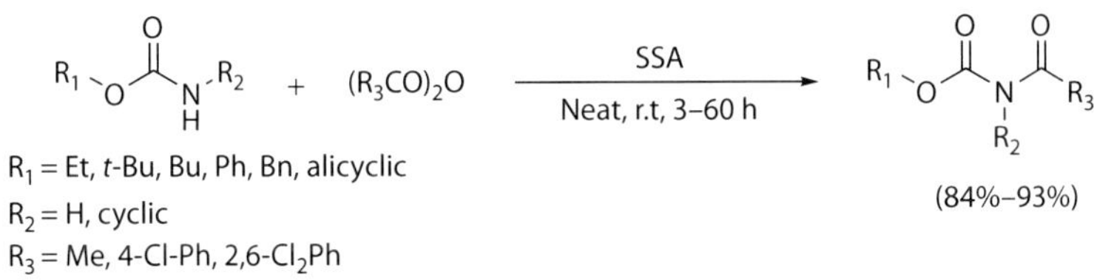

SCHEME 5.58

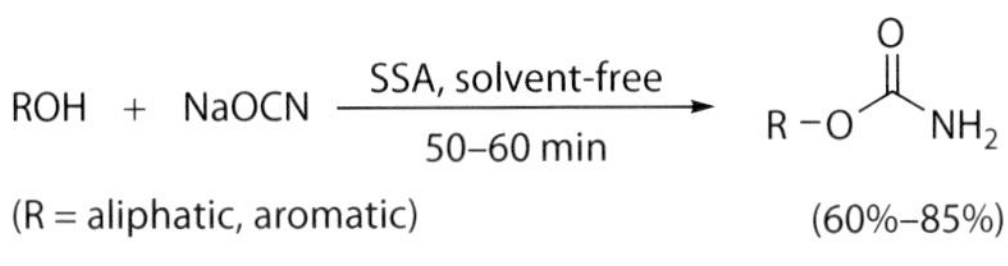

SCHEME 5.59

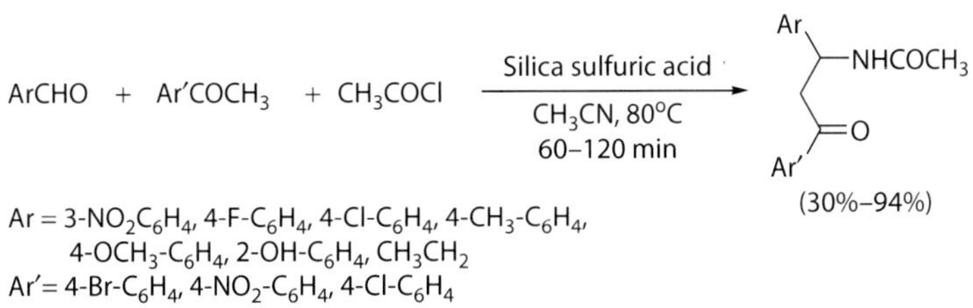

SCHEME 5.60

an inert atmosphere. Aromatic aldehydes and ketones with both electron-donating as well as electron-withdrawing substituents reacted smoothly to afford the corresponding β-acetamido ketones in excellent yields without the formation of side products. Furthermore, functional groups like nitro, chloro, bromo, hydroxyl, and methoxy were tolerated during the reaction conditions. The reactions do not take place in the absence of SSA catalyst. Also, no acetylation of the hydroxyl substrates was observed under the reaction conditions.

α,β-Unsaturated carbonyl compounds, commonly known as chalcones, are the versatile intermediates for the synthesis of a wide variety of heterocyclic compounds such as flavones, pyrazoles, benzodiazapines, and so on and also act as the precursors for aza-Michael addition, and many agrochemicals and drugs. Chalcones were obtained by the crossed-aldol condensation of 4-bromo-1-naphthyl ketones with various substituted benzaldehydes using SSA under solvent-free conditions (Thirunarayanan and Vanangamudi, 2006) (Scheme 5.61). This method has several advantages such as operational simplicity, high yield, solvent-free conditions, and so on.

The aldol reaction is one of the most significant C–C bond-forming reactions in synthetic organic chemistry. α,α′-Bis(substituted benzylidene)cycloalkanones, the starting materials for the synthesis of pyrimidines, are pharmacologically important targets. Salehi et al. (2004) reported the cross-aldol condensation of cycloalkanone with aldehydes and ketones using SSA under solvent-free conditions to afford the corresponding α,α′-bis(substituted benzylidene)cycloalkanones carbonyl compounds (Scheme 5.62). Various aldehydes and ketones that have electron-releasing as well as electron-withdrawing groups reacted with a clean profile. Furthermore, no self-condensation of the starting cyclic ketones was observed under these experimental conditions.

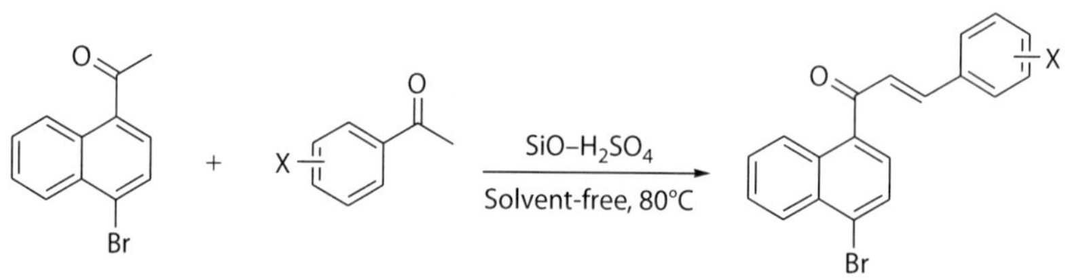

X = H, *m*-NH_2, *p*-NH_2, *m*-Br, *m*-Cl, *p*-Cl, *p*-$N(CH_3)_2$, *p*-OH, *p*-OCH_3, *o*-NO_2, *m*-NO_2, *p*-NO_2

SCHEME 5.61

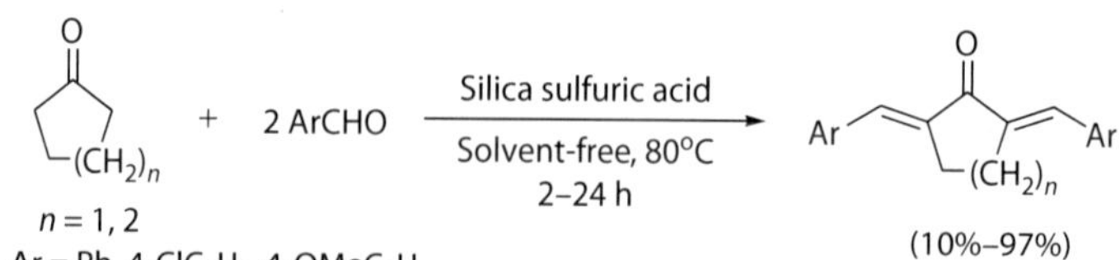

$n = 1, 2$
Ar = Ph, 4-ClC_6H_4, 4-$OMeC_6H_4$, 4-$NO_2C_6H_4$, 4-MeC_6H_4, PhCH=CH

SCHEME 5.62

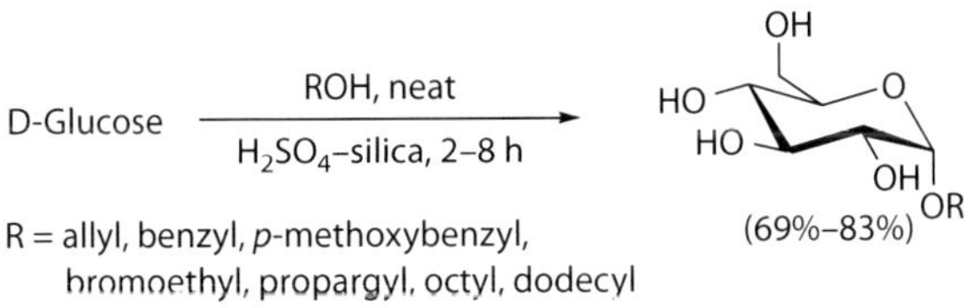

SCHEME 5.63

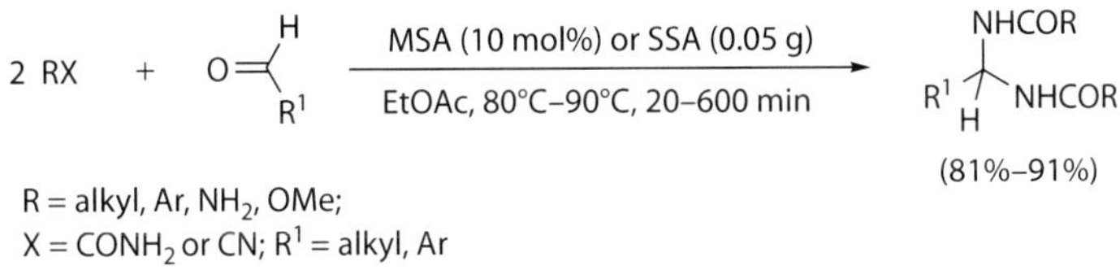

SCHEME 5.64

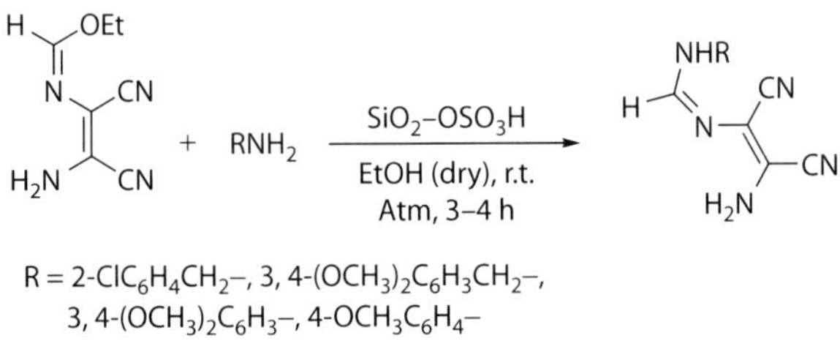

SCHEME 5.65

Glycosides of sugars are useful in synthetic and biological fields. Roy and Mukhopadhyay (2007) studied the SSA-catalyzed synthesis of alkyl, allyl, propargyl, and benzylic glycosides. Usually, the glycosylation requires use of strong mineral acids, excess alcohols, long reaction times, and high temperature (Scheme 5.63). The glycosylation reactions were successfully performed on various substrates and almost all the reactions proceeded cleanly to afford the glycosides in good yields and anomeric selectivity. The protocol was found to be applicable for large-scale preparations.

Molybdate sulfuric acid (MSA) and SSA were prepared and used as efficient solid catalysts for the synthesis of gem-bisamides and -bisurides via one-pot three-component reaction of two moles of amides, nitriles, carbamates, or urea with aldehydes (Tamaddon et al., 2013) (Scheme 5.64). SSA was used as a catalyst for the synthesis of formamidines from ethyl-(*Z*)-*N*-(2-amino-1,2-dicyanovinyl) formimidates (Yahyazadeh et al., 2012) (Scheme 5.65).

5.6 CONCLUSION

SSA is a super solid acid heterogeneous catalyst with Brønsted acidity, easily separated from the reaction mixture simply by filtration. It can be reused and shows good reusability results for several subsequent runs. It is an efficient, inexpensive, recyclable, and convenient catalyst. It shows greater selectivity, is easier to handle, and is more stable, nontoxic, and insoluble in organic solvents; thus, it has attracted considerable attention in the development of new routes for the synthesis of various heterocyclic compounds and organic intermediates with high efficiency. There is scope for future development of novel materials in which reagents are covalently bonded to silica and their applications as reusable heterogeneous catalysts in synthetic organic chemistry.

REFERENCES

Ali, K-N., Abolfath, P., Mohammad, N.S.R., Zolfigol, M.A. and Zare, A. 2007. A catalytic method for chemoselective detritylation of 5′-tritylated nucleosides under mild and heterogeneous conditions using silica sulfuric acid as a recyclable catalyst. *Tetrahedron Lett.* 48(30): 5219–5222.

Alizadeh, A., Khodaei, M.M. and Nazari, E. 2007. Silica sulfuric acid as an efficient solid acid catalyst for Friedel–Crafts acylation using anhydrides. *Bull. Korean Chem. Soc.* 28(10): 1854–1856.

Anand, M., Ranjitha, A.R. and Himaja, M. 2011. Silica sulfuric acid catalyzed microwave assisted synthesis of substituted benzoxazoles and their antimicrobial activity. *Int. Res. J. Pharm.* 2(4): 211–213.

Arash, G-C., Zolfigol, M.A. and Rastegar, T. 2009. Chemoselective oxidation of sulfides with ammonium nitrate and silica sulfuric acid catalyzed by KBr. *Chin. J. Catal.* 30(4): 273–275.

Chen, G.F., Zhang, L.Y., Jia, H.M., Chen, B.H., Li, J.T., Wang, S.X. and Bai, G.Y. 2013. Ecofriendly synthesis of 2-substituted benzothiazoles catalyzed by silica sulfuric acid. *Res. Chem. Intermed.* 39(5): 2077–2086.

Chen, W-Y., Qin, S-D. and Jin, J-R. 2007. Efficient Biginelli reaction catalyzed by sulfamic acid or silica sulfuric acid under solvent-free conditions. *Synth. Commun.* 37(1): 47–52.

Dabiri, M., Salehi, P., Baghbanzadeh, M., Zolfigol, M.A., Agheb, M. and Heydari, S. 2008. Silica sulfuric acid: An efficient reusable heterogeneous catalyst for the synthesis of 2, 3-dihydroquinazolin-4(1*H*)-ones in water and under solvent-free conditions. *Catal. Commun.* 9: 785–788.

Datta, B. and Pasha, M.A. 2011. Silica sulfuric acid: An efficient heterogeneous catalyst for the one-pot synthesis of 1,4-dihydropyridines under mild and solvent-free conditions. *Chin. J. Catal.* 32(7): 1180–1184.

Desai, U.V., Thopate, T.S., Pore, D.M. and Wadgaonkar, P.P. 2006. An efficient, solvent-free method for the chemoselective synthesis of acylals from aldehydes and their deprotection catalyzed by silica sulfuric acid as a reusable solid acid catalyst. *Catal. Commun.* 7(7): 508–511.

Du, Z., Si, C., Li, Y., Wang, Y. and Lu, J. 2012. Improved synthesis of 5-substituted 1H-tetrazoles via the [3+2] cycloaddition of nitriles and sodium azide catalyzed by silica sulfuric acid. *Int. J. Mol. Sci.* 13: 4696–4703.

Eshghi, H. and Hassankhani, A. 2007. One-pot efficient Beckmann rearrangement of ketones catalyzed by silica sulfuric acid. *J. Korean Chem. Soc.* 51(4): 361–364.

Gaikwad, D.D. and Pawar, R.P. 2010. Silica sulfuric acid: An efficient catalyst for the synthesis of substituted indazoles. *Iran. Chem. Res.* 3: 191–194.

Gawande, M.B., Polshettiwar, V., Varma, R.S. and Jayaram, R.V. 2007. An efficient and chemoselective Cbz-protection of amines using silica-sulfuric acid at room temperature. *Tetrahedron Lett.* 48(46): 8170–8173.

Hajipour, A.R., Mirjalili, B.B.F., Zarei, A., Khazdooz, L. and Ruoho, A.E. 2004. A novel method for sulfonation of aromatic rings with silica sulfuric acid. *Tetrahedron Lett.* 45(35): 6607–6609.

Joshi, U.J., Gokhale, K.M. and Kanitkar, A.P. 2011. Sulphonation of aromatics using silica sulphuric acid/$NaHSO_4$ as a novel heterogeneous system at ambient temperature. *Int. J. Pharm. Phytopharmacol. Res.* 1(3): 102–106.

Khodaei, M.M., Khosropour, A.R. and Fattahpour, P. 2005. A modified procedure for the Dakin–West reaction: An efficient and convenient method for a one-pot synthesis of β-acetamido ketones using silica sulfuric acid as catalyst. *Tetrahedron Lett.* 46(12): 2105–2108.

Khorsandi, Z., Khosropour, A.R., Mirkhani, V., Mohammadpoor-Baltork, I., Moghadam, M. and Tangestaninejad, S. 2011. A simple and efficient large-scale synthesis of 3-hydroxyphthalans via oxa-Pictet–Spengler reaction catalyzed by nanosilica sulfuric acid. *Tetrahedron Lett.* 52(11): 1213–1216.

Kiasat, A.R. and Davarpanah, J. 2013. Fe_3O_4@silica sulfuric acid nanoparticles: An efficient reusable nanomagnetic catalyst as potent solid acid for one-pot solvent-free synthesis indazolo[2,1-*b*]phthalazine-triones and pyrazolo[1,2-*b*]phthalazine-diones. *J. Mol. Catal. A Chem.* 373: 46–54.

Maleki, B., Shirvan, H.K., Taimazi, F. and Akbarzadeh, E. 2012. Sulfuric acid immobilized on silica gel as highly efficient and heterogeneous catalyst for the one-pot synthesis of 2,4,5-triaryl-1*H*-imidazoles. *Int. J. Org. Chem.* 2: 93–99.

Marmaduke, D.L. 2008. *Progress in Heterogeneous Catalysis*. New York: Nova Science.

Mirjalilia, B.B.F. and Sadeghi, B. 2009. Silica sulfuric acid: An eco-friendly and reusable catalyst for synthesis of amides via Ritter reaction. *Iran. J. Org. Chem.* 2: 76–79.

Mirjalili, B.B.F., Zolfigol, M.A. and Bamoniri, A. 2002. Deprotection of acetals and ketals by silica sulfuric acid and wet SiO_2. *Molecules* 7: 751–755.

Mirjalili, B.B.F., Zolfigol, M.A., Bamoniri, A. and Hazar, A. 2004. Acetalization of carbonyl compounds by using silica-bound sulfuric acid under green condition. *Bull. Korean Chem. Soc.* 25(6): 865–868.

Mirjalili, B.B.F., Zolfigol, M.A., Bamoniri, A., Zaghaghi, Z. and Hazara, A. 2003. Silica sulfuric acid/$KBrO_3$/wet SiO_2 as an efficient heterogeneous system for the oxidation of alcohols under mild conditions. *Acta Chim. Slov.* 50: 563–568.

Mobinikhaledi, A., Foroughifar, N., Fard, M.A.B., Moghanian, H., Ebrahimi, S. and Kalhor, M. 2009. Efficient one-pot synthesis of polyhydroquinoline derivatives using silica sulfuric acid as a heterogeneous and reusable catalyst under conventional heating and energy-saving microwave irradiation. *Synth. Commun.* 39(7): 1166–1174.

Mobinikhaledi, A., Foroughifar, N. and Khodaei, H. 2010. Synthesis of octahydroquinazolinone derivatives using silica sulfuric acid as an efficient catalyst. *Eur. J. Chem.* 1(4): 291–293.

Modarresi-Alam, A.R., Nasrollahzadeh, M. and Khamooshi, F. 2007. Solvent-free preparation of primary carbamates using silica sulfuric acid as an efficient reagent. *Arkivoc* xvi: 238–245.

Mohammadpoor-Baltork, I., Mirkhani, V., Moghadam, M., Tangestaninejad, S., Zolfigol, M.A., Abdollahi-Alibeik, M., Khosropour, A.R., Kargar, H. and Hojati, S.F. 2008. Silica sulfuric acid: A versatile and reusable heterogeneous catalyst for the synthesis of oxazolines and imidazolines under various reaction conditions. *Catal. Commun.* 9: 894–901.

Mohammadpoor-Baltorka, I., Moghadama, M., Tangestaninejad, S., Mirkhania, V., Zolfigolb, M.A. and Hojatia, S.F. 2008. Silica sulfuric acid catalyzed synthesis of benzoxazoles, benzimidazoles and oxazolo[4,5-*b*]pyridines under heterogeneous and solvent-free conditions. *J. Iran. Chem. Soc.* 5: S65–S70.

Mukhopadhyay, B. 2006. Sulfuric acid immobilized on silica: An efficient promoter for one-pot acetalation-acetylation of sugar derivatives. *Tetrahedron Lett.* 47(26): 4337–4341.

Narasimhulu, M., Srikanth, R.T., Mahesh, K.C., Prabhakar, P., Rao, C.B. and Venkateswarlu, Y. 2007. Silica supported perchloric acid: A mild and highly efficient heterogeneous catalyst for the synthesis of poly-substituted quinolines via Friedlander hetero-annulation. *J. Mol. Catal. A Chem.* 266: 114–117.

Nazeruddin, G.M., Pandharpatte, M.S. and Mulani, K.B. 2011. Heterogeneous catalyst: Silica sulfuric acid catalyzed synthesis of 9, 10-dihydro-12-aryl-8*H*-benzo[α]xanthenes-11(12*H*)-one derivatives under solvent free conditions. *Ind. J. Chem.* 50(B): 1532–1537.

Niknam, K., Karami, B. and Zolfigol, M.A. 2007. Silica sulfuric acid promoted aromatization of 1,2-dihydroquinolines by using $NaNO_2$ as oxidizing agent under mild and heterogeneous conditions. *Catal. Commun.* 8(9): 1427–1430.

Niknam, K., Zolfigol, M.A., Khorramabadi-Zad, A., Zare, R. and Shayegh, M. 2006. Silica sulfuric acid as an efficient and recyclable catalyst for the methoxymethylation of alcohols under solvent-free conditions. *Catal. Commun.* 7(7): 494–498.

Patil, D.R., Salunkhe, S.M., Sambavekar, P.P., Deshmukh, M.B., Kolekar, G.B. and Anbhule, P.V. 2011. Silica sulfuric acid: Recyclable and efficient catalyst for the 2-aryl benzothiazoles. *Der Pharma Chem.* 3(1): 189–193.

Polyakov, A.I., Eryomina, V.A., Medvedeva, L.A., Tihonova, N.I., Listratova, A.V. and Voskressensky, L.G. 2009. Silica-sulfuric acid: A highly efficient catalyst for the synthesis of imidazo[1,2-*a*]pyridines using trimethysilyl cyanide or cyanohydrins. *Tetrahedron Lett.* 50(30): 4389–4393.

Pore, D.M., Soudagar, M.S., Desai, U.V., Thopate, T.S. and Wadagaonkar, P.P. 2006. Potassium phosphate or silica sulfuric acid catalyzed conjugate addition of thiols to α,β-unsaturated ketones at room temperature under solvent-free conditions. *Tetrahedron Lett.* 47(52): 9325–9328.

Pourmousavi, S.A. and Zinati, Z. 2009. H_2SO_4-silica as an efficient and chemoselective catalyst for the synthesis of acylal from aldehydes under solvent-free conditions. *Turk. J. Chem.* 33: 385–392.

Rajput, V.K., Roy, B. and Mukhopadhyay, B. 2006. Sulfuric acid immobilized on silica: An efficient reusable catalyst for selective hydrolysis of the terminal *O*-isopropylidene group of sugar derivatives. *Tetrahedron Lett.* 47(39): 6987–6991.

Rapolu, R.K., Mukul, B.N., Bommineni, S.R., Potham, R., Mulakayalaand, N. and Oruganti, S. 2013. Silica sulfuric acid: A reusable solid catalyst for the synthesis of N-substituted amides via the Ritter reaction. *RSC Adv.* 3: 5332–5337.

Reddy, B.M., Thirupathi, B. and Patil, M.K. 2009. One-pot synthesis of substituted coumarins catalyzed by silica gel supported sulfuric acid under solvent-free conditions. *Open Catal. J.* 2: 33–39.

Rostamnia, S., Lamei, K., Mohammadquli, M., Sheykhan, M. and Heydari, A. 2012. An efficient, fast, and reusable green catalyst for the Ugi-like Groebke–Blackburn–Bienayme three-component reaction under solvent-free conditions. *Tetrahedron Lett.* 53(39): 5257–5260.

Roy, B. and Mukhopadhyay, B. 2007. Sulfuric acid immobilized on silica: An excellent catalyst for Fischer type glycosylation. *Tetrahedron Lett.* 48(22): 3783–3787.

Sadeghi, B. and Nejad, M.G. 2013. Silica sulfuric acid: An eco-friendly and reusable catalyst for synthesis of benzimidazole derivatives. *J. Chem.* 2013:Article ID 581465, 5 pages.

Salehi, P., Dabiri, M., Zolfigol, M.A. and Baghbanzadeh, M. 2005. A new approach to the facile synthesis of mono- and disubstituted quinazolin-4(3*H*)-ones under solvent-free conditions. *Tetrahedron Lett.* 46(41): 7051–7053.

Salehi, P., Dabiri, M., Zolfigol, M.A. and Fard, M.A.B. 2003. Silica sulfuric acid: An efficient and reusable catalyst for the one-pot synthesis of 3,4-dihydropyrimidin-2(1*H*)-ones. *Tetrahedron Lett.* 44(14): 2889–2891.

Salehi, P., Dabiri, M., Zolfigol, M.A. and Fard, M.A.B. 2004. Silica sulfuric acid as an efficient and reusable reagent for crossed-aldol condensation of ketones with aromatic aldehydes under solvent-free conditions. *J. Braz. Chem. Soc.* 15(5): 773–776.

Salehi, P., Dabiri, M., Zolfigol, M.A., Otokesh, S. and Baghbanzadeh, M. 2006. Selective synthesis of 2-aryl-1-arylmethyl-1*H*-1,3-benzimidazoles in water at ambient temperature. *Tetrahedron Lett.* 47(15): 2557–2560.

Seyyed Hamzeh, M., Mirzaei, P. and Bazgir, A. 2008. Solvent-free synthesis of aryl-14*H*-dibenzo[*a,j*]xanthenes and 1,8-dioxo-octahydro-xanthenes using silica sulfuric acid as catalyst. *Dyes Pig.* 76: 836–839.

Shaabani, A. and Rahmati, A. 2006. Silica sulfuric acid as an efficient and recoverable catalyst for the synthesis of trisubstituted imidazoles. *J. Mol. Catal. A Chem.* 249(1–2): 246–248.

Shaabani, A., Rahmati, A., Farhangi, E. and Badri, Z. 2007. Silica sulfuric acid promoted the one-pot synthesis of trisubstituted imidazoles under conventional heating conditions or using microwave irradiation. *Catal. Commun.* 8(7): 1149–1152.

Shaabani, A. and Rezayan, A.H. 2007. Silica sulfuric acid promoted selective oxidation of sulfides to sulfoxides or sulfones in the presence of aqueous H_2O_2. *Catal. Commun.* 8(7): 1112–1116.

Shaabani, A., Soleimani, E. and Badri, Z. 2006. Silica sulfuric acid as an inexpensive and recyclable solid acid catalyzed efficient synthesis of quinolines. *Chem. Mont.* 137: 181–184.

Shaterian, H.R., Ghashang, M. and Feyzi, M. 2008. Silica sulfuric acid as an efficient catalyst for the preparation of 2*H*-indazolo [2,1-*b*]phthalazine-triones. *Appl. Catal. A Gen.* 345: 128–133.

Sheldon, R.A., Arends, I. and Hanefeld, U. 2007. *Green Chemistry and Catalysis*. Weinheim: Wiley-VCH, pp. 1–198.

Shirini, F., Sadeghzadeh, P. and Abedini, M. 2009. Silica sulfuric acid: A versatile reagent for oxathioacetalyzation of carbonyl compounds and deprotection of 1,3-oxathiolanes. *Chin. Chem. Lett.* 20(12): 1457–1460.

Shirini, F., Zolfigol, M.A. and Mohammadi, K. 2004. Silica sulfuric acid as a mild and efficient reagent for the acetylation of alcohols in solution and under solvent free conditions. *Bull. Korean Chem. Soc.* 25(2): 325–327.

Shobha, D., Chari, M.A., Mukkanti, K. and Kim, S.Y. 2012. Synthesis and anti-neuroinflammatory activity studies of substituted 3,4-dihydroquinoxalin-2-amine derivatives. *Tetrahedron Lett.* 53(22): 2675–2679.

Tamaddon, F. and Farahi, M. 2012. A new three-component reaction catalyzed by silica sulfuric acid: Synthesis of tetrasubstituted pyrroles. *Synlett* 23: 1379–1383.

Tamaddon, F., Kargar-Shooroki, H. and Jafari, A.A. 2013. Molybdate and silica sulfuric acids as heterogeneous alternatives for synthesis of gem-bisamides and bisurides from aldehydes and amides, carbamates, nitriles or urea. *J. Mol. Catal. A Chem.* 368–369: 66–71.

Thirunarayanan, G. and Vanangamudi, G. 2006. Synthesis of some 4-bromo-1-naphthyl chalcones using silica-sulfuric acid reagent under solvent free conditions. *Arkivoc* (xii): 58–64.

Veisi, H. 2010. Silica sulfuric acid (SSA) as a solid acid heterogeneous catalyst for one-pot synthesis of substituted pyrroles under solvent-free conditions at room temperature. *Tetrahedron Lett.* 51(16): 2109–2114.

Wang, Y., Yuan, Y-Q. and Guo, S-R. 2009. Silica sulfuric acid promotes aza-Michael addition reactions under solvent-free condition as a heterogeneous and reusable catalyst. *Molecules* 14: 4779–4789.

Wu, L., Yang, X. and Yan, F. 2011. Silica sulfuric acid: A versatile and reusable heterogeneous catalyst for the synthesis of *N*-acyl carbamates and oxazolidinones under solvent-free conditions. *Bull. Chem. Soc. Ethiop.* 25(1): 151–155.

Yahyazadeh, A., Hadesinea, S. and Daneshmandi, M.S. 2012. Silica sulfuric acid as an efficient and reusable heterogeneous catalyst for the synthesis of formamidines from the diaminomaleonitrile. *Iran. J. Catal.* 2(4): 153–156.

Zolfigol, M.A., Bagherzadeh, M., Mallakpour, S., Chehardoli, G., Kolvari, E., Choghamarani, A.G. and Koukabi, N. 2007. Mild and heterogeneous oxidation of urazoles to their corresponding triazolinediones via *in situ* generation Cl^+ using silica sulfuric acid/$KClO_3$ or silica chloride/oxone system. *Catal. Commun.* 8(3): 256–260.

Zolfigol, M.A., Chehardoli, G., Dehghanian, M., Niknam, K., Shirinid, F. and Khoramabadi-Zad, A. 2008. Silica sulfuric acid and $Al(HSO_4)_3$: As efficient catalysts for the formylation of alcohols by using ethyl formate under heterogeneous conditions. *J. Chin. Chem. Soc.* 55: 885–889.

Zolfigol, M.A., Shirini, F., Choghamarani, A.G. and Mohammadpoor-Baltork, I. 2002. Silica modified sulfuric acid/$NaNO_2$ as a novel heterogeneous system for the oxidation of 1,4-dihydropyridines under mild conditions. *Green Chem.* 4: 562–564.

6 Application of Silica-Based Heterogeneous Catalysis for the Synthesis of Bioactive Heterocycles

Laishram Ronibala Devi and
Okram Mukherjee Singh

CONTENTS

6.1 INTRODUCTION

Heterogeneous catalysts have gained much importance in recent years due to economic and environmental considerations. These catalysts are advantageous over homogeneous catalysts as they can be easily recovered from the reaction mixture by simple filtration and can be reused after activation, thereby making the process economically viable. There has been a tremendous upsurge of interest in various chemical transformations promoted by catalysts under heterogeneous conditions (Smith and Notheisz, 2000; Santen and Neurock, 2006). Generally, reactions with these catalysts are clean and selective and give high yields of products. In these days of growing environmental concerns, heterogeneous catalysts have attracted much attention as catalysts for organic synthesis.

Recently, the use of silica-based heterogeneous catalysts has been reported in a number of organic transformations (Das et al., 2001; Karimi and Khalkhali, 2005; Shylesh et al., 2004; Wilson et al., 2002). The attractive features of this catalyst include (a) its surface is both thermally and chemically stable during the reaction process, (b) it is an abundant and inexpensive material, (c) it is easy to handle, (d) it has low toxicity, (e) it has a noncorrosive nature, (f) it is air tolerant, (g) it is simple to separate on completion of the reaction, and (h) it is reusable.

6.2 SYNTHESIS OF BIOACTIVE HETEROCYCLES

6.2.1 Dihydropyrimidine Synthesis

Dihydropyrimidinones (DHPMs) are found to exhibit a wide spectrum of biological activities (Singh et al., 1999; Fu et al., 2002; Atwal et al., 1991; Janis and Triggle, 1983; Godfraid et al., 1986). They also show remarkable pharmacological properties such as calcium channel antagonists and are used extensively as therapeutics in the clinical treatment of cardiovascular diseases such as hypertension, cardiac arrhythmias, and angina pectoris (Jauk et al., 1999; Rovynak et al., 1995; Atwal et al., 1990). Furthermore, they have been found to exhibit a wide spectrum of antiviral, antitumor, antibacterial, and anti-inflammatory activities (Patil et al., 1995; Snider et al., 1996).

Salehi et al. (2003) reported silica sulfuric acid (SSA) as an efficient catalyst for the three-component Biginelli reaction between an aldehyde, β-dicarbonyl compound, and urea or thiourea in refluxing ethanol to afford the corresponding DHPMs in high yields (Scheme 6.1). Several aromatic aldehydes carrying either electron-releasing or electron-withdrawing substituents in the ortho-, meta-, and para-positions afforded high yields of the products. An important feature of this procedure is the survival of

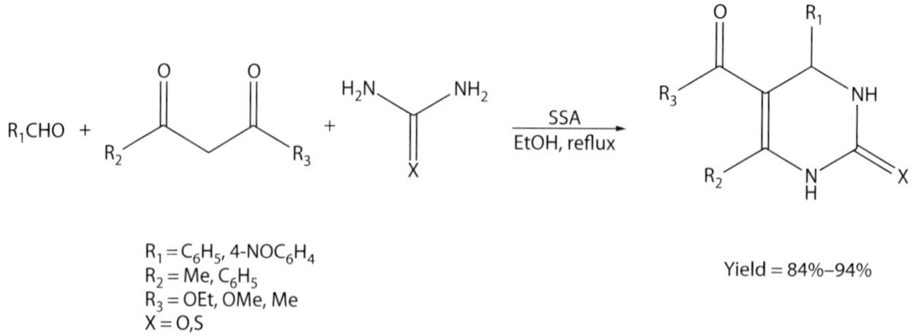

SCHEME 6.1

a variety of functional groups such as ethers, nitro groups, hydroxy groups, halides, and so on under the reaction conditions. Another advantage of this method is its efficiency for the high-yield synthesis of DHPMs from aliphatic aldehydes. Moreover, this method provides milder reaction conditions, easy workup, high yields, stability, and recyclability of the reagent.

Silica triflate efficiently catalyzed the one-pot synthesis of 3,4-dihydropyrimidin-2-(1*H*)-ones (DHPMs) and -thiones under solvent-free reaction conditions in good to high yields (Scheme 6.2). Several aliphatic and aromatic aldehydes having either electron-releasing or electron-withdrawing substituents were also afforded high yields of the products. This method offers several advantages such as mild reaction conditions, good to high yields of the products, shorter reaction time, easy purification, and an environmentally benign and simple experimental procedure (Shirini et al., 2007).

Chen et al. (2007) demonstrated a very simple, efficient, and practical method for the synthesis of DHPMs and thioderivatives through a one-pot, three-component condensation of aldehydes, 1,3-dicarbonyl compounds, and urea or thiourea catalyzed by SSA under solvent-free conditions (Scheme 6.3). The main features of this method are its operational simplicity, good yields, use of cheap, commercially available chemicals, recyclability with the comparable activity of the catalyst, and that it is cost effective and environmentally benign.

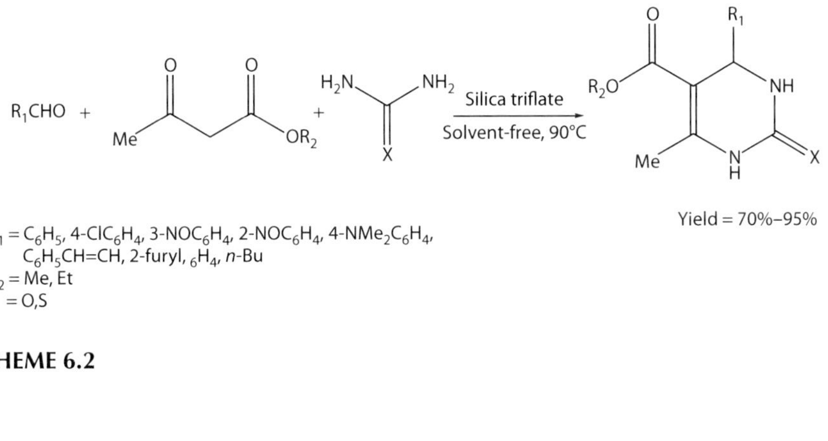

$R_1 = C_6H_5$, 4-ClC_6H_4, 3-NOC_6H_4, 2-NOC_6H_4, 4-NMe$_2$$C_6H_4$, C_6H_5CH=CH, 2-furyl, $_6H_4$, *n*-Bu
R_2 = Me, Et
X = O,S

SCHEME 6.2

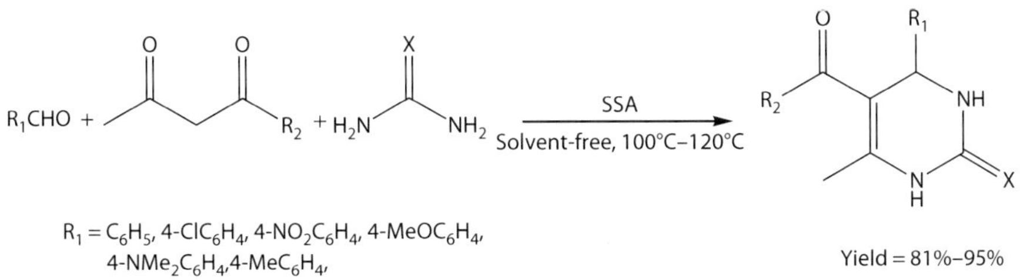

$R_1 = C_6H_5$, 4-ClC_6H_4, 4-NO$_2$$C_6H_4$, 4-MeO$C_6H_4$, 4-NMe$_2$$C_6H_4$,4-Me$C_6H_4$,
R_2 = Me, OC_2H_5
X = O, S

SCHEME 6.3

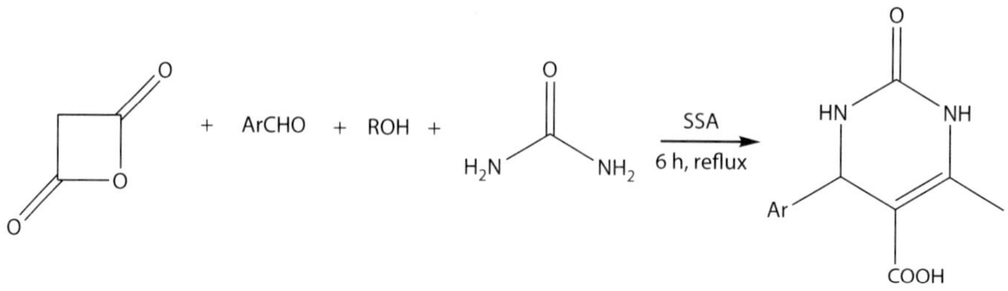

Ar = C_6H_5, 4-ClC_6H_4, 3-ClC_6H_4, 2-ClC_6H_4, 4-$NO_2C_6H_4$
R = Me, Et

Yield = 80%–91%

SCHEME 6.4

Rostamnia and Lamei (2012) have described a simple SSA-catalyzed one-pot synthesis of 3,4-dihydropyrimidin-2-(1*H*)-one derivatives via a four-component cyclocondensation reaction of diketene, alcohols, and aldehydes with urea (Scheme 6.4). The present method has several advantages, such as the reaction can be performed under one-pot neat conditions, and, in addition to the aldehyde component, the alcohols component can be modified to synthesize derivatives. The easy approach and the variability in derivatives in the presence of a heterogeneous catalyst make it an alternative to the three-component approaches.

Rafiee et al. (2009) reported silica-supported 12-tungstophosphoric (PW) acid as an efficient catalyst for the synthesis of 1,4-dihydropyridines (DHPs) under solvent-free conditions (Scheme 6.5). This catalyst has the advantages of an easy catalyst separation from the reaction medium and fewer problems with corrosion. The recycling of the catalyst and the avoidance of using harmful organic solvents are other advantages of this method. As a consequence, an eco-friendly method was developed for the preparation of DHP derivatives, which is important for the pharmaceuticals industries.

Maheswara et al. (2006) have developed a simple $HClO_4$–SiO_2-catalysed one-pot synthesis of 1,4-DHPs by Hantzsch condensation of aliphatic and heterocyclic

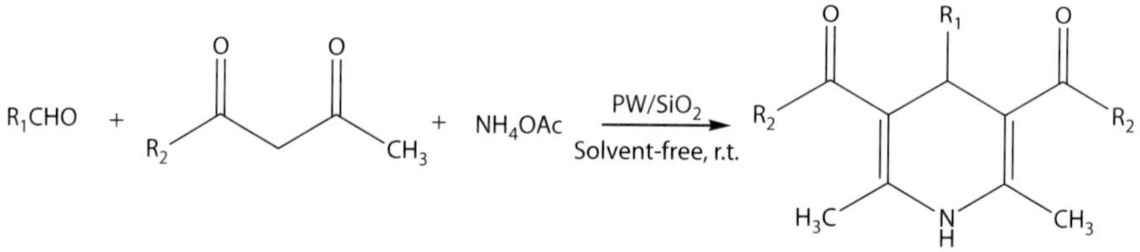

R_1 = C_6H_5, 4-ClC_6H_4, 4-MeC_6H_4, 4-$MeOC_6H_4$, 4-$NMe_2C_6H_4$, 2-furyl, 2-thionyl, $C_6H_5CH{=}CHCH_3$
R_2 = OMe, OEt, C_6H_5

Yield = 55%–94%

SCHEME 6.5

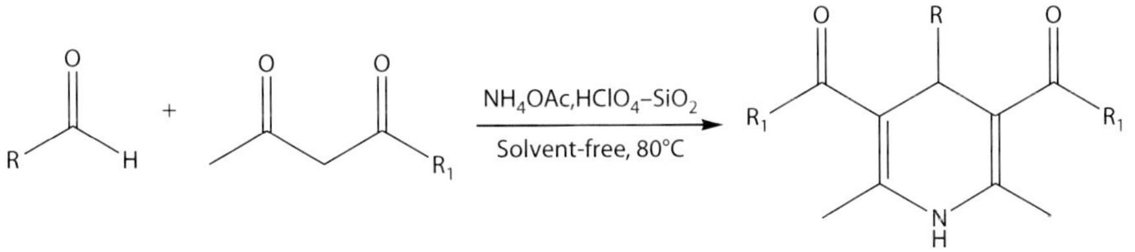

R = C_6H_5, 4-ClC_6H_4, 4-MeC_6H_4, 4-$MeOC_6H_4$, 4-$NMe_2C_6H_4$,
2-furyl, 2-thienyl, 3-pyridyl, 4-OHC_6H_5, 2-MeC_6H_4, 2-$MeOC_6H_4$,
4-BrC_6H_4, 2-BrC_6H_4, 4-MeC_6H_4, 4-MeC_6H_4, 4-MeC_6H_4
R_1 = OMe, OEt, C_6H_5

Yield = 82%–95%

SCHEME 6.6

aldehydes with ammonium acetate under solvent-free conditions (Scheme 6.6). This method offers simple workup, short reaction times, high yields, and a reusable catalyst.

Siddiqui and Farooq (2012) developed an improved methodology by using silica-supported sodium bisulfate ($NaHSO_4$–SiO_2) as a mild, highly efficient, and recyclable heterogeneous catalyst for the synthesis of pyranyl pyridine derivatives by the reaction of β-enaminone with active methylene compounds under thermal solvent-free conditions in excellent yields (Scheme 6.7). The present methodology offers very attractive features such as excellent yields of the product in a shorter reaction time, a simple workup procedure, economic viability, and reusability of the catalyst.

A simple, rapid, and efficient method for the preparation of benzoxazoles, benzimidazoles, and oxazolo[4,5-*b*]pyridines by the reaction of ortho esters with *o*-aminophenols, *o*-phenylenediamine, and 2-amino-3-hydroxypyridine in the presence of SSA under heterogeneous and solvent-free conditions is reported (Scheme 6.8). The significant features of this method are short reaction times, high yields of the products, mild reaction conditions, solvent-free reactions, cheapness, nontoxicity, and reusability of the catalyst (Baltorka et al., 2008).

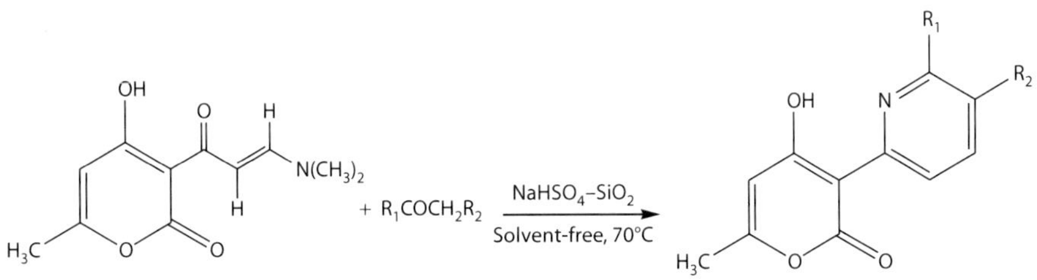

R_1 = CH_3
R_2 = $COCH_3$, $COOCH_3$

Yield = 81%–93%

SCHEME 6.7

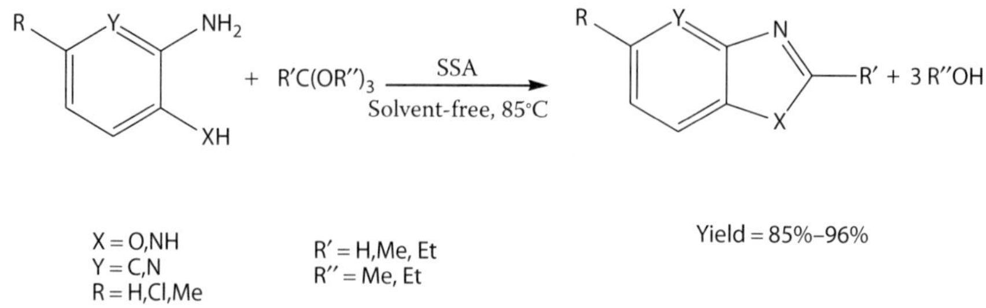

SCHEME 6.8

6.2.2 Benzimidazole Synthesis

The benzimidazole moiety plays important roles in numerous bioactive compounds. Biologically active benzimidazoles have been known for a long time and they can act as bacteriostats or bactericides (Nguyen et al., 2005; Kazimierczuk et al., 2002). Moreover, compounds containing a benzimidazole ring were found to have antifungal (Ayhan-Kilcigil and Altanlar, 2006), antitubercular (Mohamed et al., 2006), antioxidant (Kus et al., 2004), antiallergic (Nakano et al., 1999), and antiparasitic (Valdez et al., 2002) activities.

Sajjadifar et al. (2012) reported silica boron sulfonic acid (SBSA) as an efficient catalyst for the one-pot green synthesis of benzimidazole derivatives at room temperature (Scheme 6.9). Various substituted benzimidazoles were synthesized by the combination of *o*-phenylenediamines and aldehydes in the presence of SBSA with good yields in water by stirring at room temperature. This method offers several advantages such as high yields, shorter reaction times, nontoxic cost efficiency providing recyclability of the catalyst, and it is environmentally benign with simple experimental and workup procedures. A wide variety of aromatic and α,β-unsaturated aldehydes having both electron-donating and electron-withdrawing groups and substituted *o*-phenylenediamine react to give the corresponding benzimidazole in good yields.

Kumar et al. (2013) reported $NaHSO_4$–SiO_2 as an efficient catalyst for the formation of benzimidazole derivatives. The use of this inexpensive, easily available, and

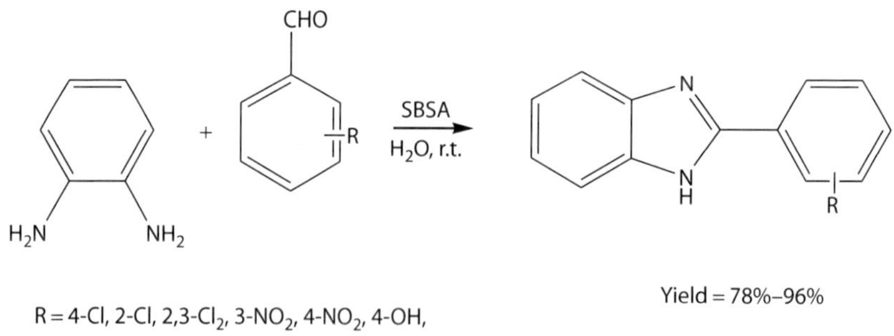

SCHEME 6.9

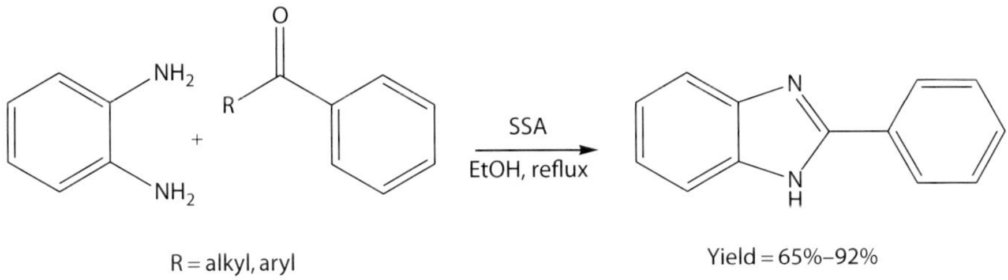

SCHEME 6.10

reusable catalyst makes this protocol practical, environment friendly, and economically attractive. The simple workup procedure, high yields of products, and nontoxic nature of the catalyst are other advantages of this method.

Sadeghi and Nejad (2013) reported SSA as an efficient catalyst for the synthesis of benzimidazole derivatives under reflux in ethanol (Scheme 6.10). The procedure is very simple and the products are isolated with an easy workup in good to excellent yields.

Paul and Basu (2012) developed an efficient and highly selective synthesis of functionalized 1,2-benzimidazoles under solvent-free conditions at ambient temperature using eco-friendly ferric sulfate soaked with silica ($Fe_2(SO_4)_3$–SiO_2). The reaction is illustrated in Scheme 6.11. This new method requires only inexpensive, eco-friendly reagents and the recyclable catalyst.

Iravani et al. (2011) reported a silica-supported catalyst as an efficient catalyst for the synthesis of 2-aryl-1-arylmethyl-1*H*-1,3-benzimidazole derivatives. The method involved the reaction of *o*-phenylenediamine and aromatic aldehydes in the presence of sulfuric acid {[3-(3-silicapropyl) sulfanyl]propyl}ester (SASPSPE) in water at 80°C (Scheme 6.12). Aryl aldehydes without a substituent give the corresponding benzimidazoles in good yields while aromatic aldehydes bearing an electron-donating substituent give the corresponding products in a shorter reaction times (20–25 min) with very good to excellent yields and aromatic aldehydes having electron-withdrawing substituents such as 4-cyano- and 4-fluorobenzaldehyde give the corresponding benzimidazoles in very good yields, whereas alkyl aldehydes such

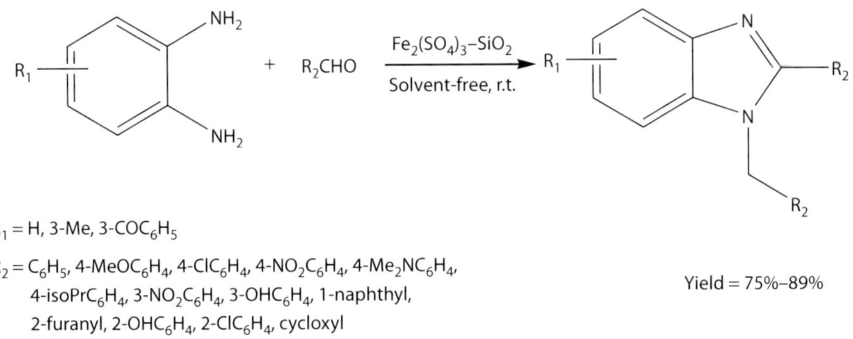

SCHEME 6.11

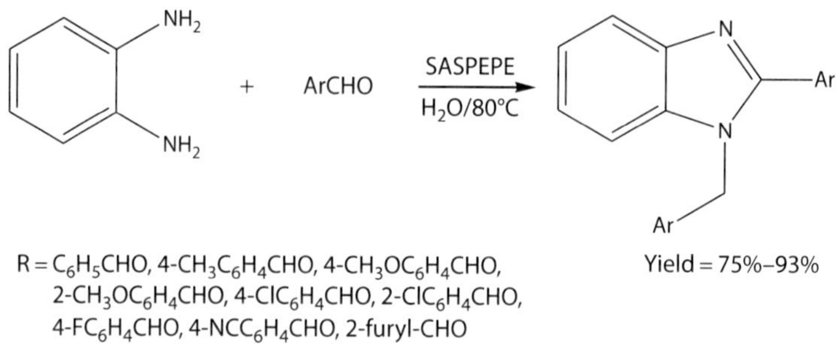

SCHEME 6.12

as butanal and octanal with *o*-phenylenediamine fail. These heterogeneous conditions offer good to excellent yields, a green solvent, and a simple and clean workup.

6.2.3 Imidazole Synthesis

Multisubstituted imidazoles, an important class of pharmaceutical compounds, exhibit a wide spectrum of biological activity (Lombardino and Wiseman, 1974). The tetrasubstituted imidazole core exists in many biological systems such as losartan and olmesartan and in some natural products and pharmacologically active compounds (Wolkenberg et al., 2004). Shaabani and Rahmati (2006) developed an environmentally benign approach for the synthesis of biologically active trisubstituted imidazoles by using SSA as an efficient heterogeneous catalyst. The reactions involved the condensation of 1,2-diketone, α-hydroxy ketone, and α-keto oxime with various aromatic aldehydes and ammonium acetate in water (Scheme 6.13).

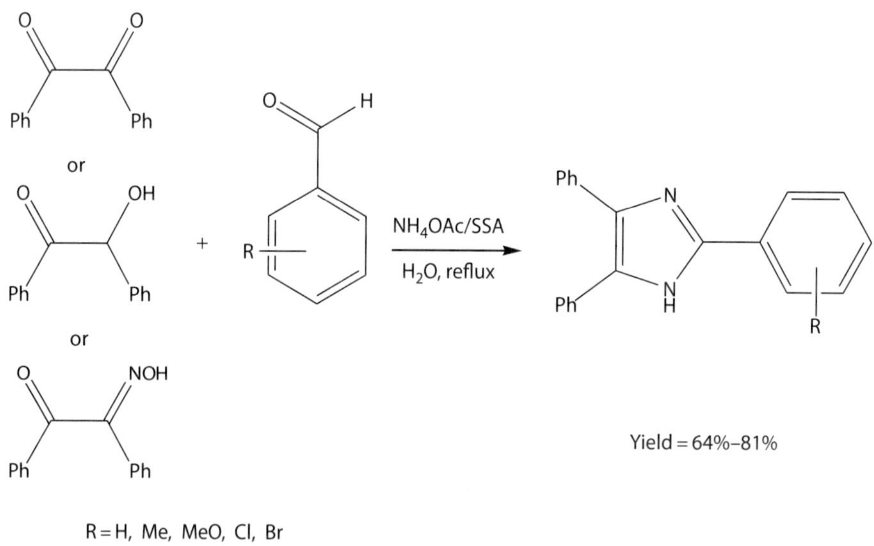

SCHEME 6.13

Aromatic aldehydes bearing either electron-releasing or electron-withdrawing substituents in the ortho-, meta-, or para-positions proceeded very efficiently, except with 4-nitrobenzaldehyde. The reactions offer noncorrosiveness, safety, low waste, ease of separation, and high yields.

Balalaie et al. (2003) reported a one-pot, three-component condensation of benzil, benzonitrile derivatives, and primary amines on the surface of silica gel with acidic character under microwave irradiation as a new and efficient method to produce 1,2,4,5-tetrasubstituted imidazoles (Scheme 6.14). This methodology offers several advantages, such as solvent-free conditions, the use of substances without any modification or activation, high yields, shorter reaction times, and reusability of solid catalysts, and it is environmentally benign compared to the existing methodologies.

Sadeghi et al. (2011) reported silica-supported antimony pentachloride ($SbCl_5 \cdot SiO_2$) as an efficient catalyst for the synthesis of 1,2,4,5-tetrasubstituted imidazoles. The reaction involved a one-pot, four-component reaction of aldehyde, amine, benzyl, and ammonium acetate under reflux conditions in ethanol with a catalytic amount of $SbCl_5 \cdot SiO_2$ to afford the corresponding 1,2,4,5-tetrasubstituted imidazoles in improved yields (Scheme 6.15). The products can also be obtained

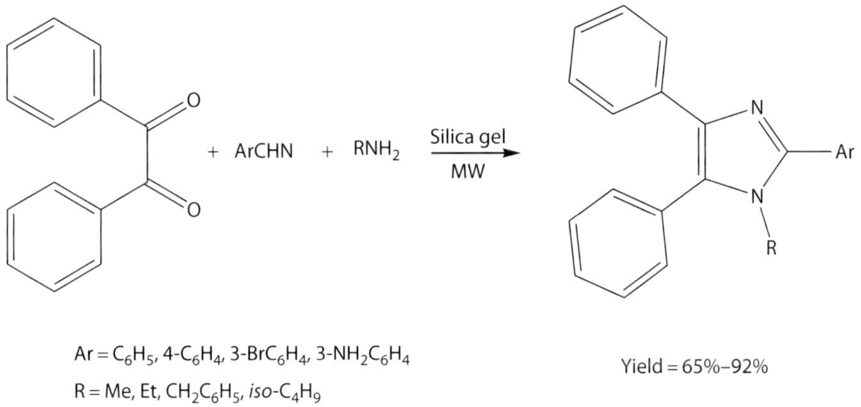

SCHEME 6.14

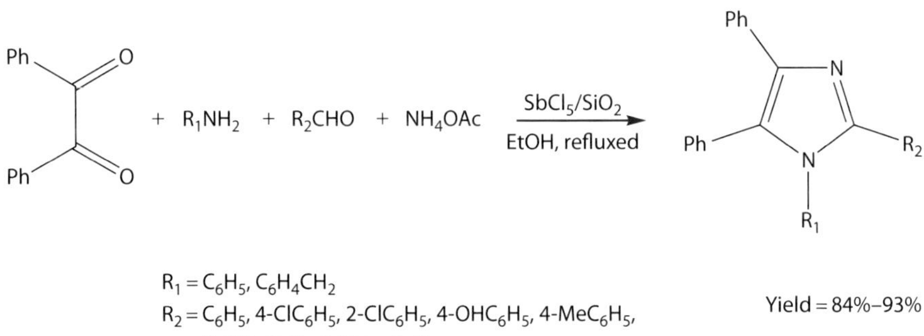

SCHEME 6.15

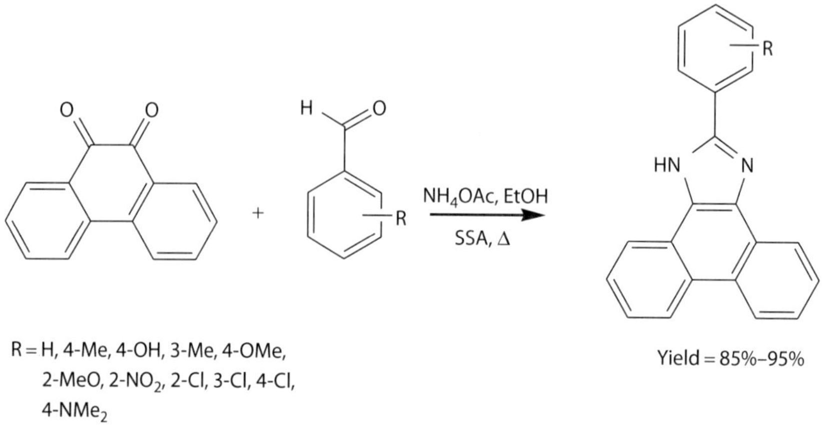

SCHEME 6.16

under solvent-free conditions. The catalyst is recoverable by simple filtration and can be used in subsequent reactions.

Behmadi et al. (2009) described a new, convenient, efficient, and cost-effective one-pot synthesis of 1*H*-phenanthro[9,10]imidazol-2-yl from phenanthraquinone and aldehydes, using sulfuric acid immobilized on silica as the catalyst (Scheme 6.16). The present methodology offers several advantages such as excellent yields, a simple procedure, shorter reaction times, the use of inexpensive reagents, ease of recovery, and it is eco-friendly. In most cases, however, such reactions are carried out in AcOH or H_2SO_4 medium by refluxing for several hours. Obviously, such reaction conditions place a heavy burden on the environment and equipment as well as being a dangerous operation.

6.2.4 Quinazolinone Synthesis

The quinazolinone derivatives form an important class of heterocyclic compounds, as they possess a broad spectrum of biological activities. They exhibit useful therapeutic and pharmacological properties such as anti-inflammatory, anticonvulsant, antihypertensive, antimalarial, and antimicrobial activities (Pereira et al., 2007, 1951; Tamaoki et al., 2007). Mobinikhaledia et al. (2010) reported SSA as an efficient catalyst for the synthesis of octahydroquinazolinone derivatives. The reaction involved a one-pot, three-component reaction of dimedone, urea, or thiourea and the corresponding aromatic aldehydes in the presence of SSA in ethanol under reflux conditions (Scheme 6.17). Several methods have been reported for the synthesis of quinazolinone derivatives. However, most of these methods have limitations including harsh reaction conditions, low yields, use of expensive and hazardous chemicals, and tedious workup procedures. This protocol offers several advantages including good yields of products, easy experimental workup procedure, and reusability of the catalyst.

Salehi et al. (2005) described SSA as an efficient catalyst for the synthesis of quinazolin-4(3*H*)-one derivatives. The reaction involved a one-pot, three-component

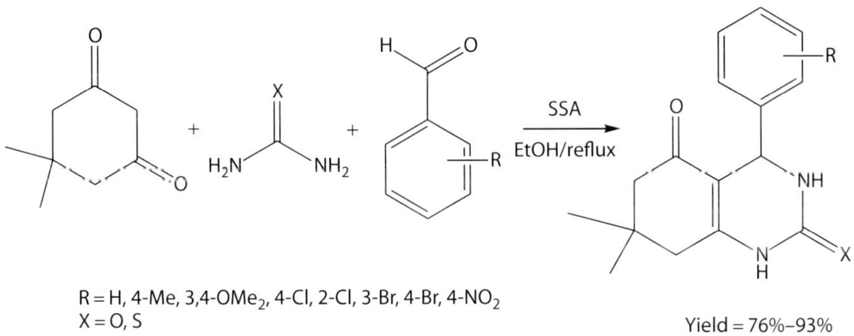

SCHEME 6.17

reaction of isatoic anhydride and an ortho ester with ammonium acetate or a primary amine under solvent-free conditions (Scheme 6.18). The most common approach toward quinazolinones involves the amidation of 2-aminobenzonitrile, 2-aminobenzoic acid, or their derivatives followed by oxidative ring closure under basic conditions (Taylor et al., 1960; Segarra et al., 1998). This approach suffers from low yields. Recently, the synthesis of 2-substituted quinazolin-4(3*H*)-ones based on the oxidative cyclization of anthranilamide with aldehydes was reported by Abdel-Jalil et al. (2004). This method also suffers from the use of a toxic catalyst. Thus, the reusability of the catalyst together with the elimination of toxic organic solvents makes the procedure green and environmentally friendly.

6.2.5 Quinoline Synthesis

Quinolines are important bioactive compound having pharmacological properties that are used as antimalarial (Sparatore et al., 2005), antiasthmatic (Dube et al., 1998), anti-inflammatory (Kalluraya and Sreevinasa, 1998), platelet-derived growth factor receptor tyrosine kinase (PDGF-RTK) inhibitor (Maguire et al., 1994), antitumor (Fryatt et al., 2004), and antiviral agents (Govindachari et al., 1974).

Khouzani et al. (2001) reported silica gel as an efficient catalyst for the synthesis of 2-ketomethylquinolines under solvent-free conditions using microwave irradiation. 2-Methylquinoline and 2,3-dimethylquinoline were reacted with acyl chlorides

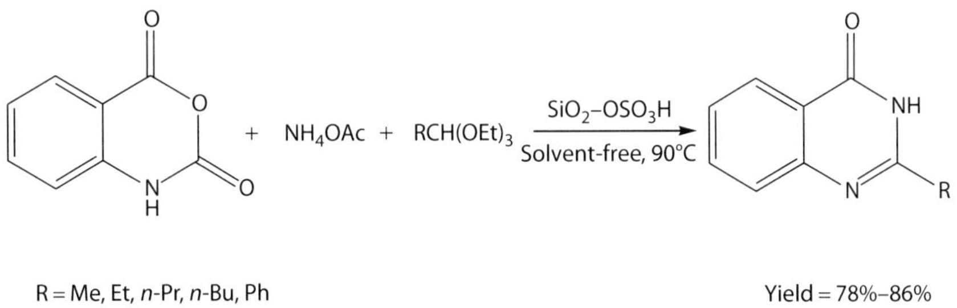

SCHEME 6.18

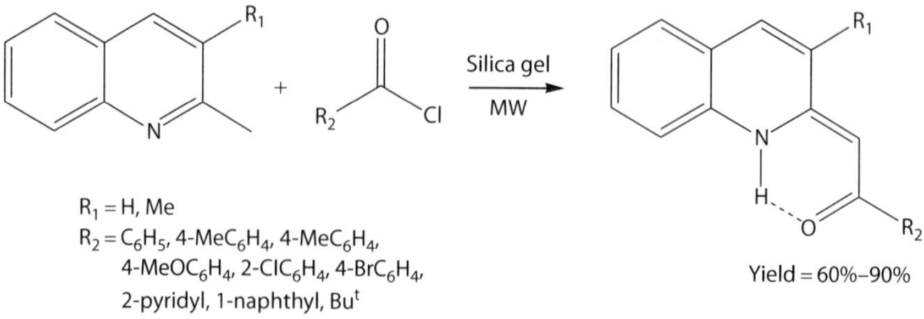

SCHEME 6.19

affording the desired 2-ketoquinolines (Scheme 6.19). This method offers high yields of the products, short reaction times, ease of workup conditions, and low cost.

Among the various synthetic methods in quinoline synthesis, Friedländer condensation is an extremely useful and versatile method for the direct construction of a quinoline ring (Friedländer, 1882; Ubeda et al., 1998). However, this condensation requires high temperature, high pressure, or long reaction conditions and hazardous solvents. Zolfigol et al. (2008) reported SSA as an efficient catalyst for the preparation of quinolines from *o*-amino aryl ketones and different ketones (including dialkyl cyclic ketones) under solvent-free conditions and microwave irradiation (Scheme 6.20). This condensation overcomes most of the mentioned disadvantages.

6.2.6 Oxadiazole Synthesis

The oxadiazole ring system is the main core of many bioactive molecules. Compounds containing this moiety have been shown to exert anti-inflammatory (Omar et al., 1996), antimicrobial (Ashour and Al Mazoroa, 1990), anticonvulsant (Tsitsa et al., 1989), and hypoglycemic (Mohamed, 1994) activities. Montazeri and Rad-Moghadam (2008) reported silica-supported sulfuric acid as an efficient catalyst for a one-pot synthesis of unsymmetrical 2,5-disubstituted-1,3,4-oxadiazoles via cyclocondensation of benzoylhydrazines with ortho esters under solvent-free and microwave conditions (Scheme 6.21). This methodology offers several advantages such as high yields, relatively short reaction times, a simple operation, and an easy workup procedure.

Dabiri et al. (2007) have developed a very simple, efficient, and eco-friendly synthesis of 2,5-disubstituted oxadiazoles by the reaction of different acyl hydrazides and ortho esters in the presence of SSA under solvent-free conditions at room

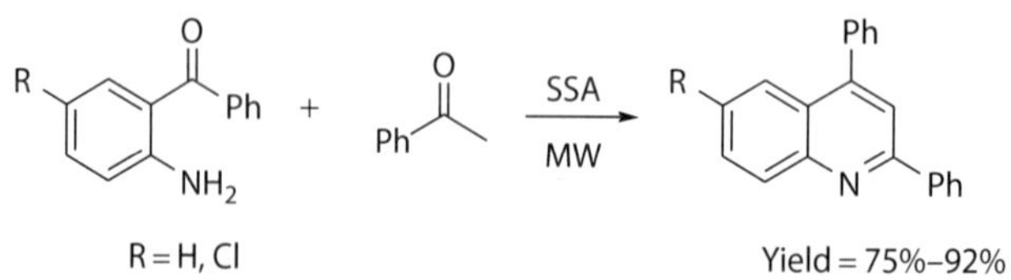

SCHEME 6.20

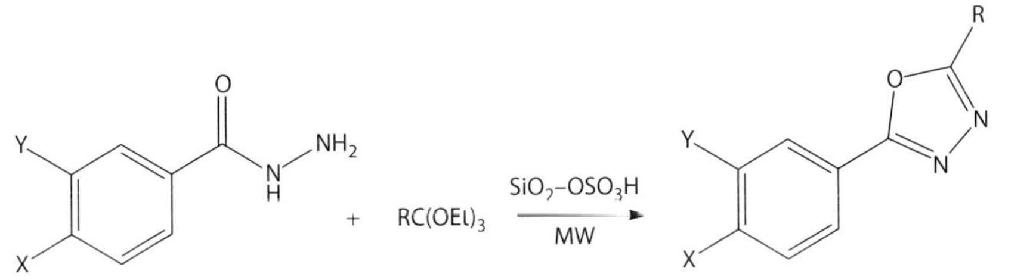

X = H or Cl
Y = H or NO_2
R = H, Me, Et, C_6H_5

Yield = 87%–94%

SCHEME 6.21

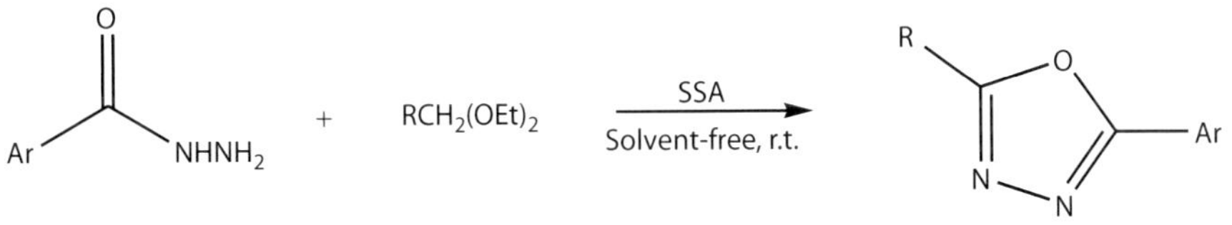

Ar = C_6H_5, 4-ClC_6H_4, 3-$NO_2C_6H_4$, 4-ClC_6H_4, 4-pyridyl
R = H, Me, Et, C_6H_5, $(CH_2)_2CH_3$, $(CH_2)_3CH_3$

Yield = 80%–95%

SCHEME 6.22

temperature (Scheme 6.22). This method offers high yields, relatively short reaction times, a simple operation, and an easy workup procedure, which are some of the advantages of this protocol.

6.2.7 Benzothiazole Synthesis

Benzothiazole and its derivatives are well known for their biological and pharmaceutical activities, such as antitumor activity (Bradshaw et al., 2002), and antimicrobial (Palmer et al., 1971) and antiglutamate/antiparkinsonism agents (Benazzouz et al., 1995). In addition, they represent one of the most promising antiamyloid therapies for the treatment of a number of a heterogeneous family of diseases referred to generically as amyloidosis, including Alzheimer's disease (AD), type II diabetes, variant Creutzfeldt-Jakob disease, painful joints associated with long-term hemodialysis, and rare cases of hereditary insomnia. Many methods are available for the synthesis of 2-arylbenzothiazoles. However, they suffer from drawbacks such as high thermal conditions, a long reaction time, and the use of acid or base catalysts and toxic metallic compounds that result in waste streams. Kodomari et al. (2004) reported an environmentally benign methodology for the synthesis of 2-substituted benzothiazoles using silica gel under microwave and solvent-free conditions (Scheme 6.23). The methodology leads to high yields, clean reactions, and shorter reaction times.

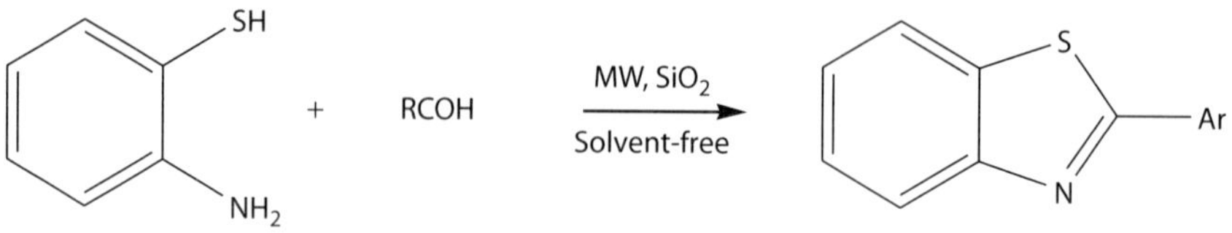

Ar = C_6H_5, 4-MeC_6H_5, 2-MeC_6H_5, 2-MeOC_6H_5,
4-OHC_6H_5, 2-OHC_6H_5, 4-ClC_6H_5, 2-ClC_6H_5,
4-pyridyl, 2-furanyl, 2-thionyl

Yield = 73%–94%

SCHEME 6.23

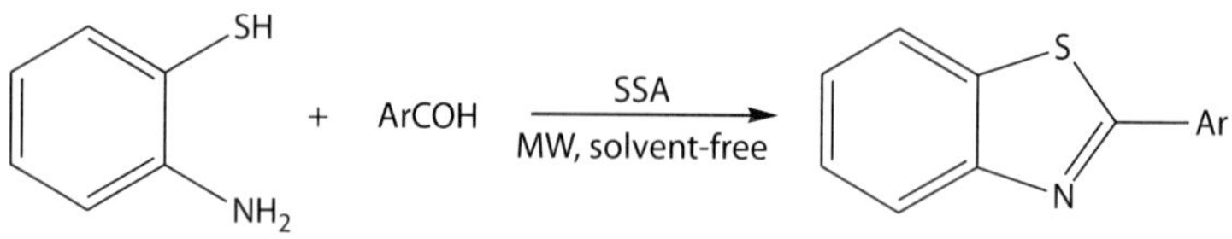

Yield = 87%–94%

Ar = 4-MeOC_6H_4, C_6H_5, 3-$NO_2C_6H_4$, 4-$NO_2C_6H_4$,2-MeOC_6H_4,
2-ClC_6H_4, 4-BrC_6H_4, 4-ClC_6H_4, 2-thienyl, 2-pyridyl

SCHEME 6.24

Niralwad et al. (2010) also reported the synthesis of 2-arylbenzothiazoles using SSA as an efficient catalyst (Scheme 6.24). The method involved the reaction of 2-aminothiophenol and several substituted aryl or heteroaryl aldehydes to afford the corresponding 2-arylbenzothiazole in good to excellent yields. The main advantages of the present synthetic protocol are mild, solvent-free conditions, an eco-friendly catalyst, and an easy reaction workup procedure.

6.2.8 Quinoxaline Synthesis

Quinoxaline and its derivatives constitute an important class of benzoheterocycles displaying a broad spectrum of biological activities, which have made them privileged structures in pharmacologically active compounds (Sakata et al., 1988; Seitz et al., 2002). Niknam et al. (2009) reported that silica-bonded *S*-sulfonic acid (SBSSA), prepared from commercially available and relatively cheap starting materials by a simple transformation, efficiently catalyzed the room temperature condensation reactions between 1,2-diamino compounds and 1,2-dicarbonyl compounds to give quinoxaline derivatives in good to excellent yield (Scheme 6.25). It could also be recovered and reused for more than 12 reaction cycles without noticeable loss of reactivity.

6.2.9 Pyrrole Synthesis

Pyrroles are an important class of heterocyclic compounds and are structural units found in a vast array of natural products, synthetic materials, and bioactive molecules,

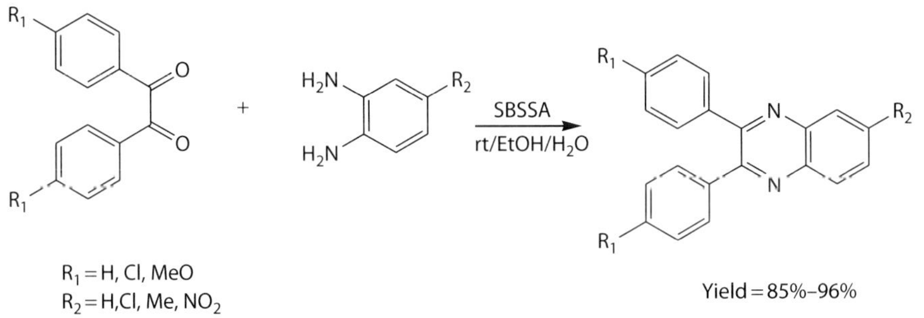

SCHEME 6.25

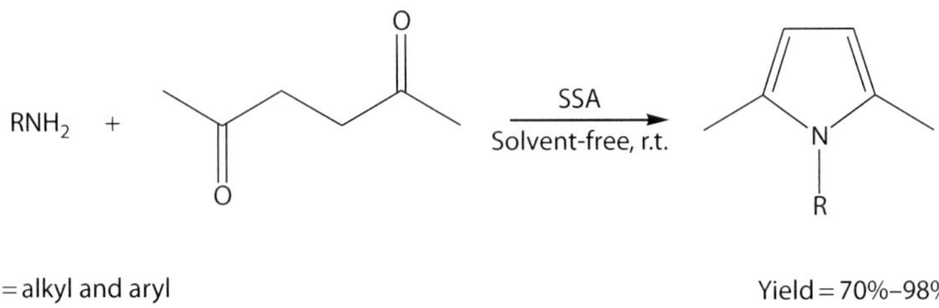

SCHEME 6.26

such as heme, vitamin B12, and cytochromes (Jones and Bean, 1977). Veisi (2010) reported SSA as an efficient heterogeneous catalyst for the one-pot synthesis of substituted pyrroles under solvent-free conditions at room temperature (Scheme 6.26). A variety of N-substituted pyrroles have been synthesized by reacting γ-diketones with amines, diamines, or triamines in the presence of SSA at room temperature under solvent-free conditions. The methodology offers simple operations and the products are isolated in high to excellent yields (70%–98%).

6.2.10 Pyrazole Synthesis

Pyrazoles are an important class of bioactive drug targets in the pharmaceutical industry, as they are the core structure of numerous biologically active compounds. They have been shown to possess important biological and pharmaceutical activities such as antianxiety, antipyretic, analgesic, antimicrobial, antiviral, antitumor, antifungal, antidepressant, anticonvulsant, and anti-inflammatory activities (McDonald et al., 2006; Elguero et al., 2002; Hatheway et al., 1978; Katayama and Oshiyama, 1997; Badawey and El-Ashmawey, 1998; Bailey et al., 1985). Silica-supported sulfuric acid has been utilized as a heterogeneous recyclable catalyst for the highly efficient regio- and chemoselective condensation of hydrazines/hydrazides, and primary amines with various β-dicarbonyl compounds at room temperature to afford pyrazoles (Scheme 6.27) (Chen et al., 2009). However, most of the methods to prepare these compounds suffer from certain drawbacks including long reaction times, unsatisfactory yields, higher temperatures, the employment of organic solvents, and

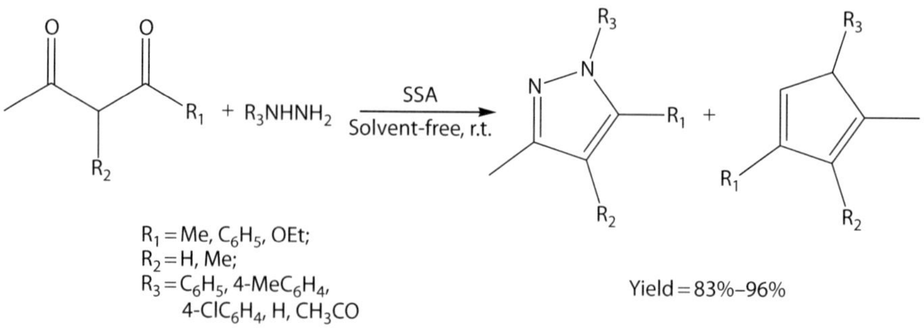

SCHEME 6.27

the use of an expensive nonreusable catalyst. Thus, there is still a need to develop greener and more efficient pathways for such synthesis.

Siddiqui and Farooq (2012) also developed an improved methodology by using silica-supported sodium bisulfate ($NaHSO_4$–SiO_2) as a mild, highly efficient, and recyclable heterogeneous catalyst for the synthesis of pyrazole by the reaction of β-enaminone with different hydrazines under a thermal solvent-free condition in excellent yields (Scheme 6.28). The present methodology offers very attractive features such as excellent yields of the product in a shorter reaction time, a simple workup procedure, and the economic viability and reusability of the catalyst. Thus, this method has provided better scope for the synthesis of pyrazole derivatives and will be a more practical alternative to the other existing methods.

6.2.11 Pyrazine, Thiazine, Piperidine, Indole, Phthalazine, Triazole, Dihydroquinolinone, Tetrazoline, and Acridine Synthesis

The imidazo[1,2-*a*]pyridine and imidazo[1,2-*a*]pyrazine moieties constitute a class of biologically active compounds. Shaabani et al. (2007) reported the synthesis of

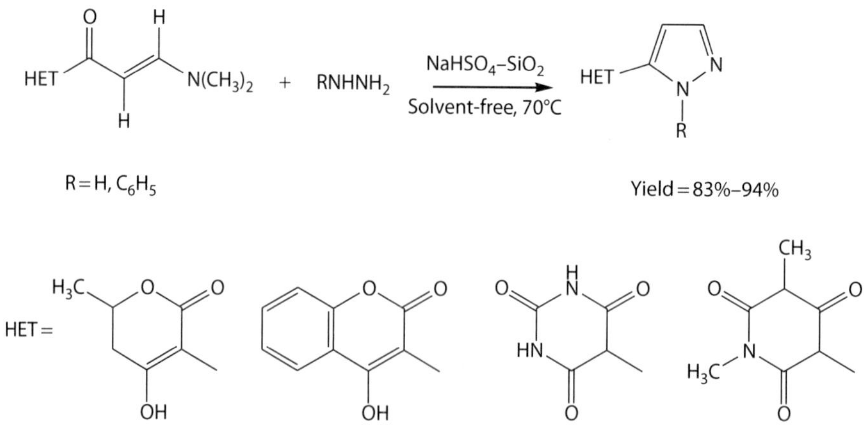

SCHEME 6.28

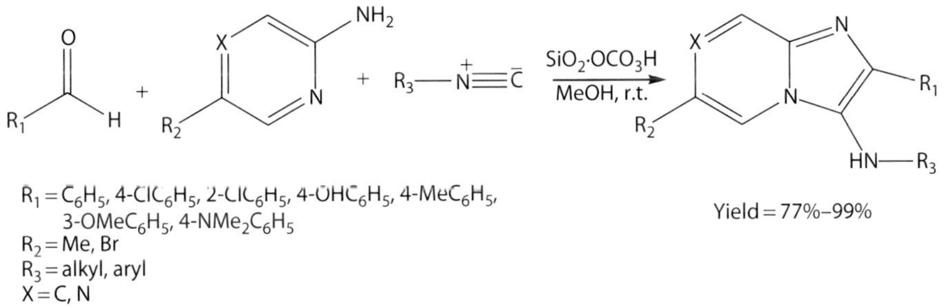

SCHEME 6.29

3-aminoimidazo[1,2-*a*]pyridines and 3-aminoimidazo[1,2-*a*] pyrazines through a condensation reaction of 2-aminopyridine or 2-aminopyrazine, aldehyde, and alkyl or arylisocyanide in high yields at room temperature in the presence of SSA (Scheme 6.29). Although many methods are available, they suffer from limitations such as the use of a corrosive reagent, a tedious workup procedure, neutralization of strong acid media producing undesired washes, and long reaction times.

1,4-Thiazine ring heterocycles play an important role in pigments, dyestuffs, and biologically active substances (Katritzky and Singh, 2002). Mihara et al. (2005) reported the synthesis of 3,4,5-triaryltetrahydro-1,4-thiazine derivatives by treating β,β-dichlorosulfides, 2:1 adducts of alkenes, and sulfur dichloride with aromatic amines in the presence of silica gel (Scheme 6.30). The silica gel can be easily recovered and reused in up to four consecutive reactions without any significant decrease in product yield.

The piperidine ring is present in many natural products such as alkaloids (Viegas et al., 2004), which are responsible for a number of unique activities including antihypertensive, anticonvulsant, and anti-inflammatory activities (Ho et al., 2001). Spiropiperidinyl compounds have attracted increasing interest as synthetic targets due to their important activity as pharmacophores in several biologically active compounds, mainly alkaloids (Kuramoto et al., 1996; Chou et al., 1996). Atar and Jeong (2013) reported a mild, efficient, and expeditious method for the synthesis of 3,5-dispirosubstituted piperidines via a three-component, one-pot cyclocondensation reaction of aromatic amines, formaldehyde, and dimedone using silica-supported tungstic acid (STA) as the heterogeneous catalyst for the first time (Scheme 6.31). The

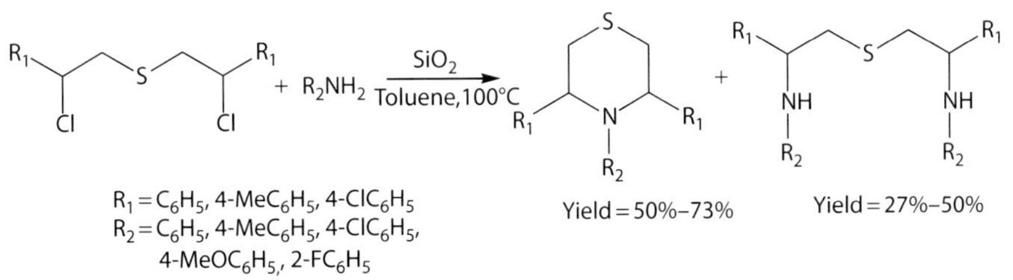

SCHEME 6.30

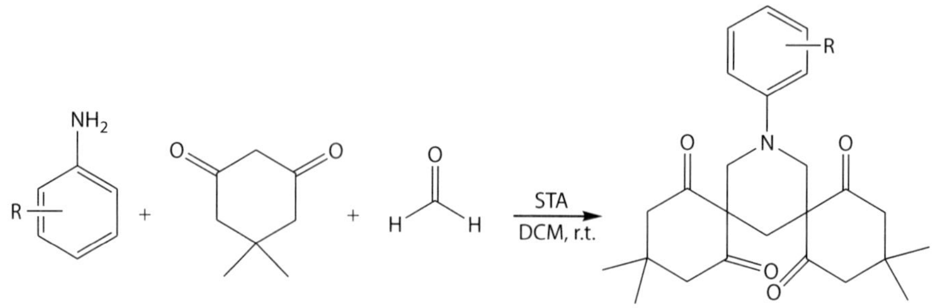

R = 4-F, 4-MeO, 2-Me, 3-Me, 4-Me, 2-MeO, 4-Cl, 3,4-(Me)$_2$, 3,5-(Me)$_2$, 4-Et, 4-OC$_6$H$_5$, 4-C$_6$H$_5$

Yield = 82%–92%

SCHEME 6.31

reaction involved the formation of six new covalent bonds conveniently promoted by STA; the catalyst could be recovered easily after the reaction and reused without any loss of its catalytic activity. The advantageous features of this methodology are its high atom economy, operational simplicity, shorter reaction time, convergence, and easy automation.

Indole derivatives exhibit various bioactivities such as anticancer, antibiotic, and anti-inflammatory activities (Gul and Hamann, 2005). Antioxidant activities were also recently reported for some analogs such as 2,2-diphenyl-1-picrylhydrazyl (DPPH) radical scavengers, highlighting an additional bioactivity in the series (Sugiyama et al., 2009). The solvent-free phosphomolybdic acid (PMA)–SiO_2-catalyzed synthesis of 3-substituted indole derivatives by a one-pot, three-component coupling reaction between aldehyde, *N*-methyl aniline, and indole is described (Scheme 6.32). This methodology offers several advantages over the existing procedures, such as simple reaction procedures, inexpensive catalysts, environmentally benignity, and single product formation (Srihari et al., 2009).

Phthalazine derivatives were reported to possess anticonvulsant (Grasso et al., 2000) and vasorelaxant (Watanabe et al., 1998) activities. Shaterian et al. (2008) described the synthesis of 2*H*-indazolo[2,1-*b*]phthalazine-1,6,11(13*H*)-trione

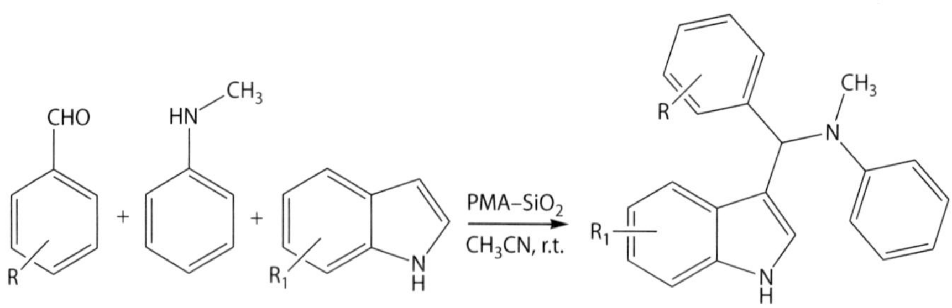

R = H, 4-CH$_3$, 4-OH, 3-OMe, 4-Cl, 4-NO$_2$
R$_1$ = H, 5-Br, 5-OMe

Yield = 85%–93%

SCHEME 6.32

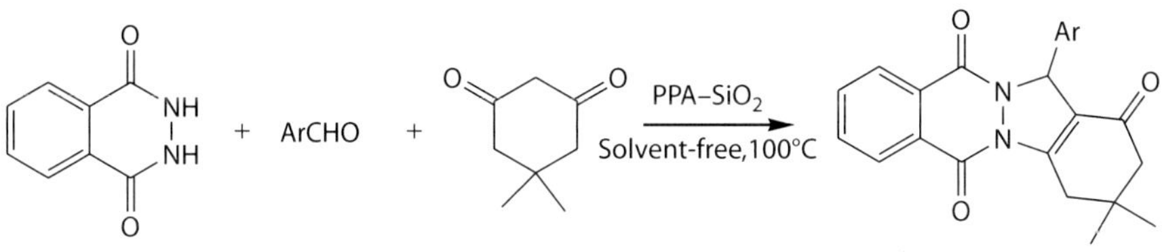

Yield = 80%–91%

Ar = C_6H_5, 4-MeC_6H_4, 4-$NO_2C_6H_4$, 3-$NO_2C_6H_5$,4-ClC_6H_4, 4-BrC_6H_4, 4-FlC_6H_4, 2-ClC_6H_4, C_6H_5

SCHEME 6.33

derivatives from the three-component condensation reaction of phthalhydrazide, dimedone, and aromatic aldehydes under solvent-free conditions in good to excellent yields and short reaction times using reusable silica-supported polyphosphoric acid (PPA-SiO_2) as the heterogeneous acid catalyst (Scheme 6.33). The methodology is safer than those using conventional catalysts such as H_2SO_4 and H_3PO_4 with respect to the amount, hazard, and reaction conditions. Also, the catalyst is safe, easy to handle, environmentally benign with fewer disposal problems and can be successfully recovered and recycled for at least five runs without significant loss of its activity.

Triazole-containing compounds exhibit various biological effects, such as antiviral (Moorhouse and Moses, 2008), antibacterial (Hou et al., 2012), antifungal (Agalave et al., 2011), and anticancer (Jordao et al., 2009). Heravia et al. (2006) developed an easy and eco-friendly method for the synthesis of thiazolo[3,2-*b*]1,2,4-triazoles using SSA as an efficient catalyst. This method involved the condensation of 3-mercapto-1,2,4-triazoles with allyl bromide. Keivanloo et al. (2013) also reported silica-supported zinc bromide as an efficient heterogeneous catalyst for the one-pot synthesis of 4,5-disubstituted-1,2,3-(NH)-triazoles. The reaction involved the cross-coupling of 1,3-dipolar cycloaddition with various acid chlorides, terminal alkynes, and sodium azide in the presence of silica-supported zinc bromide under aerobic conditions. The reaction is described in Scheme 6.34.

SSA is reported as an efficient catalyst and has been used for the preparation of 12-aryl-8,9,10,12-tetrhydro benzo[*a*]xanthene-11-one derivatives from the one-pot

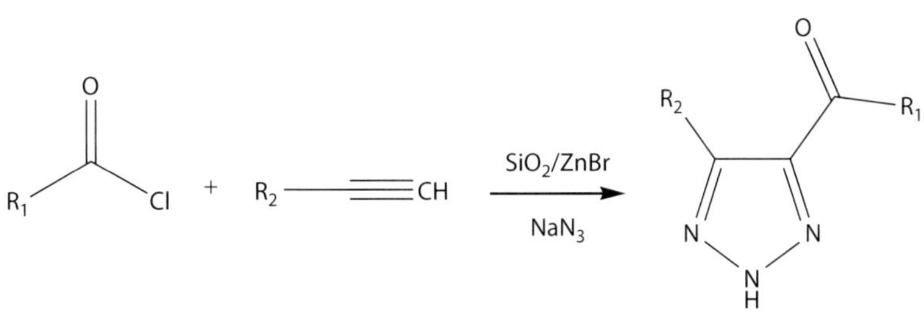

Yield = 87%–95%

R_1 = 4-BrC_6H_4, 2-ClC_6H_4, 4-ClC_6H_4, 2-MeC_6H_4, 2,4-$(Cl)_2C_6H_4$
R_2 = C_6H_5, propyl, hexyl

SCHEME 6.34

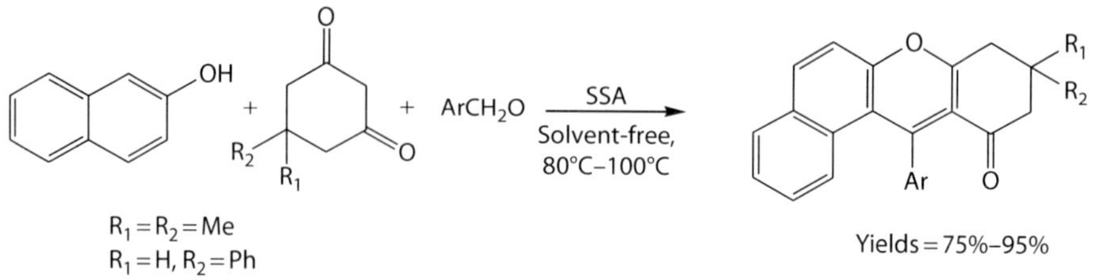

SCHEME 6.35

multicomponent reaction of aromatic aldehydes, β-naphthol, and a cyclic diketone under solvent-free conditions (Scheme 6.35) (Nemati et al., 2012). This methodology offers good yields, a simple procedure, an easy workup, short reaction times, and mild conditions.

Silica gel–supported $TaBr_5$ is an efficient catalyst for the synthesis of 2-aryl-2,3-dihydroquinolin-4(1*H*)-ones. The reaction takes place under solvent-free conditions for the easy and efficient isomerization of 20-aminochalcones to the corresponding 2-aryl-2,3-dihydroquinolin-4(1*H*)-ones (Scheme 6.36). The catalyst is easily prepared, stable, and employed under environmentally friendly conditions. The salient features of $TaBr_5$ impregnated on silica gel are rapid reaction rates, the absence of unwanted products, and improved and operational simplicity under conventional heating. This procedure has several advantages over earlier-reported features such as simplicity, fast and clean reactions, high yield, and the absence of an organic solvent (Ahmed and van Lier, 2006).

Earlier-reported methods for the synthesis of 5-substituted tetrazoles suffer from drawbacks such as the use of strong Lewis acids, or expensive and toxic metals, and the *in situ*–generated hydrazoic acid, which is highly toxic and explosive (Duncia et al., 1991; Sisido et al., 1971; Carini et al., 1991; Wittenberger and Donner, 1993). $FeCl_3$–SiO_2 is reported as an efficient and reusable heterogeneous catalyst for the synthesis of 5-substituted 1*H*-tetrazoles via [2+3] cycloaddition of nitriles and sodium azide (Scheme 6.37). The significant advantages of this methodology are high yields, the elimination of dangerous and harmful hydrazoic acid, a simple

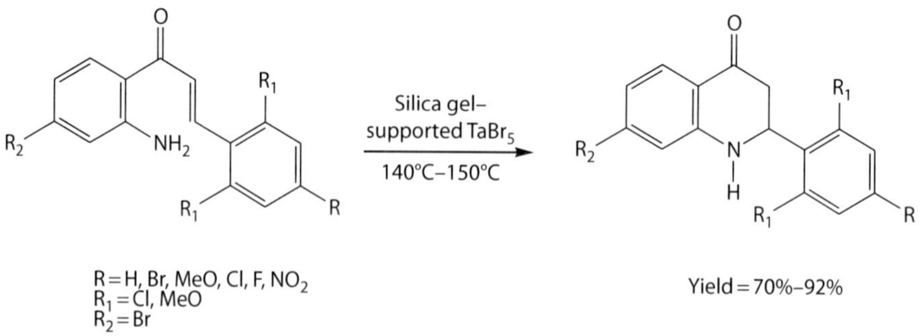

SCHEME 6.36

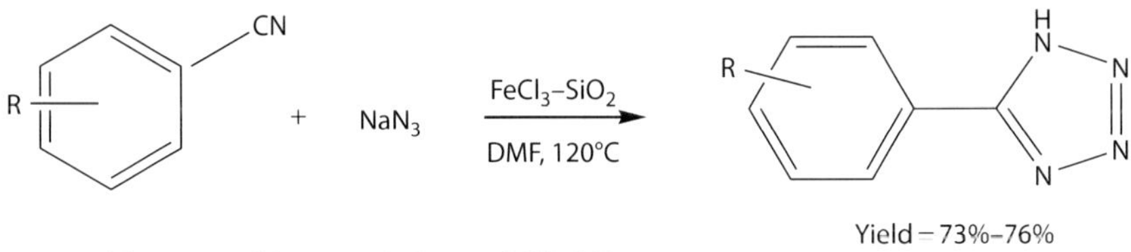

R = 2,5-$(Cl)_2C_6H_3$, 2-ClC_6H_4, 4-$NO_2C_6H_4$, 4-$OCH_3C_6H_4$, 2,4-$(CH_3)_2C_6H_3$, 4-$CH_3C_6H_4$

SCHEME 6.37

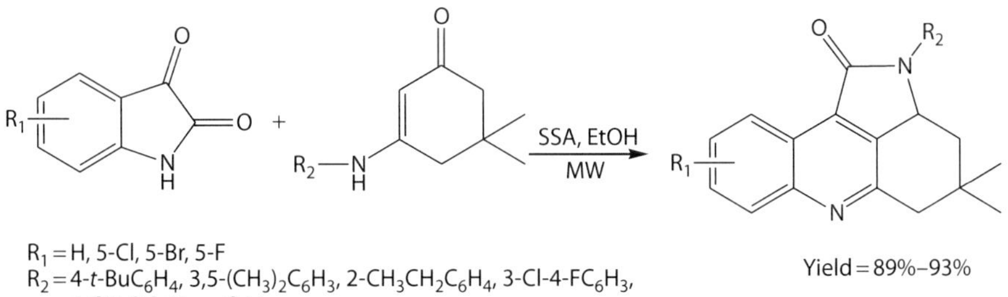

R_1 = H, 5-Cl, 5-Br, 5-F
R_2 = 4-*t*-BuC_6H_4, 3,5-$(CH_3)_2C_6H_3$, 2-$CH_3CH_2C_6H_4$, 3-Cl-4-FC_6H_3, 4-$CH_3OC_6H_4$, *n*-C_4H_9

SCHEME 6.38

workup procedure, and easy preparation and handling of the catalyst. The catalyst can be recovered by filtration and reused (Habibi and Nasrollahzadeh, 2010).

Acridine derivatives have important biological activity (Goodell et al., 2008), such as anticancer (Kapuriya et al., 2008), due to their ability to intercalate into DNA and disrupt unwanted cellular processes (Belmont and Dorange, 2008). They also possess significant biological activity toward viruses (Taraporewala, 1991), bacteria (Kavitha, 2004), parasites (Fattorusso et al., 2008; Giorgio et al., 2007), fungi (Patel et al., 2008), AD (Chauhan and Srivastava, 2001), and HIV/AIDS (Lee et al., 2008). Although several methods are available for the synthesis of these molecules (Gellerman et al., 1994; Kitahara et al., 2004), they require multiple steps and use toluene or acetic acid as solvents (Jiang et al., 2012). Cao et al. (2013) developed an improved synthesis of multifunctionalized pyrrolo[2,3,4-*kl*]acridine derivatives with different substituted patterns using SSA as a heterogeneous catalyst under microwave conditions. The reaction involved the use of readily available and inexpensive substrates within short periods of 12–15 min under microwave irradiation (Scheme 6.38). The methodology has remarkable advantages over conventional methods, such as milder reaction conditions, operational simplicity, higher yields, short reaction times, and an environmentally friendly procedure.

6.3 CONCLUSION

In this chapter, we have discussed the applications of SSA as an efficient heterogeneous catalyst for the synthesis of bioactive heterocycles. The catalyst is reusable, inexpensive, convenient, easy to handle, nontoxic, easily available, and efficient. We also believe that several unexplored, catalyzed, organic transformations can be performed using this catalyst, and its use has been growing rapidly in our laboratory.

ACKNOWLEDGMENTS

Financial supports from DBT, India (twinning project No. BT/NE/TBP/2011) and CSIR, India (project No. 01(2387)/10/EMR-II) are gratefully acknowledged.

REFERENCES

Abdel-Jalil, R.J., Volter, W. and Saeed, M. 2004. A novel method for the synthesis of 4(3*H*)-quinazolinones. *Tetrahedron Lett.* 45(17): 3475–3476.

Agalave, S.G., Maujan, S.R. and Pore, V.S. 2011. Click chemistry: 1,2,3-Triazoles as pharmacophores. *Chem. Asian J.* 6(10): 2696–2718.

Ahmed, N. and van Lier, J.E. 2006. Silica gel supported $TaBr_5$: New catalyst for the facile and rapid cyclization of 20-aminochalcones to the corresponding 2-aryl-2,3-dihydroquinolin-4(1*H*)-ones under solvent-free conditions. *Tetrahedron Lett.* 47(16): 2725–2729.

Ashour, F.A. and Al Mazoroa, S.A. 1990. Synthesis of some oxadiazole derivatives of benzimidazole as potential antimicrobial agents. *J. Pharm. Sci.* 4(1): 29–32.

Atar, A.B. and Jeong, Y.T. 2013. Silica supported tungstic acid (STA): An efficient catalyst for the synthesis of bis-spiro piperidine derivatives under milder condition. *Tetrahedron Lett.* 54(10): 1302–1306.

Atwal, K.S., Rovnyak, G.C., Kimball, S.D., Floyd, D.M., Moreland, S., Swanson, B.N., Gougoutas, J.Z., Schwarts, J., Smillie, K.M. and Malley, M.F. 1990. Dihydropyrimidine calcium channel blockers. 2. 3-Substituted-4-aryl-1,4-dihydro-6-methyl-5-pyrimidinecarboxylic acid esters as potent mimics of dihydropyridines. *J. Med. Chem.* 33(9): 2629–2635.

Atwal, K.S., Swanson, B.N., Unger, S.E., Floyd, D.M., Moreland, S., Hedberg, A. and O'Reilly, B.C. 1991. Dihydropyrimidine calcium channel blockers. 3. 3-Carbamoyl-4-aryl-1,2,3,4-tetrahydro-6-methyl-5-pyrimidinecarboxylic acid esters as orally effective antihypertensive agents. *J. Med. Chem.* 34(2): 806–811.

Ayhan-Kilcigil, G. and Altanlar, N. 2006. Synthesis and antifungal properties of some benzimidazole derivatives. *Turk. J. Chem.* 30: 223–228.

Badawey, E.A.M. and El-Ashmawey, I.M. 1998. Nonsteroidal antiinflammatory agents-Part 1: Antiinflammatory, analgesic and antipyretic activity of some new 1-(pyrimidin-2-yl)-3-pyrazolin-5-ones and 2-(pyrimidin-2-yl)-1,2,4,5,6,7-hexahydro-3*H*-indazol-3-ones. *Eur. J. Med. Chem.* 33(5): 349–361.

Bailey, D.M., Hansen, P.E., Hlavac, A.G., Baizman, E.R., Pearl, J., Defelice, A.F. and Feigenson, M.E. 1985. 3,4-Diphenyl-1*H*-pyrazole-1-propanamine antidepressants. *J. Med. Chem.* 28(2): 256–260.

Balalaie, S., Hashemib, M.M. and Akhbari, M. 2003. A novel one-pot synthesis of tetrasubstituted imidazoles under solvent-free conditions and microwave irradiation. *Tetrahedron Lett.* 44(8): 1709–1711.

Baltorka, I.M., Moghadama, M., Tangestaninejada, S., Mirkhania, V., Zolfigolb, M.A. and Hojatia, S.F. 2008. Silica sulfuric acid catalyzed synthesis of benzoxazoles, benzimidazoles and oxazolo[4,5-*b*]pyridines under heterogeneous and solvent-free conditions. *J. Iran. Chem. Soc.* 5: S65–S70.

Behmadi, H., Saadati, S.M., Roshani, M., Mohammadi, H., Razavi, A. and Ramezani, M. 2009. Sulfuric acid immobilized on silica: A versatile and reusable heterogeneous catalyst for the synthesis of phenanthrimidazole derivatives. *J. Iran. Chem. Res.* 2(3): 183–187.

Belmont, P. and Dorange, I. 2008. Acridine/acridone: A simple scaffold with a wide range of application in oncology. *Expert Opin. Ther. Pat.* 18(11): 1211–1224.

Benazzouz, A., Boraud, T., Dubédat, P., Boireau, A., Stutzmann, J.M. and Gross, C. 1995. Riluzole prevents MPTP-induced parkinsonism in the rhesus monkey: A pilot study. *Eur. J. Pharmacol.* 284(3): 299–307.

Bradshaw, T.D., Bibby, M.C., Double, J.A., Fichtner, I., Cooper, P.A., Alley, M.C., Donohue, S., et al. 2002. Preclinical evaluation of amino acid prodrugs of novel antitumor 2-(4-amino-3-methylphenyl)benzothiazoles. *Mol. Cancer Ther.* 1(4): 239–246.

Cao, C., Xu, C., Lin, W., Li, X., Hu, M., Wang, J., Huang, Z., Shi, D. and Wang, Y. 2013. Microwave-assisted improved synthesis of pyrrolo[2,3,4-*kl*]acridine and dihydropyrrolo[2,3,4-*kl*]acridine derivatives catalyzed by silica sulfuric acid. *Molecules* 18(2): 1613–1625.

Carini, D.J., Duncia, J.V., Aldrich, P.E., Chui, A.T., Johnson, A.L., Pierce, M.E., Price, W.A., et al. 1991. Nonpeptide angiotensin II receptor antagonists: The discovery of a series of *N*-(biphenylylmethyl)imidazoles as potent, orally active antihypertensives. *J. Med. Chem.* 34(8): 2525–2547.

Chauhan, P.M.S. and Srivastava, S.K. 2001. Present trends and future strategy in chemotherapy of malaria. *Curr. Med. Chem.* 8(13): 1535–1542.

Chen, W.-Y., Qin, S.-D. and Jin, J.-R. 2007. Efficient Biginelli reaction catalyzed by sulfamic acid or silica sulfuric acid under solvent-free conditions. *Synth. Commun.* 37(1): 47–52.

Chen, X., She, J., Shang, Z.-C., Wu, J. and Zhang, P. 2009. Room-temperature synthesis of pyrazoles, diazepines, β-enaminones, and β-enamino esters using silica-supported sulfuric acid as a reusable catalyst under solvent-free conditions. *Synth. Commun.* 39(6): 947–957.

Chou, T., Kuramoto, M., Otani, Y., Shikano, M., Yazawa, K. and Uemura, D. 1996. Pinnaic acid and tauropinnaic acid: Two novel fatty acids composing a 6-azaspiro[4.5]decane unit from the Okinawan bivalve *Pinna muricata. Tetrahedron Lett.* 37(22): 3871–3874.

Dabiri, M., Salehi, P. and Baghbanzadeh, M. 2007. Silica sulfuric acid: An efficient and versatile acidic catalyst for the rapid and ecofriendly synthesis of 1,3,4-oxadiazoles at ambient temperature. *Synth. Commun.* 37(7): 1201–1209.

Das, D., Lee, J.-F. and Cheng, S. 2001. Sulphonic acid functionalized mesoporous MCM-41 silica as a convenient catalyst for bisphenol-A synthesis. *Chem. Commun.* (21): 2178–2179.

Dube, D., Blouin, M., Brideau, C., Chan, C., Desmarais, S., Ethier, D., Falgueyret, J.P., et al. 1998. Quinolines as potent 5-lipoxygenase inhibitors: Synthesis and biological profile of L-746,530. *Bioorg. Med. Chem. Lett.* 8(10): 1255–1260.

Duncia, J.V., Pierce, M.E. and Santella, J.B. III. 1991. Three synthetic routes to a sterically hindered tetrazole. A new one-step mild conversion of an amide into a tetrazole. *J. Org. Chem.* 56(7): 2395–2400.

Elguero, J., Goya, P., Jagerovic, N. and Silva, A.M.S. 2002. Pyrazoles as drugs: Facts and fantasies. *Targets Heterocycl. Syst.* 6: 52–98.

Fattorusso, C., Campiani, G., Kukreja, G., Persico, M., Butini, S., Romano, M.P., Altarelli, M., et al. 2008. Design, synthesis and structure-activity relationship studies of 4-quinolinyl- and 9-acrydinylhydroazones as potent antimalarial agents. *J. Med. Chem.* 51(5): 1333–1343.

Friedländer, P. 1882. Ueber *o*-Amidobenzaldehyd. *Chem. Ber.* 15: 2572.

Fryatt, T., Pettersson, H.I., Gardipee, W.T., Bray, K.C., Green, S.J., Slawin, A.M.Z., Beall, H.D. and Moody, C.J. 2004. Novel quinolinequinone antitumor agents: Structure-metabolism studies with NAD(P)H:quinone oxidoreductase (NQO1). *Bio. Med. Chem.* 12: 1667–1687.

Fu, N.Y., Yuan, Y.F., Cao, Z., Wang, S.W., Wang, J.T. and Peppe, C. 2002. Indium(III) bromide-catalyzed preparation of dihydropyrimidinones: Improved protocol conditions for the Biginelli reaction. *Tetrahedron* 58(24): 4801–4807.

Gellerman, G., Rudi, A. and Kashman, Y. 1994. The biomimetic synthesis of marine alkaloid related pyrido- and pyrrolo[2,3,4-*kl*]acridines. *Tetrahedron* 50(45): 12959–12972.

Giorgio, C.D., Shimi, K., Boyer, G., Delmas, F. and Galy, J.P. 2007. Synthesis and antileishmanial activity of 6-mono-substituted and 3,6-di-substituted acridines obtained by acylation of proflavine. *Eur. J. Med. Chem.* 42(10): 1277–1284.

Godfraid, T., Miller, R. and Wibo, M. 1986. Calcium antagonism and calcium entry blockade. *Pharmocol. Rev.* 38: 321–416.

Goodell, J.R., Ougolkov, A.V., Hiasa, H., Kaur, H., Remmel, R., Billadeau, D.D. and Ferguson, D.M. 2008. Acridine-based agents with topoisomerase II activity inhibit pancreatic cancer cell proliferation and induce apoptosis. *J. Med. Chem.* 51(2): 179–182.

Govindachari, T.R., Ravindranath, K.R. and Viswanathan, N. 1974. Mappicine, a minor alkaloid from *Mappia foetida* miers. *J. Chem. Soc. Perkin Trans. 1* 11: 1215–1217.

Grasso, S., DeSarro, G., Micale, N., Zappala, M., Puia, G., Baraldi, M. and Demicheli, C. 2000. Synthesis and anticonvulsant activity of novel and potent 6,7-methylenedioxyphthalazin-1(2*H*)-ones. *J. Med. Chem.* 43(15): 2851–2859.

Gul, W. and Hamann, M.T. 2005. Indole alkaloid marine natural products: An established source of cancer drug leads with considerable promise for the control of parasitic, neurological and other diseases. *Life Sci.* 78(5): 442–453.

Habibi, D. and Nasrollahzadeh, M. 2010. Silica-supported ferric chloride ($FeCl_3$–SiO_2): An efficient and recyclable heterogeneous catalyst for the preparation of arylaminotetrazoles. *Synth. Commun.* 40(21): 3159–3167.

Hatheway, G.J., Hansch, C., Kim, K.H., Milstein, S.R., Schimidt, C.L., Smith, R.N. and Quin, F.R. 1978. Antitumor 1-(X-aryl)-3,3-dialkyltriazenes. 1. Quantitative structure-activity relationships vs. L1210 leukemia in mice. *J. Med. Chem.* 21(6): 563–574.

Heravia, M.M., Khademalfogharab, H.R., Sadeghia, M.M., Khaleghia, Sh. and Ghassemzadehc, M. 2006. Solvent-free regioselective cyclization of 3-allylmercapto-1,2,4-triazoles to thiazolo[3,2-*b*]1,2,4-triazoles over sulfuric acid adsorbed on silica gel. *Phosphorus Sulfur Silicon Relat. Elem.* 181(2): 377–380.

Ho, B., Crider, A.M. and Stables, J.P. 2001. Synthesis and structure–activity relationships of potential anticonvulsants based on 2-piperidinecarboxylic acid and related pharmacophores. *Eur. J. Med. Chem.* 36(3): 265–286.

Hou, J.L., Liu, X.F., Shen, J., Zhao, G.L. and Wang, P.G. 2012. The impact of click chemistry in medicinal chemistry. *Expert. Opin. Drug Discov.* 7(6): 489–501.

Iravani, N., Mohammadzade, N.S. and Niknam, K. 2011. Sulfuric acid {[3-(3-silicapropyl)sulfanyl]propyl}ester a recyclable catalyst for the synthesis of 2-aryl-1-arylmethyl-1*H*-1,3-benzimidazole derivatives. *Chin. Chem. Lett.* 22(10): 1151–1154.

Janis, R.A. and Triggle, D.J. 1983. New developments in Ca^{2+} channel antagonists. *J. Med. Chem.* 26(6): 775–785.

Jauk, B., Belaj, F. and Kappe, C.O. 1999. Synthesis and reactions of Biginelli-compounds. Part 14. A rhodium-induced cyclization-cycloaddition sequence for the construction of conformationally rigid calcium channel modulators of the dihydropyrimidine type. *J. Chem. Soc. Perkin Trans. 1.* (3): 307–314.

Jiang, B., Wang, X., Li, M.Y., Wu, Q., Ye, Q., Xu, H.W. and Tu, S.J. 2012. A domino synthetic strategy leading to two-carbon-tethered fused acridine/indole pairs and fused aceidine derivatives. *Org. Biomol. Chem.* 10(41): 8533–8538.

Jones, R.A. and Bean, G.P. 1977. *The Chemistry of Pyrroles*. Academic Press: London.

Jordao, A.K., Ferreira, V.F., Lima, E.S., de Souza, M.C.B.V., Carlos, E.C.L., Castro, H.C., Geraldo, R.B., Rodrigues, C.R., Almeida, M.C.B. and Cunha, A.C. 2009. Synthesis, antiplatelet and in silico evaluations of novel N-substituted phenylamino-5-methyl-1*H*-1,2,3-triazole-4-carbohydrazides. *Bioorg. Med. Chem.* 17(10): 3713–3719.

Kalluraya, B. and Sreevinasa, S. 1998. Synthesis and pharmacological properties of some quinoline derivatives. *Farmaco* 53(6): 399–404.

Kapuriya, N., Kapuriya, K., Zhang, X., Chou, T.C., Kakadiya, R., Wu, Y.T., Tsai, T.H., et al. 2008. Synthesis and biological activity of potent antitumor agents, aniline nitrogen mustards linked to 9-anilinoacridines via a urea linkage. *Bioorg. Med. Chem.* 16(10): 5413–5423.

Karimi, B. and Khalkhali, M. 2005. Solid silica-based sulfonic acid as an efficient and recoverable interphase catalyst for selective tetrahydropyranylation of alcohols and phenols. *J. Mol. Catal. A Chem.* 232(1–2): 113–117.

Katayama, H. and Oshiyama, T. 1997. Preparation and bioactivity of pyrazole derivatives as potential cross-linking agent. *Can. J. Chem.* 75(6): 913–919.

Katritzky, A.R. and Singh, S.K. 2002. Synthesis of C-carbamoyl-1,2,3-triazoles by microwave-induced 1,3-dipolar cycloaddition of organic azides to acetylenic amides. *J. Org. Chem.* 67(25): 9077–9079.

Kavitha, H.P. 2004. Synthesis and antimicrobial activity of 1-(9′-acridinyl)-5-(4-substituted phenyl)tetrazoles. *Asian J. Chem.* 16(2): 1191–1193.

Kazimierczuk, Z., Upcroft, J.A., Upcroft, P., Gorska, A., Starosciak, B. and Laudy, A. 2002. Synthesis, antiprotozoal and antibacterial activity of nitro- and halogeno-substituted benzimidazole derivatives. *Acta Biochim. Polon.* 49(1): 185–195.

Keivanloo, A., Bakherad, M., Taheri, S.A.N. and Samangooei, S. 2013. One-pot synthesis of 4,5-disubstituted 1,2,3-(NH)-triazoles by silica supported-zinc bromide in the aerobic condition. *C. R. Chim.* 16(3): 239–243.

Khouzani, H.L., Sadeghi, M.M., Safari, J. and Minaeifar, A. 2001. A novel method for the synthesis of 2-ketomethylquinolines under solvent-free conditions using microwave irradiation. *Tetrahedron Lett.* 42(26): 4363–4364.

Kitahara, Y., Mizuno, T. and Kubo, A. 2004. Synthetic studies of benzo[*b*]pyrrolo[4,3,2-*de*][1,10]phenanthroline. *Tetrahedron* 60(19): 4283–4288.

Kodomari, M., Tamaru, Y. and Aoyama, T. 2004. Solvent-free synthesis of 2-aryl and 2-alkylbenzothiazoles on silica gel under microwave irradiation. *Synth. Commun.* 34(16): 3029–3036.

Kumar, K.R., Satyanarayana, P.V.V. and Reddy, B.S. 2013. $NaHSO_4$–SiO_2 promoted solvent-free synthesis of benzoxazoles, benzimidazoles, and benzothiazole derivatives. *J. Chem.* 2013 (151273).

Kuramoto, M., Tong, C., Yamada, K., Chiba, T., Hayashi, Y. and Uemura, D. 1996. Halichlorine, an inhibitor of VCAM-1 induction from the marine sponge *Halichondria okadai* Kadata. *Tetrahedron Lett.* 37(22): 3867–3870.

Kus, C., Ayhan-Kilcigil, G., Can-Eke, B. and Iscan, M. 2004. Synthesis and antioxidant properties of some novel benzimidazole derivatives on lipid peroxidation in the rat liver. *Arch. Pharm. Res.* 27(2): 156–163.

Lee, Y., Hyun, S., Kim, H.J. and Yu, J. 2008. Amphiphilic helical peptides containing two acridine moieties display picomolar affinity toward HIV-1 RRE and TAR. *Angew. Chem. Int. Ed.* 47(1): 134–137.

Lombardino, J.G. and Wiseman, E.H. 1974. Preparation and antiinflammatory activity of some nonacidic trisubstituted imidazoles. *J. Med. Chem.* 17(11): 1182–1188.

Maguire, M.P., Sheets, K.R., McVety, K., Spada, A.P. and Zilberstein, A. 1994. A new series of PDGF receptor tyrosine kinase inhibitors: 3-Substituted quinoline derivatives. *J. Med. Chem.* 37(14): 2129–2137.

Maheswara, M., Siddaiah, V., Rao, Y.K., Tzeng, Y.-M. and Sridhar, C. 2006. A simple and efficient one-pot synthesis of 1,4-dihydropyridines using heterogeneous catalyst under solvent-free conditions. *J. Mol. Catal. A Chem.* 260(1–2): 179–180.

McDonald, E., Jones, K., Brough, P.A., Drysdale, M.J. and Workman, P. 2006. Discovery and development of pyrazole-scaffold Hsp90 inhibitors. *Curr. Top. Med. Chem.* 6(11): 1193–1203.

Mihara, M., Ito, T., Ishino, Y., Oderaotoshi, Y., Minakatab, S. and Komatsu, M. 2005. Silica gel-promoted synthesis of 3,4,5-triaryltetrahydro-1,4-thiazine derivatives from β,β′-dichloro sulfides and aromatic amines. *Tetrahedron Lett.* 46(47): 8105–8108.

Mobinikhaledia, A., Foroughifara, N. and Khodaeia, H. 2010. Synthesis of octahydroquinazolinone derivatives using silica sulfuric acid as an efficient catalyst. *Eur. J. Chem.* 1(4): 291–293.

Mohamed, B.G., Hussein, M.A., Abdel-Alim, A.M. and Hashem, M. 2006. Synthesis and antimicrobial activity of some new 1-alkyl-2-alkylthio-1,2,4-triazolobenzimidazole derivatives. *Arch. Pharm. Res.* 29(1): 26–33.

Mohamed, N.A. 1994. Novel wholly aromatic polyamide-hydrazides. Part V: Structure-molecular-weight-thermal-stability relationships. *Polym. Degrad.* 44(1): 33–42.

Montazeri, N. and Rad-Moghadam, K. 2008. An expeditious and one-pot synthesis of unsymmetrical 2,5-disubstituted-1,3,4-oxadiazoles under microwave irradiation and solvent-free conditions. *Chin. Chem. Lett.* 19(10): 1143–1146.

Moorhouse, A.D. and Moses, J.E. 2008. Click chemistry and medicinal chemistry: A case of "cyclo-addiction". *Chem. Med. Chem.* 3(5): 715–723.

Nakano, H., Inoue, T., Kawasaki, N., Miyataka, H., Matsumoto, H., Taguchi, T., Inagaki, N., Nagai, H. and Satoh, T. 1999. Synthesis of benzimidazole derivatives as antiallergic agents with 5-lipoxygenase inhibiting action. *Chem. Pharm. Bull.* 47(11): 1573–1578.

Nemati, F., Arghan, M. and Amoozadeh, A. 2012. Efficient, solvent-free method for the one-pot condensation of β-naphthol, aromatic aldehydes, and cyclic 1,3-dicarbonyl compounds catalyzed by silica sulfuric acid. *Synth. Commun.* 42(1): 33–39.

Nguyen, P.T.M., Baldeck, J.D., Olsson, J. and Marquis, R.E. 2005. Antimicrobial actions of benzimidazoles against oral streptococci. *Oral Microbiol. Immunol.* 20(2): 93–100.

Niknam, K., Saberi, D. and Mohagheghnejad, M. 2009. Silica bonded *S*-sulfonic acid: A recyclable catalyst for the synthesis of quinoxalines at room temperature. *Molecules* 14(5): 1915–1926.

Niralwad, K.S., Shingate, B.B. and Shingare, M.S. 2010. Solid-phase synthesis of 2-arylbenzothiazole using silica sulfuric acid under microwave irradiation. *Bull. Korean Chem. Soc.* 31(4): 981–983.

Omar, F.A., Mahfouz, N.M. and Rahman, M.A. 1996. Design, synthesis and antiinflammatory activity of some 1,3,4-oxadiazole derivatives. *Eur. J. Med. Chem.* 31(10): 819–825.

Palmer, P.J., Trigg, R.B. and Warrington, J.V. 1971. Benzothiazolines as antitubercular agents. *J. Med. Chem.* 14: 248–251.

Patel, N.A., Surti, S.C., Patel, R.G. and Patel, M.P. 2008. Synthesis, characterization, and biological activity of some new benzoic acid and thiazoloacridine derivatives. *Phosphorus Sulfur Silicon Relat. Elem.* 183(9): 2191–2203.

Patil, A.D., Kumar, N.V., Kokke, W.C., Bean, M.F., Freyer, A.J., Debrosse, C., Mai, S., Truneh, A. and Faulkner, D.J. 1995. Novel alkaloids from the sponge *Batzella* sp.: Inhibitors of HIV gp120-human CD4 binding. *J. Org. Chem.* 60(5): 1182–1188.

Paul, S. and Basu, B. 2012. Highly selective synthesis of libraries of 1,2-disubstituted benzimidazoles using silica gel soaked with ferric sulfate. *Tetrahedron Lett.* 53(32): 4130–4133.

Pereira, M.F., Chevrot, R., Rosenfeld, E., Thiery, V. and Besson, T. 2007. Synthesis and evaluation of the antimicrobial activity of novel quinazolinones. *J. Enzym. Inhib. Med. Chem.* 22(5): 577–583.

Rafiee, E., Eavani, S., Rashidzadeh, S. and Joshaghani, M. 2009. Silica supported 12-tungstophosphoric acid catalysts for synthesis of 1,4-dihydropyridines under solvent-free conditions. *Inorg. Chim. Acta* 362(10): 3555–3562.

Rostamnia, S. and Lamei, K. 2012. Diketene-based neat four-component synthesis of the dihydropyrimidinones and dihydropyridine backbones using silica sulfuric acid (SSA). *Chin. Chem. Lett.* 23(8): 930–932.

Rovynak, G.C., Kimball, S.C., Beyer, B., Cucinotta, G., Dimarco, J.D., Gougoutas, J., Hedberg, A., et al. 1995. Calcium entry blockers and activators: Conformational and structural determinants of dihydropyrimidine calcium channel modulators. *J. Med. Chem.* 38(1): 119–129.

Sadeghi, B., Mirjalili, B.B.F., Bidaki, S. and Ghasemkhani, M. 2011. $SbCl_5{\cdot}SiO_2$: An efficient alternative for one-pot synthesis of 1,2,4,5-tetrasubstituted imidazoles in solvent or under solvent-free condition. *J. Iran. Chem. Soc.* 8(3): 648–652.

Sadeghi, B. and Nejad, M.G. 2013. Silica sulfuric acid: An eco-friendly and reusable catalyst for synthesis of benzimidazole derivatives. *J. Chem.* 2013 (581465).

Sajjadifar, S., Mirshokraie, S.A., Javaherneshan, N. and Louie, O. 2012. SBSA as a new and efficient catalyst for the one-pot green synthesis of benzimidazole derivatives at room temperature. *Am. J. Org. Chem.* 2(2): 1–6.

Sakata, G., Makino, K. and Kurasama, Y. 1988. Recent progress in the quinoxaline chemistry. Synthesis and biological activity. *Heterocycles* 27(10): 2481–2515.

Salehi, P., Dabiri, M., Zolfigolc, M.A. and Baghbanzadehb, M. 2005. A new approach to the facile synthesis of mono- and disubstituted quinazolin-4(3*H*)-ones under solvent-free conditions. *Tetrahedron Lett.* 46(41): 7051–7053.

Salehi, P., Dabiri, M., Zolfigolc, M.A. and Fardb, M.A.B. 2003. Silica sulfuric acid: An efficient and reusable catalyst for the one-pot synthesis of 3,4-dihydropyrimidin-2(1*H*)-ones. *Tetrahedron Lett.* 44(14): 2889–2891.

Santen, R.A. and Neurock, M. 2006. *Molecular Heterogeneous Catalysis: A Conceptual and Computational Approach.* Wiley-VCH: Weinheim, Cambridge.

Segarra, V., Crespo, M.I., Pujol, F., Belata, J., Domenech, T., Miralpix, M., Palacios, J.M. and Castro, A. 1998. Phosphodiesterase inhibitory properties of losartan. Design and synthesis of new lead compounds. *Bioorg. Med. Chem. Lett.* 8(5): 505–510.

Seitz, L.E., Suling, W.J. and Reynolds, R.C. 2002. Synthesis and antimycobacterial activity of pyrazine and quinoxaline derivatives. *J. Med. Chem.* 45(25): 5604–5606.

Shaabani, A. and Rahmati, A. 2006. Silica sulfuric acid as an efficient and recoverable catalyst for the synthesis of trisubstituted imidazoles. *J. Mol. Catal. A Chem.* 249(1–2): 246–248.

Shaabani, A., Soleimani, E. and Maleki, A. 2007. One-pot three-component synthesis of 3-aminoimidazo[1,2-*a*]pyridines and -pyrazines in the presence of silica sulfuric acid. *Monatshefte für Chemie* 138(1): 73–76.

Shaterian, H.R., Hosseinian, A. and Ghashang, M. 2008. Silica sulfuric acid as an efficient catalyst for the preparation of 2*H*-indazolo[2,1-*b*]phthalazine-triones. *Appl. Catal. A.* 345(2): 128–133.

Shirini, F., Marjani, K. and Nahzomi, H.T. 2007. Silica triflate as an efficient catalyst for the solvent-free synthesis of 3,4-dihydropyrimidin-2(1*H*)-ones. *Arkivoc* 2007(1): 51–57.

Shylesh, S., Sharma, S., Mirajkar, S.P. and Singh, A.P. 2004. Silica functionalised sulphonic acid groups: Synthesis, characterization and catalytic activity in acetalization and acetylation reactions. *J. Mol. Catal. A Chem.* 212(1–2): 219–228.

Siddiqui, Z.N. and Farooq, F. 2012. Silica supported sodium hydrogen sulfate ($NaHSO_4$–SiO_2): A novel, green catalyst for synthesis of pyrazole and pyranyl pyridine derivatives under solvent-free condition via heterocyclic β-enaminones. *J. Mol. Catal. A Chem.* 363–364: 451–459.

Singh, K., Singh, J., Deb, P.K. and Singh, H. 1999. An expedient protocol of the Biginelli dihydropyrimidine synthesis using carbonyl equivalents. *Tetrahedron* 55: 12873–12880.

Sisido, K., Nabika, K., Isida, T. and Kozima, S. 1971. Formation of organotin-nitrogen bonds III. *N*-Trialkyltin-5-substituted tetrazoles. *J. Organomet. Chem.* 33(3): 337–346.

Smith, G.V. and Notheisz, F. 2000. *Heterogeneous Catalysis in Organic Chemistry.* Elsevier: San Diego, CA.

Snider, B.B., Chen, J., Patil, A.D. and Freyer, A. 1996. Synthesis of the tricyclic portions of batzelladines A, B and D. Revision of the stereochemistry of batzelladines A and D. *Tetrahedron Lett.* 37(39): 6977–6980.

Sparatore, A., Basilico, N., Parapini, S., Romeo, S., Novelli, F., Sparatorec, F. and Taramelli, D. 2005. 4-Aminoquinoline quinolizidinyl- and quinolizidinylalkyl-derivatives with antimalarial activity. *J. Bio. Med. Chem.* 13(18): 5338–5345.

Srihari, P., Singh, V.K., Bhunia, D.C. and Yadav, J.S. 2009. One-pot three-component coupling reaction: Solvent-free synthesis of novel 3-substituted indoles catalyzed by PMA–SiO_2. *Tetrahedron Lett.* 50(27): 3763–3766.

Sugiyama, Y., Ito, Y., Suzuki, M. and Hirota, A. 2009. Indole derivatives from a marine sponge-derived yeast as DPPH radical scavengers. *J. Nat. Prod.* 72(11): 2069–2071.

Tamaoki, S., Yamauchi, Y., Nakano, Y., Sakano, S., Asagarasu, A. and Sato, M. 2007. Pharmacological properties of 3-amino-5,6,7,8-tetrahydro-2-{4-[4-(quinolin-2-yl) piperazin-1-yl]butyl}quinazolin-4(3*H*)-one (TZB-30878), a novel therapeutic agent for diarrhea-predominant irritable bowel syndrome (IBS) and its effects on an experimental IBS model. *J. Pharm. Exp. Ther.* 322: 1315–1323.

Taraporewala, I.B. 1991. Thiazolo[5,4-*b*]acridines and thiazolo[4,5-*b*]acridines: Probable pharmacophores of antiviral and anti-tumor marine alkaloids. *Tetrahedron Lett.* 32(1): 39–42.

Taylor, E.C., Knopf, R.I. and Borror, A.L. 1960. The dimerization of 2-amino-5-nitrobenzonitrile. *J. Am. Chem. Soc.* 82(12): 3152.

Tsitsa, P., Valiraki, A.P., Papastaikoudi, T.S., Daifoiti, Z.P. and Vamvakidis, A. 1989. Synthesis and anticonvulsive activity of some new bisubstituted 1,3,4-oxadiazoles and 1*H*-1,2,4-triazoles. *Ann. Pharm. Fr.* 47(5): 296–303.

Ubeda, J.I., Villacampa, M. and Avendano, C. 1998. Friedlander synthesis of substituted quinolines from *N*-pivaloylanilines. *Synthesis* (8): 1176–1180.

Valdez, J., Cedillo, R., Hernandez-Campos, A., Yepez, L., Hernandez-Luis, F., Navarrete-Vazquez, G., Tapia, A., Cortes, R., Hernandez, M. and Castillo, R. 2002. Synthesis and antiparasitic activity of 1*H*-benzimidazole derivatives. *Bioorg. Med. Chem. Lett.* 12(16): 2221–2224.

Veisi, H. 2010. Silica sulfuric acid (SSA) as a solid acid heterogeneous catalyst for one-pot synthesis of substituted pyrroles under solvent-free conditions at room temperature. *Tetrahedron Lett.* 51(16): 2109–2114.

Viegas, C. Jr., Bolzani, V.S., Furlan, M., Barreiro, E.J., Young, M.C.M., Tomazela, D. and Eberlin, M.N. 2004. Further bioactive piperidine alkaloids from the flowers and green fruits of *Cassia spectabilis*. *J. Nat. Prod.* 67(5): 908–910.

Watanabe, N., Kabasawa, Y., Takase, Y., Matsukura, M., Miyazaki, K., Ishihara, H., Kodama, K. and Adachi, H. 1998. 4-Benzylamino-1-chloro-6-substituted phthalazines: Synthesis and inhibitory activity toward phosphodiesterase. *J. Med. Chem.* 41(18): 3367–3372.

Wilson, K., Lee, A.F., Macquarrie, D.J. and Clark, J.H. 2002. Structure and reactivity of sol-gel sulphonic acid silicas. *Appl. Catal. A Gen.* 228(1–2): 127–133.

Wittenberger, S.J. and Donner, B.G. 1993. Dialkyltin oxide mediated addition of trimethylsilyl azide to nitriles. A novel preparation of 5-substituted tetrazols. *J. Org. Chem.* 58(15): 4139–4141.

Wolkenberg, S.E., Wisnoski, D.D., Leister, W.H., Wang, Y., Zhao, Z. and Lindsley, C.W. 2004. Efficient synthesis of imidazoles from aldehydes and 1,2-diketones using microwave irradiation. *Org. Lett.* 6(9): 1453–1456.

Zolfigol, M.A., Salehib, P., Shiria, M., Faal Rastegar, T. and Ghaderi, A. 2008. Silica sulfuric acid as an efficient catalyst for the Friedländer quinoline synthesis from simple ketones and *ortho*-aminoaryl ketones under microwave irradiation. *J. Iran. Chem. Soc.* 5(3): 490–497.

7 Application of Organometallic Compounds as Heterogeneous Catalysts in Organic Synthesis

Prachi Rathi, Sudesh Kumar, K. L. Ameta, and Dharma Kishore

CONTENTS

7.1 INTRODUCTION

Organometallic chemistry deals with compounds in which one or more carbon atoms are attached to a metal. Organometallic chemistry dates back to the mid-1800s, when Frankland discovered the ethyl and methyl derivatives of zinc, tin, and mercury. Thereafter, organic transformations based on organometallic compounds continually increased. The field of organometallic chemistry is subdivided into two recognizable areas defined by the type of metal bonded to carbon. One area involves compounds containing the main group elements such as the alkali and alkaline earth metals, and the more metallic elements in the zinc, boron, carbon, nitrogen, and oxygen vertical groups in the periodic table. In these compounds, the carbon is usually bound to the metallic element by an ionic or relatively simple σ-bond. Examples include Grignard reagents, organolithium reagents, tetraethyl lead, hexadentate or tetradentate ligands, ethylenediaminetetraacetic acid (EDTA), and other compounds used routinely in the laboratory and in the chemical industry (Coates and Wade 1967). The other area includes compounds containing transition-metal elements in which a number of general compound classes may be recognized.

7.1.1 Organolithium Compounds

An organolithium reagent forms a direct bond between a carbon and a lithium atom. The electropositive nature of lithium puts most of the charge density of the bond on the carbon atom, effectively creating a carbanion; this makes organolithium compounds extremely powerful bases and nucleophiles. They are usually designated as RLi, which is a useful but incomplete representation of their composition in solution, where they exist as more complex structures. For example, simple organolithium exists mainly as a tetramer in ethers and as a hexameric structure in hydrocarbons. Its degree of aggregation as well as its reactivity can be modified by the addition of chelating agents such as tetramethylenediamine (TMEDA), *N,N*-dimethylpropyleneurea (DMPU), and hexamethylphosphoric triamide (HMPA).

7.1.1.1 Synthesis

7.1.1.1.1 By the Addition of C–C Multiple Bonds to R^{1}Li

Unsaturated compound **1** reacts with alkyl lithium and forms saturated compounds **2** or **3** (Scheme 7.1).

7.1.1.1.2 By Metalation of N-Sulfonylhydrazones to Alkyl Lithium (Shapiro)

Unsaturated compound **4** containing *N*-sulfonylhydrazone undergoes metalation with 2 mol of *n*-BuLi to give **5** (Scheme 7.2).

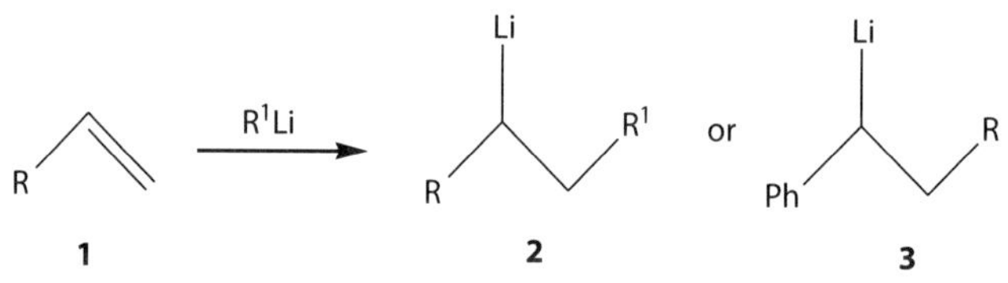

SCHEME 7.1

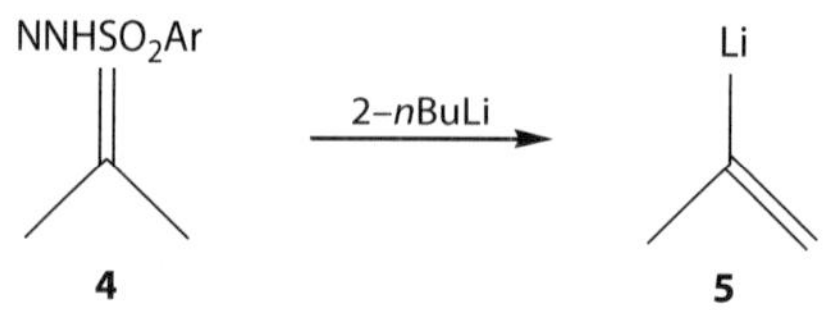

SCHEME 7.2

7.1.1.1.3 By Halogen–Metal Exchange

This reaction is often used to prepare vinyl- and aryl-lithium compounds **8** from the reaction of halo-benzene **6** and alkyl-lithium **7** species (Leroux and Schlosser 2002) (Scheme 7.3).

7.1.1.1.4 By Ortho-Metalation/Direct Metalation

Metalation of an aromatic ring near a substituent is called ortho-metalation and is shown in Scheme 7.4 (Hessler et al. 2001).

7.1.1.1.5 By Transmetalation/Metal–Metal Exchange

This reaction is preferred to form organolithium compounds from other organometallic compounds, as shown in Scheme 7.5 (Lipshutz 1994; Normant 1972).

7.1.1.2 Reactions

7.1.1.2.1 Addition Reactions

Lithio glycols **17** undergo a reaction with quinone derivatives **18** to give **19** (Parker and Georges 2000) (Scheme 7.6).

When cinnamaldehyde oxime ether (**20**) is added to RLi, it gives allylic amines **21** (Scheme 7.4). Those containing a chirality center at the carbonyl site are subject to 1,4-assymetric induction; therefore, optically active α-amino acid derivatives may be synthesized (Moody et al. 1999) (Scheme 7.7).

A bicyclic oxazolidine adduct **22** bearing an α-cyano group undergoes ring expansion on reduction with $LiAlH_4$ to form **24** (Cutri et al. 2000) (Scheme 7.8).

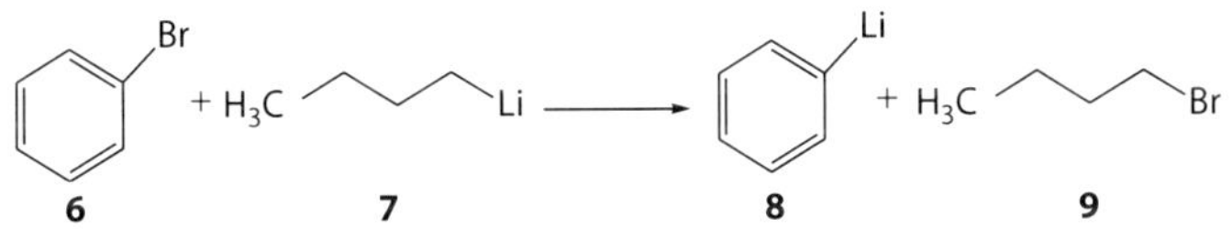

SCHEME 7.3

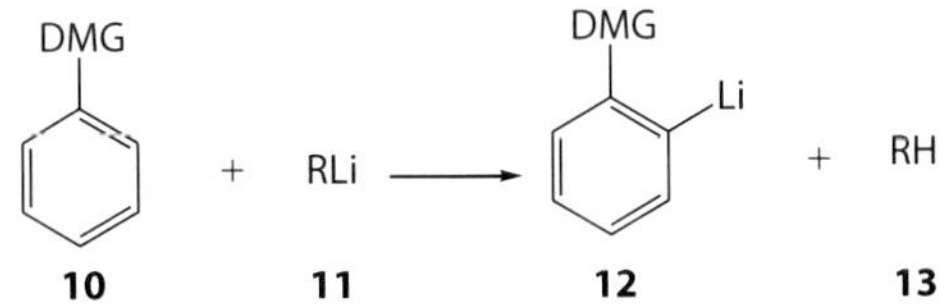

SCHEME 7.4

SCHEME 7.5

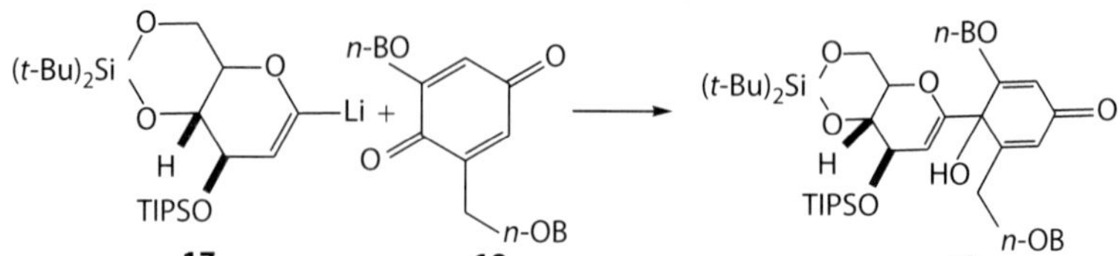

17 18 19

SCHEME 7.6

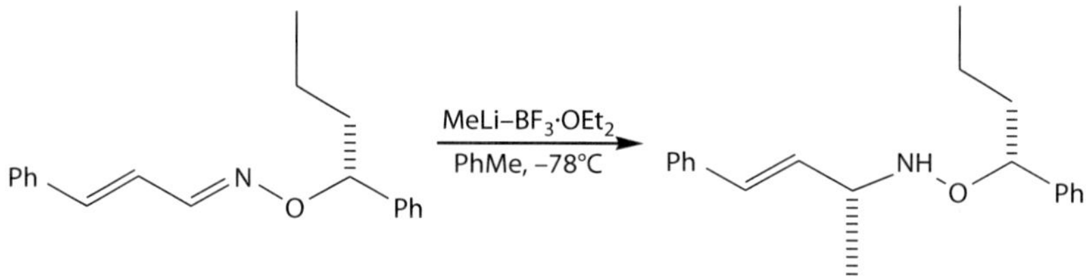

95%

20 21

SCHEME 7.7

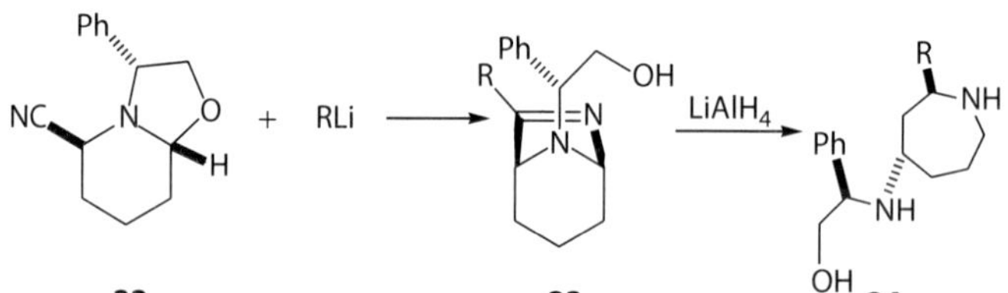

22 23 24

SCHEME 7.8

The addition of benzenesulfonylmethyllithium to nitrones **25** containing a vinyl group can lead to pyrrolidine *N*-oxides **27** after workup through the intermediacy of **26** (Hanrahan and Knight 1998) (Scheme 7.9).

7.1.1.2.2 Reaction with 1,4-Diketones

Squaric acid derivatives **28** are converted to the diketones **29** after reaction with RLi and quenching with NH_4Cl (Varea et al. 1995) (Scheme 7.10).

7.1.1.2.3 Reductive Coupling with Tosylhydrazones

Following this method, carbon chains can be assembled as in **32** from chiral components **30** and **31** (Myers and Movassaghi 1998) (Scheme 7.11).

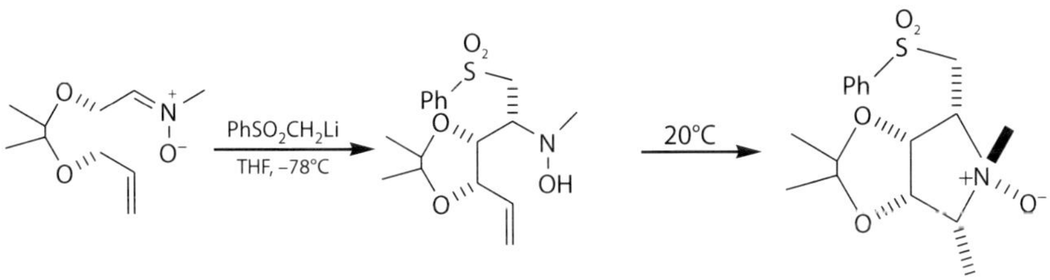

SCHEME 7.9

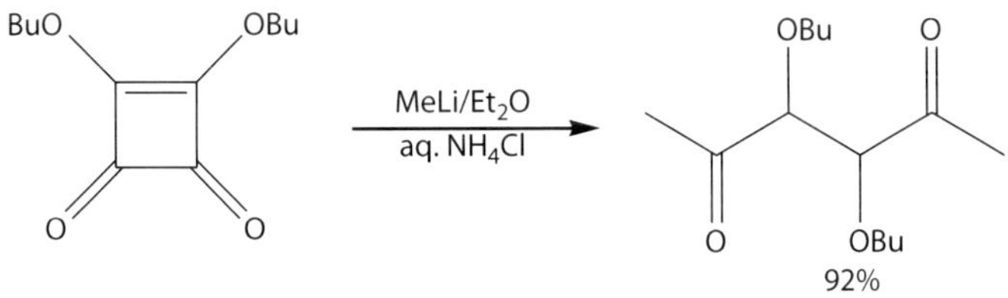

SCHEME 7.10

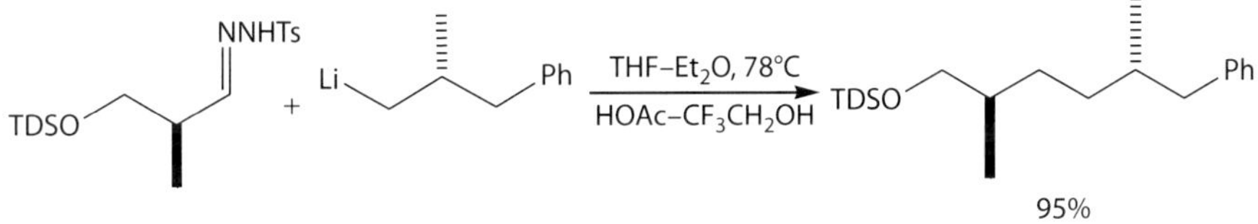

SCHEME 7.11

7.1.1.2.4 Reaction with Butenolides and 1-Isoindolinones

Enals **33** containing an iron substituent on reaction with RLi result in the formation of γ-substituted (with R) butenolides **34** (Moller et al. 2000). Cyclocarbonylation can be induced by an addition reaction (Scheme 7.12).

3-Substituted-1-isoindolinones **36** were synthesized from *N,N*-dimethylaminophthalimide **35** via reaction with organolithium (or Grignard reagent) and deoxygenation with $Et_3SiH–CF_3COOH$. N–N bond cleavage with Zn in AcOH produces **37** (Deniau and Enders 2000) (Scheme 7.13).

7.1.1.2.5 Parham Cyclization

Parham cyclization is an example of a halogen–metal exchange on an aromatic ring followed by a ring-closure reaction with the electrophile as part of a side chain. On reaction with alkyl lithium, the aromatic ring **38** forms compound **39** by halogen–metal exchange and it gives **40** via ring-closure reaction, as shown in Scheme 7.14 (Parham et al. 1975). E=CH_2–Br, CH_2–Cl, epoxide, carbonyl, and so on.

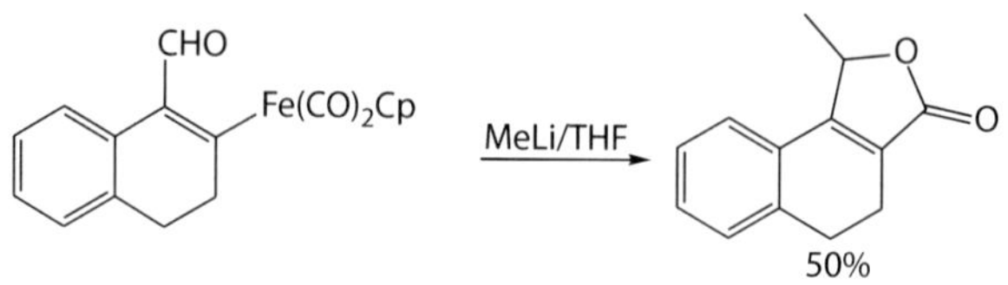

33 **34**

SCHEME 7.12

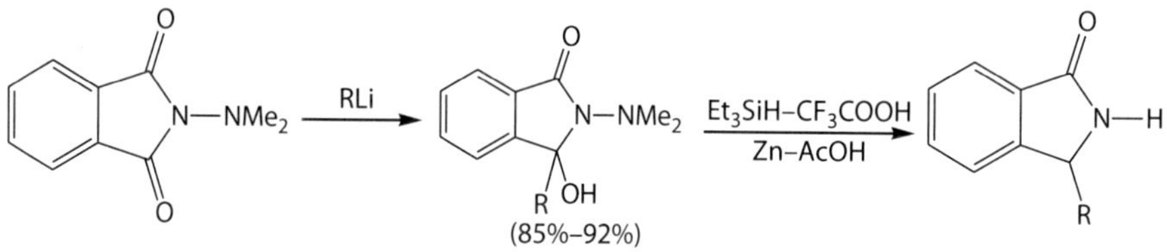

35 **36** **37**

SCHEME 7.13

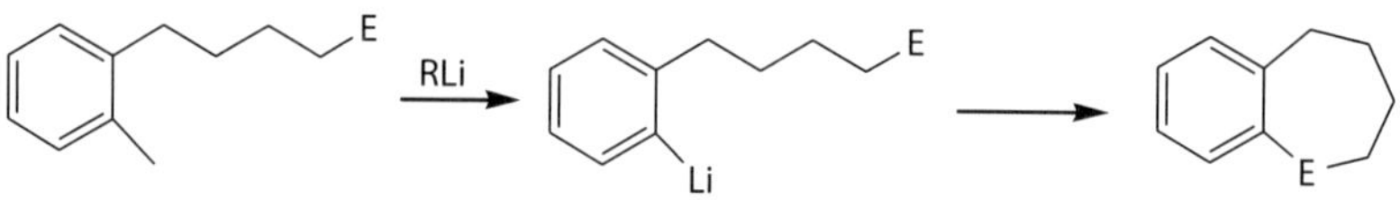

38 **39** **40**

SCHEME 7.14

7.1.1.2.6 [1,2]- and [2,3]-Wittig Rearrangement

The [1,2]-Wittig rearrangement of ethers with alkyl lithiums yields alcohols via a [1,2]-shift, as shown in Scheme 7.15. The groups R_1 and R_2 may be alkyl, aryl, or vinyl (Wittig and Löhmann 1942).

7.1.1.2.7 Ramberg–Backlund Reaction

In this reaction, an α-halogensulfone **44** is treated with a strong base to give an olefin **46**, as illustrated in Scheme 7.16 (Paquette 1968).

7.1.1.2.8 Rearrangement

A remarkable transformation observed by catalytic amounts of MeLi on an allylic alcohol that contains a silylalkyne moiety **47** is the result of a Claisen rearrangement of a 2-methylenetetrahydrofuran intermediate to yield **48** (Ovaska et al. 1998) (Scheme 7.17).

7.1.1.2.9 Reaction with Epoxy Silanes

When organolithiums containing aryl, alkenyl, alkynyl, amido, and cyano groups react with (*E*)- and (*Z*)-epoxysilanes **49**, they form adducts that, on treatment with

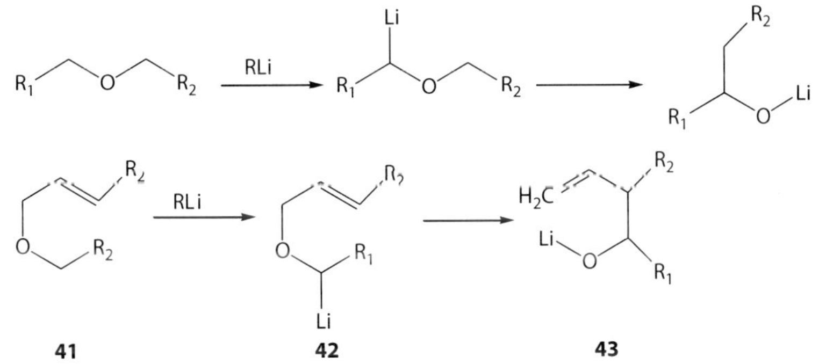

SCHEME 7.15

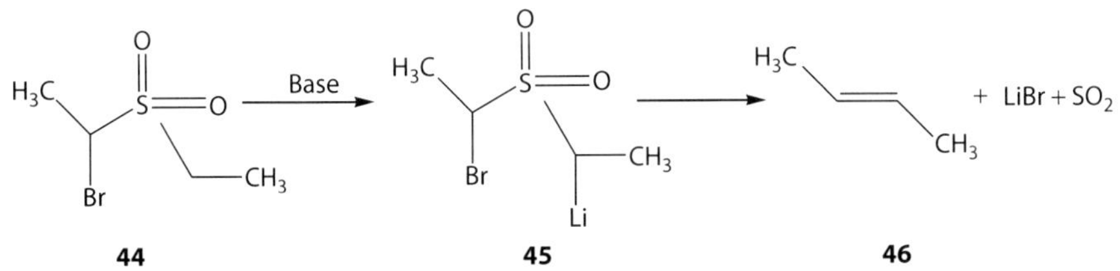

SCHEME 7.16

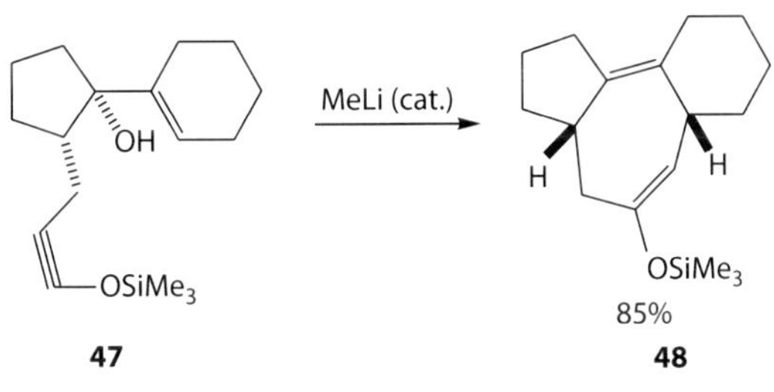

SCHEME 7.17

base (KH), are converted stereoselectively into (*E*)- and (*Z*)-alkenes **50**, respectively. Organocopper reagents may be preferred to the corresponding organolithium reagents when the carbanion is an alkenyl or aryl group (Zhang et al. 1989) (Scheme 7.18).

7.1.1.2.10 Reaction with Ketones

At low temperatures, 3,4-dihydropyranones **52** react with RLi **51** to give 1,5-diketones **53** (Harrowven and Hannam 1999) (Scheme 7.19).

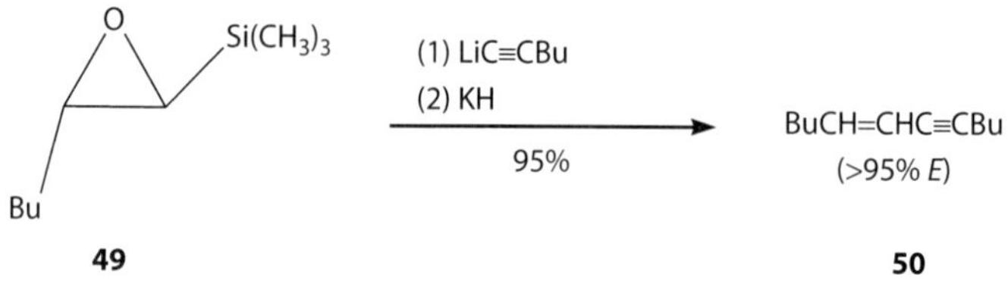

SCHEME 7.18

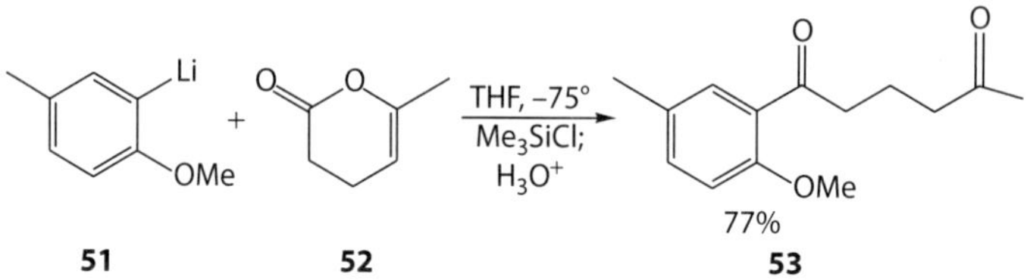

SCHEME 7.19

7.1.2 Organozinc Compounds

Organozinc compounds contain a carbon–zinc bond. Diethylzinc was the first organozinc compound, synthesized by Frankland in 1849. Organozinc compounds are either liquids, such as diethylzinc, $(C_2H_5)_2Zn$, with a boiling point of 116.8°C, or solids, such as diphenylzinc, $(C_6H_5)_2Zn$, with a melting point of 107°C. They are unstable in the presence of air, so much so that the lower members of the R_2Zn series (up to $R = C_5H_{11}$) ignite spontaneously. They are vigorously decomposed by water. Thus, their reactions are carried out in an inert atmosphere of nitrogen, argon, or carbon dioxide (Elschenbroich 2005). There are three main classes of organozinc compounds: organozinc halides (RZnX), diorganozincs (R_2Zn), and lithium or magnesium zincates (Knochel and Jones 1999).

7.1.2.1 Synthesis

Diorganozinc reagents **55** require different methods of preparation, such as iodine–zinc and boron–zinc exchange to form **56** (Langer et al. 1993, 1996). These methods are applicable to the formation of primary and secondary diorganozinc species **55** (Scheme 7.20).

Lithium or magnesium zincates are readily prepared by transmetalation reactions from the corresponding magnesium or lithium reagents and are well suited to halogenzinc exchange reactions, as shown in Scheme 7.15 (Uchiyama et al. 1996) (Scheme 7.21).

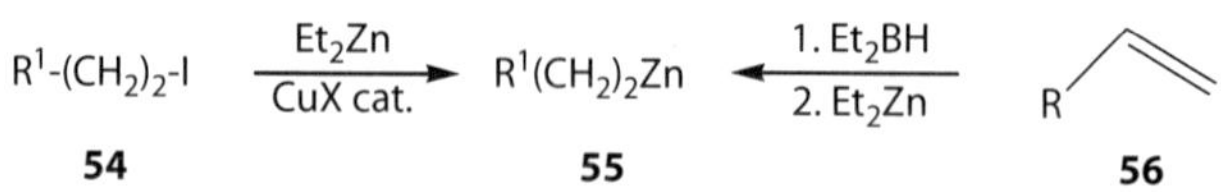

SCHEME 7.20

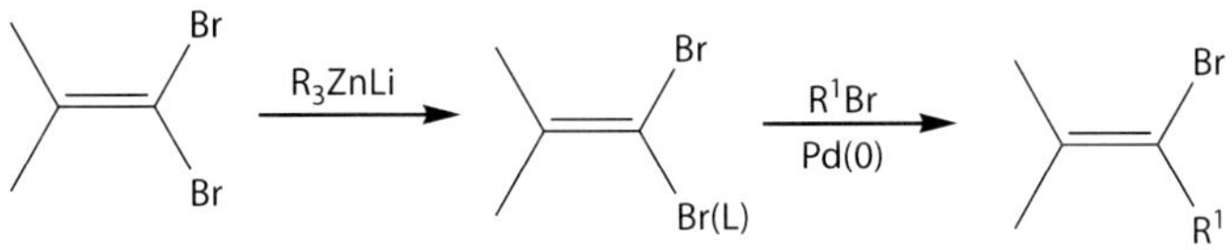

57 58 59

SCHEME 7.21

7.1.2.1.1 Carbocyclization

Vinylic ester **60** can undergo a cycloaddition reaction in the presence of zinc to produce **61** (Meyer et al. 1993) (Scheme 7.22).

7.1.2.1.2 Organozinc Compounds from Organomagnesium Halide (Reformatsky Reaction)

This reaction is shown in Scheme 7.23 (Seebach et al. 1991). Functionalized organozinc reagents **63** have been prepared from alkenes **62** via stereoselective and regioselective hydroboration and B–Zn exchange (Boudier et al. 2000) (Scheme 7.24).

7.1.2.2 Reactions

Allenylzinc reagents prepared from chiral propargylic mesylates **64** have been used to synthesize anti-homopropargylic alcohols **65** (Marshall and Adams 1999) (Scheme 7.25).

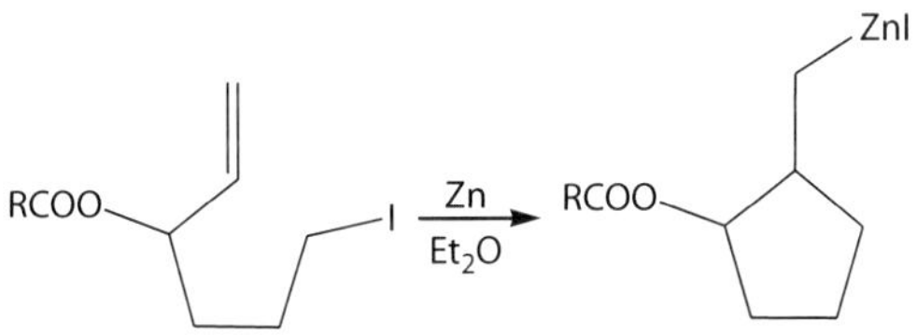

60 61

SCHEME 7.22

SCHEME 7.23

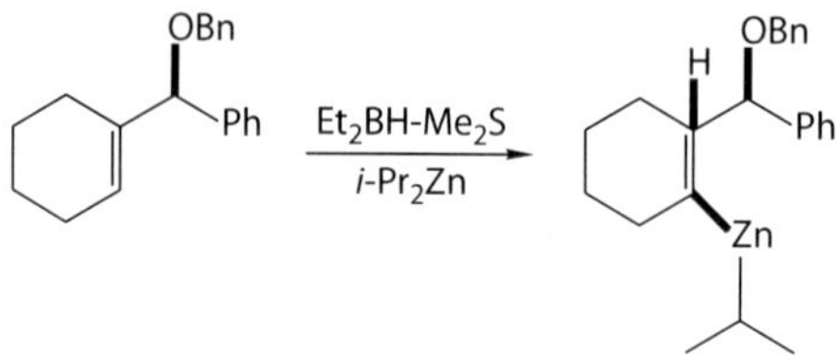

62 63

SCHEME 7.24

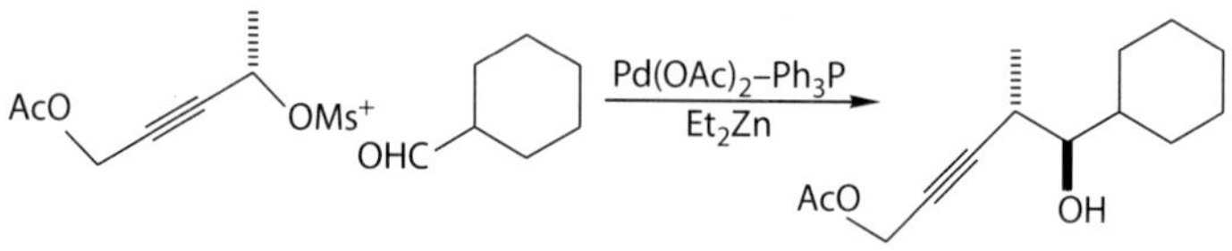

75% (anti:syn, 95:5)

64 65

SCHEME 7.25

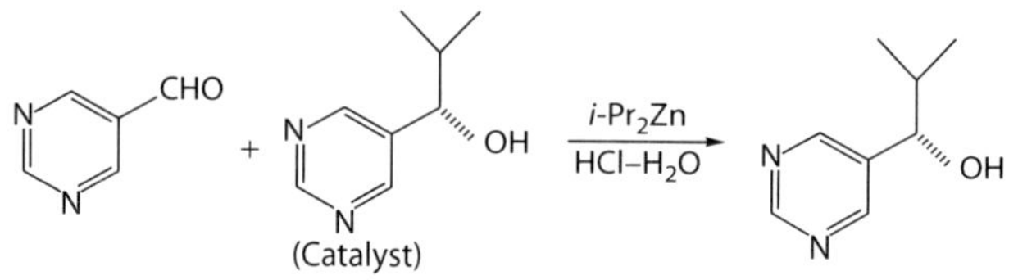

66 67 68

SCHEME 7.26

7.1.2.2.1 Addition to C=X Bond

The carbonyl group of β-keto phosphonates is attached to allylic zinc reagents (Lentsch and Wiemer 1999) with a catalytic enantioselective synthesis of chiral pyramidal carbinols; an automultiplication phenomenon is elucidated in Scheme 7.26.

7.1.2.2.2 Conjugate Additions

An enyne group is introduced to the β-position of an enone **71** in the nickel-catalyzed addition onto the product generated from alkynylzincation **70** (Scheme 7.27).

7.1.2.2.3 Addition to Alkynes and Dienes

Organozincation of alkynes **72** and dienes generates new organozinc species **73** that are exploited synthetically, as shown in Scheme 7.28, to give **74** (Qi and Montgomery 1999).

gem-Bimetalloalkenes **76** can be prepared by the addition of allylzinc bromide (allylmagnesium bromide + $ZnBr_2$) to alkynyllithium. Chlorination of the latter

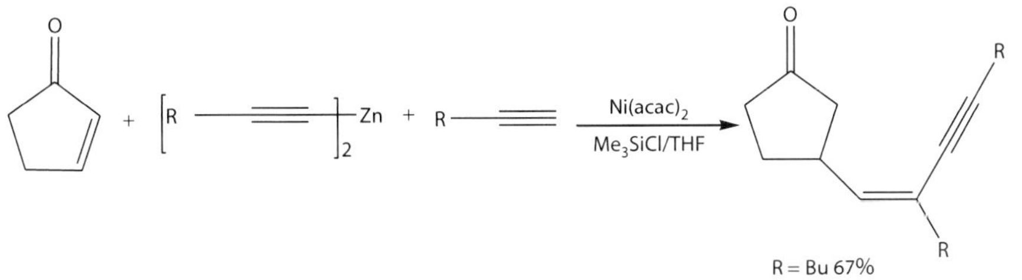

69 70 71

SCHEME 7.27

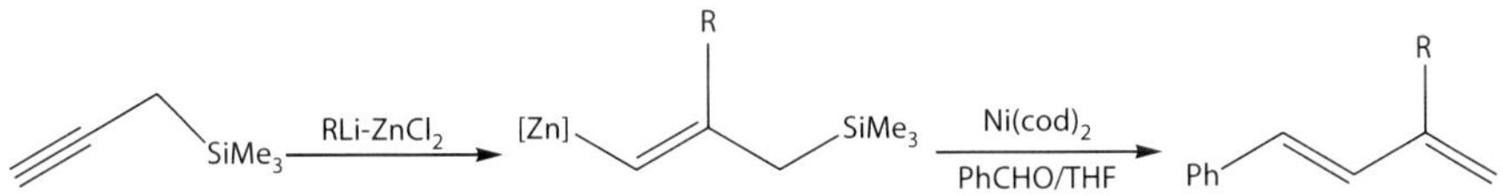

72 73 74

SCHEME 7.28

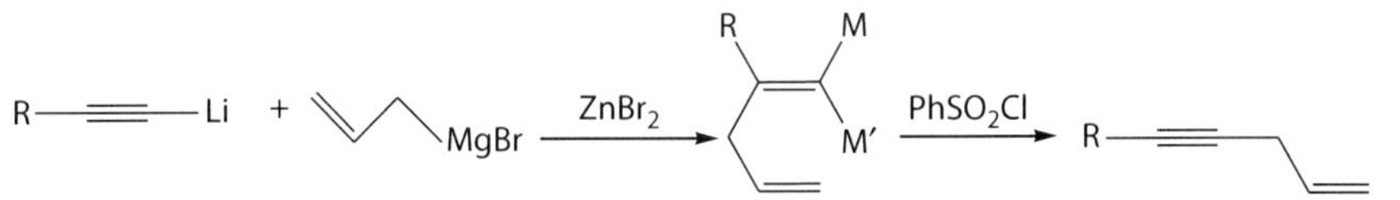

75 76

SCHEME 7.29

species with $PhSO_2Cl$ sets up a Fritsch–Buttenberg–Wiechell rearrangement to form **76** (Rezaei et al. 2000) (Scheme 7.29).

7.1.2.2.4 Coupling Reactions

Cross-coupling reactions of organozincs with activated triflates need the presence of cuprates, as shown in Scheme 7.30 (Lipshutz and Vivian 1999).

Other coupling partners to organozinc reagents include heterocycles such as 2-methylthiobenzothiazole, alkenyl aryl iodonium triflates, and aryl heteroaryl ethers (Angiolelli et al. 2000; Hinkle et al. 2000; Brigas and Johnstone 2000). Improved nickel-catalyzed cross-coupling conditions between ortho-substituted aryl iodides-nonaflates and alkyl-zinc iodides in solution and in solid phase have been defined, as shown in Scheme 7.31 (Jensen et al. 2000).

7.1.2.2.5 Decarboxylative Protonation

The formation of **85** from **84** under the influence of dialkylzinc presence is an example of carbonyl alkylation followed by hydrolysis of the anhydride to give acids (Mohr et al. 2006) (Scheme 7.32).

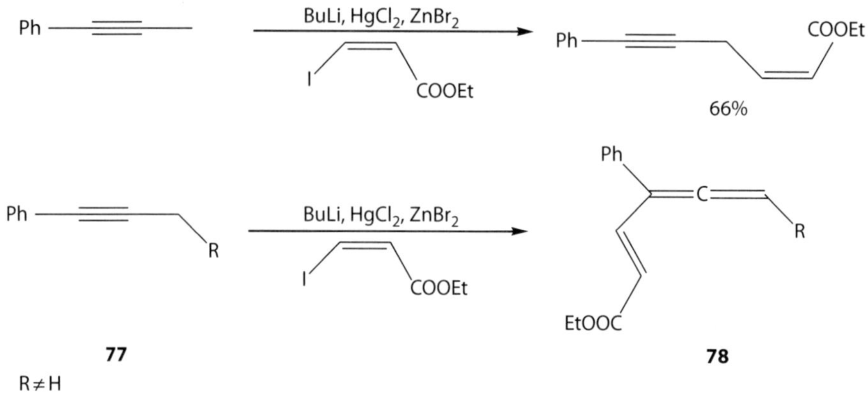

SCHEME 7.30

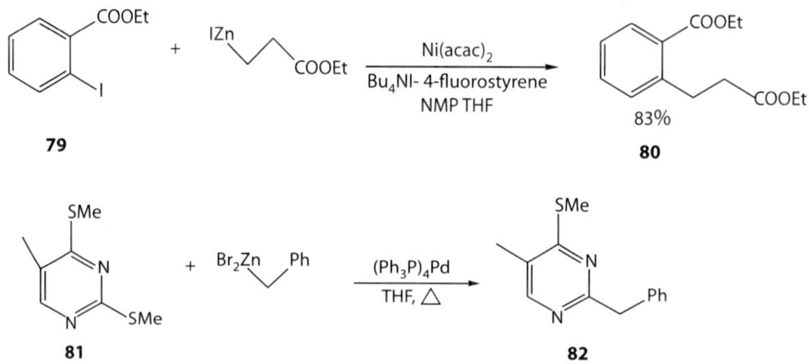

SCHEME 7.31

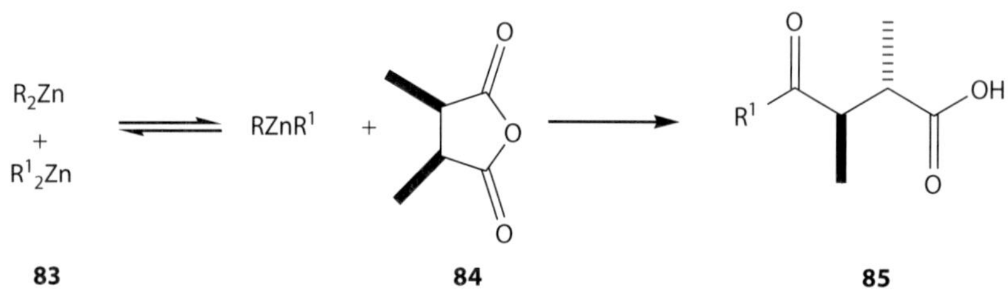

SCHEME 7.32

7.1.2.2.6 Addition of Dialkylzinc Compounds to Aldehydes

The first highly enantioselective catalytic addition of diorganozinc **87** compound to aldehyde **86** was demonstrated by Noyori et al. (1994) using (–)-3-exo-dimethyl amino isoborneol (DAIB). No reaction takes place between diorganozinc and aldehyde in toluene and hexane at room temperature. However, the addition of only 2 mol of DAIB in the reaction leads to the formation of secondary alcohol **88** with a high degree of enantioselectivity (Dimitrov and Kamenova-Nacheva 2009) (Scheme 7.33).

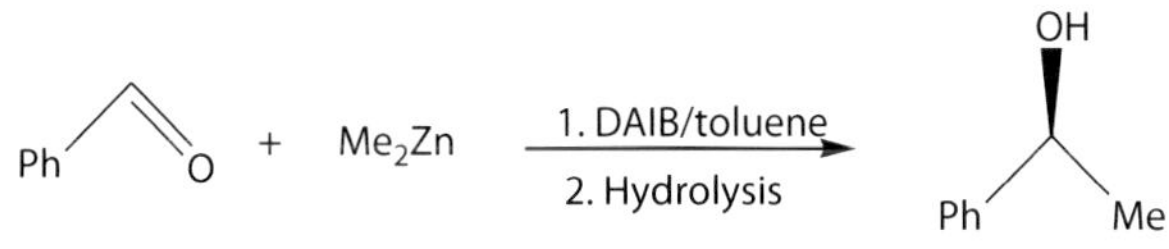

86 **87** **88**

SCHEME 7.33

7.1.2.2.7 Alkylation

Allylic oxabridge **89** diorganozincs intercept the π-allylpalladium species by donating their σ-ligands in the presence of Pd to give **90** (Lautens et al. 2000) (Scheme 7.34).

7.1.2.2.8 Elimination

An alkenyl copper species containing a sulfonylalkenyl group **92** with α-iodoalkylzinc reagent produces **93** (Ohno et al. 2000; Varghese et al. 2000, respectively) (Scheme 7.35).

7.1.2.2.9 Reaction with TsCN

The reaction of organozincs with toluenesulfonyl cyanide (TsCN) **94** gives nitriles **95** (Klement et al. 1993) (Scheme 7.36). This is interesting in the case of homopropargylic sulfonates **96**, because 1-zincoalkylidenecyclopropanes **97** are formed and it can be derivatized.

7.1.2.2.10 Formation of 1,3-Dizincs

The preparation of 1,3-dizinc species is illustrated in Scheme 7.37. It shows the reaction of the corresponding diethylborane **100** species with dialkylzinc, which in turn was prepared from the reaction of diethylhydroborane with **99** (Eick and Knochel 1996).

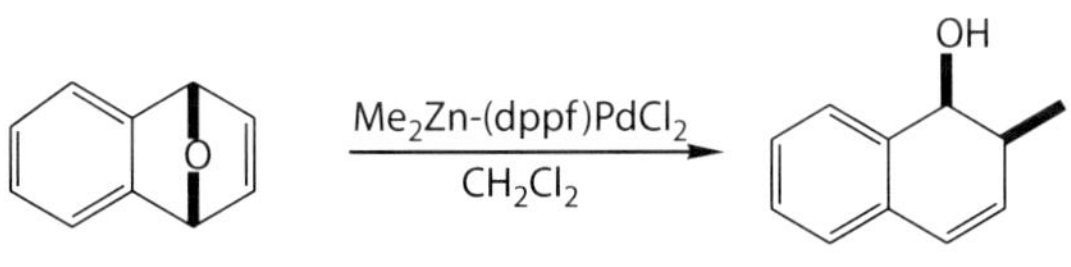

80%

89 **90**

SCHEME 7.34

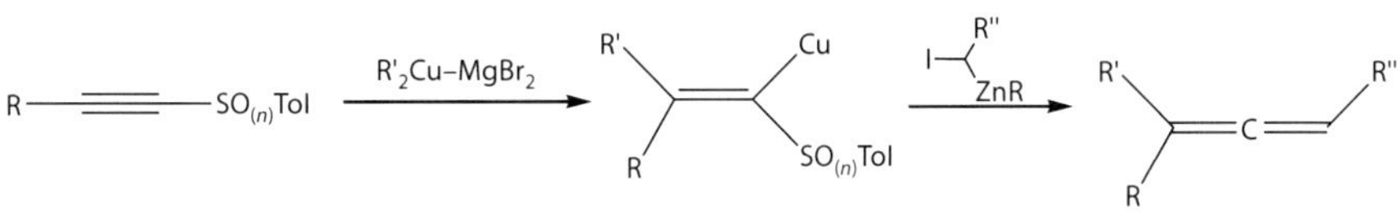

91 **92** **93**

SCHEME 7.35

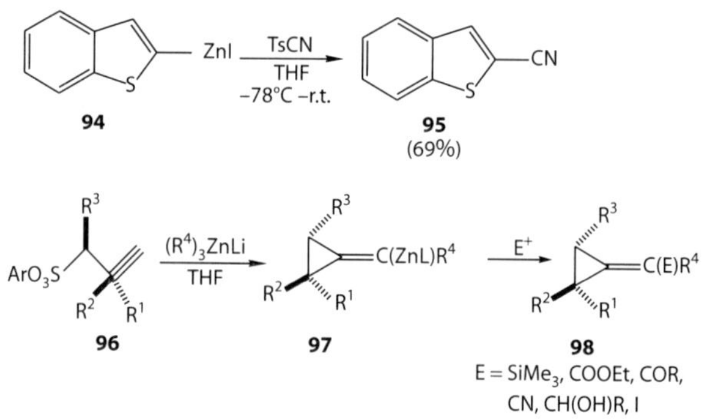

SCHEME 7.36

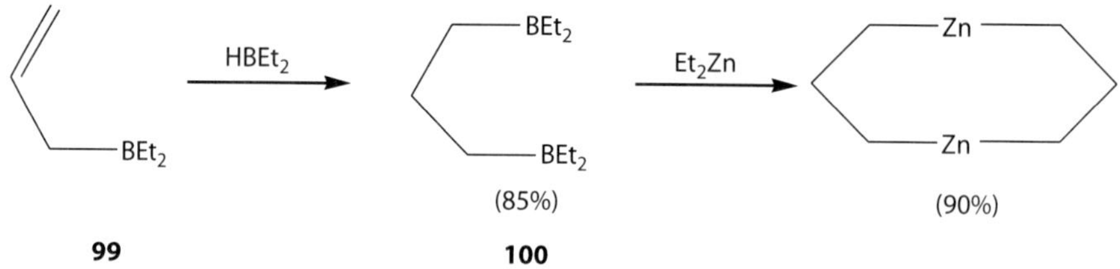

SCHEME 7.37

7.1.2.2.11 gem-Functionalization of Cyclopropanes

gem-Dibromocyclopropanes followed by metalation give 1-halocyclopropylzincates that enable stereoselective C–C bond formation, as shown in Scheme 7.38 (Harada et al. 1993).

7.1.2.2.12 Formation of γ-Keto Amino Acid

The organozinc reagent **104** couples with acid chlorides in the presence of Pd(II) catalysts to give enantiometrically pure protected l-γ-keto α-amino acids **105** (Scheme 7.39).

7.1.2.2.13 Formation of 2-Carboalkoxycyclopentenones

2-Carboalkoxycyclopentenones are formed as illustrated in Scheme 7.40 (Crimmins and Nantermet 1990).

7.1.2.2.14 Nucleophilic Displacement of Anomeric Sulfones

2-Benzenesulfonyl cyclic ethers **106**, **108**, and **110** undergo nucleophilic substitution with various organozinc reagents (Brown et al. 1989) (Scheme 7.41).

7.1.3 Organocopper Compounds

Organocopper compounds in organometallic chemistry contain a carbon-to-copper chemical bond. Organocopper chemistry is the science of organocopper compounds, describing their chemical reaction, synthesis, and physicochemical properties.

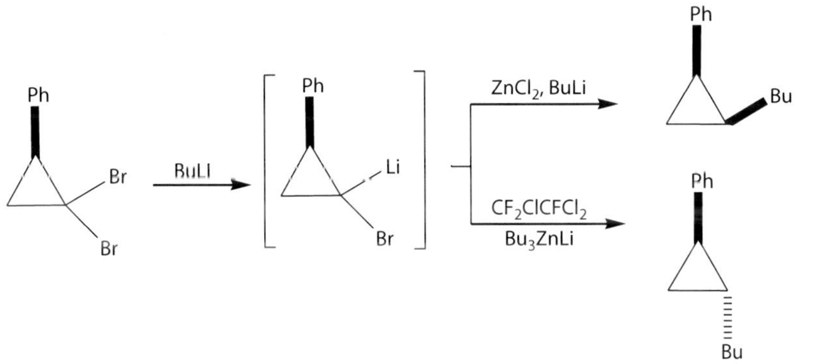

101 102 103

SCHEME 7.38

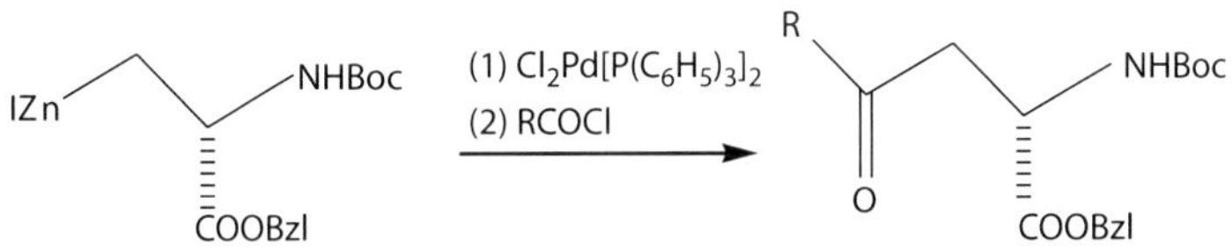

104 105

SCHEME 7.39

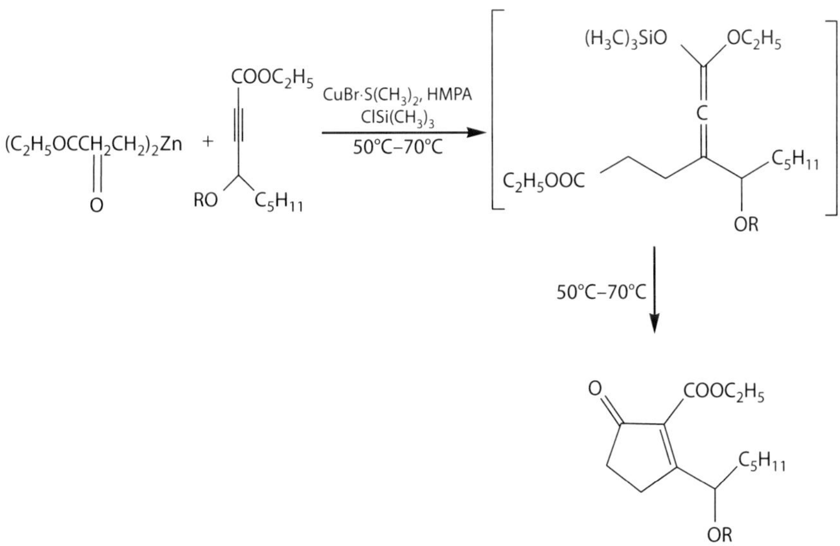

SCHEME 7.40

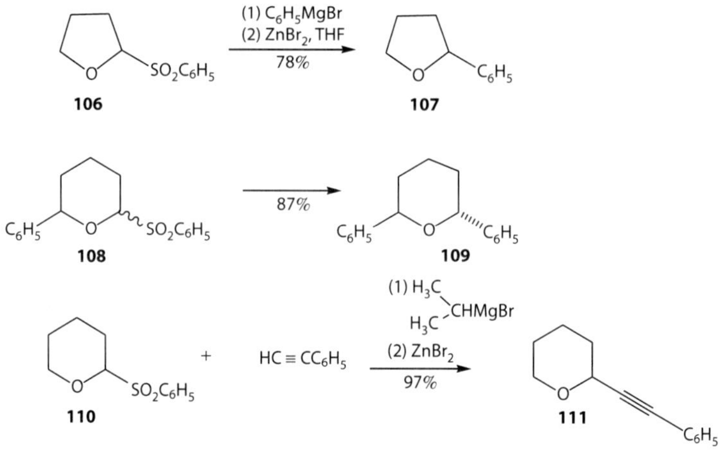

SCHEME 7.41

7.1.3.1 Synthesis

7.1.3.1.1 Transmetalation

This reaction has been used to prepare the important organocopper compounds (Scheme 7.42) and many other organometallic compounds (Lipshutz 1994; Normant 1972).

Organocopper compound **114** can be formed from the reaction of $(CH_3)_3Sn–Si(CH_3)_3$ and $Bu_2Cu(CN)Li_2$ (Lipshutz et al. 1990) (Scheme 7.43).

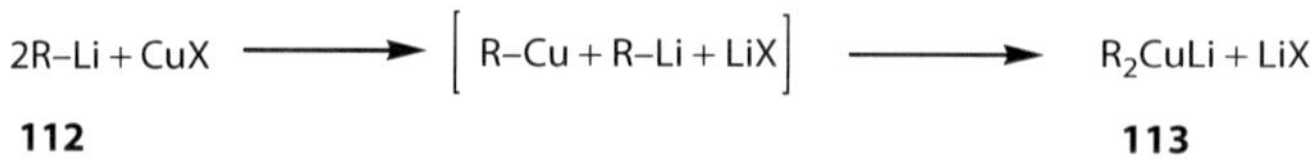

SCHEME 7.42

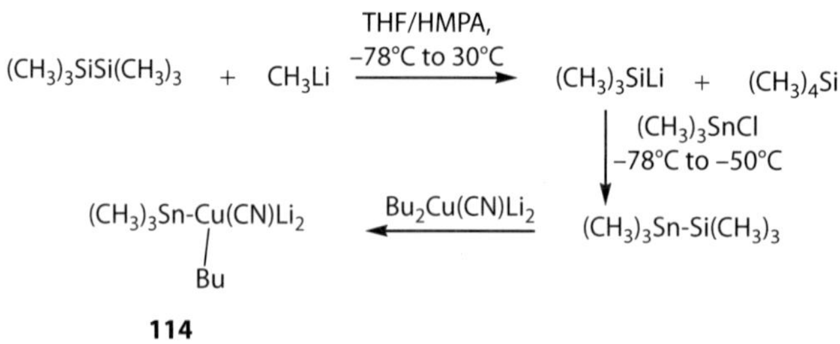

SCHEME 7.43

7.1.3.2 Reactions

7.1.3.2.1 Defunctionalization of Heteroalkenes

The C–X bonds of alkenyliodonium salts, 1,2-bis(phenyltelluro)alkenes, and ketene bis(methylthio)acetals **115**, which carry an electron-deficient substituent, are selectively replaced. Only one group of the chalcogenides is affected, and the ketene dithioacetals undergo reductive cleavage to give **116**. However, the alkenylcopper intermediates can be acetylated to form **116** (Stang et al. 1993; Hojo et al. 1994, respectively) (Scheme 7.44).

7.1.3.2.2 Epoxide Opening

Stereoselective alkene synthesis starts from the reaction of triethylsilyloxirane **117** with an organocuprate reagent and is concluded by oxidation of the formed β-silyl alcohol to the aldehyde, whose reaction with Grignard reagent and elimination of [Et_3Si/OH] leads to either the (*E*)- or the (*Z*)-alkene (**118** or **119**) by using different reagents (Scheme 7.45).

7.1.3.2.3 Allylic Displacements

Vinyl and allyl tin compounds are included in this process. Homoallylic alcohols asymmetrically substituted at the allylic position **122** are obtained from 4-bromo-2-alkenoyl derivatives of camphor sultam **120** in two steps, as shown in Scheme 7.46 (Bellina et al. 1994; Watrelot et al. 1994, respectively).

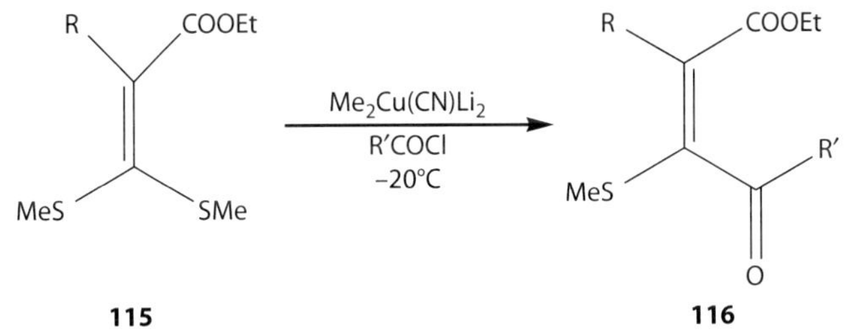

SCHEME 7.44

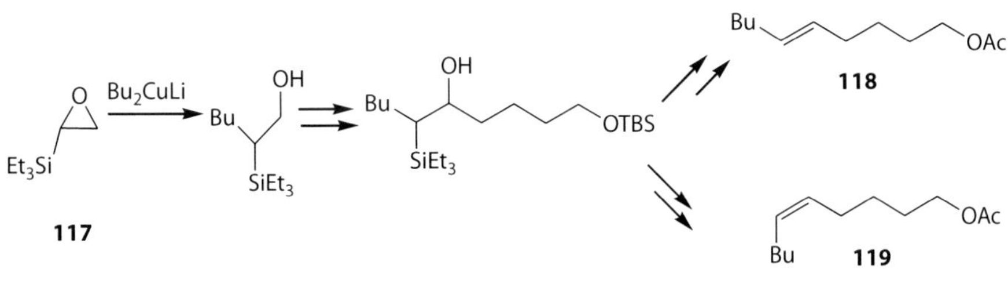

SCHEME 7.45

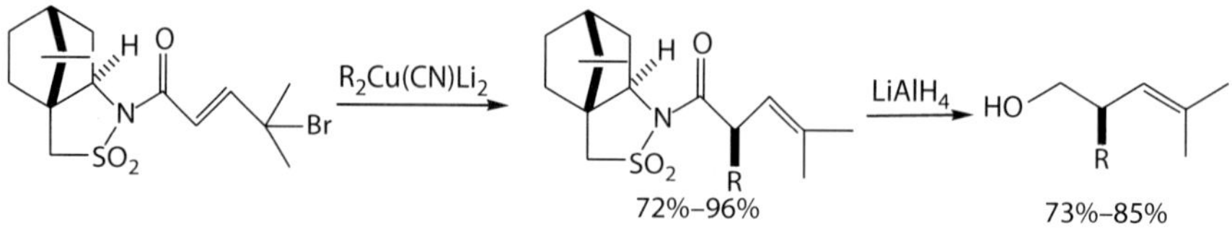

120 121 122

SCHEME 7.46

7.1.3.2.4 Conjugate Additions

1. Interesting preparations pertain to the synthesis of 3-tri-fluoromethyl-3-hydoxyalkanoic acids, 3-aminomethylcycloalkanones, and 4-vinyl-2-acetoxy-2-cyclobutanones **124**. In Step II, compound **124** gives catechol derivatives on thermolysis, as shown in Scheme 7.47 (Gautschi et al. 1994; Dieter and Alexander 1993; Gurski and Liebeskind 1993, respectively).
2. A homoallylic substituent (e.g., carbonate) **126** imparts diastereoselectivity to give **127** on the conjugate addition by its interaction with the indicated reagent (Hale and Hoveyda 1994) (Scheme 7.48).
3. A remarkable contrast in stereoselectivity of the copper and lithium reagents is unveiled in Scheme 7.49 in the formation of **129** and **130** from

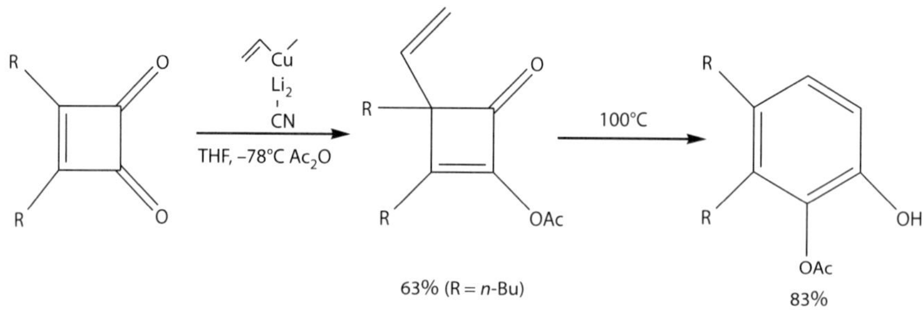

123 124 125

SCHEME 7.47

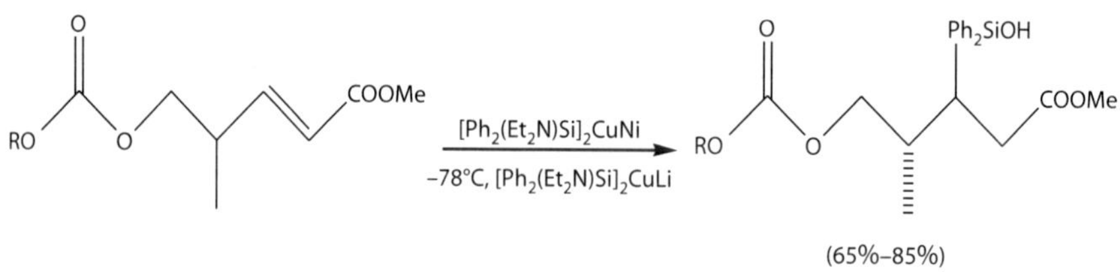

126 127

SCHEME 7.48

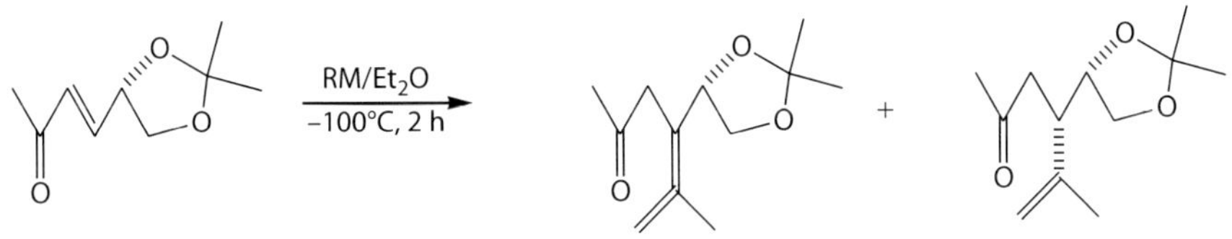

128 129 130

M = Cu 80% (5:1)
M = Li 60% (1:36)

SCHEME 7.49

128 (Leonard et al. 1995). Novel organocuprates used in conjugate additions are 2-(trimethylgermyl)allylcopper(I) [Me_2S complex], dilithium bis[2-(trimethylstannyl)vinyl] cyanocuprate, and the α-azoalkyl cuprates.

4. Alkenyl triflones have unusual reactivities. A stereoselective synthesis (Xiang and Fuchs 1996) of trisubstituted alkenyl triflones **132** involves the organocoppers and alkynyl triflones **131** (Scheme 7.50).
5. 1,6-Addition to sulfonyl enynes **133** gives allenes **134** (Hohmann and Krause 1995) (Scheme 7.51).

7.1.3.2.4 Reaction of Dialkylcuprates with α,β-Unsaturated Aldehydes

1,2-Addition to the carbonyl group takes place (Rieke 1989) (Scheme 7.52).

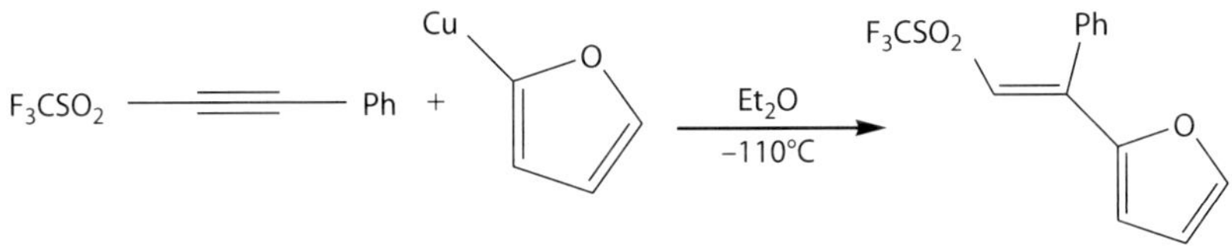

131 132

82% (*E*:*Z*, 12:1)

SCHEME 7.50

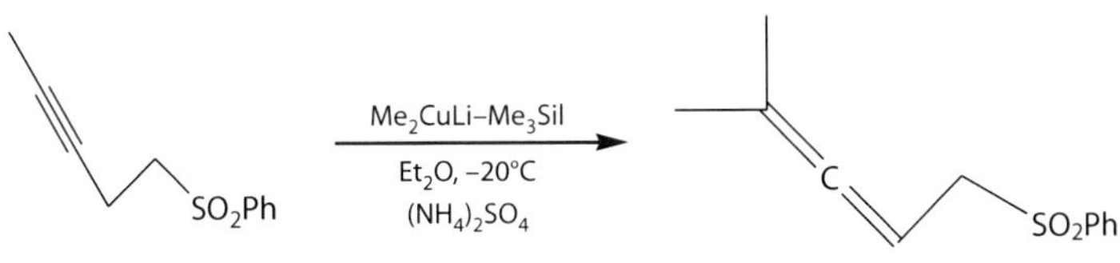

45%

133 134

SCHEME 7.51

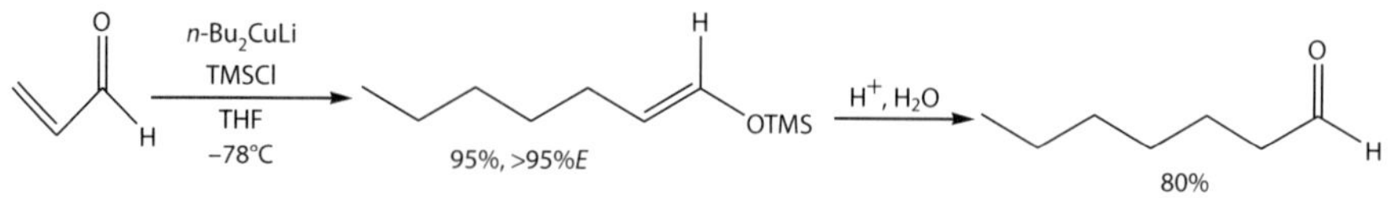

135 136 137

SCHEME 7.52

7.1.3.2.5 Acylation

Organocopper reagents react with acid chlorides **138**, leading to the formation of ketone **139** (Krasovskiy et al. 2006) (Scheme 7.53).

7.1.3.2.6 Reductive Uncoupling of Aldehydes

Reductive uncoupling of aldehydes is shown in Scheme 7.43 (Linderman and Siedlecki 1996) (Scheme 7.54).

7.1.3.2.7 Substitution Reactions

1. A methoxy group at C-4 of oxazolidin-2-ones is more reactive toward cuprates than primary chloride. The displacement of a methoxy group from tin-containing mixed acetals shows diastereoselectivity, as illustrated by the formation of **143** from **142** (Linderman and Chen 1995) (Scheme 7.55).
2. β-Amino acid derivatives of the (*R*)-series have been obtained from the β-tosylamino-γ-butyrolactone derived from l-aspartic acid, as shown in Scheme 7.56 (Jefford et al. 1996).

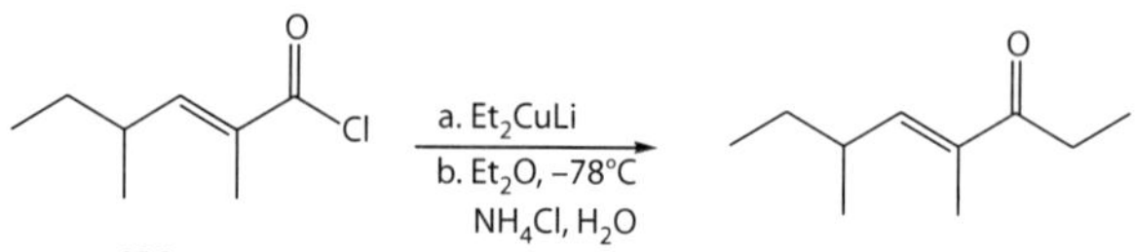

138 139

SCHEME 7.53

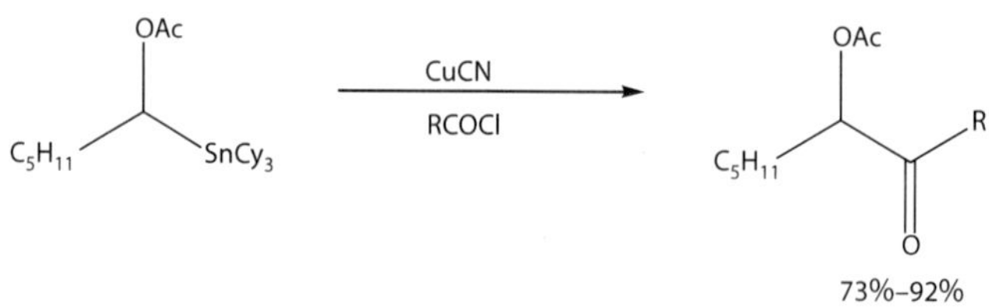

140 141

SCHEME 7.54

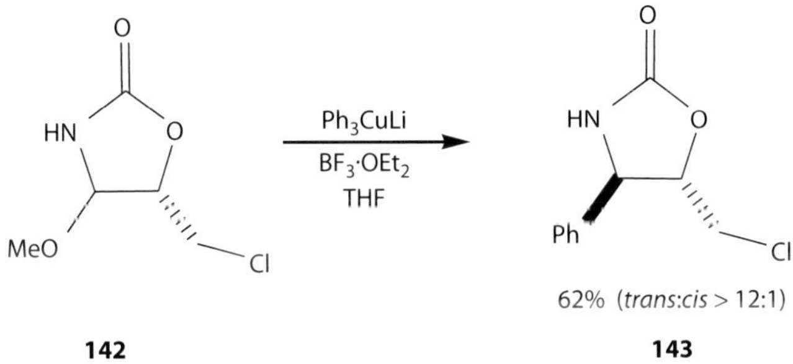

SCHEME 7.55

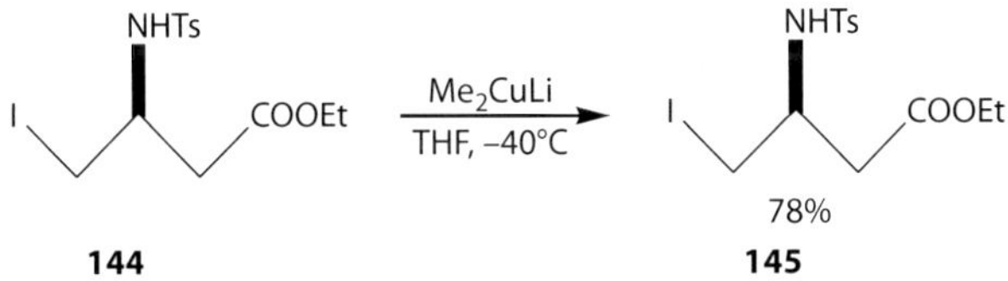

SCHEME 7.56

3. Organocopper reagents prepared from the more readily available organozincs have found much use in the synthesis of highly functionalized molecules. For example, Ni-catalyzed hydrozincation of allylic alcohol **146** initiates the preparation of α-(4-hydroxyalkyl) acrylates **147** when the cuprate intermediates are used in coupling with the α-bromomethylacrylic esters (Vettel et al. 1995) (Scheme 7.57).

7.1.3.2.7 Heterocycle Openings

Due to the attack of copper reagents at the α-carbon, α,β-epoxy silanes **148** furnish β-hydroxy silanes **149**, as shown in Scheme 7.47 (Hudrlik et al. 1996) (Scheme 7.58).

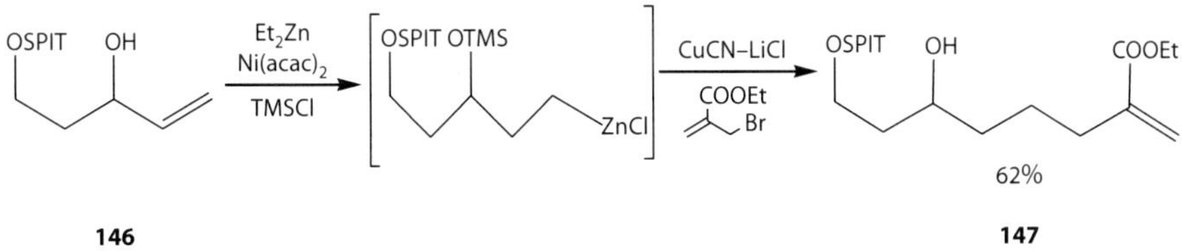

SCHEME 7.57

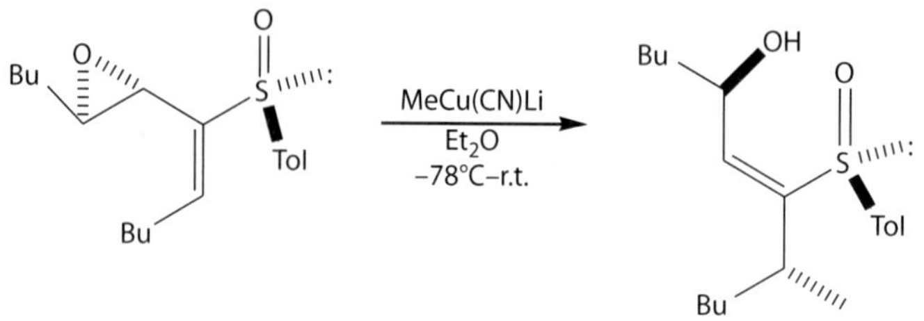

91%

148 **149**

SCHEME 7.58

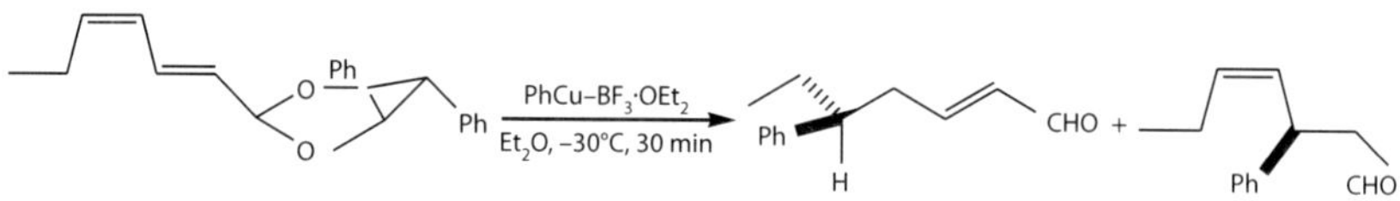

(70:30)

150 **151** **152**

SCHEME 7.59

α,β,γ,δ-Dienals derived from dioxolanes can react with organocopper reagents at both the α- and γ-position. Chiral dioxolanes of such a substitution pattern give optically active products due to remote asymmetric induction, as illustrated in Scheme 7.48 (Rakotoarisoa et al. 1996) (Scheme 7.59).

REFERENCES

Angiolelli, M. E., Casalnuovo, A. L. and Selby, T. P. 2000. Palladium-catalyzed cross-coupling of benzylzinc reagents with methylthio *N*-heterocycles: A new coupling reaction with unusual selectivity. *Synlett* 6:905–907.

Bellina, F., Carpita, A., De Santis, M. and Rossi, R. 1994. Synthesis of 2-tributylstannyl-1-alkenes from 2-tributylstannyl-2-propen-1-yl acetate. *Tetrahedron* 50:4853–4872.

Boudier, A., Hupe, E. and Knochel, P. 2000. Highly diastereoselective synthesis of monocyclic and bicyclic secondary diorganozinc reagents with defined configuration. *Angew. Chem.* 39:2294–2297.

Brigas, A. F. and Johnstone, R. A. W. 2000. Hetero aromatic ethers of phenols in nickel-catalysed dipso-replacement reactions with magnesium, zinc and tin organometallic compounds. *J. Chem. SP1*. 11:1735–1739.

Brown, D. S., Bruno, M., Davenport, R. J. and Ley, S. V. 1989. Substitution reactions of 2-benzenesulfonyl cyclic ethers with carbon nucleophiles. *Tetrahedron* 45:4293–4308.

Coates, G. E. and Wade, K. 1967. *Organometallic Compounds. The Main Group Elements*, vol. 1. Methuen, London.

Crimmins, M. T. and Nantermet, P. G. 1990. Addition of zinc homoenolates to acetylenic esters: A formal [3+2] cycloaddition. *J. Org. Chem.* 55:4235–4237.

Cutri, S., Bonin, M., Micouin, L., Froelich, O., Quirion, J. C. and Husson, H. P. 2000. Diastereoselective synthesis of 2-aryl-3-aminoazepanes via a novel ring-enlargement process. *Tetrahedron Lett.* 41:1179–1182.

Deniau, E. and Enders, D. 2000. A new simple and convenient synthesis of 3-substituted phthalimidines. *Tetrahedron Lett.* 41:2347–2350.

Dieter, R. K. and Alexander, C. W. 1993. Conjugate addition-reactions of α-aminoalkylcuprates prepared from organostannyl *tert*-butylcarbamates. *Synlett* 6:407–409.

Dimitrov, V. and Kamenova-Nacheva, M. 2009. Enantioselective organozinc-catalysed additions to carbonyl compounds: Recent developments (Review). *J. Univ. Chem. Tech. Metallurgy* 44(4):317 332.

Eick, H. and Knochel, P. 1996. The preparation of 1,3-dizinc organometallics via a boron–zinc exchange. *Angew. Chem.* 108:229–231.

Elschenbroich, C. 2005. *Organometallchemie*. Teubner Verlag, Wiesbaden.

Gautschi, M., Schweizer, W. B. and Seebach, D. 1994. Preparation of enantiomerically pure 4,4,4-trifluoro-3-hydroxy-butanoic acid-derivatives, branched in the 2-position or 3-position, from 6-trifluoromethyl-1,3-dioxan and dioxin-4-ones. *Chem. Ber.* 127:565–579.

Gurski, A. and Liebeskind, L. S. 1993. A new process for the regiocontrolled synthesis of substituted catechols and other 1,2-dioxygenated aromatics: Conjugate addition of vinyl-copper, aryl-copper, and heteroarylcopper reagents to cyclobutenediones followed by thermal rearrangement. *J. Am. Chem. Soc.* 115:6101–6108.

Hale, M. R. and Hoveyda, A. H. 1994. Diastereoselective heteroatom-directed conjugate addition of silylcuprate reagents to unsaturated carbonyls. A stereoselective route to carbonyl siloxanes. *J. Org. Chem.* 59:4370–4374.

Hanrahan, J. R. and Knight, D. W. 1998. A new strategy for the elaboration of pyrrolidine *n*-oxides using the reverse-cope elimination. *Chem. Commun.* 20:2231–2232.

Harada, T., Katsuhira, T., Hattori, K. and Oku, A. 1993. Stereoselective carbon–carbon bond-forming reaction of 1,1-dibromocyclopropanes via 1-halocyclopropylzincates. *J. Org. Chem.* 58:2958–2965.

Harrowven, D. C. and Hannam, J. C. 1999. 1,5-diketones from 3,4-dihydropyranones: An application in the synthesis of (+/−)-α-herbertenol. *Tetrahedron* 55:9333–9340.

Hessler, A., Kottsieper, K. W., Schenk, S., Tepper, M. and Stelzer, O. 2001. A novel access to tertiary and secondary *ortho*-aminophenylphosphines by protected group synthesis and palladium catalyzed P-C coupling reactions. *Z. Naturforsch. B J. Chem. Sci.* 56(4–5):347–353.

Hinkle, R. J., Leri, A. C., David, G. A. and Erwin, W. M. 2000. Addition of benzylzinc halides to alkenyl(phenyl)iodonium triflates: Stereoselective synthesis of trisubstituted alkenes. *Org. Lett.* 2:1521–1523.

Hohmann, M. and Krause, N. 1995. Synthesis of allenes by 1,6-addition of organocuprates to acceptor-substituted enynes: Scope and limitations. *Chem. Ber.* 128:851–860.

Hojo, M., Harada, H. and Hosomi, A. 1994. A novel and efficient generation of functionalized vinylcopper reagents and their reactions with electrophiles: Synthesis of β-methylthiobutenolides. *Chem. Lett.* 3:437–440.

Hudrlik, P. F., Ma, D., Bhamidipati, R. S. and Hudrlik, A. M. 1996. Ring-opening reactions of α- and β-epoxy silanes with organocopper reagents: Reaction at carbon or silicon. *J. Org. Chem.* 61:8655–8658.

Jefford, C. W., McNulty, J., Lu, Z. H. and Wang, J. B. 1996. The enantioselective synthesis of β-amino acids, their α-hydroxy derivatives, and the *n*-terminal components of bestatin and microginin. *Helv. Chim. Acta* 79:1203–1216.

Jensen, A. E., Dohle, W. and Knochel, P. 2000. Improved nickel-catalyzed cross-coupling reaction conditions. *Tetrahedron* 56:4197–4201.

Klement, I., Lennick, K., Tucker, C. E. and Knochel, P. 1993. Preparation of polyfunctional nitriles by the cyanation of functionalized organozinc halides with *p*-toluenesulfonyl cyanide. *Tetrahedron Lett.* 34:4623–4626.

Knochel, P. and Jones, P. 1999. *Organozinc Reagents. A Practical Approach*. Oxford University Press, Oxford.

Krasovskiy, A. V., Malakhov, V., Gavryushin, A. and Knochel, P. 2006. Efficient synthesis of functionalized organozinc compounds by the direct insertion of zinc into organic iodides and bromides. *Angew. Chem. Int. Ed.* 45:6040–6044.

Langer, F., Schwink, L., Devasagayaraj, A., Chavant, P. Y. and Knochel, P. 1996. Preparation of functionalized dialkylzincs via a boron-zinc exchange: Reactivity and catalytic asymmetric addition to aldehydes. *J. Org. Chem.* 61:8229–8243.

Langer, F., Waas, J. and Knochel, P. 1993. Preparation and reactions of new dialkylzincs obtained by a boron-zinc transmetalation. *Tetrahedron Lett.* 34:5261–5264.

Lautens, M., Renaud, J. L. and Hiebert, S. J. 2000. Palladium-catalyzed enantioselective alkylative ring opening. *J. Am. Chem. Soc.* 122:1804–1805.

Lentsch, L. M. and Wiemer, D. F 1999. Addition of organometallic nucleophiles to β-keto phosphonates. *J. Org. Chem.* 64:5205–5212.

Leonard, J., Mohialdin, S., Reed, D., Ryan, G. and Swain, P. A. 1995. Stereoselective conjugate addition of organolithium and organocopper reagents to δ-oxygenated α,β-unsaturated carbonyl systems derived from glyceraldehyde acetonide. *Tetrahedron* 51:12843–12858.

Leroux, F. and Schlosser, M. 2002. Der Arin-Zugang zu Biarylen mit ungewoehnlichen Substitutionsmustern. *Angew. Chem.* 114:4447–4450.

Linderman, R. J. and Chen, S. 1995. Diastereoselective additions of copper and cuprate reagents to α-stannyl substituted mixed acetals. *Tetrahedron Lett.* 36:7799–7802.

Linderman, R. J. and Siedlecki, J. M. 1996. Selective copper-catalyzed coupling reactions of (α-acetoxyhexyl)tricyclohexyltin. *J. Org. Chem.* 61:6492–6493.

Lipshutz, B. H. 1994. Synthetic procedures involving organocopper reagents. In: Schlosser, M. (ed.), *Organometallics in Synthesis*. Wiley, Chichester.

Lipshutz, B. H., Sharma, S., Reuter, D. C. 1990. A new transmetallation route to mixed trimethylstannylcuprates: $Me_3Sn(R)Cu(CN)Li_2$. *Tetrahedron Lett.* 31:7253–7261.

Lipshutz, B. H. and Vivian, R. V. 1999. Cu(I)-catalyzed substitution reactions of activated vinyl triflates with functionalized organozinc reagents. *Tetrahedron Lett.* 40:2871–2874.

Marshall, J. A. and Adams, N. D. 1999. Addition of allenylzinc reagents, prepared in situ from nonracemic propargylic mesylates, to aldehydes. A new synthesis of highly enantioenriched homopropargylic alcohols. *J. Org. Chem.* 64:5201–5204.

Meyer, C., Marek, I., Courtemanche, G. and Normant, J. F. 1993. Carbocyclization of functionalized zinc organometallics. *Synlett* 4:266–268.

Mohr, J. T., Nishimata, T., Behenna, D. C. and Stoltz, B. M. 2006. Catalytic enantioselective decarboxylative protonation. *J. Am. Chem. Soc.* 128:11348–11349.

Moller, C., Mikulas, M., Wierschem, F. and Ruck-Brawn, K. 2000. Synthesis of 5-substituted α,β-butenolides by iron-promoted intramolecular cyclocarbonylation: Addition of organometallic reagents to iron-substituted enals. *Synlett* 2:182–184.

Moody, C. J., Gallagher, P. T., Lightfoot, A. P. and Slawin, A. M. Z. 1999. Chiral oxime ethers in asymmetric synthesis. 3. Asymmetric synthesis of (*R*)-*N*-protected α-amino acids by the addition of organometallic reagents to the ROPHy oxime of cinnamaldehyde. *J. Org. Chem.* 64:4419–4425.

Myers, A. G. and Movassaghi, M. 1998. Highly efficient methodology for the reductive coupling of aldehyde tosylhydrazones with alkyllithium reagents. *J. Am. Chem. Soc.* 120:8891–8892.

Normant, J. F. 1972. Organocopper(I) compounds and organocuprates in synthesis. *Synthesis* 1972(2):63–80.

Noyori, R. 1994. *Asymmetric Catalysis in Organic Chemistry*. Wiley, New York.

Ohno, H., Toda, A., Oishi, S., Tanaka, T., Takemoto, Y., Fujii, N. and Ibuka, T. 2000. Novel synthesis of chiral terminal allenes via palladium(0)-catalyzed reduction of mesylates of 2-bromoalk-2-en-1-ols bearing a protected amino group, using diethylzinc. *Tetrahedron Lett.* 41:5131–5134.

Ovaska, T. V., Roark, J. L., Shoemaker, C. M. and Bordner, J. 1998. A convenient route to fused 5-7-6 tricyclic ring-systems. *Tetrahedron Lett.* 39:5705–5708.

Paquette, L. A. 1968. Reaction of α-halo sulfones with strong bases to yield alkenes. *Accts. Chem. Res.* 1:209–216.

Parham, W. E., Jones, L. D. and Sayed, Y. 1975. Four- to seven-membered ring annulation of aryl bromides bearing ortho side chains having an electrophilic moiety, accomplished by halogen-metal exchange and subsequent nucleophilic ring closure. *J. Org. Chem.* 40:2394–2399.

Parker, K. A. and Georges, A. T. 2000. Reductive aromatization of quinols: Synthesis of the C-arylglycoside nucleus of the papulacandins and chaetiacandin. *Org. Lett.* 2:497–499.

Qi, X. and Montgomery, J. 1999. New three-component synthesis of 1,3-dienes employing nickel catalysis. *J. Org. Chem.* 64:9310.

Rakotoarisoa, H., Perez, R. G., Mangeney, P. and Alexakis, A. 1996. Remote asymmetric induction. New mechanistic insights concerning the S_N' and S_N'' substitution in organocopper chemistry. *Organometallics* 15:1957–1959.

Rezaei, H., Yamanoi, S., Chemla, F. and Normant, J. F. 2000. Fritsch-Buttenberg-Wiechell rearrangement in the aliphatic series. *Org. Lett.* 2:419–421.

Rieke, R. D. 1989. Preparation of organometallic compounds from highly reactive metal powders. *Science* 246:1260–1264.

Seebach, D., Behrendt, L. and Felix, D. 1991. Titanate-catalyzed enantioselective addition of dialkylzinc compounds—Generated *in situ* from Grignard reagents in ether—To aldehydes. *Angew. Chem. Int. Ed.* 30:1008–1009.

Stang, P. J., Blume, T. and Zhdankin, V. V. 1993. Synthesis of enediynes by reaction of bicycloalkenyldiiodonium salts with lithium alkynyl cuprates. *Synthesis* 1:35–36.

Uchiyama, M., Koike, M., Kameda, M., Kondo, Y. and Sakamoto, T. 1996. Unique reactivities of new highly coordinate date complexes of organozinc derivatives. *J. Am. Chem. Soc.* 118:8733–8734.

Varea, T., Grancha, A. and Asensio, G. 1995. A simple and efficient route to 1,4-diketones from squaric acid. *Tetrahedron* 51:12373–12382.

Varghese, J. P., Knochel, P. and Marek, I. 2000. New allene synthesis via carbocupration-zinc carbenoid homologation and AY-elimination sequence. *Org. Lett.* 2:2849–2852.

Vettel, S., Vaupel, A. and Knochel, P. 1995. A new preparation of diorganozincs from olefins via a nickel catalyzed hydrozincation. *Tetrahedron Lett.* 36:1023–1026.

Watrelot, S., Parrain, J. L. and Quintard, J. P. 1994. Stereoselective synthesis of allyltins from vinyltins: A new route to enantioenriched α-substituted (γ-alkoxyallyl)tins from vinyltin acetals. *J. Org. Chem.* 59:7959–7961.

Wittig, G. and Löhmann, L. 1942. [1,2]-Wittig rearrangement. *Liebigs Ann. Chem.* 550:260–268.

Xiang, J. and Fuchs, P. L. 1996. Alkenylation of C–H bonds via reaction with vinyl and dienyl triflones. Stereospecific synthesis of trisubstituted vinyl triflones via organocopper addition to acetylenic triflones. *J. Am. Chem. Soc.* 118:11986–11987.

Zhang, Y., Miller, J. A. and Negishi, E. 1989. Carbon–carbon bond formation via opening of epoxysilanes with organometals containing lithium and copper. *J. Org. Chem.* 54:2043–2044.

8 Ultrasound

An Efficient Tool for the Synthesis of Bioactive Heterocycles

Essam M. Hussein

CONTENTS

8.1 INTRODUCTION

The importance of heterocycles in many fields of science (including organic, inorganic, bioorganic, agricultural, industrial, pharmaceutical, and medicinal chemistry, as well as materials science) can hardly be overstated, and justifies a long-lasting effort to work out new synthetic protocols for their production (Eicher and Hauptmann 2003). The ever-increasing awareness of the need to protect natural resources through the development of environmentally sustainable processes and the optimization of energy consumption has guided the actions of both the private and governmental sectors of society. Such a tool for this purpose that is attractive is based on ultrasound-promoted heterocyclization reactions of suitably functionalized substrates, which can allow the synthesis of highly functionalized heterocycles using readily available starting materials under mild and selective conditions. The use of ultrasound to enhance or alter chemical reactions is known as sonochemistry. Ultrasound enhances the reactivity of chemical reactions via the process of acoustic cavitation (Mason and Lorimer 2002). The assistance of ultrasonic irradiation efficiently shortens the reaction times. Simple experimental procedures, very high yields, increased selectivity, and the clean reaction of many ultrasound-induced organic transformations offer additional convenience in the field of synthetic organic chemistry (Li et al. 2012; Ghahremanzadeh et al. 2011; Kowsari and Mallakmohammadi 2011; He et al. 2011; Li et al. 2010a,b,c; Bazgir et al. 2010). Recently, ultrasound has been utilized to accelerate a large number of synthetically useful organic reactions (Suslick 1990; Fillion and Luche 1998). As well as the field of organic chemistry, sonochemistry has also been used in the preparation of micro- and nanomaterials, such as protein microspheres (Gedanken 2004). Ultrasound also has many therapeutic and diagnostic applications, such as medical ultrasonography and cleaning of teeth; however, a higher frequency (1–10 MHz) is used in these cases than in sonochemistry (20–100 kHz).

In this review, the use of ultrasound to accelerate reactions has proven to be a particularly important tool for meeting the green chemistry goals of minimization of waste and reduction of energy requirements.

8.1.1 Historical Background

The first report about the effect of ultrasound to chemical reactions is from 1927 by Richards and Loomis, involving rate studies on the hydrolysis of dimethyl sulfate and the iodine "clock" reaction (the reduction of potassium iodate by sulfurous acid) (Richards and Loomis 1927). With some exceptions (Porter and Young 1938; Renaud 1950), the field was relatively overlooked for nearly 60 years. However, in the 1980s, sonochemistry was reborn and began to be widely used in many different areas. The reason for this growth was the availability of inexpensive and appropriate laboratory equipment, such as ultrasonic cleaning baths (low intensity) and ultrasonic probes (high intensity).

8.1.2 Theory

The chemical effects of ultrasound are caused by cavitation bubbles which are generated during the rarefaction, or negative-pressure, period of sound waves. If this applied negative pressure is strong enough to break down the intermolecular van der Waals force of the liquid, small cavities or gas-filled microbubbles are formed. Cavitation is considered to be a nucleated process, meaning that these micrometer-scale bubbles will be formed at preexisting weak points in the liquid, such as gas-filled crevices in suspended particulate matter or transient microbubbles from prior cavitation events. As microbubbles are formed, they absorb energy from ultrasound waves and grow. However, they will reach a stage where they can no longer absorb energy efficiently. Without the energy input, the cavity can no longer sustain itself and explodes. It is this explosion of the cavity that creates an uncommon environment for chemical reactions (Suslick 1990). Several factors can affect the efficiency of bubble collapse, such as (Sehgal and Wang 1981): (1) temperature; (2) vapor pressure; (3) surface tension and viscosity; (4) ultrasound frequency; (5) thermal conductivity; and (6) acoustic intensity. There are three classes of sonochemical reactions (Suslick 1990; Cella and Stefani 2009):

1. *Homogeneous sonochemistry*: In the case of volatile molecules, the bubbles (or cavities) are believed to act as a microreactor; as the volatile molecules enter, the microbubbles and the high temperature and pressure produced during cavitation break their chemical bonds, and short-lived chemical species

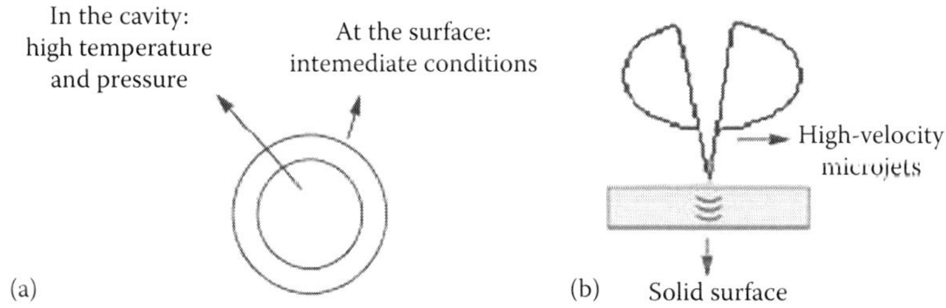

FIGURE 8.1 (a) Cavitation bubble in a homogeneous system; (b) cavitation bubble in a heterogeneous system.

are returned to the bulk liquid at room temperature, thus reacting with other species. Compounds of low volatility, which are unlikely to enter bubbles and thus be directly exposed to these extreme conditions, still experience a high energy environment resulting from the pressure changes associated with the propagation of the acoustic wave or with bubble collapse (shock waves); alternatively, they can react with radical species generated by sonolysis of the solvent. In homogeneous systems, the cavity remains spherical during collapse because its surroundings are uniform (Figure 8.1a).

2. *Heterogeneous sonochemistry*: In this case, the reaction is influenced primarily through the mechanical effects of cavitation, such as surface cleaning, particle size reduction, and improved mass transfer. When cavitation occurs in a liquid near a solid surface, the dynamics of cavity collapse change dramatically. Close to a solid boundary, cavity collapse is very asymmetric and generates high-speed jets of liquid (with velocities of approximately 400 kmh; Figure 8.1b). These jets hit the surface with massive force. This process can cause severe damage at the point of impact and produce newly exposed highly reactive surfaces.
3. *Sonocatalysis* (*overlap of homogeneous and heterogeneous sonochemistry*): These are heterogeneous reactions that include a radical and ionic mechanism. Radical reactions will be chemically enhanced by sonication, but the general mechanical effect of cavitation (described in heterogeneous sonochemistry) may very well still apply. If radical and ionic mechanisms lead to different products, ultrasound should favor the radical pathway, potentially leading to a change in the nature of the products of the reaction.

8.1.3 Sonochemical Synthesis of Bioactive Heterocycles

8.1.3.1 Sonochemical Synthesis of Three-Membered Nitrogen-Containing Heterocycles

Three-membered nitrogen-containing heterocycles (aziridines) have attracted the attention of chemists due to their association with a broad spectrum of biological activity. Many natural products possessing an aziridine moiety have been shown to have high biological activity, such as mytomycins, together with porfiromycin and mitiromycin. Aziridines mainly exhibit cytotoxic, anticancer (Hata et al. 1976), protein-inhibiting (Schirmeister and Peric 2000), antimicrobial (Harada et al. 2004), and antibiotic activities (Miller et al. 1971; Ogasawara and Liu 2009; Tsuchida et al. 1995).

The first ultrasound-assisted aziridination of olefins was reported by Chanda et al. (2001). Aziridine (**3**) was obtained in a good yield by ultrasound-assisted aziridation of styrene (**1**) in the presence of bromoamine-T (**2**) and $CuCl_2$ (10 mol%) in just 20 min at room temperature (Scheme 8.1).

Also, the authors noted that no reaction was observed when cinnamates (**4**) were subjected to conventional stirring conditions. However, when these were ultrasonically irradiated, (2*S*,3*R*)-3-phenyl-1-tosylaziridine-2-carboxylates (**5**) were obtained (Scheme 8.2).

Fluoroaziridines (**10**) were successfully synthesized by the reaction of *N*-alkyl-*N*-benzhydrylideneamines (**8**) with monofluorocarbene (**7**) that was generated *in situ* from dibromofluoromethane (**6**) in the presence of lead and tetrabutylammonium bromide (TBAB) under ultrasonic irradiation at 40°C (Konev et al. 2005). The reaction proceeded via azomethine ylide intermediates (**9**), and aziridines (**10**) were obtained (Scheme 8.3).

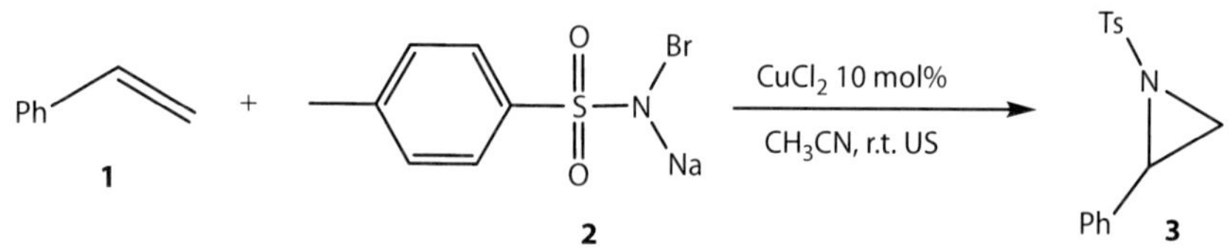

SCHEME 8.1

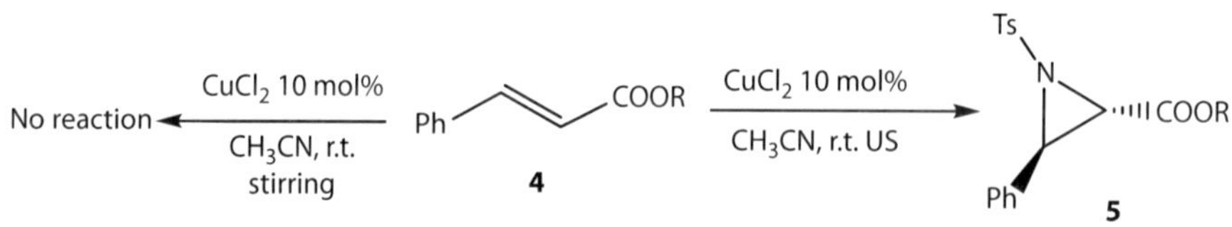

SCHEME 8.2

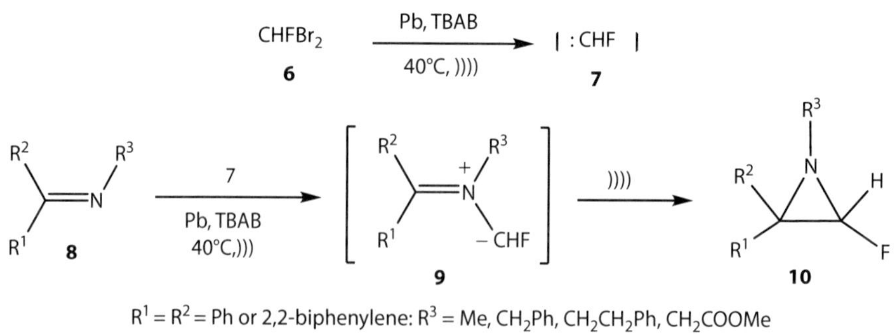

SCHEME 8.3

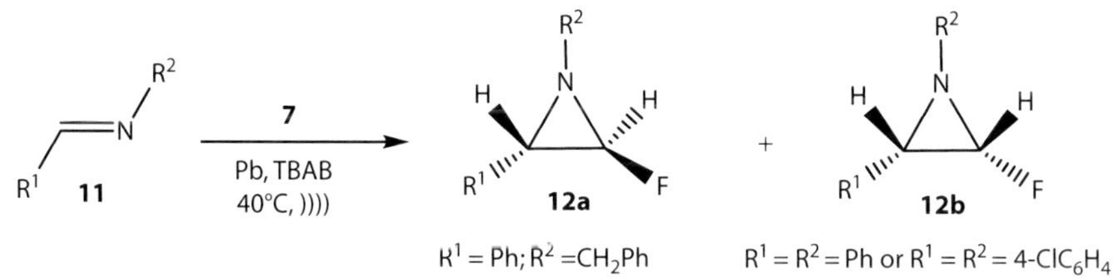

SCHEME 8.4

Also, stereoisomeric aziridines **12a** and **12b** were obtained when imine derivatives of arenecarbaldehydes (**11**) were subjected to the same conditions, in the presence of elemental lead (Pb) and TBAB under ultrasonic irradiation at 40°C (Scheme 8.4).

8.1.3.2 Sonochemical Synthesis of Five-Membered Heterocycles

Five-membered *N*-containing heterocycles and their derivatives have attracted increasing interest as key intermediates for the synthesis of biologically active compounds such as potent antioxidant, antitumor, antibacterial, antifungal, and antiviral agents.

For example, fluconazole, itraconazole, voriconazole, and posaconazole are commercially available antifungal agents that contain a triazole nucleus. Celecoxib is a nonsteroidal anti-inflammatory and analgesic agent of the pyrazole class. Isoxazole compounds such as Valdecoxib are selective COX-2 inhibitors used in the treatment of pain.

The azomethine ylide represents one of the most reactive and versatile classes of 1,3-dipoles and is trapped readily by a range of dipolarophiles, forming substituted pyrroles or pyrrolidines (Coldham and Hufton 2005). Synthesis of aziridines by reaction of a fluorocarbene with imines under ultrasonic irradiation was described previously (Konev et al. 2005). This reaction proceeds by the formation of an azomethine ylide (**9**), which can be trapped with dimethyl maleate (**13**) to produce pyrrolidines (**14**) or pyrrole rings (**15**) in moderate yields (Scheme 8.5).

Recently, Eftekhari-Sis and Vahdati-Khajeh (2013) developed a green and effective method for the preparation of 5-aryl-4-hydroxy-2-methyl-1*H*-pyrrole-3-carboxylates (**18**) via a three-component reaction of arylglyoxal hydrates (**16**) with β-dicarbonyl compounds (**17**) in the presence of ammonium acetate using water as a solvent under ultrasonic irradiation (Scheme 8.6). The reactions proceeded rapidly and afforded the corresponding pyrroles in good to high yields in very short reaction times.

Mamaghani and Dastmard (2009) reported an efficient and practical method for the regioselective synthesis of 1,5-diarylpyrazoles (**21**) in excellent yields (80%–90%) by the reaction between Baylis–Hillman adducts (**19**) and phenylhydrazine hydrochloride (**20**) in 1,2-dichloroethane (DCE) under ultrasonic irradiation with reaction times of 60–180 min at 60°C (Scheme 8.7).

1,3,5-Triaryl-2-pyrazolines (**24**) can be prepared in good to excellent yields via the condensation of chalcones (**22**) and phenylhydrazine (**23**) in acetic acid under

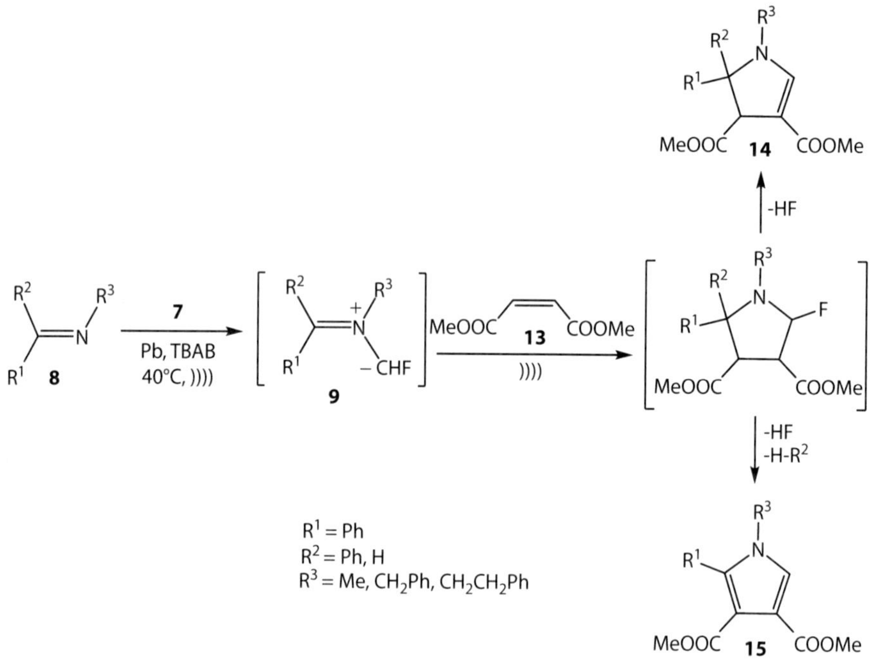

SCHEME 8.5

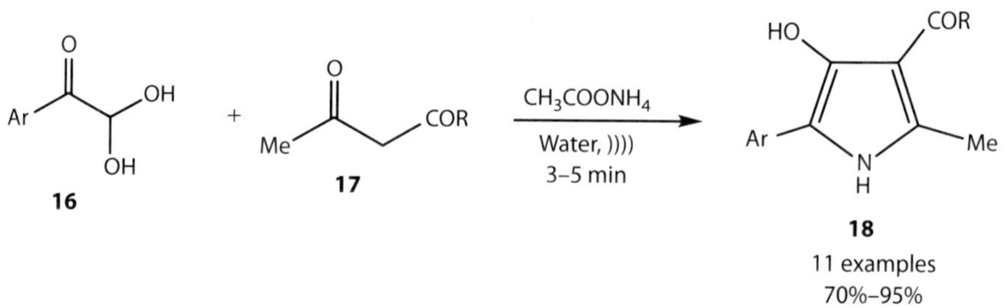

SCHEME 8.6

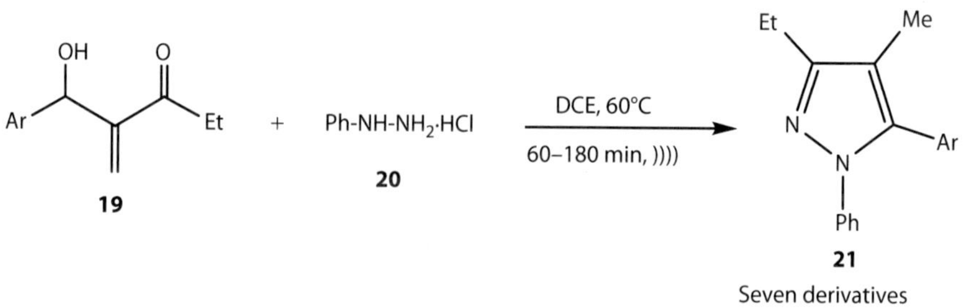

SCHEME 8.7

ultrasound irradiation at room temperature (Scheme 8.8). In this protocol, acetic acid is the reaction solvent and also acts as a catalyst (Lin and Li 2012).

Ultrasound-assisted synthesis of 1-thiocarbamoyl-3,5-diaryl-4,5-dihydropyrazoles (**27**) from chalcones (**25**) and thiosemicarbazide (**26**) catalyzed by potassium hydroxide (KOH) in high purity and yield was obtained after just 20 min at room temperature (Scheme 8.9), as reported by Pizzuti et al. (2009).

Similarly, Pizzuti et al. (2010) reported the cyclization of chalcones (**25**) with aminoguanidine hydrochloride (**28**) under essentially the same conditions (Scheme 8.10). The 4,5-dihydropyrazole derivatives (**29**) were obtained in high yields (75%–99%) in 30 min when employing sonication. The same reactions carried out under reflux without ultrasonic irradiation afforded the products in lower yields (57%–69%) and required substantially longer reaction times (3–6 h).

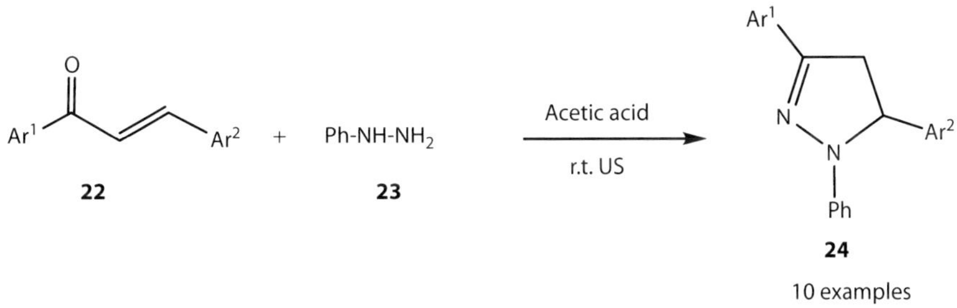

SCHEME 8.8

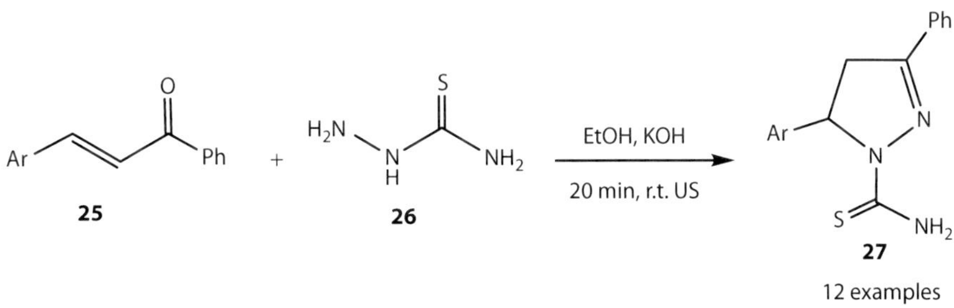

SCHEME 8.9

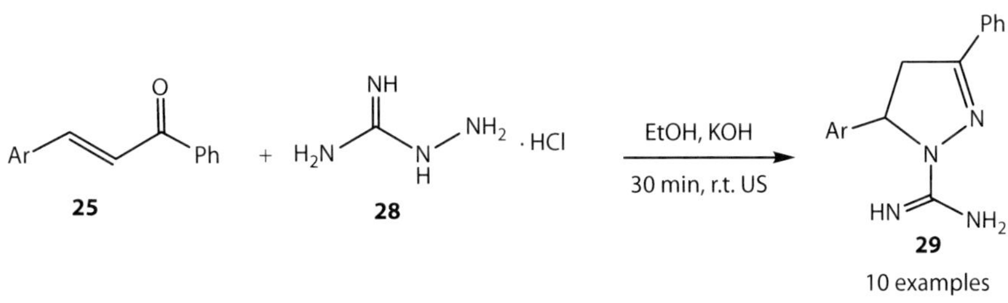

SCHEME 8.10

Analogously, Pathak et al. (2009) reported a comparative study between four activating methods for obtaining *N*-acetyl-pyrazolines (**33**), including reflux, solvent-free conditions, microwave irradiation, and ultrasonic irradiation.

Microwave irradiation was found to be the most efficient activating method, followed by ultrasound. Employing ultrasound, the reactions of 1,5-diarylpenta-1,4-dien-3-ones (**30**) with hydrazine (**31**) and acetic acid (**32**) in ethanol completed in 10–25 min and afforded the products (**33**) in excellent yields (76%–91%) (Scheme 8.11).

Ethyl benzoylacetate (**34**) was reacted with methylhydrazine (**35**) under ultrasonic irradiation to afford 1-methyl-3-phenyl-2-pyrazolin-5-one (**36**) (Scheme 8.12), which was obtained in excellent yields and shorter reaction times as compared with conventional condition. 1-Methyl-3-phenyl-1*H*-pyrazol-5(4*H*)-one (**36**) was allowed to react with phenylisothiocyanate in dimethylformamide (DMF) in the presence of KOH under ultrasonic irradiation at 25°C–30°C, followed by acidification with dilute hydrochloric acid (dil. HCl), which afforded the corresponding thiocarboxanilide (**37**).

Novel substituted 1,3,4-thiadiazoles (**40** and **41**) and bi(1,3,4-thiadiazole) (**43**) were synthesized from the reaction of thiocarboxanilide (**37**) with a series of different hydrazonyl halides (**38** and **39**) or *N,N'*-diphenyl-oxalodihydrazonoyl dichloride (**42**) in ethanol in the presence of triethylamine under ultrasonic irradiation (Schemes 8.13 and 8.14).

The reactions were carried out under both conventional and ultrasonic irradiation conditions. In general, improvements in rates and yields were observed when

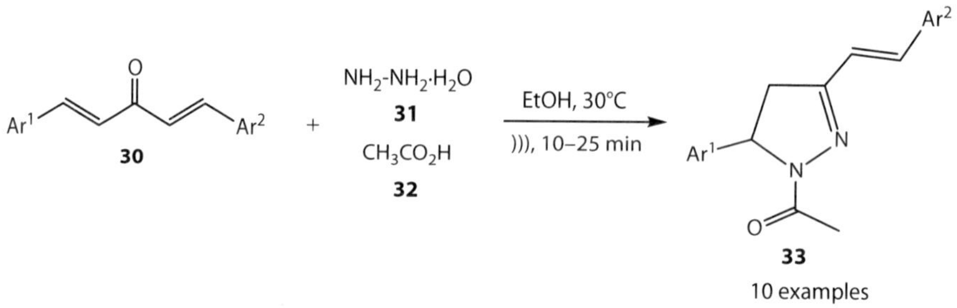

SCHEME 8.11

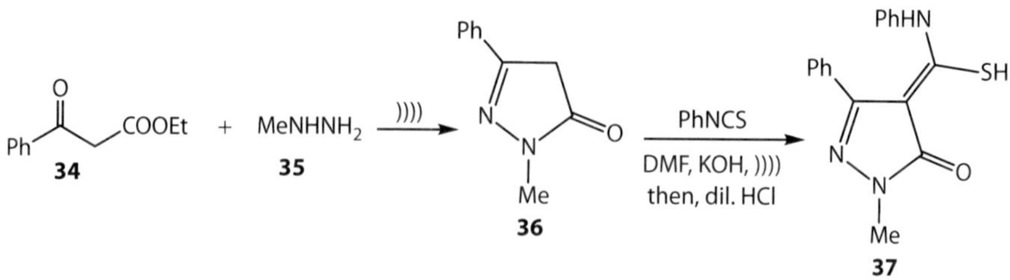

SCHEME 8.12

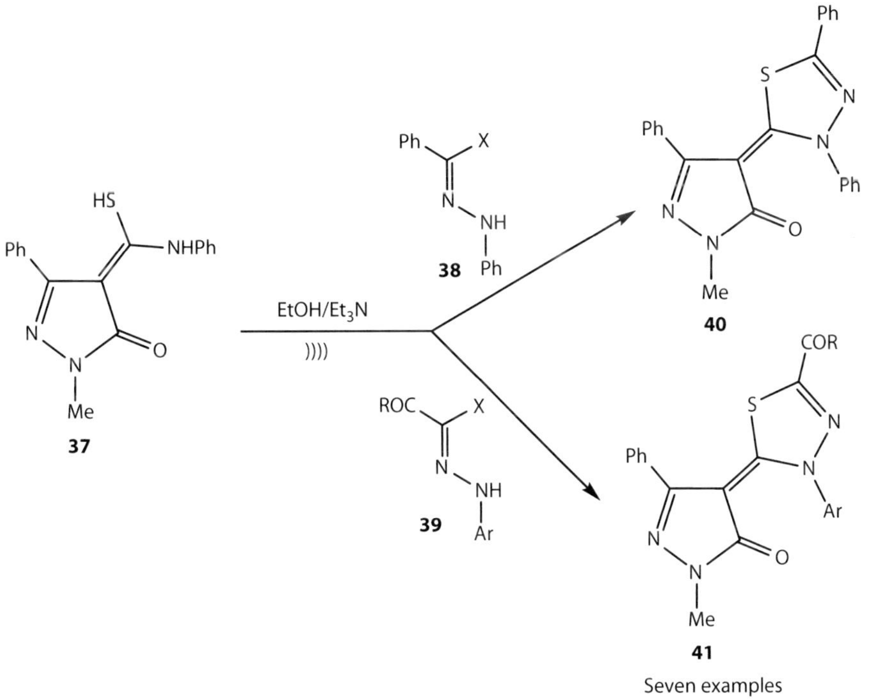

SCHEME 8.13

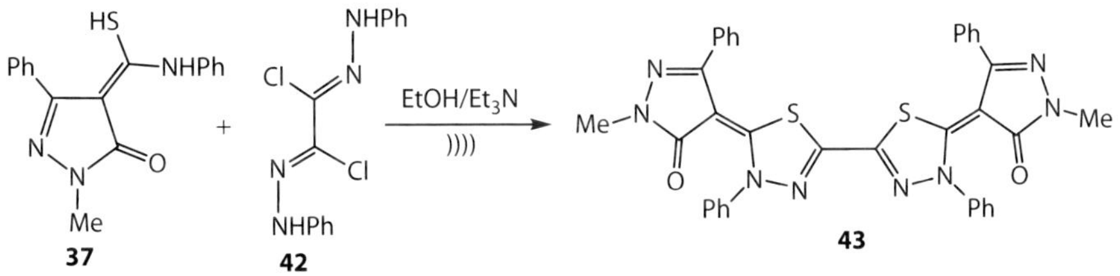

SCHEME 8.14

reactions were carried out under ultrasonic irradiation compared with traditional conditions (Abd El-Rahman et al. 2009).

3-Aryl-2,3-epoxy-1-phenyl-1-propanone (**44**) was revealed as a key intermediate for the synthesis of 1,3,5-triarylpyrazoles (**45**) in 69%–99% yields via a reaction with phenylhydrazine (**23**) catalyzed by HCl under ultrasonic irradiation at room temperature (Scheme 8.15). The same reactions performed in the absence of sonication gave substantially poorer yields (Li et al. 2010a).

Gomha and Khalil (2012) developed a successful application of ultrasound irradiation for the rapid synthesis of a novel series of 3-[1-(4-substituted-5-(aryldiazenyl) thiazol-2-yl)hydrazono)ethyl]-2*H*-chromen-2-ones (**48**) in good yields (72%–76%), via reactions of 2-(1-(2-oxo-2*H*-chromen-3-yl)ethylidene)thiosemicarbazide (**47**) and hydrazonoyl halides (**39**) in dioxane in the presence of a catalytic amount of

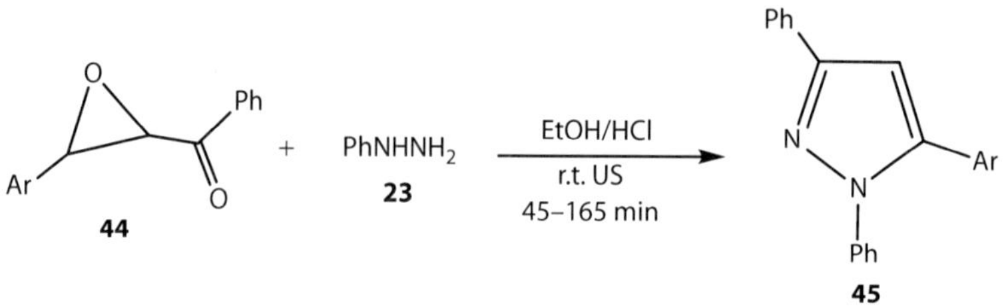

44 23 45
10 examples

SCHEME 8.15

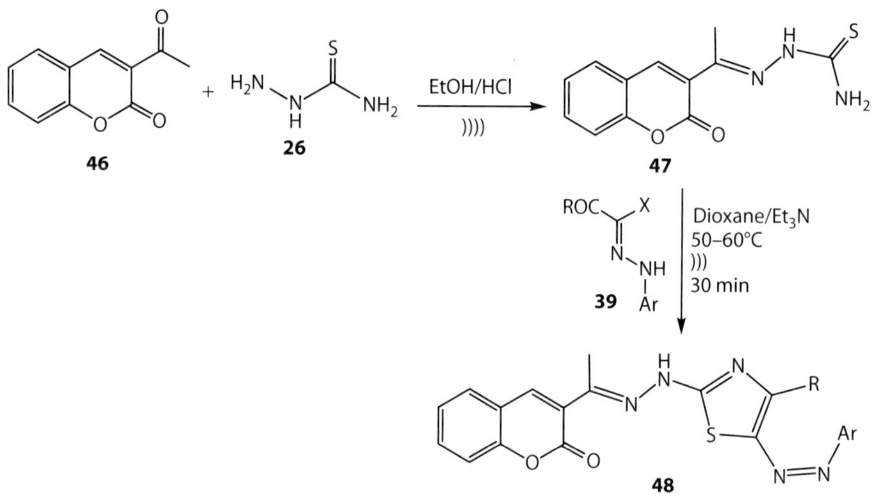

46 26 47 39 48
Eight examples

SCHEME 8.16

triethylamine (TEA) as the basic catalyst (Scheme 8.16). 2-(1-(2-oxo-2*H*-chromen-3-yl)ethylidene)thiosemicarbazide. (**47**) was previously prepared by refluxing 3-acetyl-2*H*-chromen-2-one (**46**) and thiosemicarbazide (**26**) in absolute ethanol in the presence of catalytic amounts of hydrochloric acid (Chimenti et al. 2010). Thiazolochromenone derivatives (**48**) showed potent cytotoxic activity against HaCaT cells (human keratinocytes).

Furthermore, a three-component ultrasound-assisted protocol for the synthesis of pyrazolo[1,2-*b*]phthalazine derivatives (**53**) was developed by Nabid et al. (2010). The reaction between phthalhydrazide (**49**), malononitrile (**50**) or ethyl cyanoacetate (**51**), and aromatic aldehydes (**52**) in ethanol at 50°C in the presence of triethylamine gave 1*H*-pyrazolo[1,2-*b*]phthalazine-5,10-diones (**53**) in excellent yields (85%–97%) in 60 min (Scheme 8.17).

The sonosynthesis technique was employed to promote the synthesis of pyrazolones (56) via the reaction of β-keto esters (**54**) with hydrazine derivatives (**55**) in ethanol (Al-Mutairi et al. 2010). The reactions were completed in short times (2–25 min) and afforded the products in poor to excellent yields (Scheme 8.18).

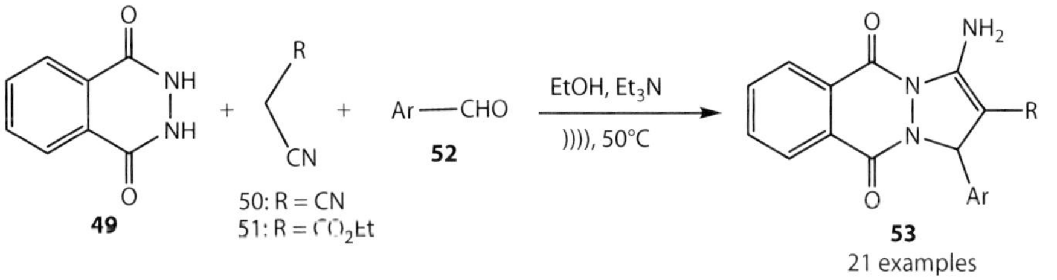

SCHEME 8.17

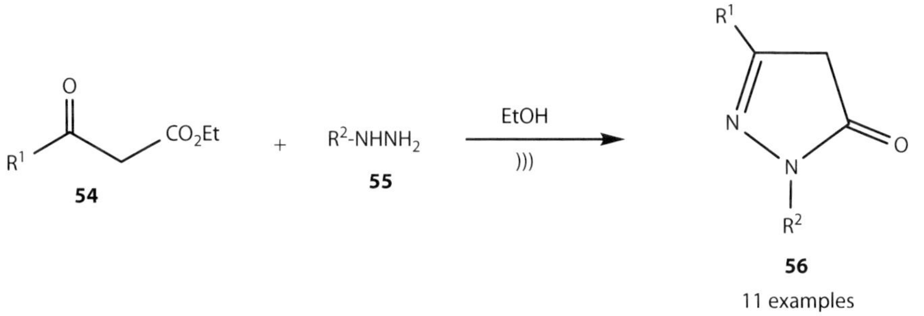

SCHEME 8.18

A multi-component one-pot reaction of ethyl acetoacetate (**57**), aromatic aldehydes (**58**), hydrazine (**31**), and malononitrile (**50**) in water afforded dihydropyrano[2,3-*c*] pyrazoles (**59**) in good yields (79%–95%) under ultrasonic irradiation in relatively short reaction times (15–40 min). In the absence of ultrasound, the products were obtained in comparatively lower yields (70%–86%) and longer reaction times (60–300 min) (Zou et al. 2011). Darandale et al. (2012) reported a simple and practical ultrasound-promoted synthetic protocol for the synthesis of dihydropyrano[2,3-*c*] pyrazoles (**59**) using sodium bisulfite ($NaHSO_3$) as a green catalyst in solvent-free conditions. The latter method provides the advantage of a shorter reaction time (30 s) and excellent yields (97%–99%) (Scheme 8.19).

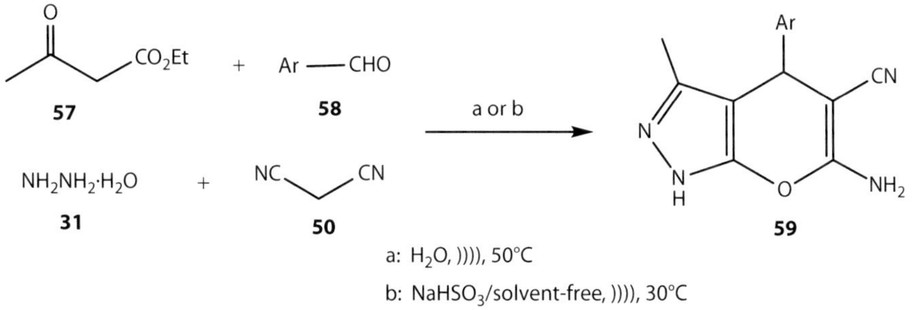

SCHEME 8.19

Recently, a catalyst-free one-pot four-component methodology was developed for the synthesis of 2*H*-indazolo[2,1-*b*]phthalazine-triones (**63**) via a reaction between dimedone (**60**), aromatic aldehyde (**61**), hydrazine hydrate (**31**), and phthalic anhydride (**62**) under ultrasonic irradiation at room temperature, using 1-butyl-3-methylimidazolium bromide ([bmim]Br) (Scheme 8.20) as a neutral reaction medium (Shekouhy and Hasaninejad 2012). A broad range of structurally diverse aldehydes (aromatic aldehydes bearing electron-withdrawing and/or electron-donating groups as well as heteroaromatic aldehydes) were applied successfully, and corresponding products were obtained in good to excellent yields without any by-product.

Khosropour (2008) developed an ultrasound-promoted, simple, efficient, and green methodology for the synthesis of 2,4,5-trisubstituted imidazoles (**66**) catalyzed by zirconium(IV) acetylacetonate ($Zr(acac)_4$). The reactions were carried out at room temperature for 20–85 min by taking a 1:1:10 mol ratio mixture of aldehyde (**64**), benzyl (**65**) and ammonium acetate, respectively, in the presence of 20 mol% of $Zr(acac)_4$ using EtOH as the solvent under sonication at 24 kHz (Scheme 8.21).

Shelke et al. (2009) reported the synthesis of 2,4,5-triaryl-1*H*-imidazoles (**70**) from the three-component one-pot condensation of benzil (**67**)/benzoin (**68**), aldehydes (**69**), and ammonium acetate in aqueous media under ultrasound at room temperature (Scheme 8.22). BO_3H_3 (5 mol%) was used as the catalyst. The reaction, performed under conventional stirring without ultrasound, required a reaction time (180 min) clearly longer than that needed when ultrasound was used (30–95 min).

However, Joshi et al. (2010) reported a chemoselective method for the synthesis of 1,3-imidazoles (**73**) by the reaction of substituted aldehydes (**72**) with *o*-phenylenediamine (**71**) catalyzed by 5 mol% of tetrabutylammonium fluoride (TBAF) as an inexpensive and relatively nontoxic phase-transfer catalyst in water as a green solvent under ultrasonic irradiation at room temperature (Scheme 8.23). High yields

SCHEME 8.20

SCHEME 8.21

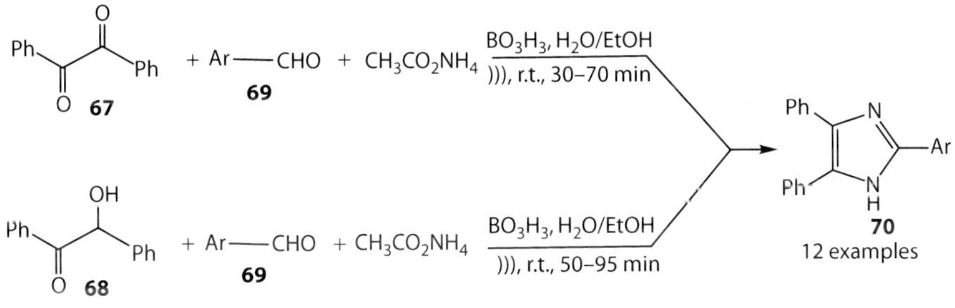

SCHEME 8.22

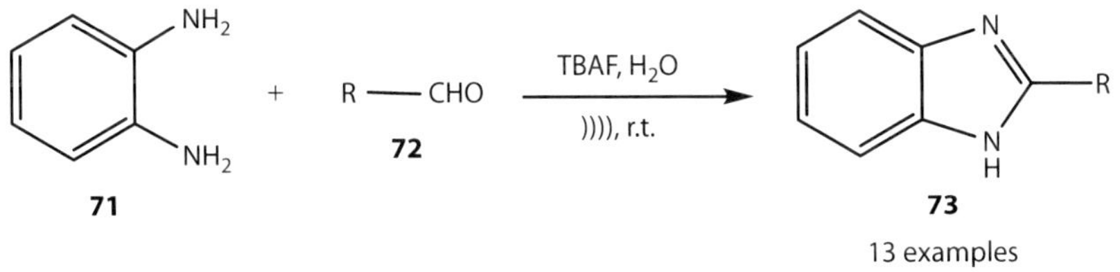

SCHEME 8.23

(82%–94%), short reaction times (30–65 min), and easy workup are the advantages of this sonocatalyzed procedure.

Arani and Safari (2011) mentioned a very efficient high-yield (96%–98%) synthesis of 5,5-diphenylhydantoin (**76**) and 5,5-diphenyl-2-thiohydantoin (**77**) derivatives via condensation of benzils (**74**) and substituted ureas (**75**) under mild conditions (Scheme 8.24). The reactions were accomplished in DMSO/H_2O with ultrasonic irradiation and catalyzed by KOH.

Synthesis of aminooxazole derivatives via thermal and ultrasonic methods using deep eutectic solvent as the medium was recently reported by Singh et al. (2013). It was observed that the ultrasound-assisted method gave 90% yield in just 8 min as against 3.5 h required to achieve 69% yield by the thermal method. 2-Amino-4-aryloxazoles (**80**) were obtained via the reaction of phenacyl bromide derivatives (**78**) and amide derivatives (**79**) in a deep eutectic solvent (Scheme 8.25). The deep eutectic solvent used in this protocol was easily prepared from choline chloride (1 eq) and urea (2 eq) at 80°C by a previously reported method (Toukoniitty et al. 2006) with 100% atom economy.

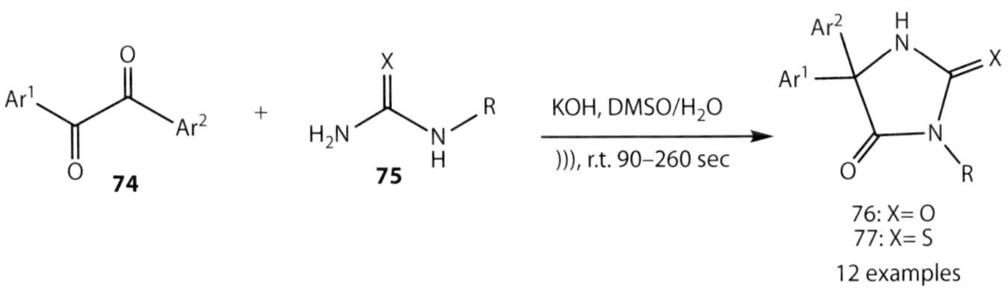

SCHEME 8.24

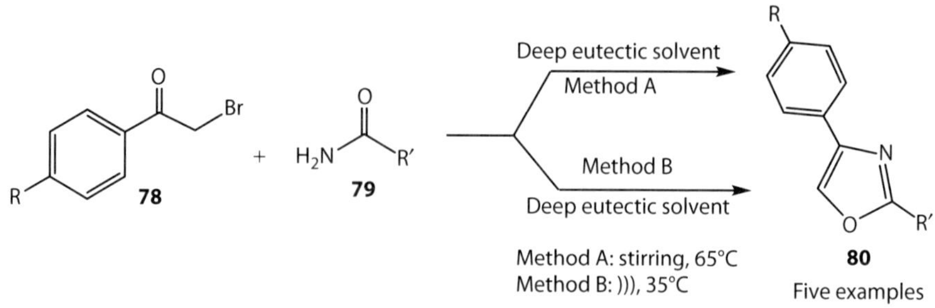

SCHEME 8.25

Ultrasound-promoted synthesis of imidazolidine-2-thione (**83**) was reported by Entezari et al. (2008) and the heterocycle was obtained from the reaction between ethylenediamine (**81**) and carbon disulfide (**82**) in a methanol and water solvent system (Scheme 8.26). The reaction occurs in the presence of an acid catalyst (HCl) or in the absence of any catalyst, and the yield depends on the reaction temperature.

A convenient one-pot protocol for the synthesis of 2-imino-1,3-thiazolidin-4-ones (**86** and **87**) was developed by the reaction of amines (**84**), isocyanates (**85**), aldehydes, and chloroform in the presence of 1,8-diazabicycloundec-7-ene (DBU)/sodium hydroxide as the basic medium under ultrasonic irradiation in high yields (75%–91%) and shorter reaction times (12–15 min) (Scheme 8.27) (Mamaghani et al. 2011).

Neuenfeldt et al. (2011) used ultrasonic power to promote the synthesis of thiazolidinones (**91**). 2-Aryl-3-(piperonylmethyl)-1,3-thiazolidin-4-ones (**91**) were obtained in good yields (60%–92%) from the multicomponent reaction of aromatic

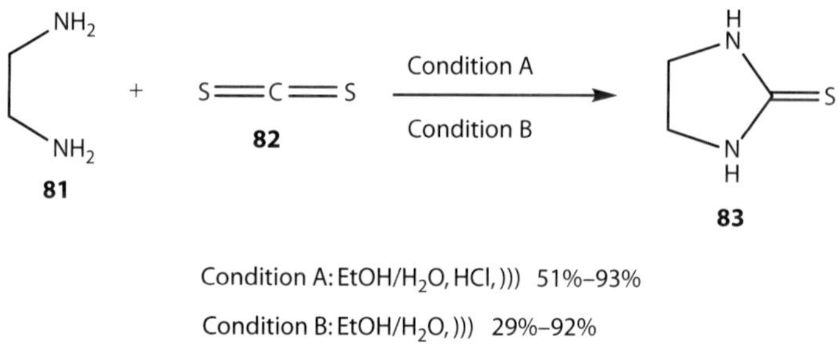

SCHEME 8.26

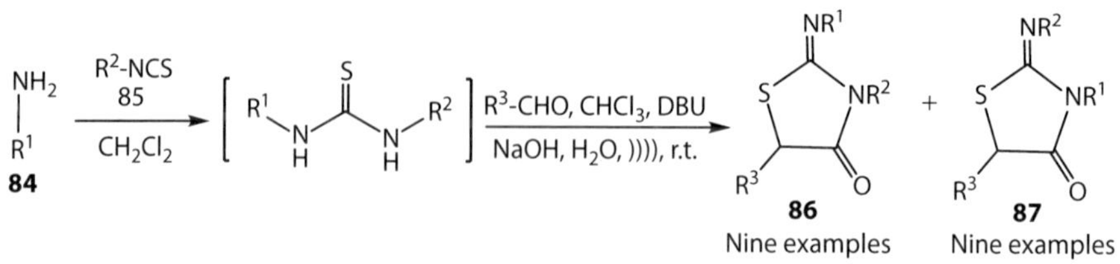

SCHEME 8.27

aldehydes (**88**), mercaptoacetic acid (**90**), and piperonilamine (**89**) in toluene under sonication for 5 min (Scheme 8.28). The corresponding conventional thermal reactions in the absence of ultrasound also furnished similar yields of these heterocycles, but required much longer times (960 min).

Previously, the synthesis of five-membered heterocycles containing sulfur, selenium, and tellurium was reported by Šibor and Pazdera (1996). Two different kinds of heterocycles (**93–95**) can be prepared by the reaction of substrates (**92**), a chalcogen atom, and isothiocyanate or ketones in the presence of triethylamine under ultrasonic conditions (Scheme 8.29). Beyond the utilization of ultrasound, the authors studied the use of conventional conditions, as well as microwave irradiation,

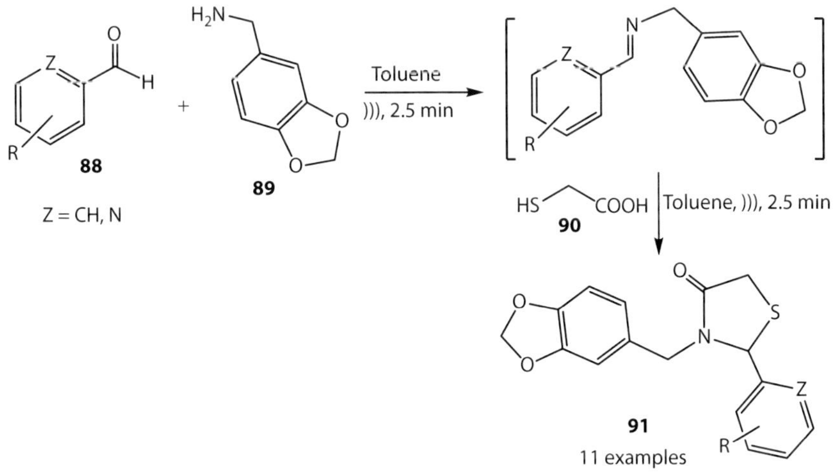

SCHEME 8.28

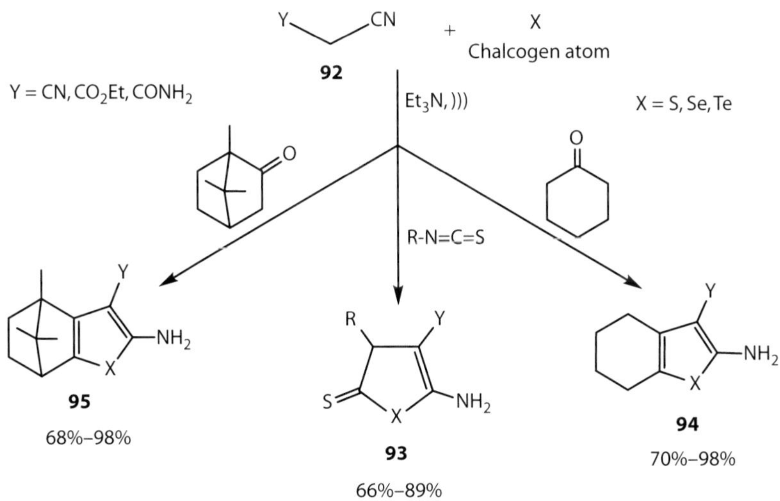

SCHEME 8.29

and found that although ultrasound accelerated the reaction described, the effect of microwaves was more significant.

8.1.3.3 Sonochemical Synthesis of Six-Membered Heterocycles

Six-membered heterocycles comprise an important class of bioactive heterocycles. They attract considerable interest by medicinal chemists because of their wide spectrum of therapeutic and pharmacological properties. They are recognized for possessing antibacterial, antiviral, antitumor, antiplatelet aggregation, antihypertensive, analgesic, and anti-inflammatory activities (Matsuda and Hirao 1965; Hurst and Hull 1961; Bokaeva 1967; Kato 1984; Sadanandam et al. 1992; Kurono et al. 1987).

Zhidovinova et al. (2003) showed that the classical Biginelli reaction (EtOH and HCl) (Biginelli 1893) is accelerated by a factor of more than 40 times as a result of ultrasonic irradiation. The three-component reaction among aldehydes (**96**), ethyl acetoacetate (**97**), and urea (**98**) or thiourea (**99**) in the presence of a catalytic amount of HCl was completed within 2–5 min at room temperature, and dihydropyrimidin-2-ones (thiones) (DHPM) (**100**) were obtained in excellent yields (90%–95%) (Scheme 8.30).

Li et al. (2003) used ultrasound to promote the Biginelli reaction between aldehydes (**96**), β-keto esters (**101**), and urea (**98**) to obtain DHPM (**102**) in good to excellent yields (Scheme 8.31). The reaction was catalyzed by aminosulfonic acid and it was very tolerant of aromatic aldehydes carrying either electron-withdrawing or electron-donating substituents.

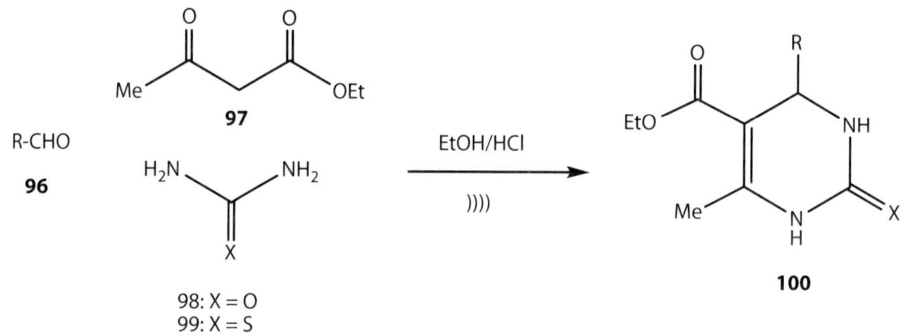

SCHEME 8.30

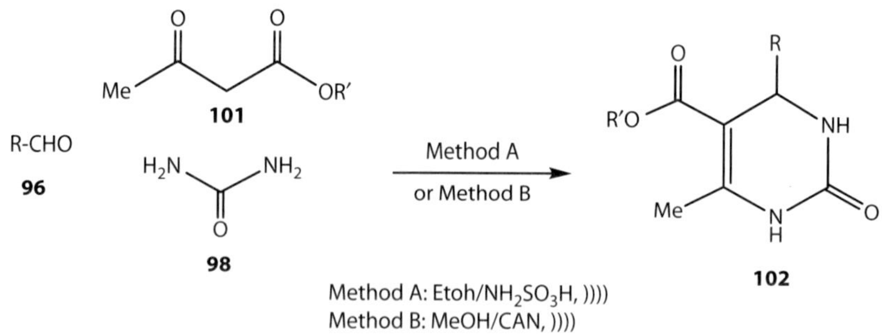

SCHEME 8.31

Yadav et al. (2001) showed that ceric ammonium nitrate (CAN) may also be used as a catalyst in an ultrasonic-promoted Biginelli reaction (Scheme 8.31). The reaction was carried out in methanol under ultrasonic irradiations. Heteroaryl, aromatic (electron-poor or electron-rich), aliphatic, and α,β-unsaturated aldehydes were used and, in all cases, compounds (**102**) were obtained in high yields and with high purity.

The authors suggest a radical mechanism for the reaction, in which a single-electron transfer from CAN to the β-keto esters and latter radical adds to the imine intermediate (Scheme 8.32).

DHPM(s) (**103**) have been synthesized by the utilization of inexpensive ammonium chloride as a mediator of the reaction under ultrasonic irradiation (Scheme 8.33). The reaction was carried out in methanol and irradiated for 3–5 h in a cleaning bath.

The antioxidant activity of these DHPM(s) (**103**) was evaluated, and some of these compounds exhibited potent activity against lipid peroxidation induced by Fe and ethylenediaminetetraacetic acid (EDTA) (Stefani et al. 2006).

Li et al. (2004) extended their earlier study of the Biginelli reaction and achieved ultrasound-mediated synthesis of 4-oxo-2-thioxohexahydropyrimidines (**105**). One-pot condensation of aromatic aldehydes (**96**) with ethyl cyanoacetate (**104**) and thiourea (**99**) was performed in the presence of potassium carbonate and ethanol, leading to compounds (**105**) in low to high yields (20%–90%) (Scheme 8.34).

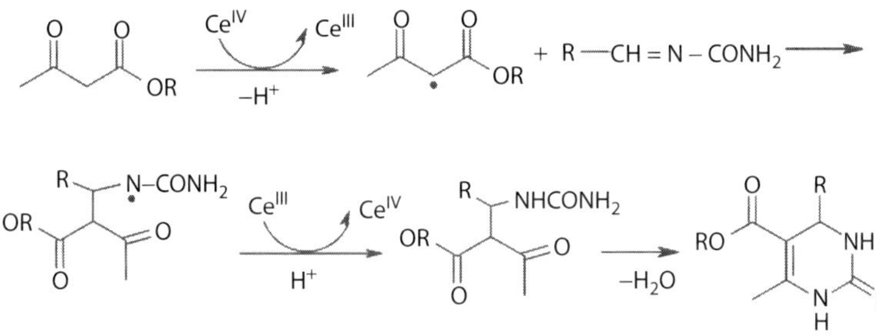

SCHEME 8.32

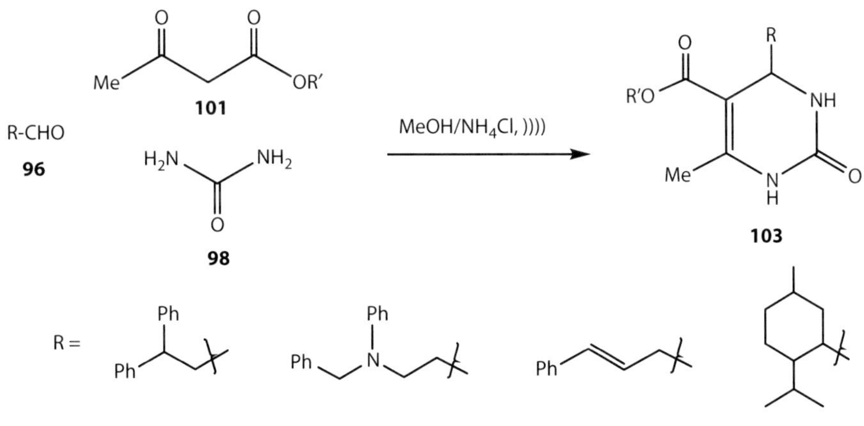

SCHEME 8.33

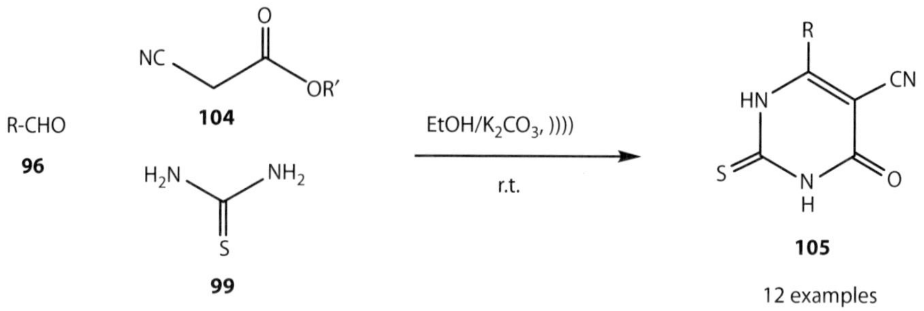

SCHEME 8.34

As well as an efficient synthesis of novel 4-(2-phenyl-1,2,3-triazol-4-yl)-3,4-dihydropyrimidin-2(1*H*)-(thio)ones (**108**) from 1,3-dicarbonyl compounds (**107**), the synthesis of 2-phenyl-1,2,3-triazole-4-carbaldehyde (**106**) and urea (**98**) or thiourea (**99**) under ultrasound irradiation and using samarium perchlorate as a catalyst was described by Liu and Wang (2010). Compared with conventional methods, the main advantages of this methodology are milder conditions, shorter reaction times, and higher yields (Scheme 8.35).

An optimized procedure for the preparation of pyrimidine-2-thione derivatives (**110**) under mild and clean conditions was achieved by Safaei-Ghomi and Ghasemzadeh (2011). Chalcone derivatives (**109**) were prepared by the condensation of various substituted aryl aldehydes and acetophenone in alkaline ethanol, while pyrimidine-2-thione derivatives (**110**) were prepared by the combination of chalcones (**109**) and thiourea (**99**) under conventional and ultrasonic conditions. Advantages of the ultrasound effect were observed and high yields (73%–82%) of the products were obtained after 20–30 min sonication (Scheme 8.36).

Niralwad et al. (2010) reported an efficient synthesis of octahydroquinazolinone derivatives (**112**) under ultrasound irradiation via the reaction of dimedone (**60**), aromatic aldehyde (**111**), and ureas (**98** or **99**) using [tbmim]Cl_2/$AlCl_3$ as an acidic ionic liquid catalyst (Scheme 8.37). This protocol has advantages in terms of (i) a short reaction time, (ii) a solvent-free reaction, (iii) high yield, (iv) easy workup, (v) being environmentally friendly, and (vi) recyclability of ionic liquid.

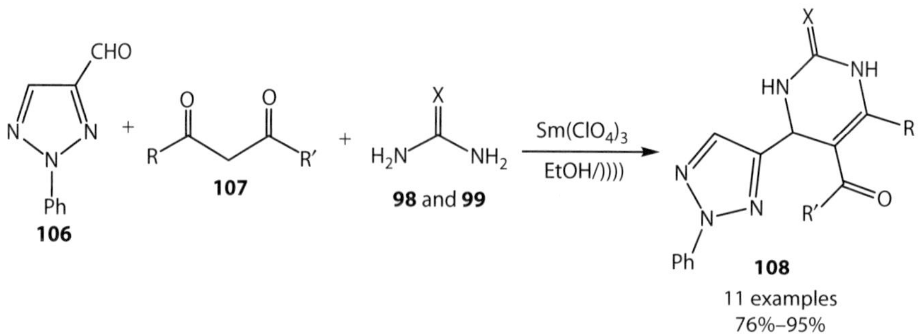

SCHEME 8.35

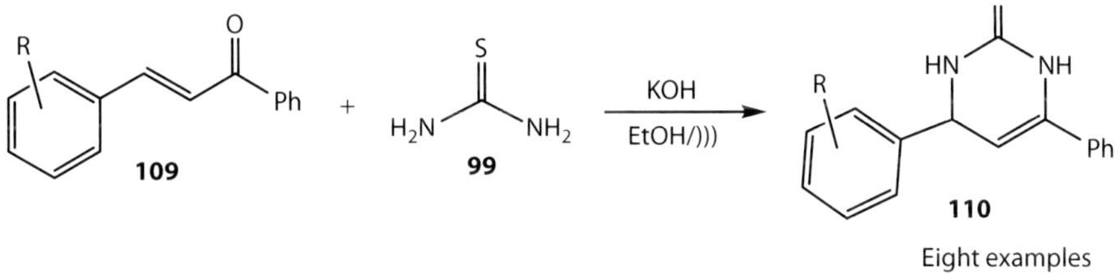

SCHEME 8.36

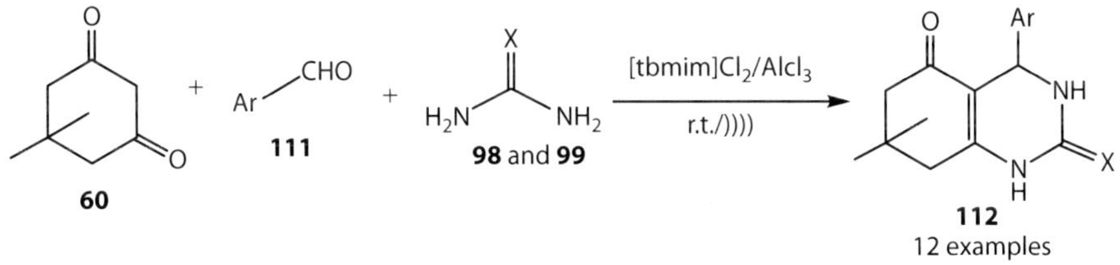

SCHEME 8.37

The pyridine ring system is an important motif in naturally occurring products as well as in many synthetic compounds of pharmaceutical interest (Murata et al. 2003; Mantri et al. 2008; Deng et al. 2007). Ruiz et al. (2011) have developed an efficient and straightforward procedure for the synthesis of 3,4-dihydropyridone derivatives (**117**) by the condensation of Meldrum's acid (**114**), aromatic aldehydes (**116**), alkyl acetoacetates (**115**), and ammonium acetate in glacial acetic acid under ultrasonic irradiation at room temperature (Scheme 8.38). Compared with conventional methods, the main advantages of this procedure are milder conditions, shorter reaction times (15–20 min) and higher yields (85%–96% yields).

A green and convenient approach to the synthesis of 2-amino-4,6-diphenylnicotinonitriles via a four-component reaction of malononitrile, aromatic aldehydes, acetophenone derivatives, and ammonium acetate in water under ultrasound irradiation has been developed by Safari et al. (2012). The combinatorial synthesis was achieved for this methodology by applying ultrasound irradiation while making use of water

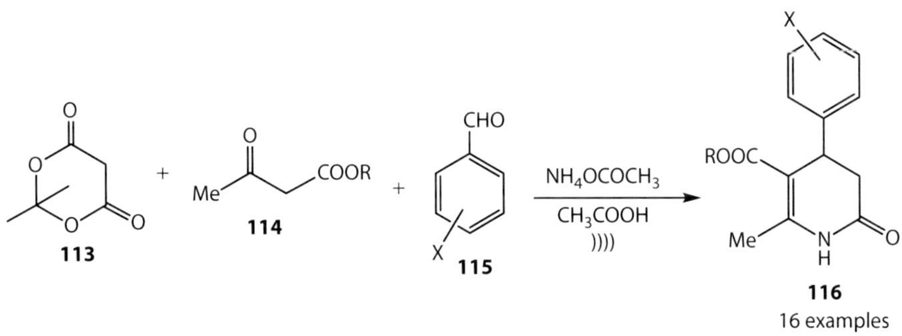

SCHEME 8.38

as a green solvent (Scheme 8.39). In comparison to conventional methods, experimental simplicity, good functional group tolerance, excellent yields, short routine, and selectivity without the need for a transition metal or base catalyst are prominent features of this sonocatalyzed procedure.

Pyridazines are useful compounds with a broad array of biological activities. They possess notable hypertensive (Betti et al. 2003), platelet aggregation inhibition (Sotelo et al. 2002), phosphodiesterase (Van Der May et al. 2001), antiasthmatic (Yamaguchi et al. 1993), antisecretory, antiulcer (Yamada et al. 1983), antidepressant, antibacterial (Ishida et al. 1990), antifungal, α-adrenoceptor antagonists (Betti et al. 2006), analgesic (Pinna et al. 1985), anti-inflammatory (Cignarella et al. 1982), antianemic (Black et al. 2000), nephrotropic (Okushima et al. 1987), and cardiotonic properties (Narimatsu et al. 1987). As mentioned previously, Eftekhari-Sis and Vahdati-Khajeh (2013) developed an efficient method for the preparation of 5-aryl-4-hydroxy-2-methyl-1*H*-pyrrole-3-carboxylates (**18**). Also, the authors reported the same protocol for the synthesis of 6-aryl-3-methylpyridazine-4-carboxylates (**120**) via a three-component reaction of arylglyoxal hydrates (**16**) with β-dicarbonyl compounds (**17**) in the presence of hydrazine hydrate (Scheme 8.40).

The first ultrasound-promoted multicomponent synthesis of pyridazinones (**124**) and phthalazinones (**125**) from arenes, cyclic anhydrides, and $ArNHNH_2$ in the presence of an efficient recyclable catalyst, [bmim]Br/$AlCl_3$, in high yield and a short reaction time was reported by Zare et al. (2012) (Scheme 8.41).

Tetrahydrobenzo[*c*]xanthenes-11-ones (**128**) were synthesized through effective shorter reaction durations, mild reaction conditions, and a low-cost protocol by an

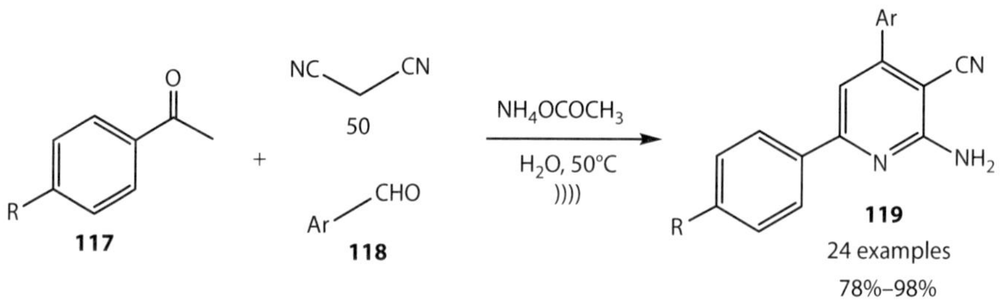

SCHEME 8.39

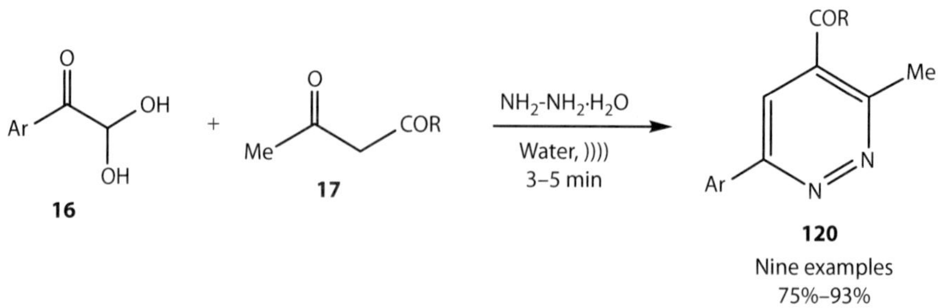

SCHEME 8.40

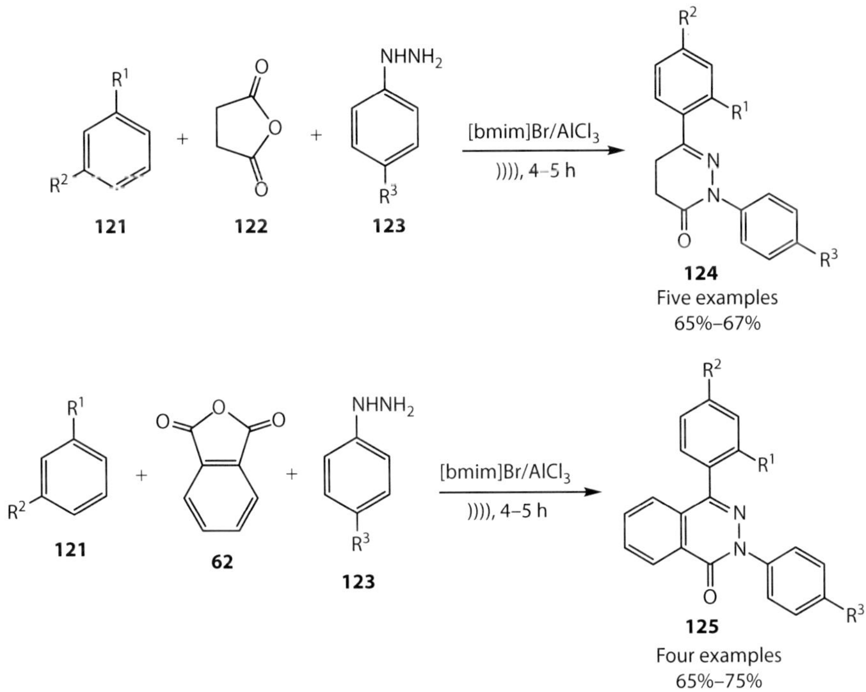

SCHEME 8.41

ultrasound-assisted three-component reaction of α-naphthol (**126**), aromatic aldehyde (**127**), and dimedone (**60**). Ceric ammonium nitrate (CAN) acts as a suitable eco-friendly catalyst for this method (Scheme 8.42).

The formation of tetrahydrobenzo[*c*]xanthene-11-ones (**128**) from α-naphthol, benzaldehyde (**127**), and dimedone (**60**) in the presence of CAN may be explained by a tentative mechanism presented in Scheme 8.43. The ultrasonic cavitation induced shear forces and the jets produced near the surface of the vessel and the catalyst may activate the passive α-naphthol through sonolysis of the O–H bond. The reaction between the activated α-naphthol and the aromatic aldehyde (activated by CAN) may facilitate the formation of the corresponding *o*-quinone methides (*o*-QM) under sonic

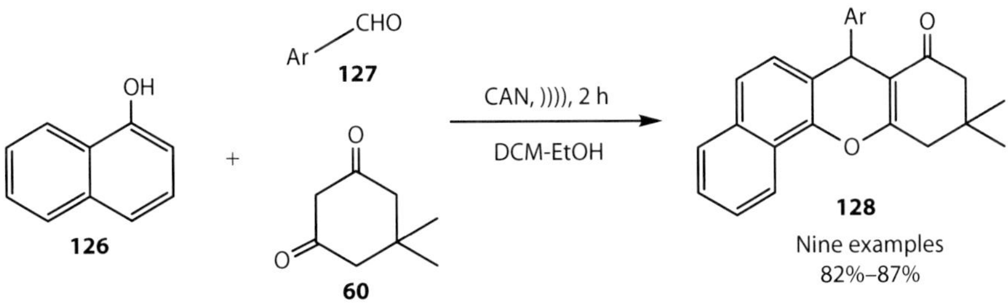

SCHEME 8.42

conditions, and this *o*-QM intermediate may then react with the active methylene or dimedone (**60**), followed by cyclization to give tetrahydrobenzo[*c*]xanthene-11-one (**128**) with the removal of a molecule of water (Scheme 8.43) (Sudha and Pasha 2012).

Very recently, an environmentally benign method for the synthesis of 2-amino-4,8-dihydropyrano[3,2-*b*]pyran-3-carbonitrile derivatives (**131**) via a three-component reaction of arylaldehydes (**130**), malononitrile (**60**), and kojic acid (**129**) in water at 50°C under ultrasound irradiation was achieved by Banitaba et al. (Scheme 8.44). In comparison to conventional methods, this method provides several advantages such as experimental simplicity, good functional group tolerance, excellent yields (85%–98%), a short routine (5–30 min), and selectivity, without the need for a transition metal or base catalyst (Banitaba et al. 2013).

Pyrido[2,3-*d*]pyrimidines have diverse pharmacological activity such as antitumor (Gangjee et al. 1999; Grivsky et al. 1980), cardiotonic (Heber et al. 1993; Furuya and Ohtaki 1994), hepatoprotective, antihypertensive (Heber et al. 1993), antibronchitic (Sakuma et al. 1989), antifungal (Singh et al. 2002), antibacterial (Zakharov et al. 1994), and antifolate activity (DeGraw et al. 1992).

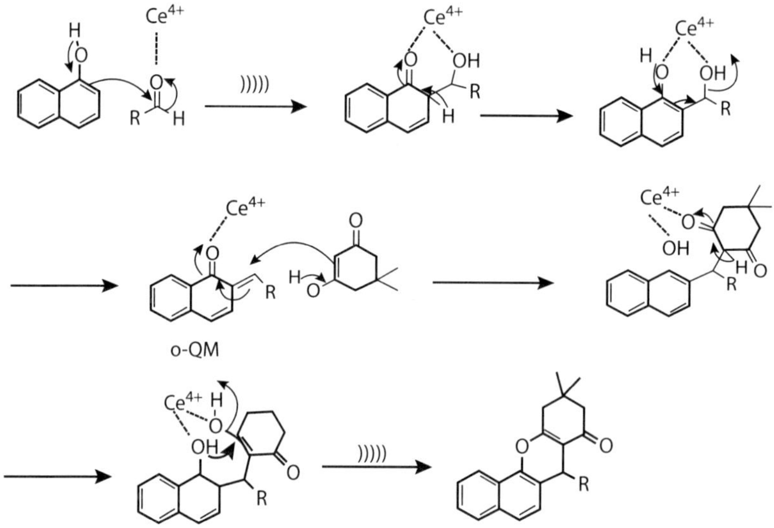

SCHEME 8.43

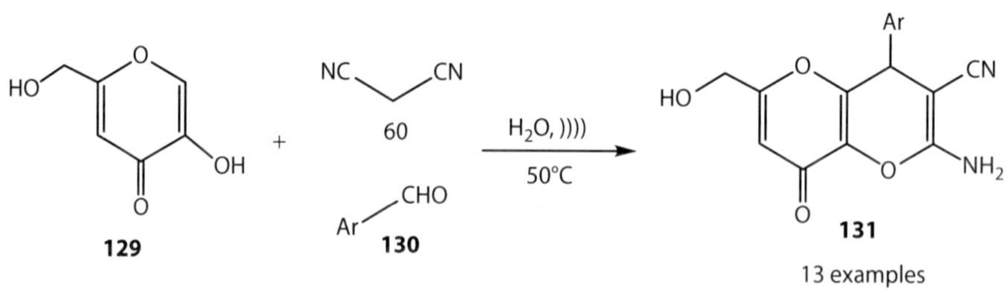

SCHEME 8.44

Tetrabutyl ammonium bromide (TBAB) was found to be an efficient phase-transfer catalyst for the synthesis of pyrido[2,3-*d*]pyrimidines by the one-pot reaction of 6-aminouracils (**132**), aromatic aldehydes (**134**), and malononitrile or ethyl cyanoacetate in water under ultrasonic irradiation (Scheme 8.45). The advantages of this method are the use of an inexpensive and readily available catalyst, short reaction times, easy workup, improved yields, and the use of water as a solvent that is environmentally benign (Hussein, 2012).

A series of novel 5-amino-7-aryl-7,8-dihydro-[1,2,4]triazolo[4,3-*a*]-pyrimidine-6-carbonitriles (**138**) were synthesized by a concise, mild, one-pot three-component reaction of 3-amino-1,2,4-triazole, malononitrile, and aryl aldehydes in the presence of NaOH (20 mol%) in ethanol under heating or ultrasonic irradiation as shown in Scheme 8.46 (Ablajan et al. 2012).

Saleh et al. (2012) have developed a novel and efficient three-component condensation reaction of an aldehyde, amino azoles, and β-keto sulfone for the synthesis of pyrimido[1,2-*a*]benzimidazoles (**143**) and pyrazolo[3,4-*b*]pyridines (**146**) under ultrasonic irradiation via utilization of β-keto sulfone as an important synthon (Schemes 8.47 and 8.48).

The simple one-pot nature of the reaction makes it an interesting alternative to other multistep approaches. In general, improvements in rates and yield of reactions

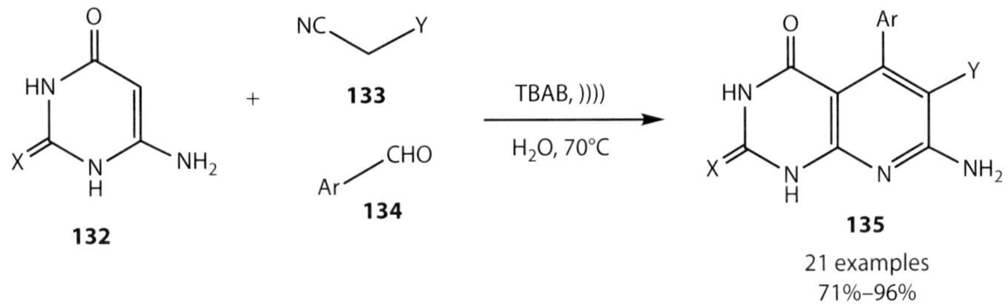

SCHEME 8.45

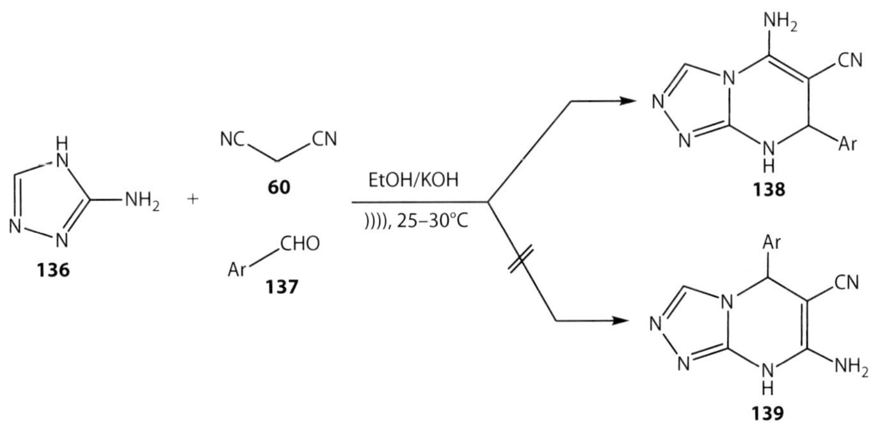

SCHEME 8.46

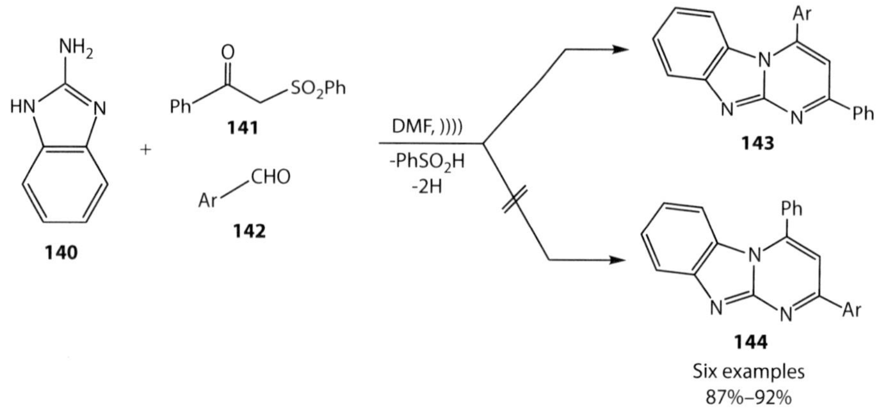

SCHEME 8.47

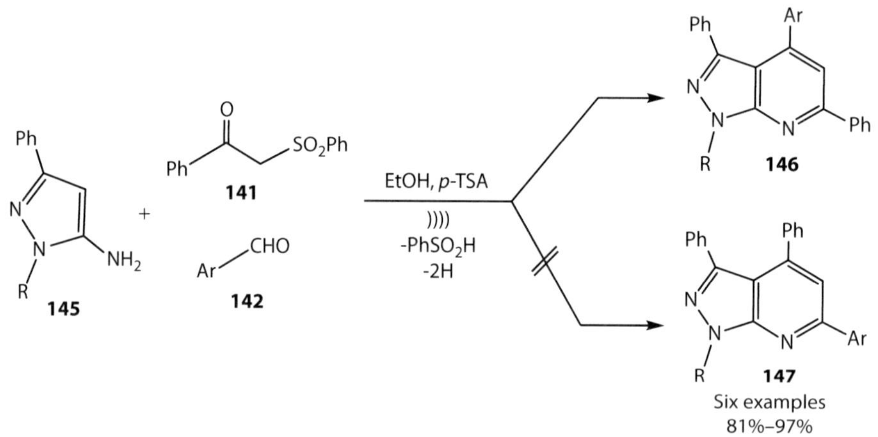

SCHEME 8.48

were observed when reactions were carried out under sonication, compared with classical conditions.

8.1.3.4 Sonochemical Synthesis of Spiroheterocycles

Spiro compounds represent an important class of naturally occurring substances characterized by highly pronounced biological properties. The spiro[pyrrolidin-3,3′-indole] ring system is a recurring structural motif in a number of natural products such as vinblastine and vincristine, which function as cytostatics and are of prime importance in cancer chemotherapy (Cordell 1998). The derivatives of spirooxindole find very wide biological application as antimicrobial (Okita and Isobe 1994), antitumor, and antibiotic agents, and inhibitors of the human NK-1 receptor (Kornet and Thio 1976) and potent nonpeptide inhibition of the p53–MDM2 interaction (Skiles and McNeil 1990). The highly substituted 4-piperidone (**148**) was employed as a synthon for spiro heterocycles (**150**). When the compound (**148**) was treated with thiohydrazide derivatives (**149**) with the presence of a catalytic amount of piperidine under ultrasonic irradiation, the corresponding piperidinothiadiazoles (**150**) were obtained in good yields (Scheme 8.49) (Padmavathi et al. 2007).

However, Li et al. (2010c) described a simple and economical procedure for the synthesis of azaspiro[4.5]decane derivatives (**154**). 3-Aza-6,10-diaryl-2-oxaspiro[4.5]decane-1,4,8-triones (**154**) were obtained in moderate to good yields (49% 91%) from one-pot synthesis of 1,5-diaryl-1,4-pentadien-3-one (**151**), dimethyl malonate (**152**), and hydroxylamine hydrochloride in the presence of sodium hydroxide under ultrasound irradiation at 50°C (Scheme 8.50).

A library of spiro[indole-thiazolidinones] (**158**) was prepared sonochemically by a three-component reaction in an aqueous medium in the presence of cetyltrimethylammonium bromide (CTAB) as a phase-transfer catalyst (Dandia et al. 2011b). The reaction of isatins (**155**), aryl or heteroaryl amines (**156**), and α-mercaptocarboxylic acids (**157**) under ultrasound for 40–50 min afforded the target molecules in good to excellent yields (80%–98%) (Scheme 8.51).

1,3-Dipolar cycloaddition reactions are considered to be one of the most useful processes for the construction of five-membered heterocyclic ring systems in a highly regio- and stereoselective manner. Novel substituted pyrrolidines (**162**) were synthesized from the 1,3-dipolar cycloaddition of azomethine ylides generated *in situ* by heating a solution of acenaphtenequinone (**159**) and proline (**160**) with various dipolarophiles (**161**) (Scheme 8.52). The reactions were highly diastereoselective and regioselective and were carried out under reflux and ultrasonic conditions at room temperature. Generally, milder conditions and moderate improvements in

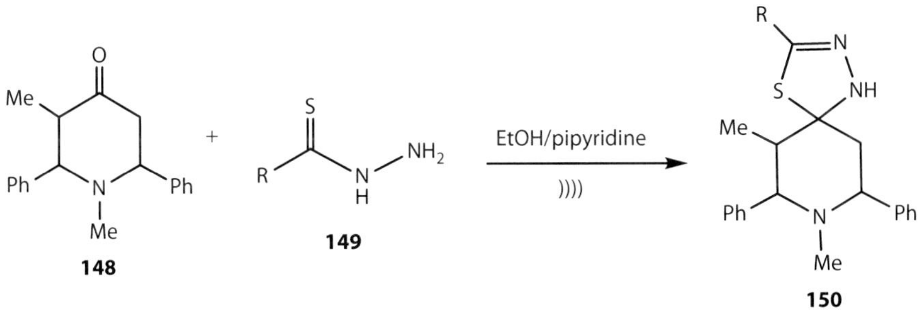

SCHEME 8.49

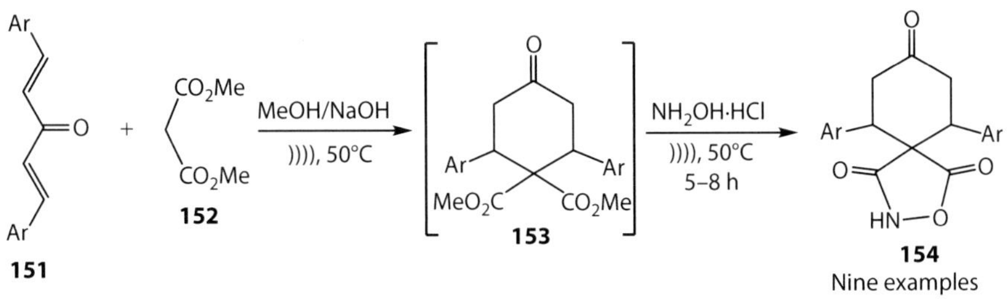

SCHEME 8.50

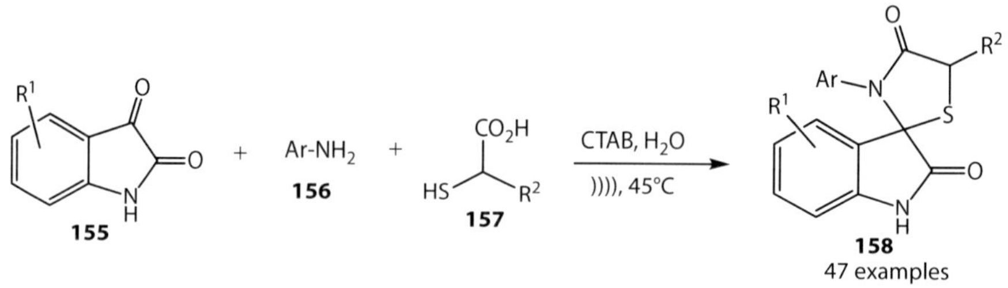

SCHEME 8.51

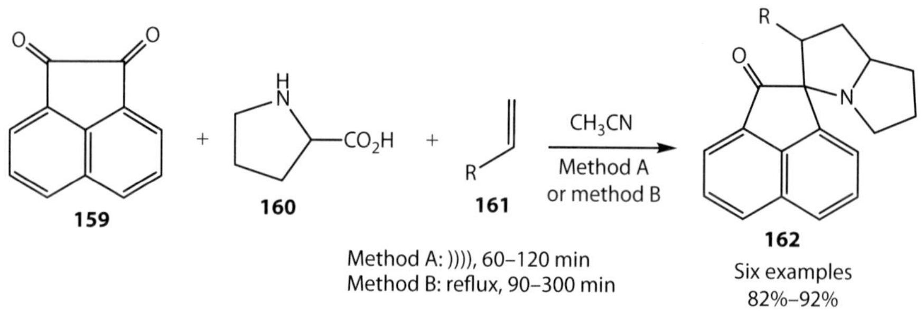

SCHEME 8.52

rates and reaction times were observed when ultrasonic conditions were used. The products were obtained in high yields (82%–92%) (Jadidi et al. 2008).

Novel dicyano-functionalized spiropyrrolizidine (**166** and **167**) and spiropyrrolidine (**168** and **169**) were synthesized in a regio- and stereoselective manner from the reaction of various arylidenemalononitriles (**163**) with nonstabilized azomethine ylides generated from isatin (**155**)/acenaphthenequinone (**159**) and proline (**160**)/sarcosine (**164**)/*N*-phenylglycine (**165**). The reactions were carried out under both conventional heating and ultrasonic irradiation conditions (Scheme 8.53). In general, improvement in rates and yields were observed when the reactions were carried out under sonication compared with classical conditions (Rezaei et al. 2011).

A series of novel dispirooxindolecyclo[pyrrolo[1,2-*c*]thiazole-6,5′-thiazolidine] derivatives (**172**) were obtained regioselectively by the 1,3-dipolar cycloaddition reaction of 5-arylidene-2-thioxothiazolidin-4-ones (**171**) as dipolarophiles with non-stabilized azomethine ylides, generated *in situ* via decarboxylative condensation of isatin (**155**) and thiazolidine-4-carboxylic acid (**170**) in ethanol under both classical refluxing and ultrasound irradiation (Scheme 8.54). Because of the advantages of ultrasonic irradiation of mild reaction conditions, short reaction times, and high efficiency, it is quite profitable to develop the 1,3-dipolar cycloaddition of azomethine ylides under these conditions (Hua et al. 2012).

Epoxidation of 3-aroylmethylene-indole-2-ones (**173**) with 30% aqueous hydrogen peroxide using cetyltrimethyl ammonium bromide (CTAB) as a phase-transfer catalyst under ultrasound irradiation afforded spiro[indole-3,2′-oxiranes] derivatives (**174**) in 90%–97% yields (Scheme 8.55). The lead compounds showed potent antimicrobial activity and antioxidant properties (Dandia et al. 2011a).

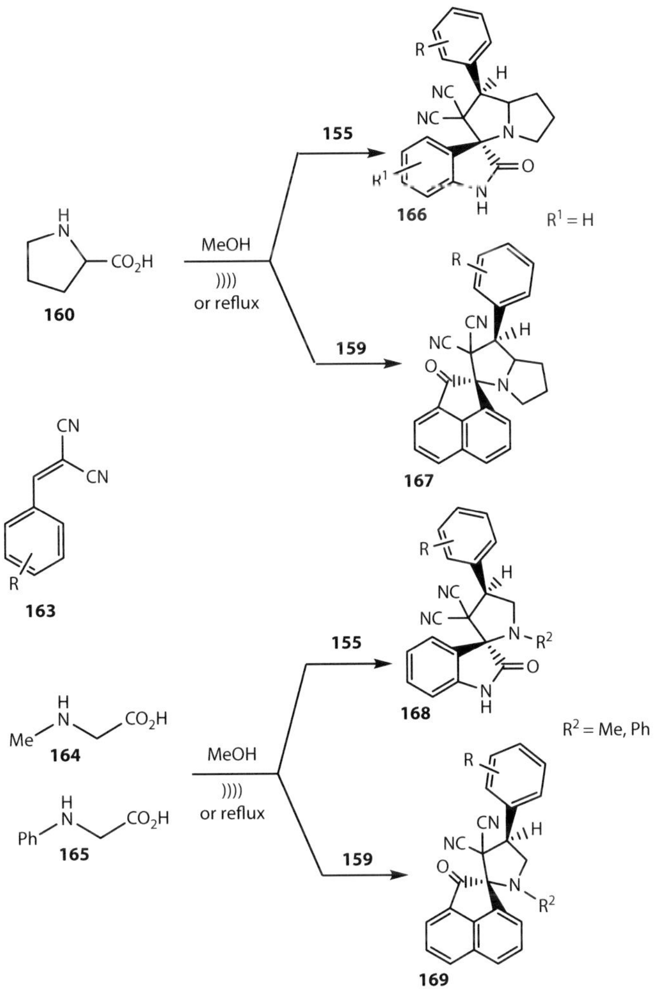

SCHEME 8.53

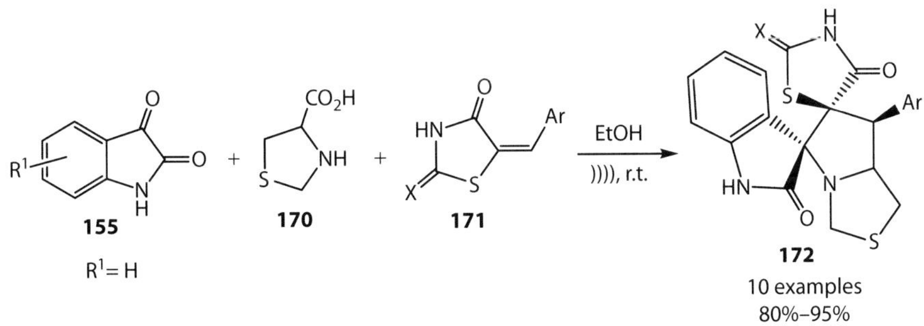

SCHEME 8.54

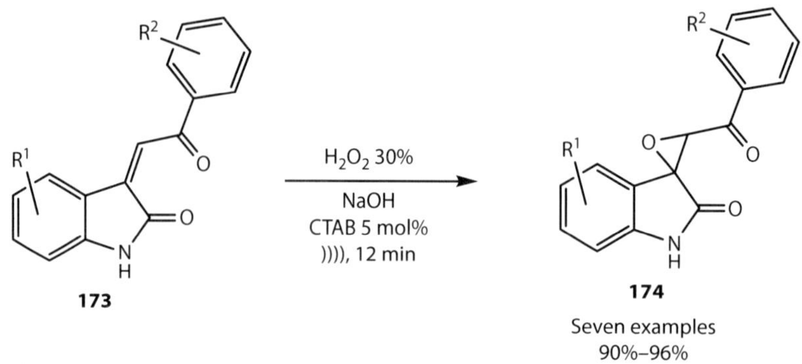

SCHEME 8.55

The reaction of spiro[indole-3,2′-oxiranes] (**174**) with thioacetamide in the presence of LiBr as a catalyst in an aqueous medium produced substituted 2′-amino-4′-benzoyl-2′-methylspiro[indole-3,5′-[1,3]oxathiolane]-2(1*H*)-ones (**175**). The reaction was carried out under both microwaves and sonication, and the results were also compared with the conventional method (Dandia et al. 2010). In general, improvement in rate and yields were observed when reaction was carried out under sonication as compared to microwave irradiation and the conventional method (Scheme 8.56).

The possible mechanism for present investigation involves the nucleophilic attack of a bromide ion at the less substituted *C*-3 position of the spiro epoxide leading to the formation of an intermediate (**I1** and **I2**). Further, as sulfur is a better nucleophile than nitrogen, in the next step cyclization leads to formation of an oxathiolane ring (Scheme 8.57).

Bazgir et al. (2010) developed a simple, facile, and efficient three-component procedure for the synthesis of spiro[indoline-3,4′-pyrazolo[3,4-*b*]pyridine]-2,6′(1′*H*)-diones (**179**) by the reaction of 4-hydroxycoumarins (**176**), isatins (**177**), and 5-amino-1*H*-pyrazoles (**178**) in water as an environmentally benign solvent, using *p*-toluenesulfonic acid as an inexpensive and readily available catalyst under ultrasonic irradiation (Scheme 8.58).

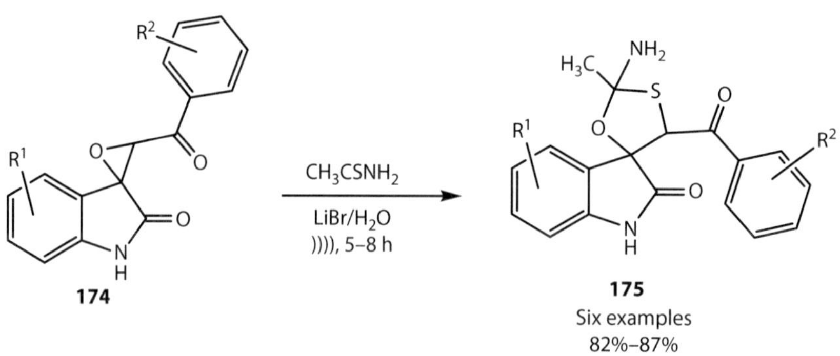

SCHEME 8.56

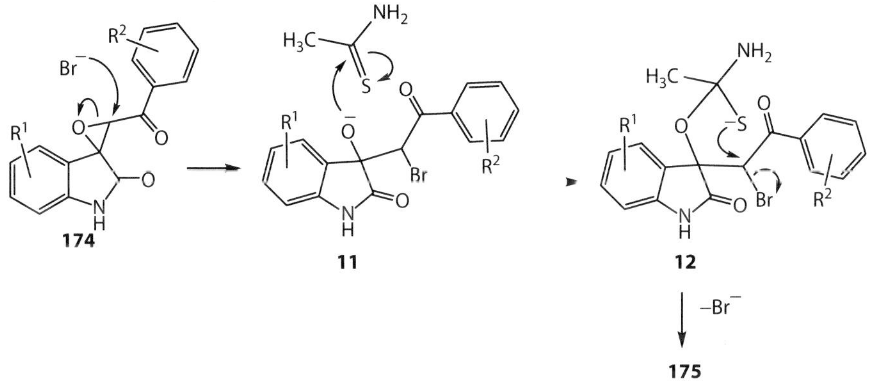

SCHEME 8.57

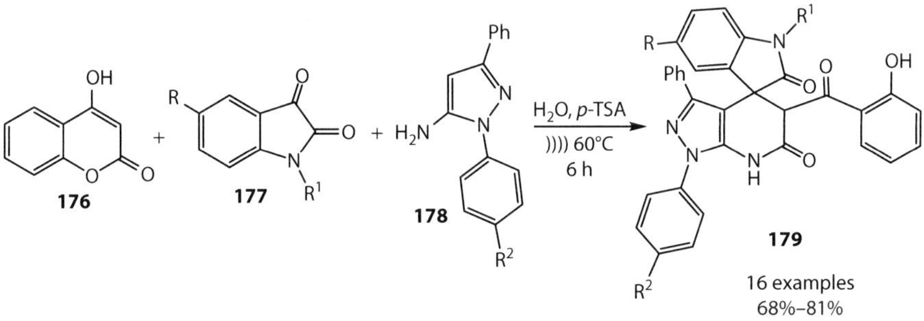

SCHEME 8.58

A series of 3′-aminospiro[indoline-3,1′-pyrazolo[1,2-*b*]phthalazine]-2,5′,10′-trione derivatives (**181**) have been synthesized by a one-pot three-component reaction of isatin (**180**), malononitrile (**50**) or ethyl cyanoacetate (**51**), and phthalhydrazide (**49**) catalyzed by piperidine under ultrasound irradiation (Scheme 8.59). For comparison, the reactions were carried out under both conventional and ultrasonic conditions. As a whole, improvement in rates and yields were observed when the reactions were carried out under sonication compared with classical conditions (Wang et al. 2012).

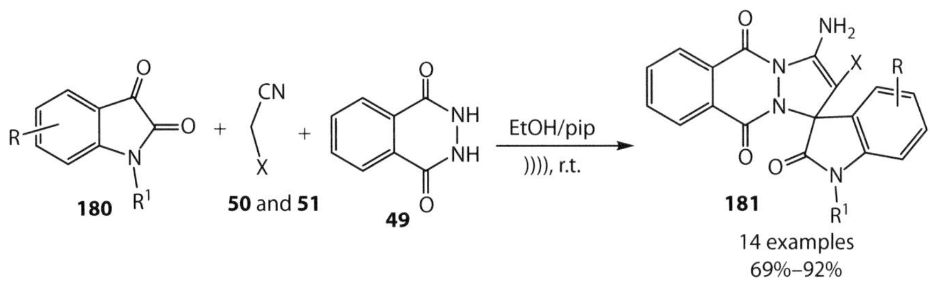

SCHEME 8.59

Also, Rezaei et al. reported a facile and efficient one-pot three-component ultrasound-promoted synthesis of spiroacenaphthylene pyrazolotriazoles (**184**) and pyrazolophthalazines (**185**) via a three-component reaction of various nitrile-contained acidic C–H (**182**), acenaphthylene-1,2-dione (**159**), and phthalhydrazide (**49**) or 4-phenyl-1,2,4-triazolidine-3,5-dione (**183**) in the presence of Et_3N as the catalyst (Scheme 8.60).

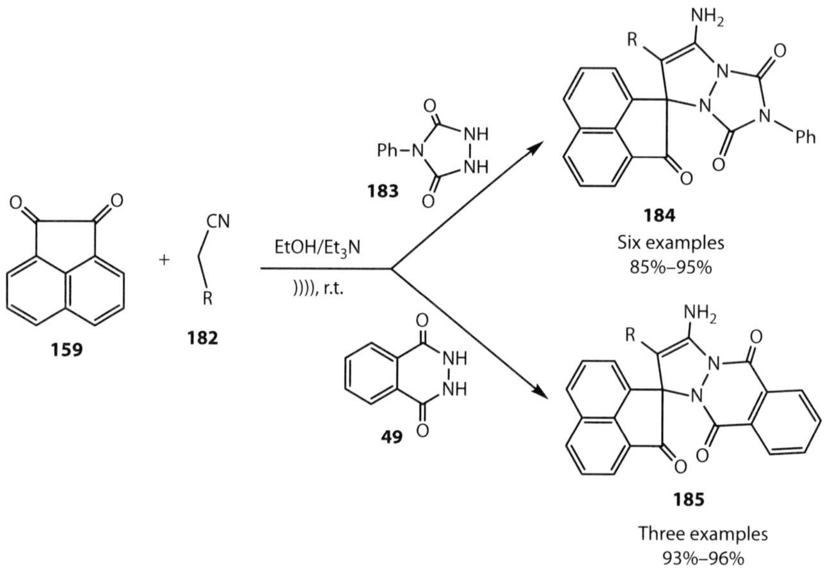

SCHEME 8.60

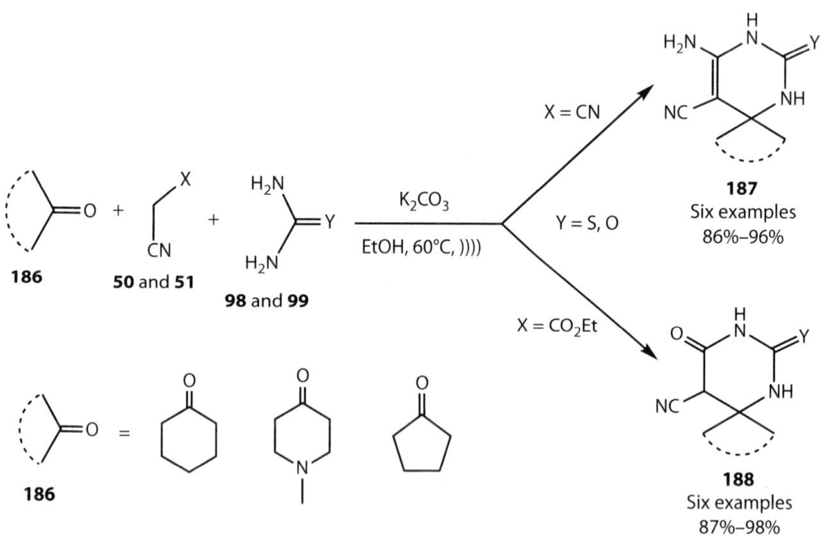

SCHEME 8.61

Employing ethanol as a solvent, which is assumed to be relatively environmentally benign, and use of an inexpensive and readily available catalyst combined with the advantages of sonochemistry, such as mild reaction conditions, excellent yield, and short reaction times (30–40 min), highlight this synthetic route (Rezaei et al. 2012). Very recently, a novel and efficient protocol was devised for the synthesis of various spirocyanopyrimidines (**187** and **188**) via the three-component condensation of cyclic ketones (**186**), nitrile derivatives (**50** and **51**), and thiourea (**99**) or urea (**98**) in ethanol at 60°C using potassium carbonate as a basic catalyst under ultrasonic irradiation. In comparison with conventional methods, high yields, easy workup, and short reaction times are the advantages of this sonocatalyzed procedure (Scheme 8.61) (Hussein 2013).

REFERENCES

Abd El-Rahman, N. M., Saleh, T. S. and Mady, M. F. 2009. Ultrasound assisted synthesis of some new 1,3,4-thiadiazole and bi(1,3,4-thiadiazole) derivatives incorporating pyrazolone moiety. *Ultrason. Sonochem.* 16:70–74.

Ablajan, K., Kamil, W., Tuoheti, A. and Wan-Fu, S. 2012. An efficient three component one-pot synthesis of 5-amino-7-aryl-7,8-dihydro[1,2,4]triazolo[4,3-*a*]pyrimidine-6-carbonitriles. *Molecules* 17:1860–1869.

Al-Mutairi, A. A., El-Baih, F. E. M. and Al-Hazimi, H. M. 2010. Microwave versus ultrasound assisted synthesis of some new heterocycles based on pyrazolone moiety. *J. Saudi Chem. Soc.* 14:287–299.

Arani, N. M. and Safari, V. J. 2011. A rapid and efficient ultrasound-assisted synthesis of 5,5-diphenylhydantoins and 5,5-diphenyl-2-thiohydantoins. *Ultrason. Sonochem.* 18:640–643.

Banitaba, S. H., Safari, J. and Khalili, S. D. 2013. Ultrasound promoted one-pot synthesis of 2-amino-4,8-dihydropyrano[3,2-*b*]pyran-3-carbonitrile scaffolds in aqueous media: A complementary "green chemistry" tool to organic synthesis. *Ultrason. Sonochem.* 20:401–407.

Bazgir, A., Ahadi, S., Ghahremanzadeh, R., Khavasi, H. R. and Mirzaei, P. 2010. Ultrasound-assisted one-pot, three-component synthesis of spiro[indoline-3,4′-pyrazolo[3,4-*b*]pyridine]-2,6′(1′*H*)-diones in water. *Ultrason. Sonochem.* 17:447–452.

Betti, L., Floridi, M., Giannacini, G., Manetti, F., Strappaghetti, G., Tafi, A. and Botta, M. 2003. α1-Adrenoceptor antagonists. 5. Pyridazinone-arylpiperazines. Probing the influence on affinity and selectivity of both *ortho*-alkoxy groups at the arylpiperazine moiety and cyclic substituents at the pyridazinone nucleus. *Bioorg. Med. Chem. Lett.* 13:171–173.

Betti, L., Zanelli, M., Giannacini, G., Manetti, F., Schenone, S. and Strappaghetti, G. 2006. Synthesis of new piperazine-pyridazinone derivatives and their binding affinity toward α1-, α2-adrenergic and 5-HT1A serotoninergic receptors. *Bioorg. Med. Chem. Lett.* 14:2828–2836.

Biginelli, P. 1893. Derivati aldeiduredici degli eteri acetil- e dossal-acetico. *Gazz. Chim. Ital.* 23:360–416.

Black, L. A., Basha, A., Kolasa, T., Kort, M. E., Liu, H., MacCarty, C. M., Patel, M. V., Rohde, J. J., Coghlan, M. J. and Stewart, A. O. 2000. Preparation of arylpyridazinones as prostaglandin endoperoxide H synthase biosynthesis inhibitors. *Int. Appl. Chem. Abstr.* 132:321867.

Bokaeva, S. S. 1967. Effects of some pyrimidine derivates on the growth of transplanted tumors in animals. *Tr. Kaz. Nauch.-Issled. Inst. Onkol. Radiol.* 3:305–309.

Cella, R., Stefani, H. A. 2009. Ultrasound in heterocycles chemistry. *Tetrahedron* 65:2619–2641.

Chanda, B. M., Vyas, R. and Bedekar, A. V. 2001. Investigations in the transition metal catalyzed aziridination of olefins and insertion reactions with bromamine-T as the source of nitrene. *J. Org. Chem.* 66:30–34.

Chimenti, F., Bolasco, A., Secci, D., Chimenti, P., Granese, A., Carradori, S., Yáñez, M., Orallo, F., Ortuso, F. and Alcaro, S. 2010. Investigations on the 2-thiazolylhydrazyne scaffold: Synthesis and molecular modeling of selective human monoamine oxidase inhibitors. *Bioorg. Med. Chem.* 18:5715–5723.

Cignarella, G., Loriga, M., Pinna, G. A., Pirisi, M. A., Schiatti, P. and Selva, D. 1982. Unexpected anti-inflammatory activity of rigid structures derived from antihypertensive 6-arylpyridazinones. III. Synthesis and activity of 7-fluoro- and 5-keto-5*H*-indeno[1,2-*c*]pyridozines. *Farmaco-Ed. Sci.* 37:133–144.

Coldham, I. and Hufton, R. 2005. Intramolecular dipolar cycloaddition reactions of azomethine ylides. *Chem. Rev.* 105:2765–2810.

Cordell, G. A. 1998. *The Alkaloids: Chemistry and Biology*, vol. 5, pp. 259–311. Academic: San Diego.

Dandia, A., Singh, R. and Bhaskaran, S. 2010. Ultrasound promoted greener synthesis of spiro[indole-3,5′-[1,3]oxathiolanes in water. *Ultrason. Sonochem.* 17:399–402.

Dandia, A., Singh, R. and Bhaskaran, S. 2011a. Facile stereoslective synthesis of spiro[indole-oxiranes] by combination of phase transfer catalyst and ultrasound irradiation and their bioassay. *Ultrason. Sonochem.* 18:1113–1117.

Dandia, A., Singh, R., Bhaskaran, S. and Samant, S. D. 2011b. Versatile three component procedure for combinatorial synthesis of biologically relevant scaffold spiro[indolethiazolidinones] under aqueous conditions. *Green Chem.* 13:1852–1859.

Darandale, S. N., Sangshetti, J. N. and Shinde, D. B. 2012. Ultrasound mediated, sodium bisulfite catalyzed, solvent free synthesis of 6-amino-3-methyl-4-substitued-2,4-dihydropyrano[2,3-*c*]pyrazole-5-carbonitrile. *J. Korean Chem. Soc.* 56:328–333.

DeGraw, J. I., Christie, P. H., Clowell, W. T. and Sirotnak, F. M. 1992. Synthesis and antifolate properties of 5,10-ethano-5,10-dideazaaminopterin. *J. Med. Chem.* 35:320–324.

Deng, J., Sanchez, T., Al-Mawsawi, L. Q., Dayam, R., Yunes, R. A., Garofalo, A., Bolger, M. B. and Neamati, N. 2007. Discovery of structurally diverse HIV-1 integrase inhibitors based on a chalcone pharmacophore. *Bioorg. Med. Chem.* 15:4985–5002.

Eftekhari-Sis, B. and Vahdati-Khajeh, S. 2013. Ultrasound-assisted green synthesis of pyrroles and pyridazines in water via three-component condensation reactions of arylglyoxals. *Curr. Chem. Lett.* 2:85–92.

Eicher, T. and Hauptmann, S. 2003. *The Chemistry of Heterocycles: Structure, Reactions, Syntheses, and Applications*, 2nd edn. Wiley-VCH: Weinheim.

Entezari, M. H., Asghari, A. and Hadizadeh, F. 2008. Sono-synthesis of imidalidine-2-thione as a base compound of some pharmaceutical products. *Ultrason. Sonochem.* 15:119–123.

Fillion, H. and Luche, J.-L. 1998. In: Luche, J.-L. (ed.), *Synthetic Organic Sonochemistry*. Plenum: New York, NY.

Furuya, S. and Ohtaki, T. 1994. Pyrido[2,3-*d*]pyrimidines and their use as endothelin anatagonists. Eur. Pat. Appl. EP. 608565. *Chem. Abstr.* 1994. 121:205395.

Gangjee, A., Adair, O. and Queener, S. F. 1999. Pneumocystis carinii and Toxoplasma gondii dihydrofolate reductase inhibitors and antitumor agents: Synthesis and biological activities of 2,4-diamino-5-methyl-6-[(monosubstitutedanilino)methyl] pyrido[2,3-*d*]pyrimidines. *J. Med. Chem.* 42:2447–2455.

Gedanken, A. 2004. Using sonochemistry for the fabrication of nanomaterials. *Ultrason. Sonochem.* 11:47–55.

Ghahremanzadeh, R., Fereshtehnejad, F., Mirzaei, P. and Bazgir, A. 2011. Ultrasound-assisted synthesis of 2,2′-(2-oxoindoline-3,3-diyl)bis(1*H*-indene-1,3(2*H*)-dione) derivatives. *Ultrason. Sonochem.* 18:415–418.

Gomha, S. M. and Khalil, Kh. D. 2012. A convenient ultrasound-promoted synthesis of some new thiazole derivatives bearing a coumarin nucleus and their cytotoxic activity. *Molecules* 17:9335–9347.
Grivsky, E. M., Lee, S., Sigel, C. W., Duch, D. S. and Nichol, C. A. 1980. Synthesis and antitumor activity of 2,4-diamino-6-(2,5-dimethoxybenzyl)-5-methylpyrido[2,3-*d*]pyrimidine. *J. Med. Chem.* 23:327–329.
Harada, K. I., Tomita, K., Fujii, K., Masuda, K., Mikami, Y., Yazawa, K. and Komaki, H. 2004. Isolation and structure characterization of siderophore, madurastatins produced by a pathogenic *Actinomadura madurae*. *J. Antibiot.* 57:125–135.
Hata, Y., Watanabe, M., Matasubara, T. and Touchi, A. 1976. Fragmentation reaction of ylide 5. A new metabolic reaction of aziridine derivative. *J. Am. Chem. Soc.* 98:6033–6036.
He, J. Y., Xin, H. X., Yan, H., Song, X. Q. and Zhong, R. G. 2011. Convenient ultrasound-mediated synthesis of 1,4-diazabutadienes under solvent-free conditions. *Ultrason. Sonochem.* 18:466–469.
Heber, D., Heers, C. and Ravens, U. 1993. Positive inotropic activity of 5-amino-6-cyano-1,3-dimethyl-1,2,3,4-tetrahydropyrido[2,3-*d*]pyrimidine-2,4-dione in cardiac muscle from guinea-pig and man. Part 6: Compounds with positive inotropic activity. *Pharmazie* 48:537–541.
Hua, Y., Zou, Y., Wub, H. and Shi, D. 2012. A facile and efficient ultrasound-assisted synthesis of novel dispiroheterocycles through 1,3-dipolar cycloaddition reactions. *Ultrason. Sonochem.* 19:264–269.
Hurst, E. W. and Hull, R. I. 1961. Two new synthetic substances active against viruses of the psittacosis-lymphogranuloma-trachoma group. *J. Med. Pharm. Chem.* 3:215–229.
Hussein, E. M. 2012. Enviro-economic, ultrasound-assisted one-pot, three-component synthesis of pyrido[2,3-*d*]pyrimidines in aqueous medium. *Z. Naturforsch.* 67b:231–237.
Hussein, E. M. 2013. Ultrasound-promoted efficient domino reaction for one-pot synthesis of spiro-5-cyanopyrimidines: A rapid procedure. *Monatsh. Chem.* 144(11):1691–1697.
Ishida, A., Homma, K., Kono, H., Tamura, K. and Sasaki, Y. 1990. Eur. Pat. Appl. *Chem. Abstr.* 1994. 121:35618.
Jadidi, K., Gharemanzadeh, R., Mehrdad, M., Darabi, H. R., Khavasi, H. R. and Asgari, D. 2008. A facile synthesis of novel pyrrolizidines under classical and ultrasonic conditions. *Ultrason. Sonochem.* 15:124–128.
Joshi, R. S., Mandhane, P. G., Dabhade, S. K. and Gill, C. H. 2010. Tetrabutylammonium fluoride (TBAF) catalysed synthesis of 2-arylbenzimidazole in water under ultrasound irradiation. *J. Chin. Chem. Soc.* 57:1227–1231.
Kato, T. 1984. Japan Kokai Tokkyo Koho JP Patent 59,190,974. *Chem. Abstr.* 102, 132067.
Khosropour, A. R. 2008. Ultrasound-promoted greener synthesis of 2,4,5-trisubstituted imidazoles catalyzed by Zr(acac)4 under ambient conditions. *Ultrason. Sonochem.* 15:659–664.
Konev, A. S., Novikov, M. S. and Khlebnikov, A. F. 2005. The first example of the generation of azomethine ylides from a fluorocarbene: 1,3-Cyclization and 1,3-dipolar cycloaddition. *Tetrahedron Lett.* 46:8337–8340.
Kornet, M. J. and Thio, A. P. 1976. Oxindole-3-spiropyrrolidines and -piperidines. Synthesis and local anesthetic activity. *J. Med. Chem.* 19:892–898.
Kowsari, E. and Mallakmohammadi, M. 2011. Ultrasound promoted synthesis of quinolines using basic ionic liquids in aqueous media as a green procedure. *Ultrason. Sonochem.* 18:447–454.
Kurono, M., Hayashi, M., Miura, K., Isogawa, Y. and Sawai, K. 1987. Sanwa Kogaku Kenkyusho Co., Japan Kokai, Tokkyo Koho JP 62, 267, 272.
Li, J.-T., Han, J.-F., Yang, J.-H. and Li, T.-S. 2003. An efficient synthesis of 3,4-dihydropyrimidin-2-ones catalyzed by NH_2SO_3H under ultrasound irradiation. *Ultrason. Sonochem.* 10:119–122.

Li, J.-T., Li, Y. W., Song, Y. L. and Chen, G. F. 2012. Improved synthesis of 2,2′-arylmethylene bis(3-hydroxy-5,5-dimethyl-2-cyclohexene-1-one) derivatives catalyzed by urea under ultrasound. *Ultrason. Sonochem.* 19:1–4.

Li, J.-T., Lin, Z.-P., Han, J.-F. and Li, T.-S. 2004. One-pot synthesis of 4-oxo-2-thioxohexahydropyrimidines catalyzed by potassium carbonate under ultrasound. *Synth. Commun.* 34:2623–2631.

Li, J.-T., Yin, Y., Li, L. and Sun, M.-X. 2010a. A convenient and efficient protocol for the synthesis of 5-aryl-1,3-diphenylpyrazole catalyzed by hydrochloric acid under ultrasound irradiation. *Ultrason. Sonochem.* 17:11–13.

Li, J.-T., Yin, Y. and Sun, M. X. 2010b. An efficient one-pot synthesis of 2,3-epoxyl-1,3-diaryl-1-propanone directly from acetophenones and aromatic aldehydes under ultrasound irradiation. *Ultrason. Sonochem.* 17:363–366.

Li, J.-T., Zhai, X.-L. and Chen, G.-F. 2010c. Ultrasound promoted one-pot synthesis of 3-aza-6,10-diaryl-2-oxa-spiro[4.5]decane-1,4,8-trione. *Ultrason. Sonochem.* 17:356–358.

Lin, Z. and Li, J. 2012. A convenient and efficient protocol for the synthesis of 1,3,5-triaryl-2-pyrazolines in acetic acid under ultrasound irradiation. *E-J. Chem.* 9:267–271.

Liu, C.-J. and Wang, J.-D. 2010. Ultrasound-assisted synthesis of novel 4-(2-phenyl-1,2,3-triazol-4-yl)-3,4-dihydropyrimidin-2(1*H*)-(thio)ones catalyzed by $Sm(ClO_4)_3$. *Molecules* 15:2087–2095.

Mamaghani, M. and Dastmard, S. 2009. One-pot easy conversion of Baylis-Hillman adducts into arylpyrazoles under ultrasound irradiation. *Arkivoc* ii:168–173.

Mamaghani, M., Loghmanifar, A. and Taati, M. R. 2011. An efficient one-pot synthesis of new 2-imino-1,3-thiazolidin-4-ones under ultrasonic conditions. *Ultrason. Sonochem.* 18:45–48.

Mantri, M., De Graaf, O., Van Veldhoven, J., Göblyös, A., Von Frijtag Drabbe Künzel, J. K., Mulder-Krieger, T., Link, R., et al. 2008. 2-Amino-6-furan-2-yl-4-substituted nicotinonitriles as A_{2A} adenosine receptor antagonists. *J. Med. Chem.* 51:4449–4455.

Mason, T. J. and Lorimer, J. P. 2002. *Applied Sonochemistry: Uses of Power Ultrasound in Chemistry and Processing*. Wiley-VCH: Weinheim.

Matsuda, T. and Hirao, I. 1965. Antibacterial activity of 5-nitrofuran derivatives. *Nippon Kagaku Zasshi* 86:1195–1197.

Miller, T. W., Tristram, E. W. and Wolf, F. J. 1971. Azirinomycin II. Isolation and chemical characterization as 3-methyl-2(H)aziridinecarboxylic acid. *J. Antibiot.* 24:48–50.

Murata, T., Shimada, M., Sakakibara, S., Yoshino, T., Kadono, H., Masuda, T., Shimazaki, M., et al. 2003. Discovery of novel and selective IKK-beta serine-threonine protein kinase inhibitors. Part 1. *Bioorg. Med. Chem. Lett.* 13:913–918.

Nabid, M. R., Rezaei, S. J. T., Ghahremanzadeh, R. and Bazgir, A. 2010. Ultrasound-assisted one-pot, three-component synthesis of 1H-pyrazolo[1,2-*b*]phthalazine-5,10-diones. *Ultrason. Sonochem.* 17:159–161.

Narimatsu, A., Kitada, Y., Satoh, N., Suzuki, R. and Okushima, H. 1987. Cardiovascular pharmacology of 6-[4-(4′-pyridyl)aminophenyl]-4,5-dihydro-3(2*H*)-pyridazinone hydrochloride, a novel and potent cardiotonic agent with vasodilator properties. *Arzneimittelforschung* 37:398–406.

Neuenfeldt, P. D., Duval, A. R., Drawanz, B. B., Rosales, P. F., Gomes, C. R. B., Pereira, C. M. P. and Cunico, W. 2011. Efficient sonochemical synthesis of thiazolidinones from piperonilamine. *Ultrason. Sonochem.* 18:65–67.

Niralwad, K. S., Shingate, B. B. and Shingare, M. S. 2010. Ultrasound-assisted one-pot synthesis of octahydroquinazolinone derivatives catalyzed by acidic ionic liquid [tbmim] $Cl_2/AlCl_3$. *J. Chin. Chem. Soc.* 57:89–92.

Ogasawara, Y. and Liu, H.-W. 2009. Biosynthetic studies of aziridine formation in azicemicins. *J. Am. Chem. Soc.* 131:18066–18068.

Okita, T. and Isobe, M. 1994. Synthesis of the pentacyclic intermediate for dynemicin a and unusual formation of spiro-oxindole ring. *Tetrahedron* 50:11143–11152.

Okushima, H., Narimatsu, A., Kobayashi, M., Furuya, R., Tsudo, K. and Kitada, Y. 1987. A novel class of cardiotonics. Synthesis and pharmacological properties of [4-(substituted-amino)phenyl]pyridazinones and related derivatives. *J. Med. Chem.* 30:1157–1161.

Padmavathi, V., Ramana Reddy, T. V. and Audisesha Reddy, K. 2007. 4-Piperidone- A synthon for spiroheterocycles. *Indian J. Chem.* 46B:818–822.

Pathak, V. N., Joshi, R., Sharma, J., Gupta, N. and Rao, V. M. 2009. Mild and ecofriendly tandem synthesis, and spectral and antimicrobial studies of N1-acetyl-5-aryl-3-(substituted styryl)pyrazolines. *Phosphorus Sulfur Silicon* 184:1854–1865.

Pinna, G. A, Curzu, M. M, Loriga, M., Cignarella, G., Barlocco, D. and Malandrino, S. 1985. Rigid congeners of arylpyridazinones. IV. Synthesis and activity of derivatives of the new heterocyclic system 9*H*-indeno[2,1-*c*]pyridazine. *Farmaco-Ed. Sci.* 40:979–986.

Pizzuti, L., Martins, P. L. G., Ribeiro, B. A., Quina, F. H., Pinto, E., Flores, A. F. C., Venzke, D. and Pereira, C. M. P. 2010. Efficient sonochemical synthesis of novel 3,5-diaryl-4,5-dihydro-1*H*-pyrazole-1-carboximidamides. *Ultrason. Sonochem.* 17:34–37.

Pizzuti, L., Piovesan, L. A., Flores, A. F. C., Quina, F. H. and Pereira, C. M. P. 2009. Environmentally friendly sonocatalysis promoted preparation of 1-thiocarbamoyl-3,5-diaryl-4,5-dihydro-1*H*-pyrazoles. *Ultrason. Sonochem.* 16:728–731.

Porter, C. W. and Young, L. 1938. A molecular rearrangement induced by ultrasonic waves. *J. Am. Chem. Soc.* 60:1497–1500.

Renaud, P. 1950. Note de laboratoire sur L'application des ultra-sons a la preparation de compose organo-metalliques. *Bull. Soc. Chim. Fr.* 1044–1048.

Rezaei, S. J. T., Bide, Y. and Nabid, M. R. 2012. An efficient ultrasound-promoted one pot synthesis of spiroacenaphthylene pyrazolotriazole and pyrazolophthalazine derivatives. *Tetrahedron Lett.* 53:5123–5126.

Rezaei, S. J. T., Nabid, M. R., Yari, A. and Weng, Ng. S. 2011. Ultrasound-promoted synthesis of novel spirooxindolo/spiroacenaphthen dicyano pyrrolidines and pyrrolizidines through regioselective azomethine ylide cycloaddition reaction. *Ultrason. Sonochem.* 18:49–53.

Richards, W. T. and Loomis, A. L. 1927. The chemical effects of high frequency sound waves I. A preliminary survey. *J. Am. Chem. Soc.* 49:3086–3100.

Ruiz, E., Rodriguez, H., Coro, J., Salfrán, E., Suárez, M., Martinez-Alvarez, R. and Martin, N. 2011. Ultrasound-assisted one-pot, four component synthesis of 4-aryl 3,4-dihydropyridone derivatives. *Ultrason. Sonochem.* 18:32–36.

Sadanandam, Y. S., Shetty, M. M. and Diwan, P. V. 1992. Synthesis and biological evaluation of new 3,4-dihydro-6-methyl-5-*N*-methyl-carbamoyl-4-(substituted phenyl)-2(1*H*)-pyrimidinones and pyrimidinethiones. *Eur. J. Med. Chem.* 27:87–92.

Safaei-Ghomi, J. and Ghasemzadeh, M. A. 2011. Ultrasound-assisted synthesis of dihydropyrimidine-2-thiones. *J. Serb. Chem. Soc.* 76:679–684.

Safari, J., Banitaba, S. H. and Khalili, S. D. 2012. Ultrasound-promoted an efficient method for one-pot synthesis of 2-amino-4,6-diphenylnicotinonitriles in water: A rapid procedure without catalyst. *Ultrason. Sonochem.* 19:1061–1069.

Sakuma, Y., Hasegawa, M., Kataoka, K., Hoshina, K., Yamazaki, N., Kadota, T. and Yamaguchi, H. 1989. *Chem. Abstr.* 115:71646. PCT Int. Appl. WO 9105785.

Saleh, T. S., Eldebss, T. M. A. and Albishri, H. M. 2012. Ultrasound assisted one-pot, three-components synthesis of pyrimido[1,2-*a*]benzimidazoles and pyrazolo[3,4-*b*]pyridines: A new access via phenylsulfone synthon. *Ultrason. Sonochem.* 19:49–55.

Schirmeister, T. and Peric, M. 2000. Aziridinyl peptides as inhibitors of cysteine proteases: Effect of a free carboxylic acid function on inhibition. *Bioorg. Med. Chem.* 8:1281–1291.

Sehgal, C. M. and Wang, S. Y. J. 1981. Threshold intensities and kinetics of sonoreaction of thymine in aqueous solutions at low ultrasonic intensities. *J. Am. Chem. Soc.* 103:6606–6611.

Shekouhy, M. and Hasaninejad, A. 2012. Ultrasound-promoted catalyst-free one-pot four component synthesis of 2*H*-indazolo[2,1-*b*]phthalazine-triones in neutral ionic liquid 1-butyl-3-methylimidazolium bromide. *Ultrason. Sonochem.* 19:307–313.

Shelke, K. F., Sapkal, S. B., Sonar, S. S., Madje, B. R., Shingate, B. B. and Shingare, M. S. 2009. An efficient synthesis of 2,4,5-triaryl-1*H*-imidazole derivatives catalyzed by boric acid in aqueous media under ultrasound-irradiation. *Bull. Korean Chem. Soc.* 30:1057–1060.

Šibor, J. and Pazdera, P. 1996. Synthesis of some new five-membered heterocycles containing selenium and tellurium. *Molecules* 1:157–162.

Singh, B. S., Lobo, H. R., Pinjari, D. V., Jarag, K. J., Pandit, A. B. and Shankarling, G. S. 2013. Ultrasound and deep eutectic solvent (DES): A novel blend of techniques for rapid and energy efficient synthesis of oxazoles. *Ultrason. Sonochem.* 20:287–293.

Singh, G., Singh, G., Yadav, A. K. and Mishraa, K. 2002. Synthesis and antimicrobial evaluation of some new pyrido[2,3-*d*]pyrimidines and their ribofuranosides. *Indian J. Chem.* 41B:430–432.

Skiles, J. W. and Mc Neil, D. 1990. Spiro indolinone β-lactams inhibitors of poliovirus and rhinovirus 3C-proteinases. *Tetrahedron Lett.* 31:7277–7280.

Sotelo, E., Fraiz, N., Yanez, M., Terrades, V., Laguna, R., Cano, E. and Ravina, E. 2002. Pyridazines. Part XXIX: Synthesis and platelet aggregation inhibition activity of 5-substituted-6-phenyl-3(2*H*)-pyridazinones novel aspects of their biological action. *Bioorg. Med. Chem.* 10:2873–2882.

Stefani, H. A., Oliveira, C. B., Almeida, R. B., Pereira, C. M. P., Braga, R. C., Cella, R., Borges, V. C., Savegnago, L. and Nogueira, C. W. 2006. Dihydropyrimidine-2(*H*)-ones obtained by ultrasound irradiation: A new class of potential antioxidant agents. *Eur. J. Med. Chem.* 41:513–518.

Sudha, S. and Pasha, M. A. 2012. Ultrasound assisted synthesis of tetrahydro-benzo[*c*]xanthene-11-ones using CAN as catalyst. *Ultrason. Sonochem.* 19:994–998.

Suslick, K. S. 1990. Sonochemistry. *Science* 247:1439–1445.

Toukoniitty, B., Toukoniitty, E., Maki-Arvela, P., Mikkola, J.-P., Salmi, T., Murzim, D., Yu, P. and Kooyman, J. 2006. Effect of ultrasound in enantioselective hydrogenation of 1-phenyl-1,2-propanedione: Comparison of catalyst activation, solvents and supports. *Ultrason. Sonochem.* 13:68–75.

Tsuchida, T., Inuma, H., Kinoshita, N., Ikeda, T., Sawa, T., Hamada, M. and Takenchi, T. 1995. Azicemicins A and B, a new antimicrobial agent produced by Amycolatopsis. I. Taxonomy, fermentation, isolation, characterization and biological activities. *J. Antibiot.* 48:217–221.

Van Der May, M., Hatzelmann, A., Van Der Laan, I. J., Sterk, G. J., Thibaut, V. and Timmerman, H. 2001. Novel selective PDE4 inhibitors. 1. Synthesis, structure-activity relationships, and molecular modeling of 4-(3,4-dimethoxyphenyl)-2*H*-phthalazin-1-ones and analogues. *J. Med. Chem.* 44:2511–2522.

Wang, J., Bai, X., Xu, C., Wang, Y., Lin, W., Zou, Y. and Shi, D. 2012. Ultrasound-promoted one-pot, three-component synthesis of spiro[indoline-3,1′-pyrazolo[1,2-*b*]phthalazine] derivatives. *Molecules* 17:8674–8686.

Yadav, J. S., Subba Reddy, B. V., Bhaskar Reddy, K., Raj, K. S. and Prasad, A. R. 2001. Ultrasound-accelerated synthesis of 3,4-dihydropyrimidin-2(1*H*)-ones with ceric ammonium nitrate. *J. Chem. Soc. Perkin Trans. 1* 1939–1941.

Yamada, T., Shimamura, H., Tsukamoto, Y., Yamaguchi, A. and Ohki, M. 1983. Pyridazinones. 3. Synthesis, antisecretory, and antiulcer activities of 2-cyanoguanidine derivatives. *J. Med. Chem.* 26:1144–1149.

Yamaguchi, M., Kamei, K., Koga, T., Akima, M., Kuroki, T. and Ohi, N. 1993. Novel antiasthmatic agents with dual activities of thromboxane A2 synthetase inhibition and bronchodilation. 1. 2-[2-(1-Imidazolyl)alkyl]-1(2*H*)-phthalazinones. *J. Med. Chem.* 36:4052–4060.

Zakharov, A. V., Gavrilov, M. Y., Demina, L. M., Novoselova, G. N., Gornova, N. A. and Konshin, M. Y. 1994. Derivatives of 1-aryl-2-vinylpyrido[2,3-*d*]pyrimidine-4-ones: Synthesis and antimicrobial activity. *Kim. Farm. Zh.* 28:24–26.

Zare, L., Mahmoodi, N. O., Yahyazadeh, A. and Nikpassand, M. 2012. Ultrasound-promoted regio and chemoselective synthesis of pyridazinones and phthalazinones catalyzed by ionic liquid [bmim]Br/$AlCl_3$. *Ultrason. Sonochem.* 19:740–744.

Zhidovinova, M. S., Fedorova, O. V., Rusinov, G. L. and Ovchinnikova, I. G. 2003. Sonochemical synthesis of Biginelli compounds. *Russ. Chem. Bull.* 52:2527–2528.

Zou, Y., Wu, H., Hu, Y., Liu, H., Zhao, X., Ji, H. and Shi, D. 2011. A novel and environment friendly method for preparing dihydropyrano[2,3-*c*]pyrazoles in water under ultrasound irradiation. *Ultrason. Sonochem.* 18:708–712.

9 Nano–Zinc Oxide

An Efficient Heterogeneous Catalyst for the Synthesis of Heterocyclic Compounds

Sunil U. Tekale, Ambadas B. Rode, K. L. Ameta, and Rajendra P. Pawar

CONTENTS

9.1 INTRODUCTION

Heterogeneous catalysts have many advantages over homogenous ones, such as easy separation, recycling ability in successive runs with minimization of metal traces in the product, high selectivity, and so on (Ross 2011). Currently more attention is being paid to the development and catalytic applications of novel and reusable heterogeneous catalysts due to their environmental and economic considerations. Heterogeneous catalysis is being explored in fine chemical industries due to the need for environmentally friendly technology. The development of novel heterogeneous catalytic materials for the synthesis of various heterocyclic compounds has become an intensely studied area in organic and medicinal chemistry (Kulkarni and Torok 2012; Magar et al. 2013). In short, the advantages of heterogeneous catalysts are briefly outlined below.

1. Easy handling
2. Easy separation after completion of reaction by filtration or centrifugation
3. Safe to store for longer time
4. Tolerance at higher temperature and pressure range
5. Lower probability of products being contaminated
6. Formation of inorganic salts avoided

Consequently, heterogeneous catalysts are extensively employed by researchers in the development of novel synthetic routes for the construction of various heterocyclic compounds.

For the development of efficient synthetic processes and minimization of by-products using heterogeneous catalysts, nanomaterials have come into the limelight during the past decade. Nanocatalysts have larger surface areas, on account of their high surface area-to-volume ratio as compared to the classical bulk catalysts, and reactive morphologies, which assist them to work as efficient catalysts. Since the properties of nanomaterials are dependent on size and shape, they are associated with diverse catalytic activity, and hence they can be used in organic synthesis as effective catalysts over a wide range of temperatures. With nanoparticulate catalysts, satisfactory results are obtained in short reaction times, and high yields are produced, with operational simplicity in the experimental procedure. Furthermore, reusable heterogeneous catalysts can be employed with high atom efficiency, selectivity, and high product yield. With the use of nanocatalysts, reactions usually occur under mild reaction conditions rather than harsh ones and require easier workup procedures (Polshettiwar and Asefa 2013).

Recently, many inorganic metal oxide nanoparticles have emerged as sustainable alternatives to conventional bulk catalytic materials and have attracted the interest of synthetic organic chemists due to their characteristic properties and application as heterogeneous catalysts. Nanoparticulate metal oxides have high surface-to-volume ratio and fewer coordinated parts; this results in a larger number of active sites per unit area as compared to their bulk heterogeneous counterparts. Metal oxides exhibit both Lewis acid and Lewis base characteristics at their surfaces, which makes them excellent adsorbents and activators for organic compounds. They have a good level of stability under reaction conditions and their reusability makes them highly useful as catalysts for organic transformations. Consequently, these metal oxide nanoparticles have received significant attention as efficient catalysts and have become promising inexpensive, nontoxic, and biocompatible catalytic materials in different organic transformations, including C–C bond–forming reactions, and for the synthesis of various heterocyclic compounds (Astruc 2008).

Essentially, nano-ZnO is an inorganic material of significant interest in nanotechnology, used particularly in gas sensors, solar cells, and luminescent materials. It is a nonvolatile, nonhygroscopic, odorless, and white crystalline solid with versatile properties. It can be easily synthesized in the laboratory by various methods including sol-gel, precipitation, thermal, and pyrolysis techniques and is characterized by sophisticated instrumental analytical techniques such as scanning electron microscopy (SEM), transmission electron microscopy (TEM), x-ray diffraction (XRD), electron dispersion spectroscopy (EDS), and so on.

Nano–zinc oxide acts as a mild Lewis acid that is a reusable, heterogeneous, noncorrosive (compared to other Lewis acids), and nontoxic catalyst. It has a high surface area and low coordinating sites which give it excellent catalytic activity. ZnO provides a better support than many catalysts such as Al_2O_3, SiO_2, clays, and so on. In many cases, it efficiently catalyzes organic transformations involving solvent-free conditions or reactions in an aqueous medium and helps in the development of green and economically competitive protocols for the synthesis of heterocyclic compounds of biological interest. It offers chemoselectivity and environmental compatibility as well as simplicity of operation and, in many cases, solvent-free conditions are required for the reactions when nano-ZnO is used as a catalyst. In short, reactions using nano-ZnO catalysts have the following characteristics:

1. Easy isolation of the product and catalyst
2. Mild reaction conditions
3. Use of solvent-free conditions or aqueous medium in many cases
4. Good chemoselectivity
5. Short reaction time
6. Clean reaction profile

Thus, on account of its peculiar characteristics, such as large surface area, heterogeneous nature, reusability, thermal stability, and sustainability, nano-ZnO has attracted the attention of synthetic organic chemists as an efficient and versatile heterogeneous catalyst in organic chemistry. This chapter describes different organic transformations catalyzed by nano-ZnO as a valuable heterogeneous catalyst.

9.1.1 Nano-ZnO-Catalyzed Synthesis of Heterocyclic Compounds

- Synthesis of nitrogen-containing heterocyclic compounds
- Synthesis of oxygen-containing heterocyclic compounds
- Synthesis of sulfur-containing heterocyclic compounds

9.1.2 ZnO Nanoparticle–Catalyzed Miscellaneous Reactions

- C–C bond–forming reactions
- Alkylation/arylation
- Other reactions catalyzed by nano-ZnO

9.1.3 Nano-ZnO-Catalyzed Synthesis of Various Heterocyclic Compounds

Heterocyclic compounds are cyclic compounds containing one or more atoms of elements other than carbon—commonly sulfur, oxygen, or nitrogen—in their ring structure. Heterocyclic compounds constitute an important class of organic compounds useful and essential in human and animal life, including the discovery of life-saving new candidates as drugs for the welfare of mankind. They are also present in many antibiotics, hormones, amino acids, alkaloids, vitamins, hemoglobin, and so on. These heterocyclic compounds may occur naturally or be synthesized artificially in the laboratory. Besides these applications, heterocyclic compounds are also recruited as valuable intermediates in organic synthesis. This section of the chapter is concerned with the synthesis of various heterocyclic compounds of biological interest using nano-ZnO as a valuable heterogeneous catalyst.

9.1.3.1 Synthesis of Nitrogen-Containing Heterocyclic Compounds

Many nitrogen-containing heterocyclic compounds with potent biological activities, including imidazoles, pyridines, pyrimidines, triazoles, and so on, are present in nature or may be synthesized artificially. Nano-ZnO may be used as a catalyst for the synthesis of such heterocyclic compounds.

Hantzsch's reaction is a multicomponent organic reaction between an aldehyde, a β-keto ester, and a nitrogen donor such as ammonium acetate or ammonia. It provides rapid access to dihydropyridine and polyhydroquinoline derivatives, which constitute an important class of calcium channel blockers. Zinc oxide nanoparticles have been used for one-pot four-component synthesis of Hantzsch's polyhydroquinolines (**7**) and 1,4-dihydropyridines (**5**) in higher yields and short reaction times by the condensation of aldehydes (**1**), dimedone (**2**), active methylene compounds (**4**, **6**), and ammonium acetate (**3**) under solvent-free conditions at room temperature (Kassaee et al. 2010) (Scheme 9.1).

Nanorods of ZnO were employed as heterogeneous catalysts for the one-pot synthesis of imidazo[1,2-*a*]azines and diazines (**11**) from aldehydes (**8**), amine (**9**), and trimethyl silylcyanide (TMSCN) (**10**) in short reaction times (20–40 min) (Sadjadi and Eskandari 2012) (Scheme 9.2). The synthesized products were purified by crystallization from ethanol. The synthesis of substituted 2,4,5-triarylimidazoles (**14**) by the one-pot three-component condensation reaction of benzil (**12**), aldehydes (**13**), and ammonium acetate (**3**) under solvent-free conditions has also been reported (Reza et al. 2011) (Scheme 9.3). 2-Amino-3,5-dicyano-4-phenyl-6-(phenylthio)pyridines (**18**) were synthesized (Safaei-Ghomi and Ghasemzadeh 2012) in excellent yields by the multicomponent reaction of aldehydes (**15**), malononitrile (**16**), and thiols (**17**) using ZnO nanoparticles over short times by heating the reaction mixture to 50°C (Scheme 9.4), with good recoverability of the catalyst. After completion of

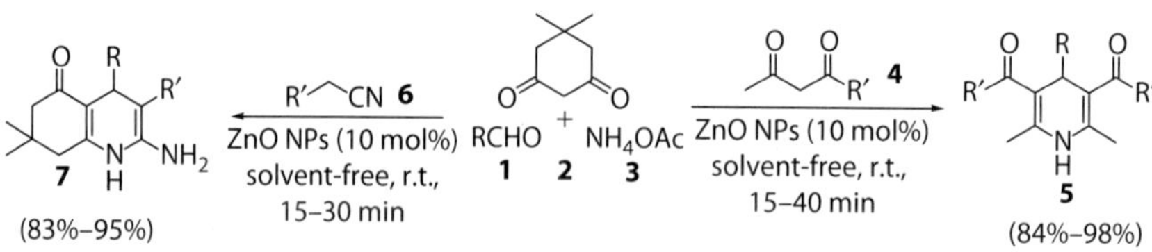

R = Ph, 4-$OMeC_6H_4$, 4-MeC_6H_4, 4-ClC_6H_4, 4-ClC_6H_4, 3-ClC_6H_4, 2-Cl, 2,4-$Cl_2C_6H_3$, 3,4-$Cl_2C_6H_3$, 4-BrC_6H_4, 4-FC_6H_4, 4-$NO_2C_6H_4$, 3-$NO_2C_6H_4$, 2-$NO_2C_6H_4$, 4-OHC_6H_4, 4-OH,3-$OMeC_6H_3$, etc.
R′ = OEt, OMe, CN, COOEt

SCHEME 9.1

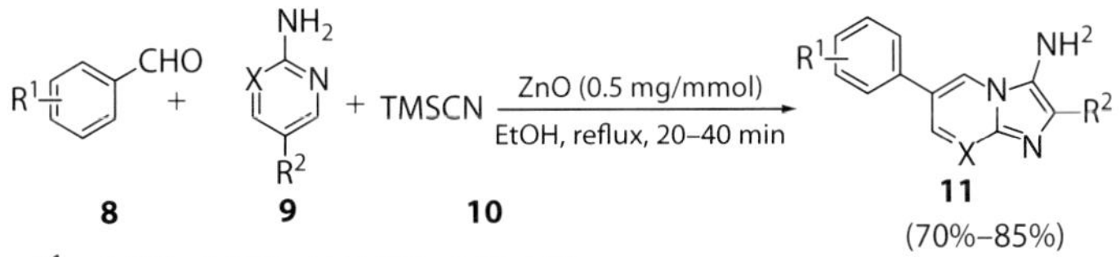

(70%–85%)

R^1 = 4-NO_2, 4-OMe, H, 3-NO_2, 4-Cl, 4-Me
R^2 = H, Br, Me; X = N, CH

SCHEME 9.2

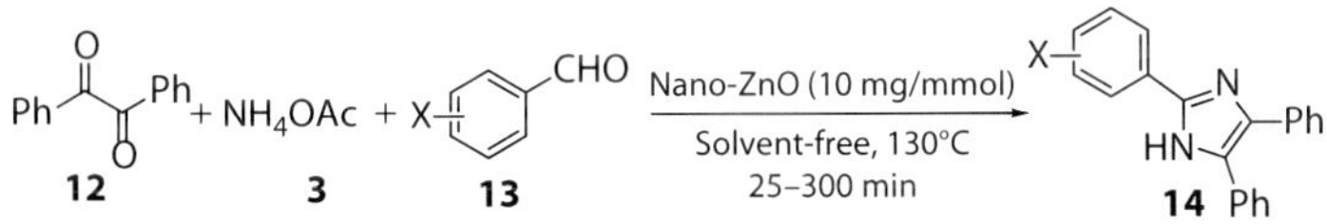

(56%–90%)

(X = H, 4-OH, 4-OMe, 4-Me, 4-*t*-Bu, 2-Me,
4-F, 2-Cl, 4-Cl, 2, 4-Cl_2, 2-NO_2, 4-NO_2)

SCHEME 9.3

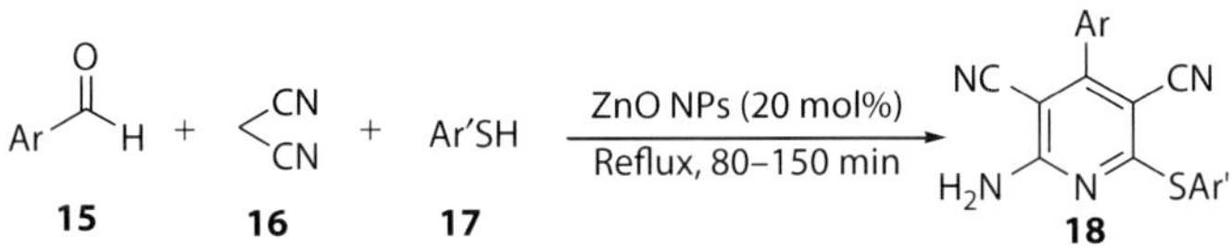

(75%–94%)

(Ar = Ph, *m*-MeC_6H_4, *p*-MeC_6H_4, *m*-OHC_6H_4, *p*-OHC_6H_4, *p*-$OMeC_6H_4$,
m-$NO_2C_6H_4$, *p*-$NO_2C_6H_4$, *p*-ClC_6H_4, *p*-BrC_6H_4, *p*-CNC_6H_4
Ar′ = Ph, *p*-MeC_6H_4)

SCHEME 9.4

the reaction, the reaction mass was cooled to room temperature and centrifuged to separate the catalyst. Purification of the corresponding products was carried out by simple crystallization from ethanol.

ZnO-CuO hybrid nanoparticles were used as a catalyst for cycloaddition reactions between azides (**19**) and alkyne (**20**) under ultrasound-assisted conditions to afford substituted triazole derivatives (**21**) (Park et al. 2012) (Scheme 9.5). A green and efficient ZnO nanoparticle–catalyzed synthesis of *bis*-isoquinolinones (**26**) from 3-substituted isocoumarins (**24**) and heptadiamines (**25**) is also reported in the literature (Scheme 9.6) (Krishnakumar et al. 2012). The 3-substituted isocoumarins (**24**) required were synthesized from homophthalic acid (**22**) and various acid chlorides (**23**). Khazaei et

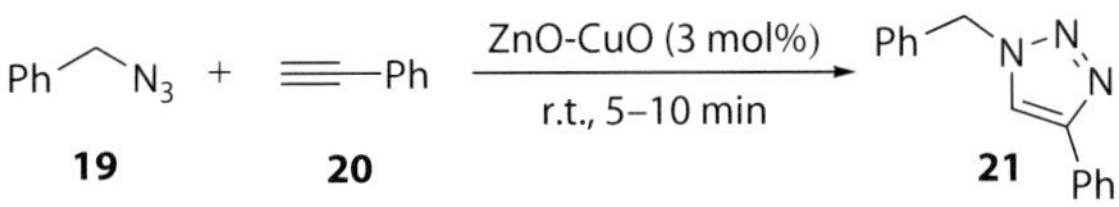

SCHEME 9.5

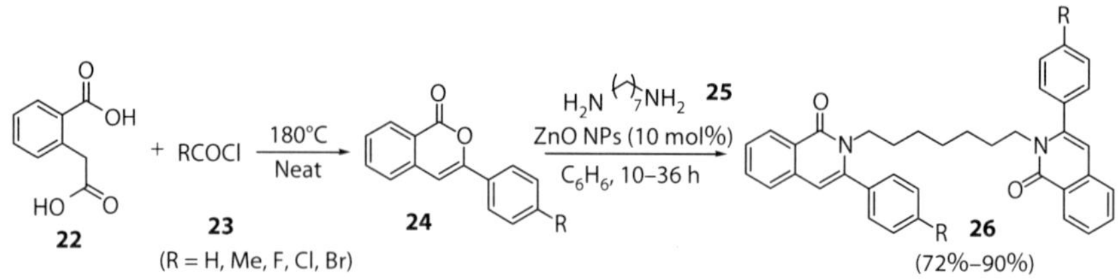

SCHEME 9.6

al. reported one-pot synthesis of dihydropyridine derivatives (**29**, **30**) by the one-pot multicomponent reaction of aldehydes (**15**), ethyl acetoacetate (**28**), cyanoacetamide (**27**), and ammonium acetate (**3**) using nano-ZnO catalyst (Scheme 9.7) (Khazaei et al. 2012). The transformation was carried out in ethanol under reflux conditions to afford the corresponding products in excellent yields. Among the various substituents of the aldehydes used, the product (**29**) was predominantly formed and, interestingly, with 4-nitrobenzaldehyde products (**29**) and (**30**) were formed in yields of 63% and 37% respectively. A solvent-free protocol for the synthesis of bis(indolyl)methanes (**32**) from the condensation of aldehydes (**15**) and indoles (**31**) at 80°C using the environmentally friendly and reusable nano-ZnO as a solid catalyst (Scheme 9.8) was documented (Hosseini-Sarvaria et al. 2008). After completion of the reaction, ethyl acetate was added to the reaction mixture and the catalyst was separated by filtration. The catalyst was then washed with water, dried, and reused in subsequent runs for recycle studies. The catalyst could be recycled at least five times without much loss of catalytic activity. The mechanism for nano-ZnO-catalyzed synthesis of bis(indolyl)methanes is depicted in Figure 9.1. Initially, the catalyst polarizes the oxygen atom of the carbonyl compound to form an intermediate adduct (**33**) that reacts with the first molecule of indole to afford the 3-substituted hydroxyl indole (**34**). The hydroxyl compound (**34**) undergoes

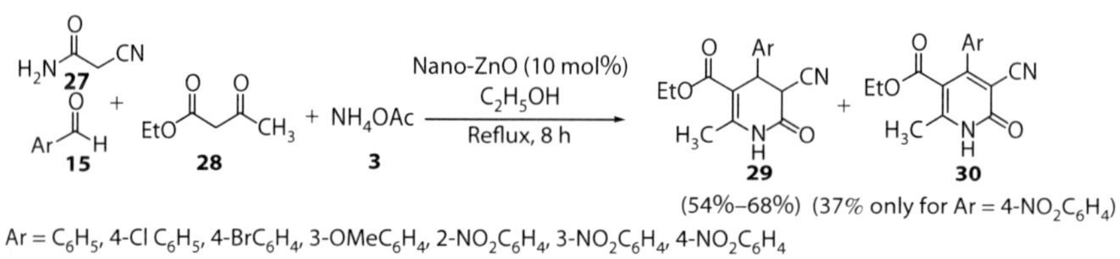

SCHEME 9.7

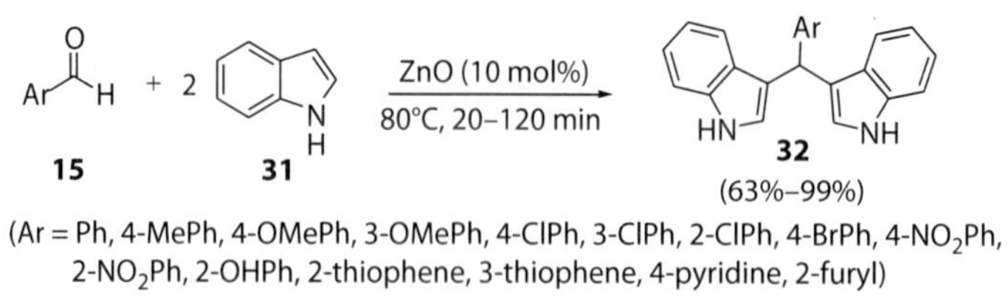

SCHEME 9.8

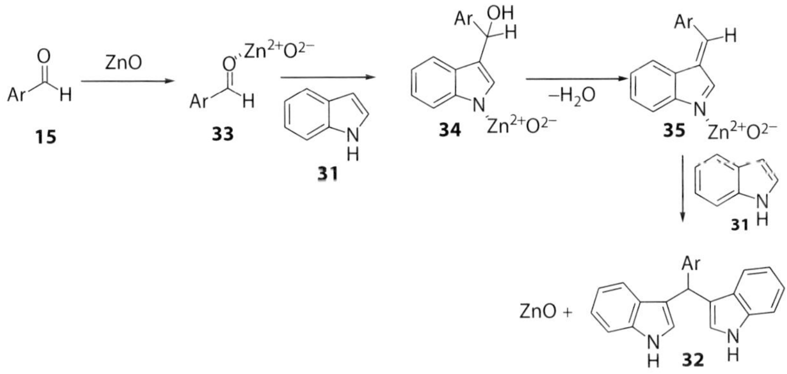

FIGURE 9.1 Mechanism for ZnO nanoparticle–catalyzed synthesis of bisindolylmethanes.

dehydration in forming the alkene which in adduct form with the catalyst (**35**) reacts with another molecule of indole (**31**) to afford the desired bis(indolyl)methane (**32**).

The Biginelli reaction is an important one-pot three-component condensation which allows a facile and easy synthesis of 3,4-dihydropyrimidin-2(1*H*)-ones and their thio analogs. The reaction was discovered by the Italian chemist Pietro Biginelli in 1891 and is being modified even today by different researchers using different acidic as well as basic catalysts. 3,4-Dihydropyrimidin-2-(1*H*)-ones (**37**) were synthesized via a Biginelli reaction from aldehydes (**1**), urea/thiourea (**36**), and active methylene compounds (**4**) such as ethyl acetoacetate in the presence of nano-ZnO as a reusable catalyst under solvent-free conditions at 80°C (Scheme 9.9) (Bahrami et al. 2009). High catalytic activity, reusability, solvent-free conditions, and good yields make this protocol an attractive process for the synthesis of dihydropyrimidinones and their sulfur analogs. Hekmatshoar et al. (2010) documented an efficient method for the synthesis of 4-amino-5-pyrimidinecarbonitriles (**39**) from three-component synthesis using aldehydes (**1**), malononitrile (**16**), and amidines (**38**) in the presence of nano-ZnO in an aqueous medium (Scheme 9.10). The reactions were completed in short times producing products in higher yield (91%–98%). Furthermore, since water was used as the reaction solvent, this constituted a green method for the synthesis of 4-amino-5-pyrimidinecarbonitriles (**39**). Mechanistically, the reaction involves

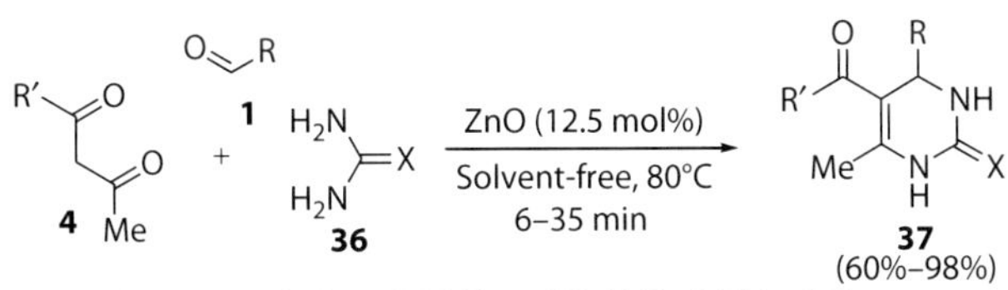

R = C_6H_5, 4-MeC_6H_4, 4-OHC_6H_4, 4-$NO_2C_6H_4$, 4-$NMe_2C_6H_4$, 4-$OMeC_6H_4$, 4-ClC_6H_4, CH_3, CH_2CH_3, 4-FC_6H_4, 2-ClC_6H_4
R′ = OEt, OMe, Me; X = O, S

SCHEME 9.9

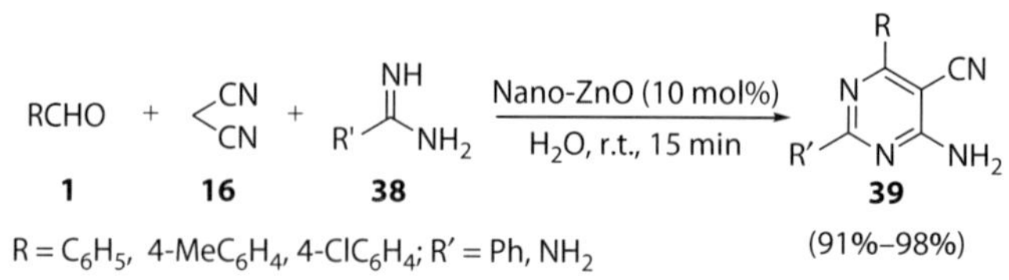

SCHEME 9.10

Knoevenagel condensation between the aldehyde (**1**) and malononitrile (**16**) to form cyanoolefins (**40**), followed by a Michael addition with amidine (**38**), cycloaddition, isomerization, and aromatization to afford the 4-amino-5-pyrimidinecarbonitriles (**39**) (Figure 9.2).

The synthesis of 1,4-dihydropyridines (**45**) by four-component condensation reaction of 1,3-dicarbonyls (**43**), aldehydes (**1**), and ammonium carbonate (**44**) in an aqueous medium was reported at 60°C within 1.5–2.5 h (Scheme 9.11) (Tamaddon and Moradi 2012). ZnO nanoparticles can be used as a heterogeneous catalyst (Yavari and Beheshti 2011) for the one-pot three-component condensation of isatoic anhydride (**46**), amines (**47**), and aldehydes (**48**) to afford the corresponding 2,3-disubstituted quinazolin-4(1*H*)-ones (**49**) (Scheme 9.12). A short reaction time (3 h), solvent-free conditions, and an atom-economic nature are the significant advantages of the present method. The catalyst showed good recyclability without much loss of its catalytic activity.

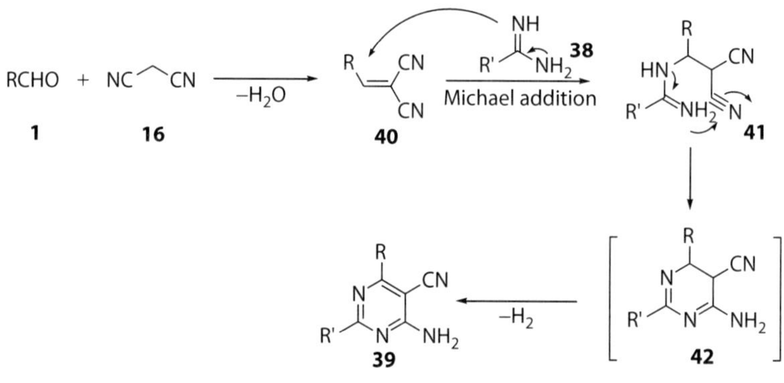

FIGURE 9.2 Mechanism for the synthesis of 4-amino-5-pyrimidinecarbonitriles using ZnO nanoparticles.

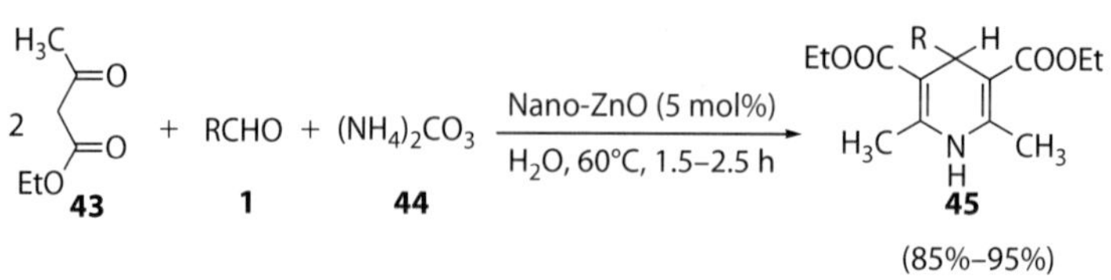

SCHEME 9.11

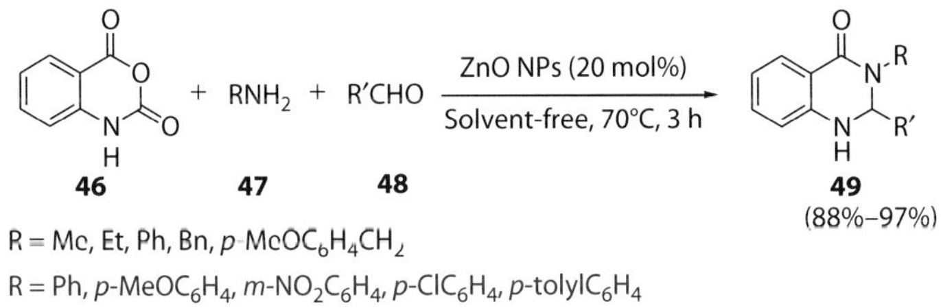

SCHEME 9.12

Ameta et al. (2012) reported one-pot three-component reactions of chalcones (**50**) with *S*-benzylthiouronium chloride (SBT) (**51**) and aliphatic heterocyclic secondary amines such as morpholine (**52**), pyrrolidine (**53**), and piperidine (**54**) in the presence of ZnO heterogeneous catalyst to afford 2-substituted-4,6-diarylpyrimidines (**55**, **56**, **57**) (Scheme 9.13). The method is highly economic, simple, and environmentally friendly in nature. Furthermore, the catalyst showed good reusability up to three successive runs.

Hosseini-Sarvaria (2011b) reported the synthesis of quinoline derivatives (**60**) by the condensation of 2-aminoaryl ketones (**58**) with methylene carbonyl compounds (**59**) catalyzed by nanoflake ZnO as a reusable heterogeneous catalyst under solvent-free conditions involving Friedlander heteroannulation (Scheme 9.14). The employed catalyst was synthesized from zinc acetate dihydrate and urea. It constitutes a simple, environmentally benign, and cost-effective method for the synthesis of quinolines. Tamaddon and Moradi (2013) reported the application of nano-ZnO as a reusable heterogeneous catalyst for the synthesis of Biginelli dihydropyrimidines (**61**) and Hantzsch 1,4-dihydropyridines (**62**) (Schemes 9.15, 9.16). ZnO catalyzed the Biginelli reaction at 60°C while the synthesis of Hantzsch's 1,4-dihydropyridines (**62**) required a higher temperature of 120°C–140°C in an aqueous medium or in

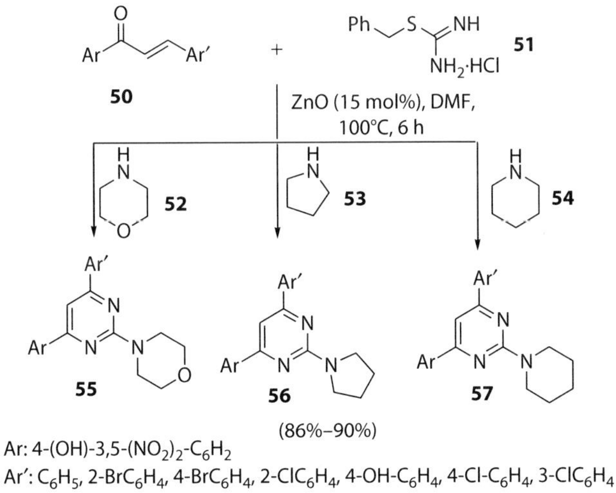

SCHEME 9.13

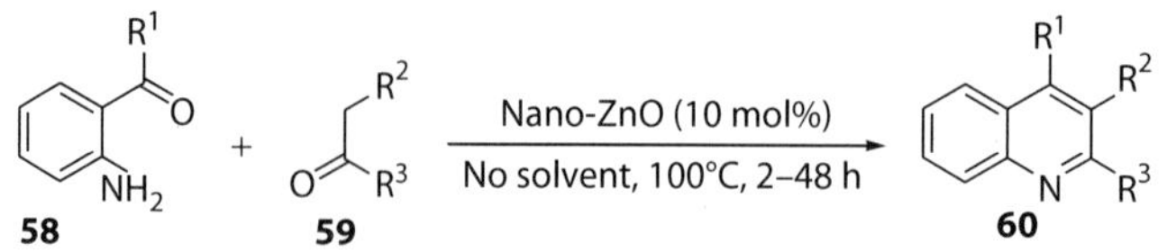

R^1 = Me, Ph
R^2 = COOEt, COOMe, CO_2CH_2Ph, COMe, COcycloalkyl, $-COCF_3$
R^3 = Me, $-CF_3$, cycloalkyl

SCHEME 9.14

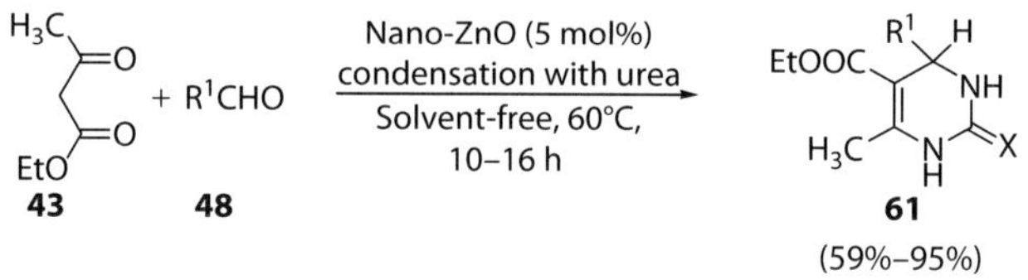

SCHEME 9.15

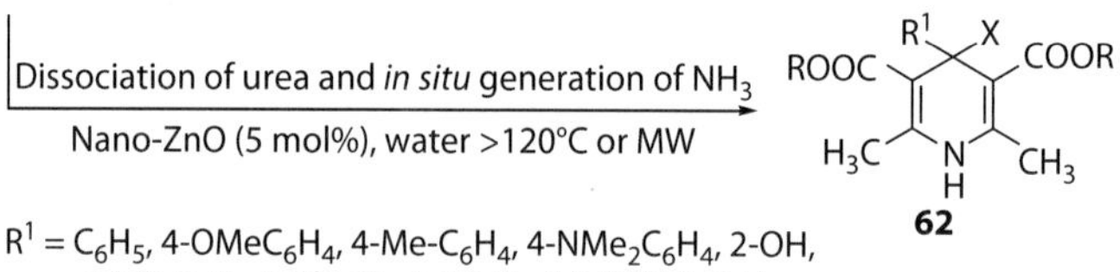

R^1 = C_6H_5, 4-OMeC_6H_4, 4-Me-C_6H_4, 4-$NMe_2C_6H_4$, 2-OH, 4-$NO_2C_6H_4$, 4-ClC_6H_4, 4-FC_6H_4, 3,4-$(OMe)_2C_6H_3$
X = O, S, NH

SCHEME 9.16

microwave irradiation. It was suggested that the NH_3 required for the synthesis of Hantzsch's 1,4-dihydropyridines (**62**) was formed due to *in situ* dissociation of urea.

Sadjadi and Eskandari (2013) reported a rapid protocol for the ultrasound-promoted synthesis of imidazoazines (**63**) by three-component reaction of aldehydes (**48**), 2-aminoazines (**49**), and TMSCN (**10**) (Scheme 9.17). Short reaction times, high yields, and environmentally friendly conditions are some of the advantages of this method. Nano-ZnO/Co_3O_4 can be used as a novel heterogeneous catalytic system (Agawane and Nagarkar 2012) for the cycloaddition between different aromatic nitriles (**64**) and sodium azide (**65**) for the synthesis of 5-substituted 1*H*-tetrazoles (**66**) in excellent yield and high efficiency (Scheme 9.18). It was observed that the electron-donating group at the para-position of the nitrile group resulted in excellent yields of the corresponding tetrazoles (**66**). Furthermore, the synthesized catalyst showed better catalytic activity than merely using either ZnO or Co_3O_4. A similar transformation was also documented in the literature using porous nano-ZnO (Scheme 9.19) (Giri et al. 2013).

Nanoparticles of ZnO were utilized as an efficient catalyst for the synthesis of benzimidazoles (**71**) from *o*-diamines (**69**) and formic acid (**70**) under solvent-free

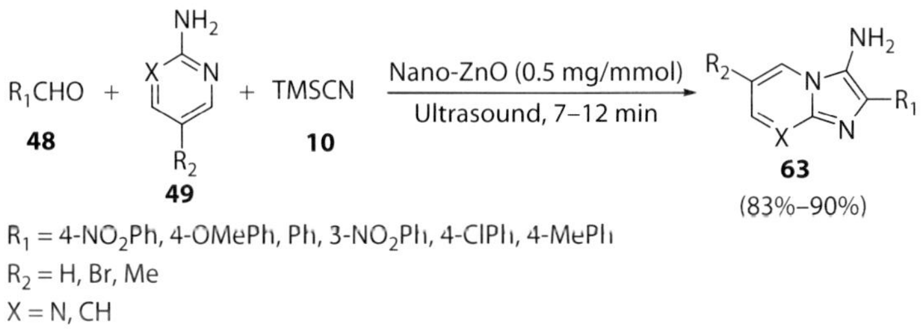

SCHEME 9.17

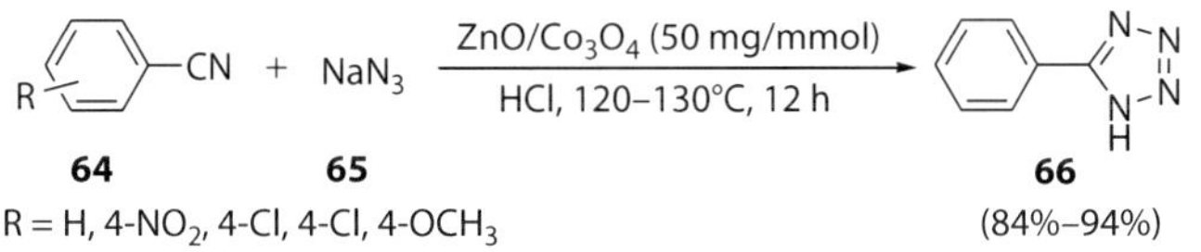

SCHEME 9.18

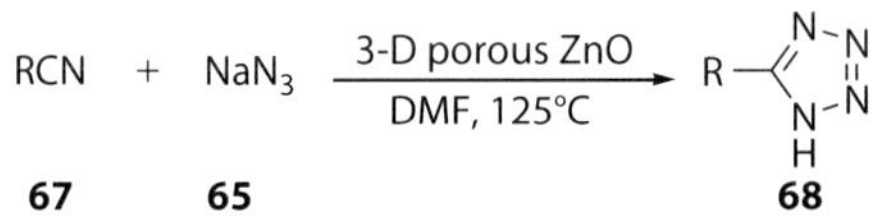

SCHEME 9.19

conditions at 70°C. Since the method uses solvent-free conditions for performing the transformation, it constitutes an environmentally benign method for the synthesis of benzimidzoles (**71**) (Scheme 9.20) (Alinezhad et al. 2012). Some of the advantages of the protocol are solvent-free conditions, easy workup, and an environmentally benign nature. ZnO was employed in the form of nanorods as a reusable heterogeneous catalyst for the reaction between isatins (**72**) and indoles (**73**) by a Friedel–Craft reaction in an aqueous medium at 80°C to afford the corresponding 3-indolyl-3-hydroxy oxindoles (**74**) (Scheme 9.21) (Hosseini-Sarvari and Tavakolian 2012b). The method has distinct advantages, such as atom efficiency, high reaction rate, simple operational procedure, easy recovery of the catalyst, and so on. Ziarati et al. (2013) documented a one-pot three-component synthesis of *N*-cyclohexyl-3-aryl-quinoxaline-2-amines (**78**) from *o*-phenylenediamine (**75**), aldehydes (**76**), and cyclohexyl isocyanide (**77**) in the presence of nano-ZnO under reflux conditions in ethanol as the reaction medium (Scheme 9.22). High yield, an environmentally

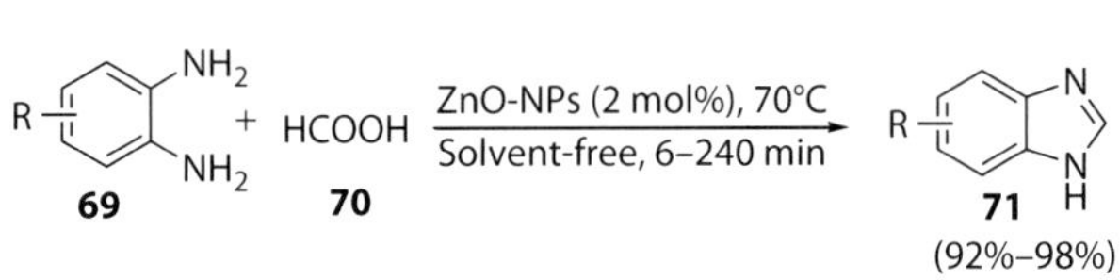

SCHEME 9.20

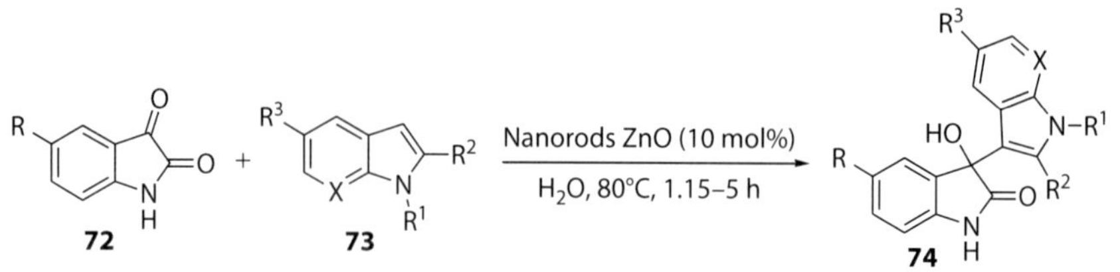

R = H, F; R^1 = H, Me; R^2 = H, Me; R^3 = H, Br, Me
X = C(H), N

SCHEME 9.21

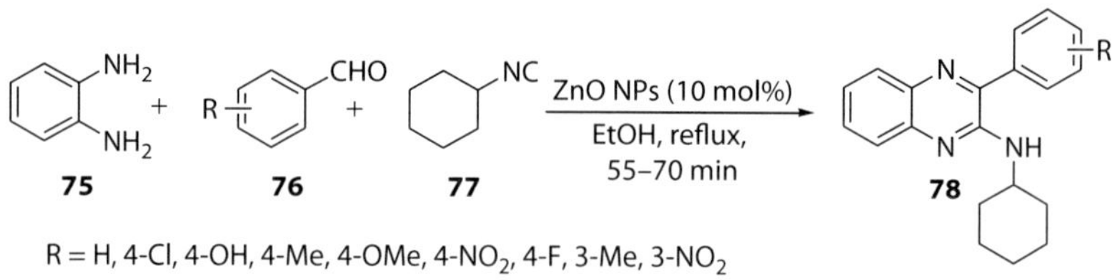

R = H, 4-Cl, 4-OH, 4-Me, 4-OMe, 4-NO_2, 4-F, 3-Me, 3-NO_2

SCHEME 9.22

benign nature, easy purification, and reusability of the heterogeneous catalyst indicate the significance of this protocol.

9.1.3.2 Synthesis of Oxygen-Containing Heterocyclic Compounds

Tetrahydrobenzopyrans, chromenes, coumarins, xanthenes, and so on are some of the most important oxygen-containing heterocyclic compounds. Several synthetic methods are available for these compounds. This part of the chapter deals with the synthesis of these heterocyclic compounds using nano-ZnO as a reusable heterogeneous catalyst.

Tetrahydrobenzo[*b*]pyrans exhibit many important biological and pharmacological properties. Hosseini-Sarvari and Shafiee-Haghighi (2012) documented the synthesis of tetrahydrobenzo[*b*]pyrans (**81**) by one-pot three-component reaction of arylaldehydes (**15**), active methylene compounds (**79**), and 1,3-diketones (**80**) in the presence of nano-ZnO in an aqueous medium (Scheme 9.23).

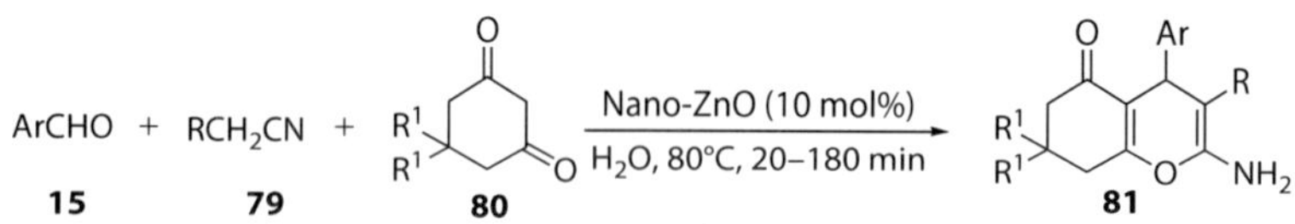

(R = CN, COOEt; R^1 = Me, H; Ar = 4-ClC_6H_4, 4-$O_2NC_6H_4$, 4-HOC_6H_4, 4-$MeOC_6H_4$, 4-MeC_6H_4, Ph, 3-ClC_6H_4, 2-ClC_6H_4, 2-thienyl)

SCHEME 9.23

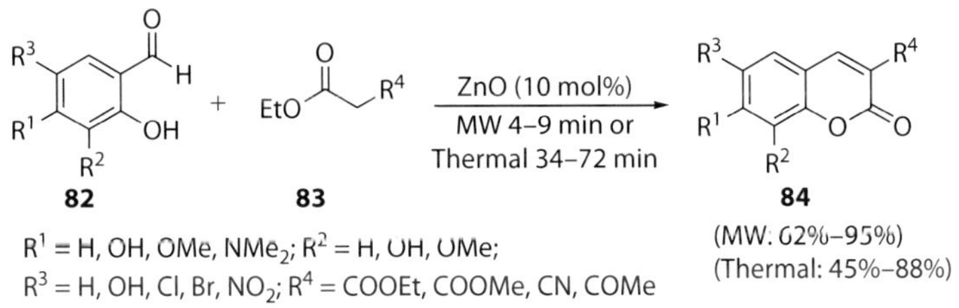

SCHEME 9.24

Nanoparticulate ZnO was used as an efficient catalyst for the synthesis of coumarins (**84**) by the reaction of *o*-hydroxy benzaldehydes (**82**) and 1,3-dicarbonyl compounds (**83**) via Knoevenagel condensation under microwaves and thermal conditions (Scheme 9.24) in moderate to excellent yields (Kumar et al. 2011). This protocol differs from the previous methods for the synthesis of coumarins (**84**) in terms of simplicity and effectiveness. The application of ZnO/MgO in ionic liquid [bmim][BF_4] was carried out successfully for the synthesis of 4*H*-pyrans (**85**) and coumarins (**88**) at ambient temperature via Knoevenagel condensation reaction of aldehydes (**8**) or 2-hydroxybenzaldehyde derivatives (**86**) with active methylene compounds (**16**, **43**, **87**) (Schemes 9.25 and 9.26) (Valizadeha and Azimib 2011). The method has several advantages in terms of mild reaction conditions, reusability of the catalyst, high yields of the products, and short reaction times. In comparison with methods mentioned in the literature for the synthesis of 4*H*-pyrans (**85**) and coumarins (**88**), this protocol has better yield and eco-friendly advantages.

Nanoparticulate ZnO was used as an environmentally benign reusable heterogeneous catalyst for the synthesis of a series of biologically important

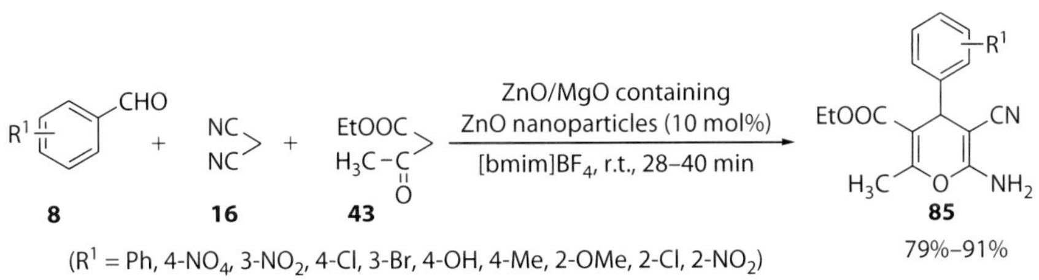

SCHEME 9.25

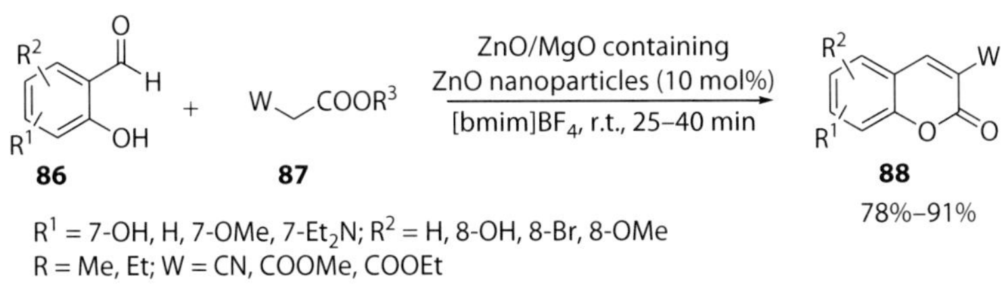

SCHEME 9.26

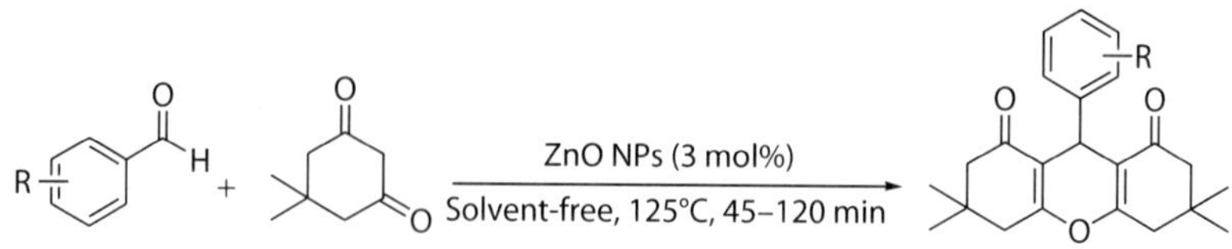

76 2 89

R = 4-Cl, 3-OH, 2-Cl, 3-OCH_3, 4-NO_2, 4-CH_3, 4-OCH_3, 3-NO_2, 3-OC_2H_5-4-OH, 5-Br-3-OCH_3, 1,3-bis-C_6H_5

80%–93%

SCHEME 9.27

1,8-dioxo-octahydroxanthenes (**89**) by the condensation of aldehydes (**76**) with 5,5-dimethyl-1,3-cyclohexanedione (**2**) (Scheme 9.27) in excellent yields and short reaction times with a clean reaction profile (Rao et al. 2012). A rapid, green, and high yielding one-pot three-component synthesis of dihydropyranochromene derivatives (**91**) from 3-hydroxy coumarin (**90**), aldehydes (**15**), and malononitrile (**16**) in an aqueous medium (Scheme 9.28) was explored (Paul et al. 2011). After completion of the reaction, the reaction mass was filtered in hot conditions to separate the catalyst and the filtrate was cooled to room temperature. The solid that reappeared was filtered, washed with cold aqueous ethanol, and recrystallized from ethanol to afford the corresponding products in high yields and good purity. Magnesium oxide–impregnated zinc oxide (MgO/ZnO) was studied as an effective catalyst for the Claisen–Schmidt condensation between *o*-hydroxyacetophenone (**92**) and benzaldehyde (**93**) to form *o*-hydroxychalcone (**94**) which was subsequently cyclized to afford the flavanone (**95**) under solvent-free conditions (Scheme 9.29) (Saravanamurugan et al. 2005). This catalyst showed better reactivity as compared to the other screened catalysts such as ZnO, SiO_2, Al_2O_3, and so on.

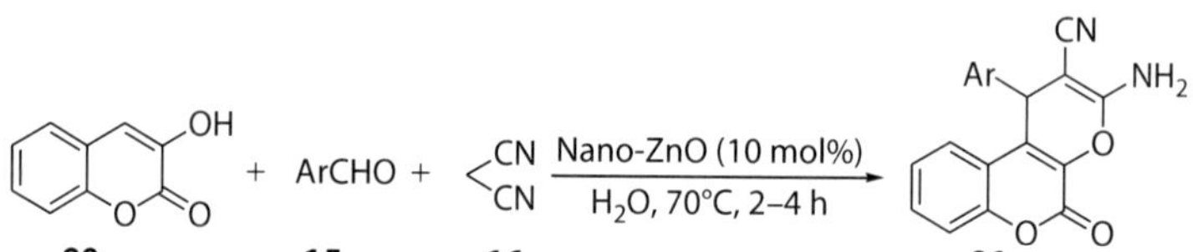

90 15 16 91

(65%–91%)

SCHEME 9.28

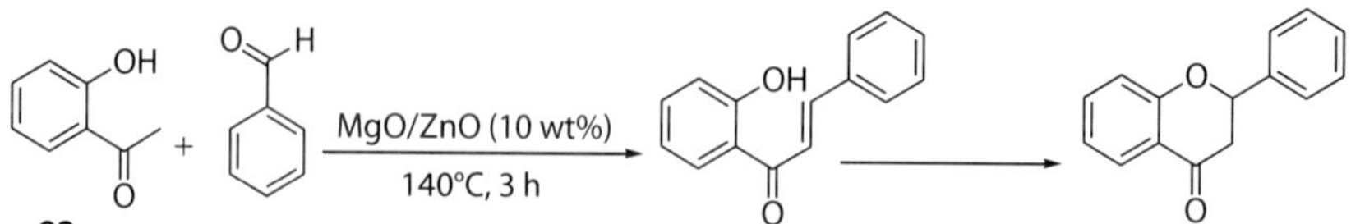

92 93 94 (39.55%) 95 (41.37%)

SCHEME 9.29

In continuation of our efforts in the development of new synthetic routes for the synthesis of heterocyclic compounds using nanocatalysts, we have recently reported a novel synthesis of 3,4,5-trisubstituted furan-2(5*H*)-one derivatives by the one-pot three-component condensation of aldehydes, amines, and dimethyl acetylenedicarboxylate (DMAD) by using nanoparticulate ZnO as a catalyst in EtOH:H_2O (1:1) at 90°C (Scheme 9.30) (Tekale et al. 2013). Almost all the employed aldehydes and amines reacted smoothly to afford excellent yields of the products, irrespective of the nature of the substituent present on the aldehyde or amine. The plausible mechanism for the synthesis of furan-2(5*H*)-ones using nano-ZnO is depicted in Figure 9.3. The catalyst promotes the formation of enamines (**99**) from amines (**97**) and DMAD (**96**). ZnO polarizes the carbonyl group of aldehydes to form a polarized adduct (**100**) which reacts with the enamines, followed by cyclization with the elimination of methanol molecules to afford the corresponding trisubstituted furanone derivatives (**98**).

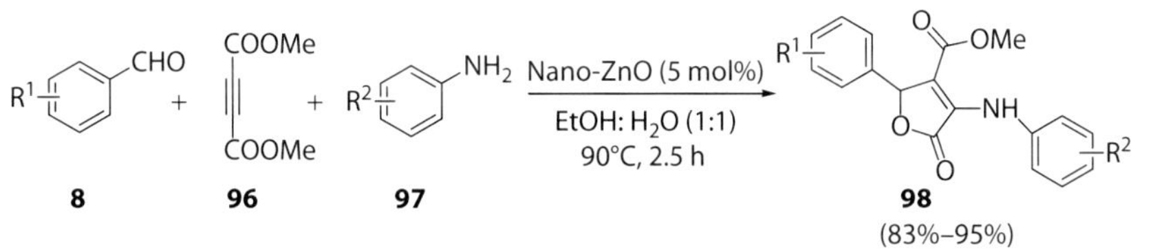

SCHEME 9.30

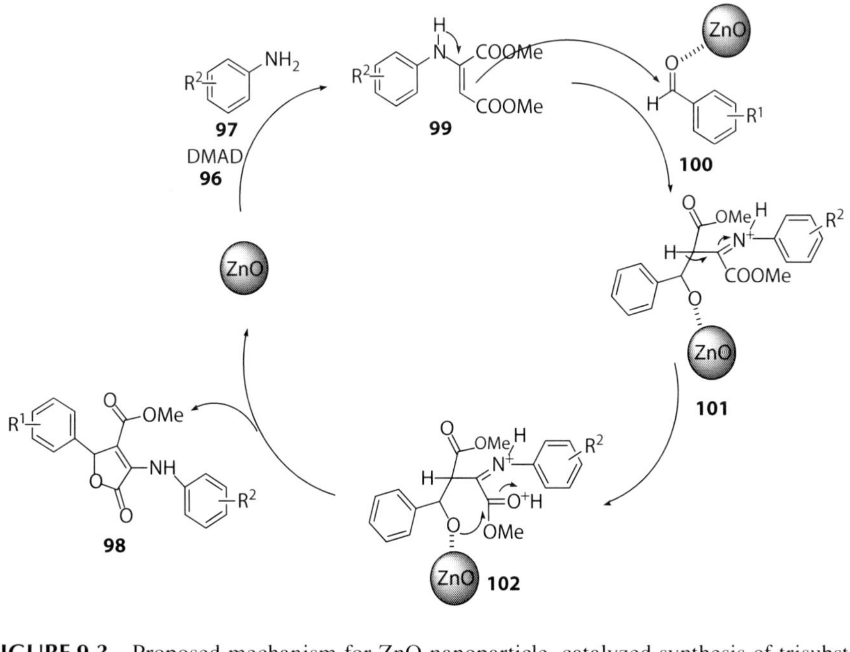

FIGURE 9.3 Proposed mechanism for ZnO nanoparticle–catalyzed synthesis of trisubstituted furan-2(5*H*)-ones.

9.1.3.3 Synthesis of Sulfur-Containing Heterocyclic Compounds

The Gewald reaction involves synthesis of 2-aminothiophenes via multicomponent condensation of α-methylene carbonyl compounds, cyano compounds, and sulfur. Recently, Tayebee et al. (2013) have successfully accomplished a rapid and efficient synthesis of 2-aminothiophenes (**105**) from ketones or aldehydes (**103**), malononitrile (**16**), and sulfur (**104**) via a one-pot three-component Gewald reaction in the presence of a catalytic amount of ZnO nanoparticles (Scheme 9.31). The catalyst required for the reaction was synthesized through sedimentation of zinc acetate dihydrate in ethanol.

9.1.4 ZnO Nanoparticle–Catalyzed Miscellaneous Reactions

9.1.4.1 C–C Bond–Forming Reactions

The element carbon is considered as the backbone of organic chemistry. Carbon has the unique characteristic property to form long chains of its own atoms, called catenation power. Heterocyclic, as well as alicyclic and open-chain organic compounds, contain several C–C bonds. Consequently, C–C bond–forming reactions are of great significance in organic as well as heterocyclic chemistry. This section of the chapter is concerned with the significant C–C bond–forming organic transformations catalyzed by nano-ZnO.

Gold nanoparticles on ZnO were used as a catalyst for the rapid one-pot three-component condensation of amines (**106**), aldehydes (**107**), and phenyl acetylene (**20**), affording the selective synthesis of propargylamines (**108**) at room temperature (Scheme 9.32) (Gonzalez-Bejar et al. 2013). Propargylic alcohols constitute important building blocks for many biologically active compounds and natural products.

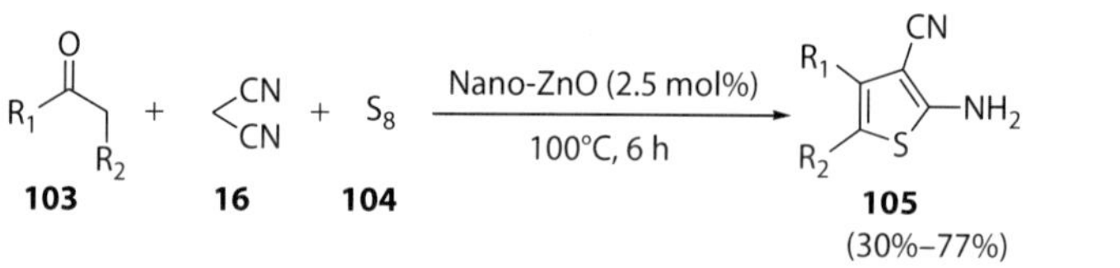

2-Pentanone, acetaldehyde, propiopaldehyde, butyraldehyde, cyclopentanone, cyclohexanone, cycloheptanone, acetophenone, 4-Me-acetophenone, 4-NO_2-acetophenone

SCHEME 9.31

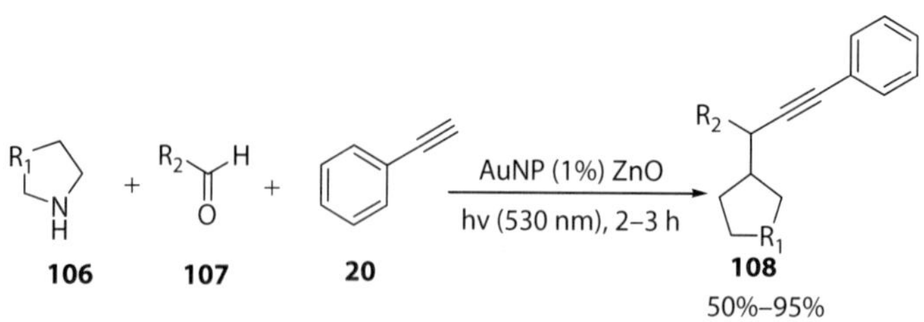

$R_1 = CH_2, C_2H_4, OCH_2$; $R_2 = H, C_6H_5, CH(CH_3)_2$

SCHEME 9.32

A new protocol for the alkynylation of various aromatic as well as aliphatic aldehydes (**48**) and alkynes (**109**) using ZnO under solvent- and base-free conditions was developed by Hosseini-Sarvari and Mardaneh (2011b) (Scheme 9.33). This demonstrates an important strategy for C–C bond formation in organic chemistry.

The Suzuki reaction is an important C–C bond–forming reaction of an aryl- or vinylboronic acid (**112**) with an aryl- or vinylhalide (**111**, **115**), catalyzed by a palladium(0) complex. Recently, Akira Suzuki won the 2010 Nobel Prize in chemistry for his significant contribution and development of this reaction. The literature survey reveals several modifications to the reaction. The application of various ZnO-supported metal combinations such as Pd-Ag, Pd-Cu, Pd-Ni, and Pd (Kim and Choi 2009) were reported for the cross-coupling Suzuki reaction between aryl halides (**111**, **115**) and phenyl boronic acid (**112**) (Schemes 9.34, 9.35). These metal combinations with ZnO resulted in excellent yields of the biaryl products.

Recently the addition of terminal alkynes (**109**) to acid chlorides (**117**) was reported using nano-ZnO under solvent-free conditions at room temperature to afford the corresponding (*Z*)-β-Cl-α,β-unsaturated ketones (**118**) selectively (Scheme 9.36) (Hosseini-Sarvari and Mardaneh 2011a). An important aspect of this method was

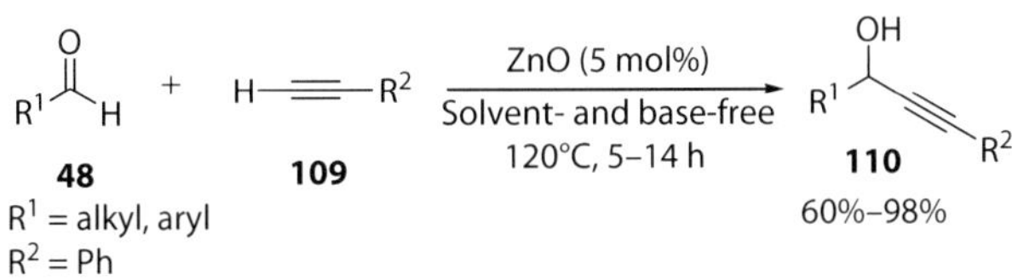

SCHEME 9.33

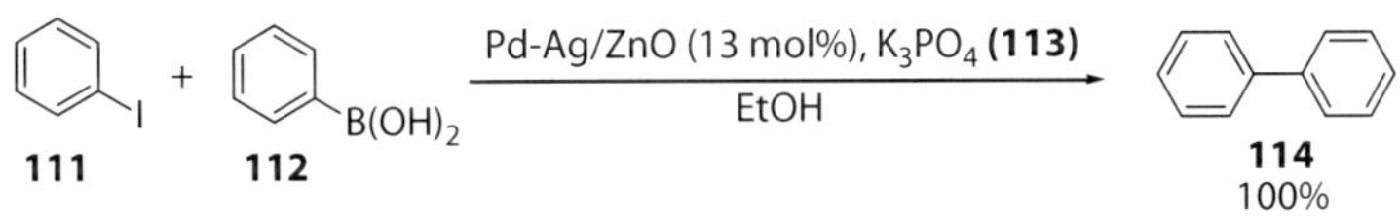

SCHEME 9.34

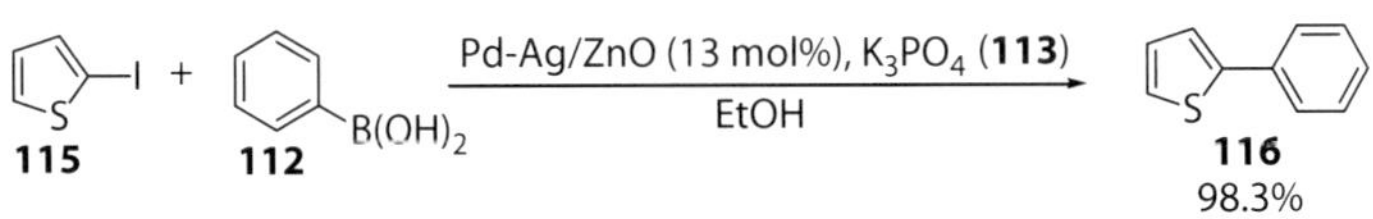

SCHEME 9.35

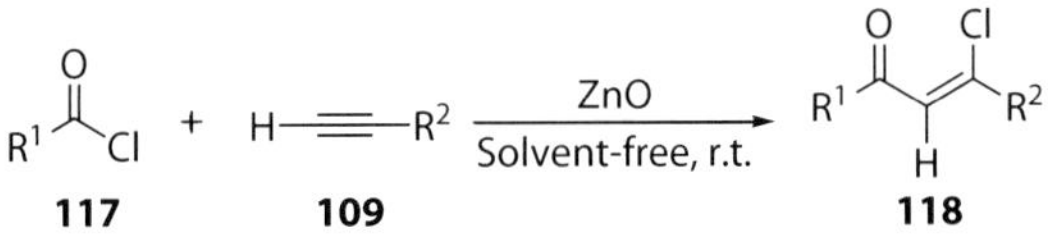

SCHEME 9.36

that the reaction occurs without decarbonylation in short reaction time and high yield. Chemoselectivity and solvent-free conditions make this process an economic and widely accepted method for the synthesis of β-Cl-α,β-unsaturated ketones (**118**).

Knoevenagel condensation is the addition of a nucleophile from active methylene compound to a carbonyl group followed by dehydration to form a β-conjugated enone. The Knoevenagel condensation between different aldehydes (**15**), including various aliphatic, aromatic, and heterocyclic aldehydes with active methylene compounds (**119**) in the presence of nano-ZnO under solvent-free conditions (Scheme 9.37) has been reported (Hosseini-Sarvari et al. 2008). Most of the aldehydes investigated reacted smoothly to afford the corresponding products in excellent yields (90%–98%) in a reaction time of 5 min to 3 h.

Chalcones (**123**) are commonly prepared by Claisen–Schmidt or aldol condensation. An important way to synthesize chalcones (**123**) is the Friedel–Crafts acylation involving treatment of acid chlorides (**121**) with arenes (**122**). Such a protocol for the synthesis of chalcones (**123**) was developed by More et al. (2012) using nano-ZnO heterogeneous catalyst under solvent-free conditions at room temperature (Scheme 9.38). Arenes (**122**) of all sorts, activated as well as unactivated, reacted smoothly to afford the chalcones (**123**) in excellent yield. High regioselectivity was observed during the course of reaction, which occurred selectively at the para-position of OMe, Br, Me, and Cl.

Along with the attempts to reduce environmental pollution and economic problems, solvent-free reactions, as a characteristic of green chemistry, are preferred by the scientific community. The solvent-free condensation of an equimolar mixture of aldehyde (**1**) with 2,4-thiazolidinedione/rhodanine (**124**) in the presence of nano-ZnO involving the Knoevenagel condensation reaction has been documented (Scheme 9.39) (Suresh and Sandhu 2012). After completion of the reaction, the reaction mass was cooled and stirred with ethanol. This was followed by centrifugation

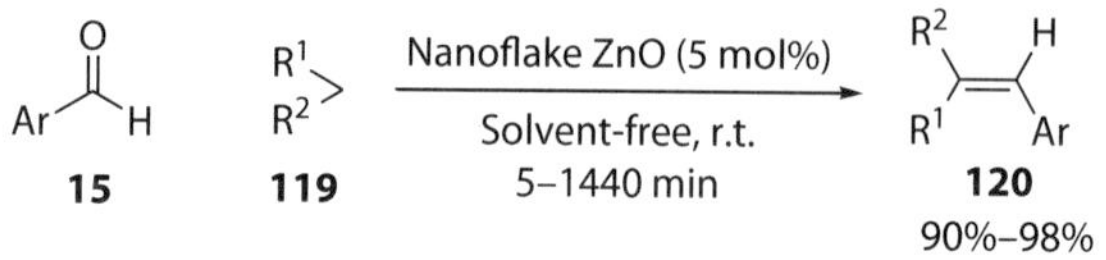

Ar = Ph, 4-MeC_6H_4, 4-OMeC_6H_4, 2-OMeC_6H_4, 4-ClC_6H_4, 4-OHC_6H_4, 4-$NO_2C_6H_4$, 3-ClC_6H_4, 2-ClC_6H_4, 2-thienyl, 4-pyridyl, 2-furyl

R^1 = CN, COOEt; R^2 = CN, COOEt, OMe, Cl

SCHEME 9.37

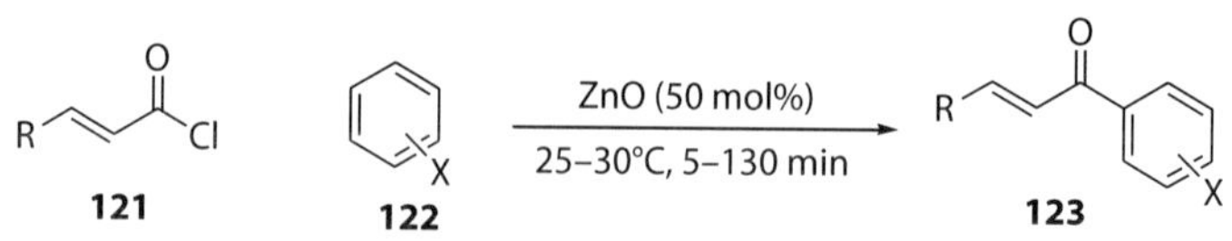

R = C_6H_5, 4-MeOC_6H_5, 4-OHC_6H_5
X = OMe, Me, Cl, Br

SCHEME 9.38

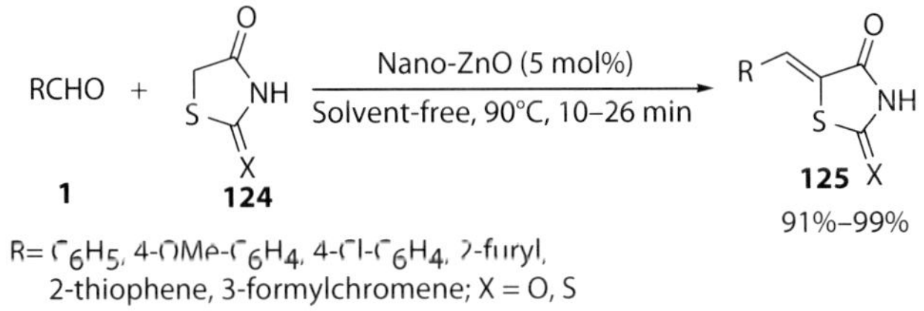

SCHEME 9.39

for a few minutes, and finally by filtration to separate the catalyst. The filtrate was concentrated and the products were purified using 3:2 EtOH:DMF. After drying at 100°C, the catalyst could be reused several times. The nature of the substituents did not show any remarkable effect on the yield of products during the reaction. The synthesis of 4-arylmethylidene-2-phenyl-5(4*H*)-oxazolones (**127**) from the condensation of aldehydes (**76**) and hippuric acid (**126**) using a catalytic amount of zinc oxide was developed at room temperature in ethyl alcohol as the solvent in a short reaction time and excellent yield of the products (Scheme 9.40) (Pasha et al. 2007). The synthesized compounds were evaluated for antibacterial activity against various bacterial species such as *Bacillus subtilis* and *Escherichia coli*. Some of these exhibited remarkable antimicrobial activity against these bacteria as compared to standard drugs like streptomycin and ampicillin.

9.1.4.2 Alkylation/Arylation

Many reactions, including alkylation/arylation of C/O/N of arenes, alcohols, phenols, and amines catalyzed by nano-ZnO, are discussed in this part of the chapter.

The montmorillonite clay-encapsulated ZnO nanoparticle–catalyzed N-benzylation of amines (**47**) (Scheme 9.41) has been studied, as documented in the literature (Dhakshinamoorthy et al. 2011). Although conversions up to 96% were observed, good selectivity was difficult to achieve in many cases. In many methodologies for the benzylation of amines (**47**), hazardous wastes are formed, but with the

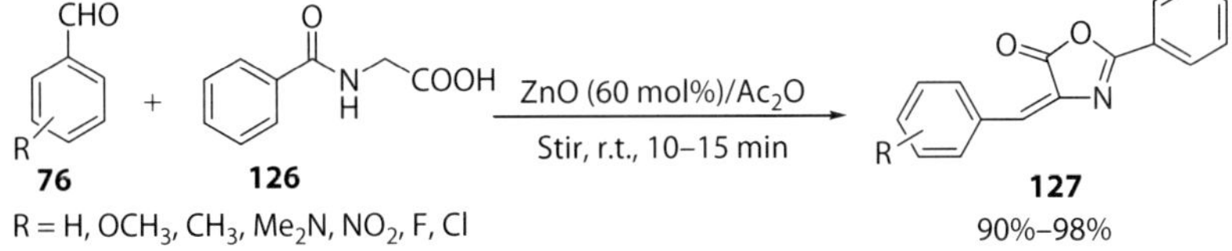

SCHEME 9.40

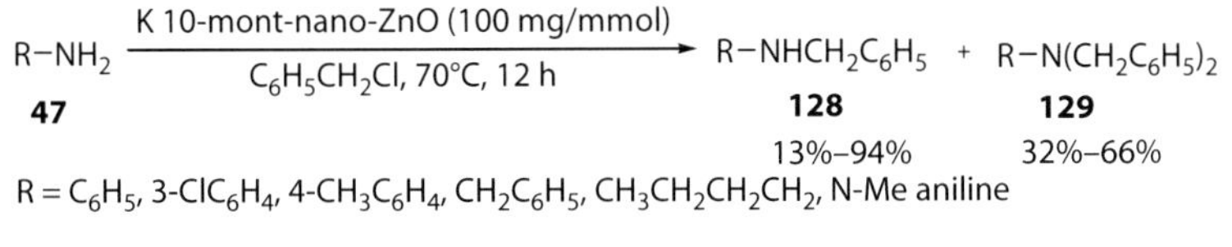

SCHEME 9.41

use of ZnO, this problem is minimized. Due to mild basic oxyanions (O^{2-}) and the polar surface of the Zn^{2+} cation of ZnO, this combination resulted in an acid–base bifunctional catalytic system in which the surface of the montmorillonite allowed close proximity of the reactants. The resulting catalyst showed good reusability up to five successive runs without much loss of catalytic activity. ZnO was used as an efficient catalyst for *N*-alkyl derivatives (**132**, **134**) of phthalimide (**130**) and saccharin (**133**). The Michael addition of phthalimide (**130**) and saccharin (**133**) to acrylic acid esters (**131**) was studied using a catalytic amount of zinc oxide and tetrabutylammonium bromide (TBAB) (Zare et al. 2007) under microwave and solvent-free thermal conditions in a short reaction time (Schemes 9.42 and 9.43).

The Friedel–Crafts reaction involves an electrophilic aromatic substitution that facilitates the alkylation or acylation of arenes (**135**) and heterocyclic compounds catalyzed by acidic catalysts. Zinc oxide has been found to be an effective catalyst for the Friedel–Crafts acylation of activated and nonactivated aromatic compounds (**135**) (Hosseini-Sarvari and Sharghi 2004) under solvent-free and room temperature conditions (Scheme 9.44). The catalyst provides a large surface area for the reaction. This Friedel–Crafts reaction is a safe and environmentally benign method which requires simple workup, mild reaction conditions and a short reaction time.

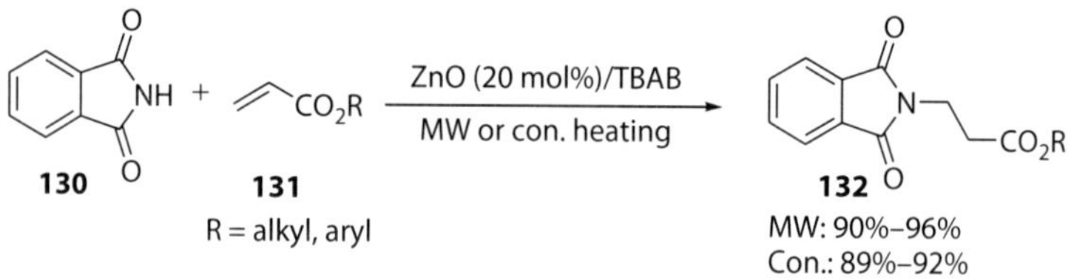

SCHEME 9.42

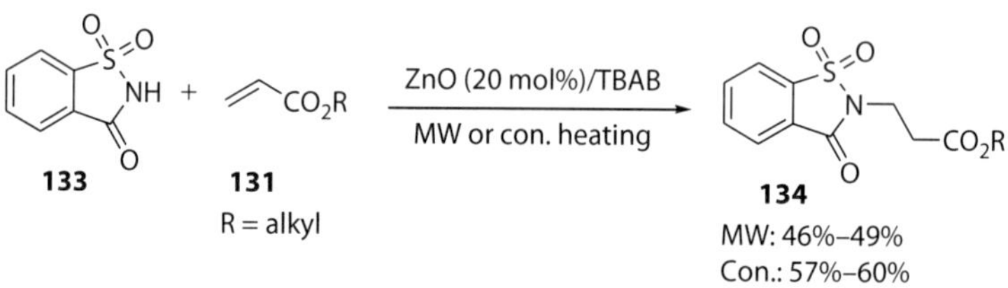

SCHEME 9.43

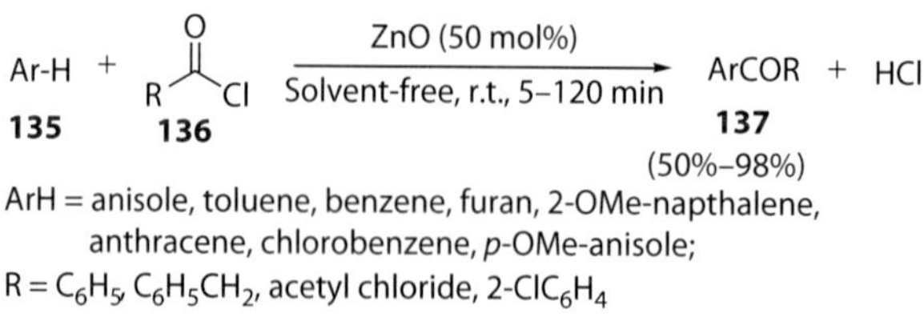

SCHEME 9.44

Protection and deprotection strategies are important elements in multistep organic synthesis. Among the different methods of protecting functional groups, acylation is the most simple and common method for protecting alcohols, phenols, and amines. Thus, the development of simpler and novel acetylation methods is desirable. Hosseini-Sarvari and Sharghi (2005) documented ZnO as a catalyst for the acylation of various compounds such as alcohols, phenols, and amines (**138**) under solvent-free conditions (Scheme 9.45). This method is applicable to a wide variety of organic compounds such as primary, secondary, tertiary, allylic and benzylic alcohols, diols, and phenols (**138**) possessing electron-donating as well as electron-withdrawing substituents. Due to the application of solvent-free conditions used during the reaction, the method is highly viable yet with a green approach.

9.1.4.3 Other Reactions Catalyzed by Nano-ZnO

Finally, several important reactions catalyzed by nano-ZnO are discussed in this last portion of the chapter.

The one-pot three-component condensation reaction of aldehydes (**48**), amines (**140**), and terminal alkynes (**141**) to form propargylamines (**142**) in the presence of ZnO nanoparticles as a reusable heterogeneous catalyst has been reported (Scheme 9.46) (Satyanarayana et al. 2012). Aldehydes (**48**) having substituents at the para-position reacted to afford the corresponding products in a short reaction time and excellent yields, the order of reactivity for various substituents on aldehydes being Br > Cl > OMe > NO_2. Among amines (**140**), piperidine gave better results as compared to morpholine. Higher yields were obtained under solvent-free conditions without the use of any activator or cocatalyst and the recycling ability of the catalyst was found to be good up to 10 recycles. An important C–C and C–P bond–forming reaction involving the three-component condensation of isatins (**143**), dialkyl phosphites (**144**) and malononitriles (**16**) catalyzed by ZnO nanorods under solvent-free and room temperature conditions has been reported (Scheme 9.47) (Hosseini-Sarvari and Tavakolian 2012a).

$$\underset{\mathbf{138}}{\text{RXH}} \xrightarrow[\text{ZnO (50 mol\%), solvent-free, r.t., 8–300 min}]{\text{RCOCl or (R'CO)}_2\text{O}} \underset{\substack{\mathbf{139}\\ 53\%\text{–}96\%}}{\text{RXCOR'}}$$

R = alkyl and aryl; R' = Ph, Me; X = O, NH

SCHEME 9.45

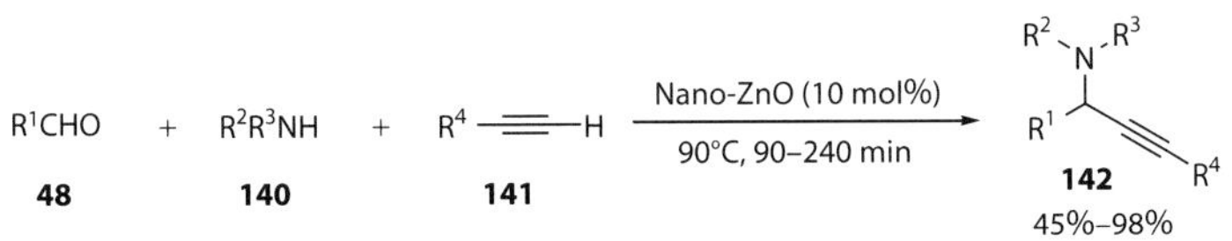

$R^1 = C_6H_5$, 2-ClC_6H_4, 4-ClC_6H_4, 4-$NO_2C_6H_4$, 4-$OCH_3C_6H_4$, 4-$CH_3C_6H_4$, 4-BrC_6H_4
R^2, R^3 = piperidine, morpholine

SCHEME 9.46

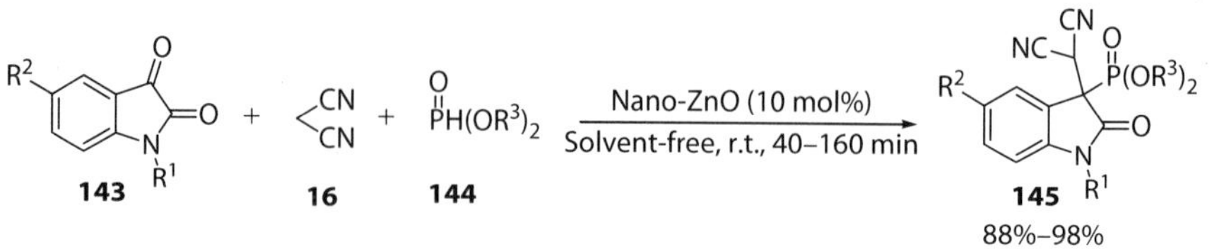

SCHEME 9.47

The Mannich reaction involves the condensation of ammonia or primary or secondary amines with formaldehyde. MaGee et al. (2011) reported the one-pot three-component synthesis of β-amino carbonyl compounds using amines (**146**), aromatic aldehydes (**13**, **152**), and cyclic ketones (**148**) via the Mannich reaction in the presence of ZnO nanoparticles to afford the corresponding products (**149**, **150**) in good yields (74%–93%) with moderate diastereoselectivity (Scheme 9.48). The reactions were carried out in an aqueous medium, which is an important aspect of green chemistry. Aldehydes (**147**) and amines (**146**) with both electron-withdrawing and electron-donating groups reacted smoothly to afford the corresponding products with antiselectivity (**150**). Mirjafary et al. (2008) documented a one-pot multicomponent reaction of aromatic aldehydes (**144**), enolizable ketones (**151**) or β-keto esters, acetonitrile, and acetyl chloride using ZnO nanoparticles for the synthesis of β-acetamido ketones/esters (**153**, **155**, **156**) at room temperature in a short reaction time (1–1.5 h) (Schemes 9.49, 9.50). Electron-donating substituents promoted the reaction to accomplish the desired products in high yields, whereas electron-withdrawing substituents decreased the reaction rate and afforded poor yields.

Organometallic compounds possess a prominent position in synthetic organic chemistry. The ferrocene compound was first prepared in 1951 by Pauson and Kealy, and ever since then the chemistry of these compounds has been developed. Recently,

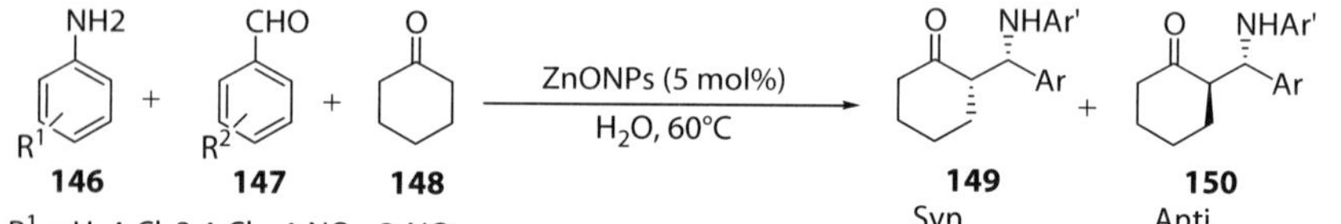

R^1 = H, 4-Cl, 3,4-Cl_2, 4-NO_2, 3-NO_2
R^2 = H, 4-NO_2, 4-Cl, 3-NO_2, 4-Me, 3-F, 2-Cl, 4-OMe, 2,4-Cl_2

SCHEME 9.48

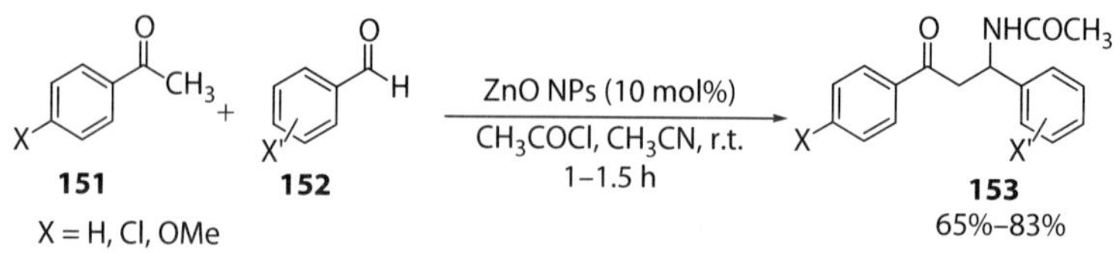

X = H, Cl, OMe
X' = H, 4-Cl, 2-Cl, 4-OCH_3, 4-NO_2, 3-NO_2

SCHEME 9.49

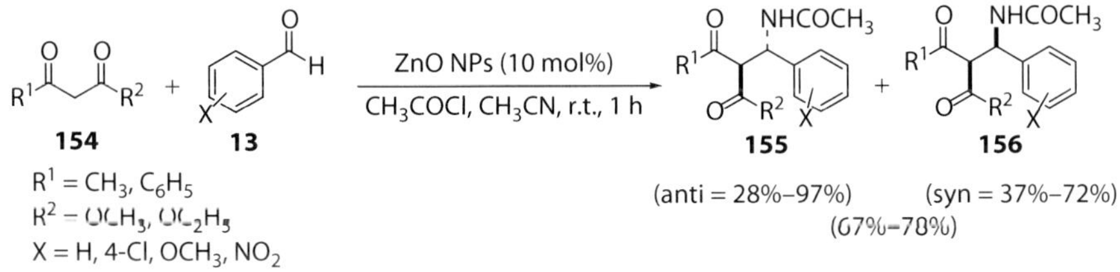

SCHEME 9.50

Hosseini-Sarvari (2011a) reported solvent-free selective synthesis of ferrocenyl aminophosphonic esters (**159**) by treating ferrocene carbaldehyde (**157**) with different aromatic and aliphatic amines (**47**) and dialkyl phosphate (**158**) in the presence of nano-ZnO (Scheme 9.51). For these reactions, several other catalysts such as CuO, MgO, basic Al_2O_3, nano-TiO_2, $Mg(ClO_4)_2$, and so on were screened, but the results showed excellent activity for the nano-ZnO catalyst. No formation of imines or α-OH phosphonates was observed. After completion of the reaction, the reaction mass was diluted with ethyl acetate and centrifuged to separate the catalyst. The separated organic layer was concentrated and the products were purified by chromatography on cellulose.

A green synthetic protocol under solvent-free conditions for the synthesis of phosphono malonates (**162**) from alkenes (**160**) and phosphorous nucleophiles (**161**) using nanoflake ZnO at 50°C has been documented (Scheme 9.52) (Hosseini-Sarvari and Etmad 2008). The reaction is an example of one of the most powerful and important C–P bond–forming reactions, which is the phospha-Michael addition. The advantages of this method include mild reaction conditions, a simple set up, high yields, and so on. The formylation of amines (**163**) with formic acid (**70**) under solvent-free

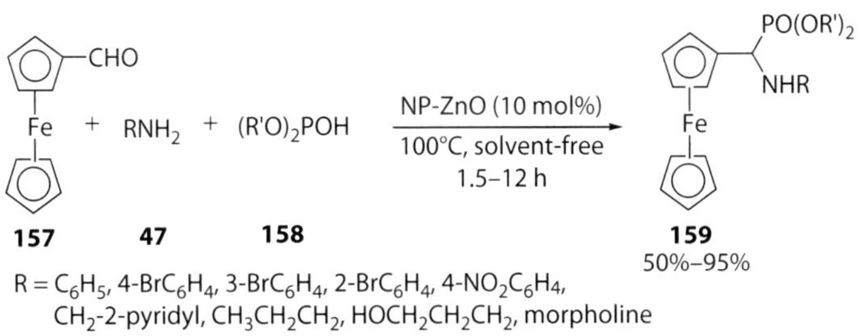

SCHEME 9.51

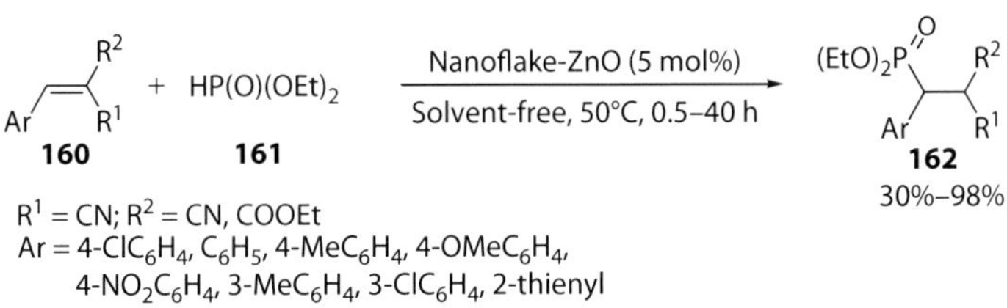

SCHEME 9.52

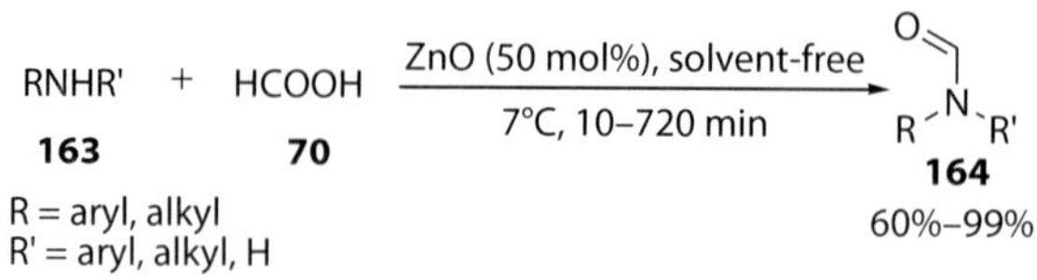

SCHEME 9.53

conditions was successfully accomplished using nano-ZnO (Scheme 9.53) (Hosseini-Sarvari and Sharghi 2006). Under these conditions, a wide variety of amines (**163**), including aromatic, aliphatic, and heterocyclic—both primary and secondary—were successfully formylated. This method produces higher yields and has a short reaction time, which makes it a more acceptable protocol than other methods of formylation. The method is chemoselective and does not require any specialized equipment.

The N-benzylation of imidazole-4-carboxaldehyde (**165**) and 4-cyanoimidazole (**168**) was studied using nano-ZnO and the triethylamine catalytic system under solvent-free conditions to afford the N-benzylated derivatives (**166**, **167**, **169**, **170**) (Schemes 9.54, 9.55) (Oresmaa et al. 2007).

Cyano functionality is an important functional group in multistep organic synthesis. It facilitates the interconversion of many functional groups. The successful conversion of the carbonyl group of aldehydes (**171**) into oximes (**173**) followed by dehydration to nitriles (**67**) was documented by Pasha (Scheme 9.56) (Reddy and Pasha 2010).

A report on the one-pot synthesis of α-aminophosphonates (**176**) by the reaction of aldehydes (**15**), amines (**175**), and dialkyl phosphites (**174**) at room temperature

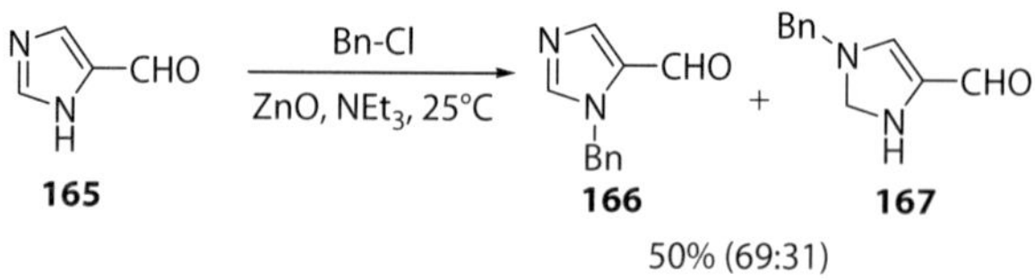

SCHEME 9.54

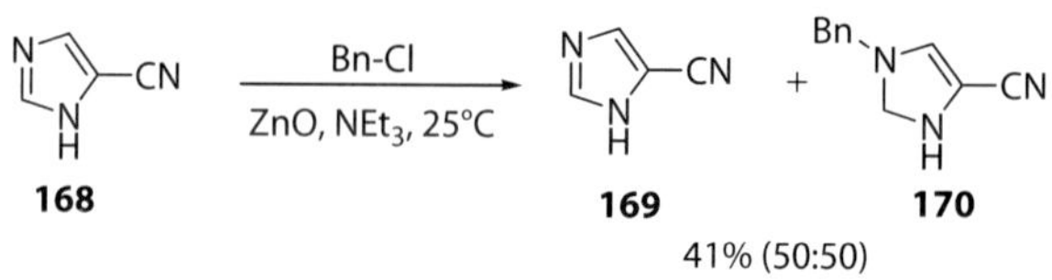

SCHEME 9.55

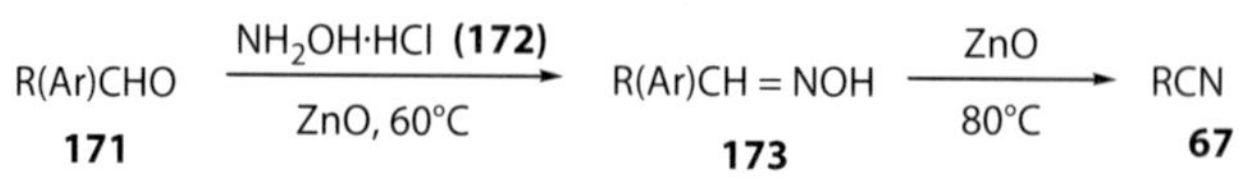

SCHEME 9.56

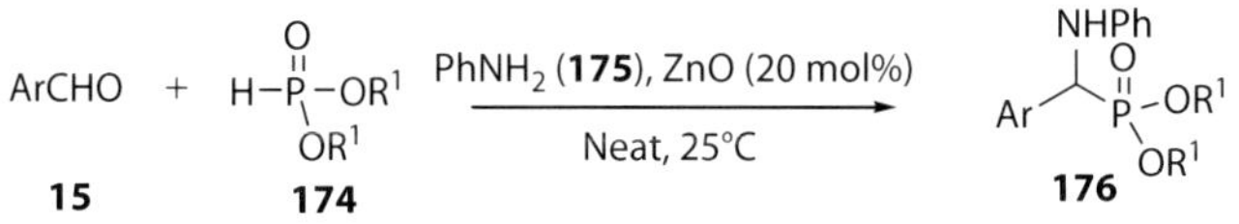

SCHEME 9.57

under solvent-free conditions was made by Kassaee et al. (2009) (Scheme 9.57). The reaction requires mild reaction conditions, short reaction times, and an easy workup procedure. ZnO nanofluids were used as a pseudohomogeneous catalyst for the synthesis of amides (**179**) from aliphatic carboxylic acids (**177**) with primary aliphatic, as well as aromatic, amines (**178**) under solvent-free conditions (Scheme 9.58) (Tamaddon et al. 2011). The acid–base properties of the catalyst activated both COOH and NH_2 groups for the synthesis of amides (**179**). This method has several advantages, such as high chemoselectivity, low effect on the environment, simplicity, and so on.

The use of conventional strong Lewis/Brønsted acids is not suitable from the environmental, corrosion, and operational points of view. Furthermore, such Lewis acids may decompose the products formed during the reaction if this is continued for a prolonged period. The Beckmann rearrangement is one of the fundamental reactions for the synthesis of amides (**179**) from oximes in the presence of acidic catalysts. This reaction, which normally requires strong Lewis/Brønsted acids, can be carried out with nano-ZnO under solvent-free conditions. Sharghi et al. studied the rearrangement for various aldehydes (**1**) and ketones (**180**), which reacted smoothly to afford the (*Z*)-isomer of the oximes (**181**) in excellent yields (60%–95%) and short times (5–15 min) with lower formation of the (*E*)-isomer (**182**) (10%–20%) (Schemes 9.59, 9.60) (Sharghi and Hosseini 2002).

N-Sulfonylimines are the electron-deficient imines and stable synthetically important intermediates in organic synthesis. A simple method for the synthesis of *N*-sulfonylaldimines (**184**) using aldehydes (**48**) and sulfonamides (**183**) under solvent-free conditions in the presence of ZnO has been reported in the literature (Scheme 9.61) (Hosseini-Sarvari and Sharghi 2007). This protocol has several distinct

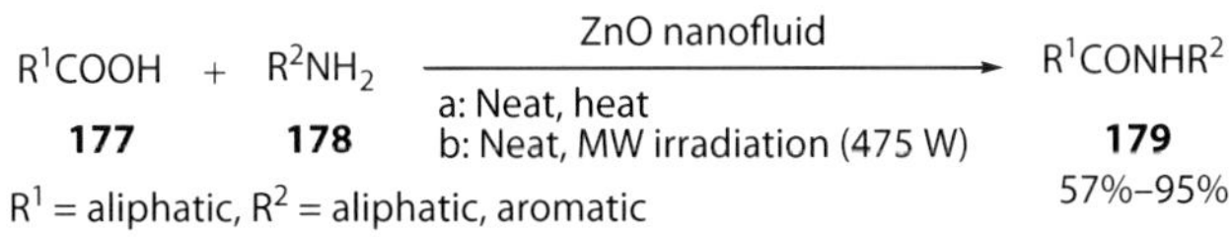

SCHEME 9.58

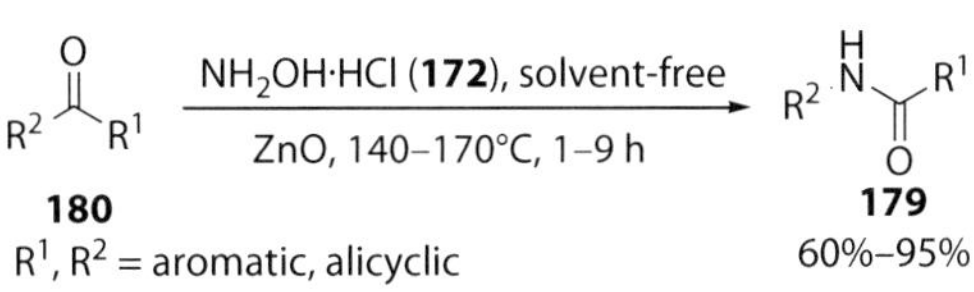

SCHEME 9.59

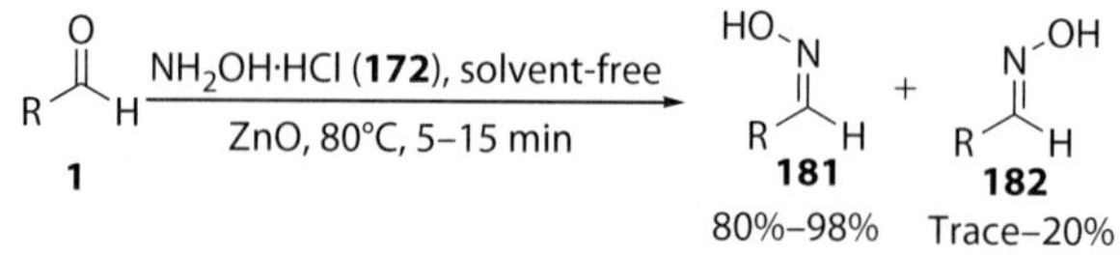

(R = *p*-MeC_6H_4, *p*-OHC_6H_4, *p*-ClC_6H_4, *p*-$OMeC_6H_4$, *m*-MeC_6H_4, *m*-OHC_6H_4, *m*-ClC_6H_4, *m*-$OMeC_6H_4$, *o*-OHC_6H_4, *o*-ClC_6H_4, Ph)

SCHEME 9.60

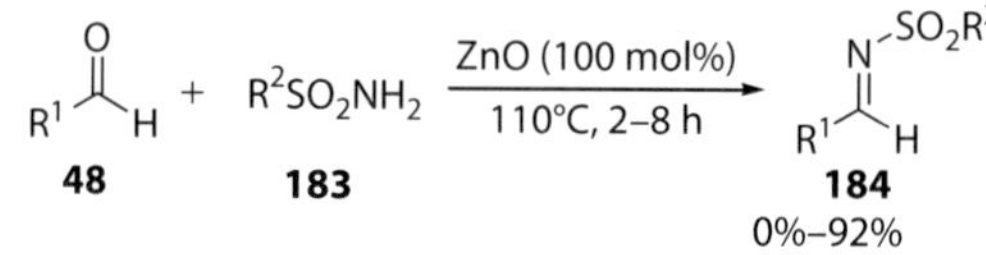

R^1 = C_6H_5, 4-MeC_6H_4, 3-MeC_6H_4, 4-$OMeC_6H_4$, 3-$OMeC_6H_4$, 4-ClC_6H_4, 2-ClC_6H_4, 3-ClC_6H_4, 2-$NO_2C_6H_4$, 4-pyridyl, 2-thienyl, 3-pyridyl, acetophenone, diphenylketone

R^2 = C_6H_5, 4-MeC_6H_4

SCHEME 9.61

salient features such as solvent-free conditions, lack of toxic waste formation during the reaction, simple workup, and so on. The catalyst showed good reusability up to three recycles. The substituents on the aldehydes (**48**) did not produce any significant effect, but in the case of sulfonamides (**183**), *p*-toluenesulfonamide reacted faster than benzensulfonamide. When ketones were used as the substrates, lower yields were obtained, probably due to steric hindrance. The synthesis of β-amino alcohols (**186**) via the opening of epoxide rings (**185**) by the nucleophilic attack of amines (**47**) under solvent-free conditions in the presence of nano-ZnO was investigated by Hosseini-Sarvari (2008a, 2008b) (Scheme 9.62). Various amines (**47**)—aromatic, aliphatic, and cyclic—reacted smoothly to afford the corresponding products in good to excellent yields. Both primary and secondary amines reacted rapidly. The reaction proceeds in *trans*-stereoselectivity for the cyclic epoxides (**185**). It was observed that anilines (**47**) possessing electron-withdrawing substituents and sterically hindered anilines reacted slowly and the reactions required prolonged reaction times.

Oxidation of thio compounds (**187**) is an important and challenging task in synthetic organic chemistry. Suryanarayana et al. reported the application of ZnO-dispersed polyaniline solid-phase (ZnO/PANI) composites for the oxidation of sulfides (**187**) to sulfoxides (**189**) using H_2O_2 (**188**) under solvent-free conditions

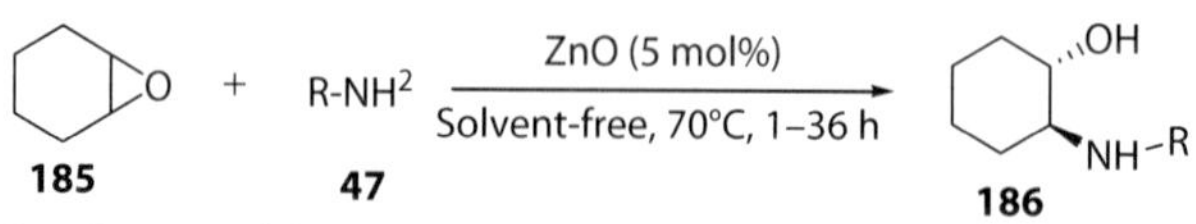

R = CH_2-2-pyridyl, C_6H_5, 4-Me C_6H_4, 3-MeC_6H_4, etc. (50%–98%)

SCHEME 9.62

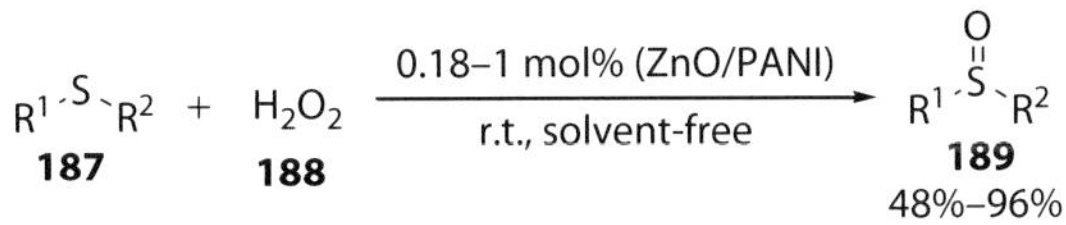

$R^1 = CH_3CH_2CH_2, C_6H_5$, 4-Me$C_6H_4$, 4-Cl$C_6H_4$, $CH_2C_6H_5$, 4-OMeC_6H_4, 4-$NO_2C_6H_4$, $CH_2{-}CHCH_2$, $C_{11}H_{23}$
$R^2 = CH_3CH_2CH_2, CH_2CH_2Cl$, Me, C_6H_5, $CH_2C_6H_5$, $CH_2CH{=}CH_2$, $C_{11}H_{29}$, $C_{11}H_{23}$

SCHEME 9.63

(Scheme 9.63) (Shiv et al. 2009). The reaction was also studied with raw ZnO, which revealed more selectivity and high activity for the nanocomposite. In case of aryl methyl sulfides, the chlorine substituent at the para-position lowered the yield of corresponding sulfoxides (**189**) due to its electron-withdrawing nature, whereas no substituent effect was observed for the methyl group. The reactions were carried out under mild conditions and the isolation of the catalyst was easy due to its heterogeneous nature. Bandgar et al. (2009) reported a new method for the synthesis of thioesters (**191**) from acid chlorides (**117**) and catalytic amounts of ZnO under solvent-free conditions and at room temperature. The reactions required short reaction times and mild conditions to afford the corresponding thioester products (**191**) in excellent yields (Scheme 9.64).

A green and efficient enamination of 1,3-dicarbonyl compounds (**192**) with various amines (**146**) can be successfully accomplished using nano-ZnO (Scheme 9.65) (Indulkar et al. 2012). Aryl amines (**138**) possessing electron-donating substituents reacted to give good yield of the products. Aliphatic and cyclic amines reacted smoothly to afford the desired products in short reaction times (1–2 h). *tert*-Butyl

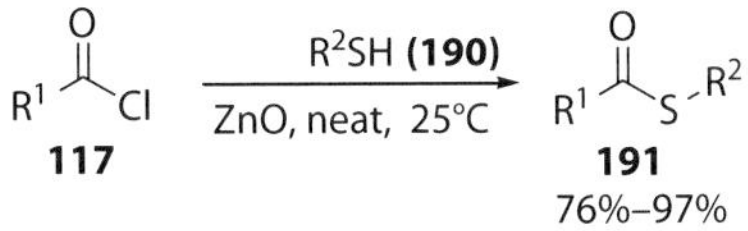

R^1 = Ph, *t*-Bu, Me, *p*-ClC_6H_4, *p*-OMe-C_6H_4, CH_2Ph, *p*-$NO_2C_6H_4$; R^2 = Ph, Et

SCHEME 9.64

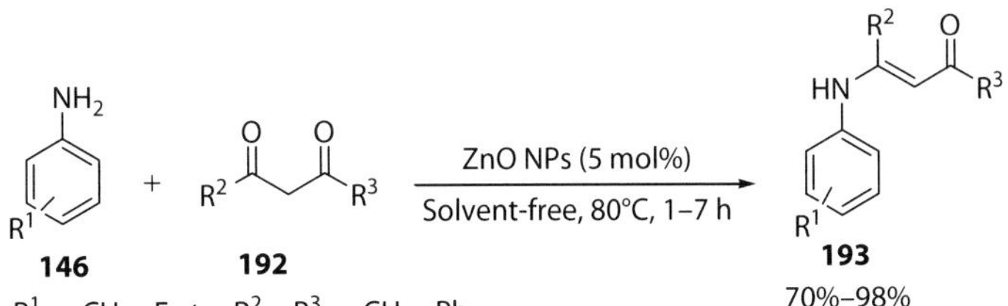

R^1 = -CH_3, -F etc.; $R^2 = R^3$ = -CH_3, -Ph

SCHEME 9.65

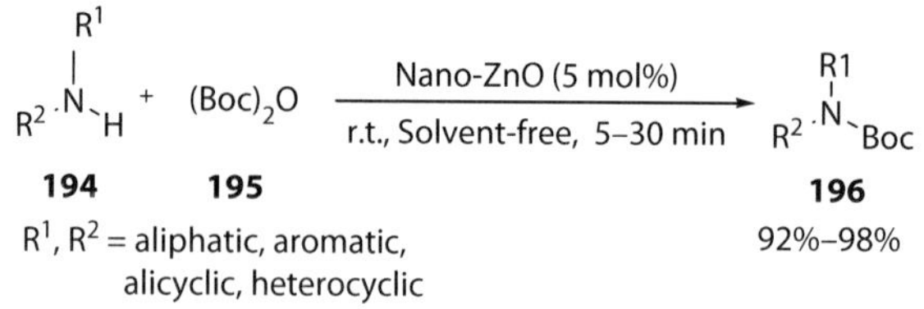

SCHEME 9.66

dicarbonate (**195**) is the common protecting group for amines (**194**) in organic synthesis due to its ease of protection and deprotection. The use of strong Lewis or Brønsted acids for the *N*-Boc protection is not favored because the product formed may decompose when the reaction is continued for a prolonged period. A rapid and chemoselective *N*-Boc protection of various aliphatic, aromatic, alicyclic, and heterocyclic amines (**194**) was carried out using ZnO nanorods as a highly active catalyst (Scheme 9.66) (Nouria et al. 2011). The catalyst showed good recycle results up to five times without much loss of catalytic activity. The reaction was also attempted in various solvents such as MeOH, EtOH, CH_3CN, DCM, DMF, and so on, but solvent-free conditions afforded good results.

9.2 CONCLUSION

In summary, ZnO is a green and efficient reusable heterogeneous catalyst for several environmentally friendly and atom-economic organic transformations in modern organic synthesis. Recently it has emerged as a promising heterogeneous catalyst and viable alternative to conventional materials for the construction of many heterocyclic compounds. It has several advantages such as its heterogeneous nature, reusability, and sustainability. It exhibits better activity in several organic reactions due to its smaller particle size; this helps to contribute a great deal to synthetic organic chemistry. This chapter will be helpful and provide future guidelines to researchers in synthetic organic chemistry and nanocatalysis to explore the utility of nano-ZnO for synthesis of novel biologically active heterocyclic compounds.

REFERENCES

Agawane, S.M. and Nagarkar, J.M. 2012. Synthesis of 5-substituted 1H-tetrazoles using a nano ZnO/Co_3O_4 catalyst. *Catal. Sci. Tech.* 2: 1324–1327.

Alinezhad, H., Salehian, F. and Biparva, P. 2012. Synthesis of benzimidazole derivatives using heterogeneous ZnO nanoparticles. *Synth. Commun.* 42: 102–108.

Ameta, K.L., Kumar, B. and Rathore, N.S. 2012. ZnO catalyzed efficient synthesis of some new 2-substituted-4,6-diarylpyrimidines. *Org. Chem.* 2012, ID 242569.

Astruc, D. 2008. *Nanoparticles and Catalysis*. Wiley-VCH Verlag GmbH & Co. KGaA: Weinheim.

Bahrami, K., Khodaei, M.M. and Farrokhi, A. 2009. Highly efficient solvent-free synthesis of dihydropyrimidinones catalyzed by zinc oxide. *Synth. Commun.* 39: 1801–1808.

Bandgar, B.P., More, P.E., Kamble, V.T. and Sawant, S.S. 2009. Convenient and efficient synthesis of thiol esters using zinc oxide as a heterogeneous and eco-friendly catalyst. *Austr. J. Chem.* 61: 1006–1010.

Dhakshinamoorthy, A., Visuvamithiran, P., Tharmaraj, V. and Pitchumani, K. 2011. Clay encapsulated ZnO nanoparticles as efficient catalysts for N-benzylation of amines. *Catal. Commun.* 16: 15–19.

Giri, A.K., Sinhamahapatra, A., Prakash, S., Chaudhari, J., Shahi, V.K. and Panda, A.B. 2013. Porous ZnO microtubes with excellent cholesterol sensing and catalytic properties. *J. Mat. Chem. A* 1: 814–822.

Gonzalez-Bejar, M., Peters, K., Hallett-Tapley, G.L., Grenier, M. and Scaiano, J.C. 2013. Rapid one-pot propargylamine synthesis by plasmon mediated catalysis with gold nanoparticles on ZnO under ambient conditions. *Chem. Commun.* 49: 1732–1734.

Hekmatshoar, R., Kenary, G.N., Sadjadi, S. and Beheshtiha, Y.S. 2010. ZnO nanoparticles: A mild and efficient reusable catalyst for the one-pot synthesis of 4-amino-5-pyrimidine-carbonitriles under aqueous conditions. *Synth. Commun.* 1: 2007–2013.

Hosseini-Sarvari, M. 2008a. Synthesis of bis(indolyl)methanes using a catalytic amount of ZnO under solvent-free conditions. *Synth. Commun.* 38: 832–840.

Hosseini-Sarvari, M. 2008b. Synthesis of β-aminoalcohols catalyzed by ZnO. *Acta Chim. Slov.* 55: 440–447.

Hosseini-Sarvari, M. 2011a. An efficient and eco-friendly nanocrystalline zinc oxide catalyst for one-pot, three component synthesis of new ferrocenyl aminophosphonic esters under solvent-free. *Catal. Lett.* 141: 347–355.

Hosseini-Sarvari, M. 2011b. Synthesis of quinolines using nano-flake ZnO as a new catalyst under solvent-free conditions. *J. Iran. Chem. Soc.* 8: S119–S128.

Hosseini-Sarvari, M. and Etmad, S. 2008. Nanosized zinc oxide as a catalyst for the rapid and green synthesis of β-phosphono malonates. *Tetrahedron* 64: 5519–5523.

Hosseini-Sarvari, M. and Mardaneh, Z. 2011a. Selective and CO-retentive addition reactions of acid chlorides to terminal alkynes in synthesis of β-chloro-α, β-unsaturated ketones using ZnO. *Bull. Chem. Soc. Japan* 84: 778–782.

Hosseini-Sarvari, M. and Mardaneh, Z. 2011b. Solvent-free synthesis of propargylic alcohols using ZnO as a new and reusable catalyst by direct addition of alkynes to aldehydes. *Bull. Korean Chem. Soc.* 32: 4297–4303.

Hosseini-Sarvari, M. and Shafiee-Haghighi, S. 2012. Nano-ZnO as heterogeneous catalyst for three-component one-pot synthesis of tetrahydrobenzo[b]pyrans in water. *Chem. Hetero. Comp.* 48: 1307–1313.

Hosseini-Sarvari, M. and Sharghi, H. 2004. A simple, economical and efficient Friedel–Crafts acylation reaction over zinc oxide (ZnO) as a new catalyst. *J. Org. Chem.* 69: 6953–6956.

Hosseini-Sarvari, M. and Sharghi, H. 2005. Zinc oxide (ZnO) as a new, highly efficient, and reusable catalyst for acylation of alcohols, phenols and amines under solvent free conditions. *Tetrahedron* 61: 10903–10907.

Hosseini-Sarvari, M. and Sharghi, H. 2006. ZnO as a new catalyst for N-formylation of amines under solvent-free conditions. *J. Org. Chem.* 71: 6652–6654.

Hosseini-Sarvari, M. and Sharghi, H. 2007. A novel method for the synthesis of *N*-sulfonylaldimines by ZnO as a recyclable neutral catalyst under solvent-free conditions. *Phosphorus Sulfur Silicon Relat. Elem.* 182: 2125–2130.

Hosseini-Sarvari, M., Sharghi, H. and Etemad, S. 2008. Nanocrystalline ZnO for Knoevenagel condensation and reduction of the carbon, carbon double bond in conjugated alkenes. *Helv. Chim. Acta* 91: 715–724.

Hosseini-Sarvari, M. and Tavakolian, M. 2012a. P–C bond formation via direct and three-component conjugate addition catalyzed by ZnO nano-rods for the synthesis of 2-oxindolin-3-yl-phosphonates under solvent-free conditions. *New J. Chem.* 36: 1014–1021.

Hosseini-Sarvari, M. and Tavakolian, M. 2012b. Preparation, characterization, and catalysis application of nano-rods zinc oxide in the synthesis of 3-indolyl-3-hydroxy oxindoles in water. *Appl. Catal. A Gen.* 441–442: 65–71.

Indulkar, U.U., Kale, S.R., Gawande, M.B. and Jayaram, R.V. 2012. Ecofriendly and facile nano ZnO catalyzed solvent-free enamination of 1, 3-dicarbonyls. *Tetrahedron Lett.* 53: 3857–3860.

Kassaee, M.Z., Masrouri, H. and Movahedi, F. 2010. ZnO-nanoparticle-promoted synthesis of polyhydroquinoline derivatives via multicomponent Hantzsch reaction. *Chem. Month.* 141: 317–322.

Kassaee, M.Z., Movahedi, F. and Masrouri, H. 2009. ZnO nanoparticles as an efficient catalyst for the one-pot synthesis of α-amino phosphonates. *Synlett* 8: 1326–1330.

Khazaei, M., Anary-Abbasinejad, M., Hassanabadi, A. and Sadeghi, B. 2012. ZnO nanoparticles: An efficient reagent, simple and one-pot procedure for synthesis of highly functionalized dihydropyridine derivatives. *E-J. Chem.* 9: 615–620.

Kim, M-R. and Choi, S-H. 2009. One-step synthesis of Pd-M/ZnO (M = Ag, Cu, and Ni) catalysts by γ-irradiation and their use in hydrogenation and Suzuki reaction. *J. Nanomat.* 2009, ID 302919.

Krishnakumar, V., Kumar, K.M., Mandal, B.K. and Nawaz Khan, F-R. 2012. Zinc oxide nanoparticles catalyzed condensation reaction of isocoumarins and 1,7-heptadiamine in the formation of bis-isoquinolinones. *Sci. World J.* 2012, ID 619080.

Kulkarni, A. and Torok, B. 2012. Environmentally benign synthesis of heterocyclic compounds by combined microwave-assisted heterogeneous catalytic approaches. *Green Chem.* 14: 17–37.

Kumar, B.V., Naik, H.S.B., Girija, D. and Kumar, B.V. 2011. ZnO nanoparticle as catalyst for efficient green one-pot synthesis of coumarins through Knoevenagel condensation. *J. Chem. Sci.* 123: 615–621.

Magar, R.L., Thorat, P.B., Jadhav, V.B., Tekale, S.U., Dake, S.A., Patil, B.R. and Pawar, R.P. 2013. Silica gel supported polyamine: A versatile catalyst for one pot synthesis of 2-amino-4*H*-chromene derivatives. *J. Mol. Catal. A Chem.* 374–375: 118–124.

MaGee, D.I., Dabiri, M., Salehi, P. and Torkian, L. 2011. Highly efficient one-pot three-component Mannich reaction catalyzed by ZnO-nanoparticles in water. *ARKIVOC* 2011(11): 156–164.

Mirjafary, Z., Saeidian, H., Sadeghi, A. and Moghaddam, F.M. 2008. ZnO nanoparticles: An efficient nanocatalyst for the synthesis of β-acetamido ketones/esters via a multi-component reaction. *Catal. Commun.* 9: 299–306.

More, P.E., Bandgar, B.P. and Kamble, V.T. 2012. Zinc oxide as a regioselective and heterogeneous catalyst for the synthesis of chalcones at room temperature. *Catal. Commun.* 27: 30–32.

Nouria, A., Akbari, J., Heydaric, A. and Nouri, A. 2011. ZnO nanorods as an efficient and heterogeneous catalyst for *N*-Boc protection of amines and amine derivatives. *Lett. Org. Chem.* 8: 38–42.

Oresmaa, L., Taberman, H., Haukka, M., Vainiotalo, P. and Aulaskari, P. 2007. Regiochemistry of N-substitution of some 4(5)-substituted imidazoles under solvent-free conditions. *J. Heter. Chem.* 44: 1445–1451.

Park, J.C., Kim, A.Y., Kim, J.Y., Park, S., Park, K.H. and Song, H. 2012. ZnO-CuO core-branch nanocatalysts for ultrasound-assisted azide–alkyne cycloaddition reactions. *Chem. Commun.* 48: 8484–8486.

Pasha, M.A., Jayashankara, V.P., Venugopala, K.N. and Rao, G.K. 2007. Zinc oxide (ZnO): An efficient catalyst for the synthesis of 4-arylmethylidene-2-phenyl 5(4*H*)-oxazolones having antimicrobial activity. *J. Pharmaco. Toxico.* 2: 264–270.

Paul, S., Bhattacharyya, P. and Das, A.R. 2011. One-pot synthesis of dihydropyrano[2,3-*c*] chromenes via a three component coupling of aromatic aldehydes, malononitrile, and 3-hydroxycoumarin. *Tetrahedron Lett.* 52: 4636–4641.

Polshettiwar, V. and Asefa, T, (eds). 2013. *Nanocatalysis: Synthesis and Applications*, p. 1–10. Wiley.

Rao, G.B.D., Kaushik, M.P. and Halve, A.K. 2012. Zinc oxide nanoparticles: An environmentally benign and reusable catalyst for the synthesis of 1,8-dioxooctahydroxanthene derivatives under solvent-free conditions. *Heter. Lett.* 2: 411–418.

Reddy, M.B.M. and Pasha, M.A. 2010. Environment friendly protocol for the synthesis of nitriles from aldehydes. *Chin. Chem. Lett.* 21: 1025–1028.

Reza, M., Shafiee, M., Cheraghipoor, M. and Ghashang, M. 2011. ZnO nanopowder: An efficient catalyst for the preparation of 2,4,5-triaryl imidazoles under solvent-free condition. *Int. Conf. Nanotech. Biosens. IPCBEE.* 25: 90–93.

Ross, J.R.H. 2011. *Heterogeneous Catalysis: Fundamentals and Applications*, pp. 1–232. Elsevier, Technology & Engineering: Oxford.

Sadjadi, S. and Eskandari, M. 2012. ZnO nanorods as an efficient catalyst for the synthesis of imidazo[1,2-*a*]azines and diazines. *Chem. Month.* 143: 653–656.

Sadjadi, S. and Eskandari, M. 2013. Ultrasonic assisted synthesis of imidazo[1,2-*a*]azine catalyzed by ZnO nanorods. *Ultrason. Sonochem.* 20: 640–643.

Safaei-Ghomi, J. and Ghasemzadeh, M.A. 2012. ZnO nano particles as new and efficient catalyst for the one-pot synthesis of polyfunctionalized pyridines. *Acta Chim. Slov.* 59: 697–702.

Saravanamurugan, S., Palanichamy, M., Arabindoo, B. and Murugesan, V. 2005. Solvent free synthesis of chalcone and flavanone over zinc oxide supported metal oxide catalysts. *Catal. Commun.* 6: 399–403.

Satyanarayana, K.V.V., Ramaiah, P.A., Murty, Y.L.N., Chandra, M.R. and Pammi, S.V.N. 2012. Recyclable ZnO nano particles: Economical and green catalyst for the synthesis of A3 coupling of propargylamines under solvent free conditions. *Catal. Commun.* 25: 50–53.

Sharghi, H. and Hosseini, M. 2002. Solvent-free and one-step Beckmann rearrangement of ketones and aldehydes by zinc oxide. *Synthesis* 8: 1057–1060.

Shiv, P., Sharma, M.V.S., Suryanarayana, A.K., Nigam, A.S., Chauhan, A.S. and Tomar, L.N.S. 2009. PANI/ZnO composite: Catalyst for solvent-free selective oxidation of sulfides. *Catal. Commun.* 10: 905–912.

Suresh and Sandhu, J.S. 2012. ZnO nanobelts: An efficient catalyst for synthesis of 5-arylidine-2,4-thiazolidinediones and 5-arylidine-rhodanine. *Int. J. Org. Chem.* 2: 305–310.

Tamaddon, F., Aboee, F. and Nasiri, A. 2011. ZnO nanofluid as a structure base catalyst for chemoselective amidation of aliphatic carboxylic acids. *Catal. Commun.* 16: 194–197.

Tamaddon, F. and Moradi, S. 2012. Nano ZnO as an efficient and reusable catalyst for the preparation of 1,4-DHPs via Hantzsch reaction. *Iran. J. Catal.* 2: 101–106.

Tamaddon, F. and Moradi, S. 2013. Controllable selectivity in Biginelli and Hantzsch reactions using nano-ZnO as a structure base catalyst. *J. Mol. Catal. A Chem.* 370: 117–122.

Tayebee, R., Javadi, F. and Argi, G. 2013. Easy single-step preparation of ZnO nano-particles by sedimentation method and studying their catalytic performance in the synthesis of 2-aminothiophenes via Gewald reaction. *J. Mol. Catal. A Chem.* 368–369: 16–23.

Tekale, S.U., Kauthale, S.S., Pagore, V.P., Jadhav, V.B. and Pawar, R.P. 2013. ZnO nanoparticle catalyzed efficient one pot three component synthesis of 3,4,5-trisubstituted furan-2(5*H*)-ones. *J. Iran. Chem. Soc.*

Valizadeha, H. and Azimib, A.A. 2011. ZnO/MgO containing ZnO nanoparticles as a highly effective heterogeneous base catalyst for the synthesis of 4*H*-pyrans and coumarins in [bmim]BF_4. *J. Iran. Chem. Soc.* 8: 123–130.

Yavari, I. and Beheshti, S. 2011. ZnO nanoparticles catalyzed efficient one-pot three-component synthesis of 2,3-disubstituted quinalolin-4(1*H*)-ones under solvent-free conditions. *J. Iran. Chem. Soc.* 8: 1030–1035.

Zare, A., Hasaninejad, A., Khalafi-Nezhad, A., Zare, A.R.M., Parhami, A. and Nejabat, G.R. 2007. A green solventless protocol for Michael addition of phthalimide and saccharin to acrylic acid esters in the presence of zinc oxide as a heterogeneous and reusable catalyst. *Arkivoc* 2007(i): 58–69.

Ziarati, A., Safaei-Ghomi, J. and Rohani, S.A. 2013. One-pot multi-component synthesis of *N*-cyclohexyl-3-arylquinoxaline-2-amines using ZnO nanoparticles as a heterogeneous reusable catalyst. *Lett. Org. Chem.* 10: 47–52.

10 Application of Heterogeneous Catalysts for the Synthesis of Bioactive Coumarins

Lokesh A. Shastri and Manohar V. Kulkarni

CONTENTS

10.1 INTRODUCTION

Chromene[2*H*]-2-one (**1**), commonly referred to as coumarin, represents the parent compound of a group of naturally occurring lactones exhibiting a wide range of biological activities (Kulkarni et al. 2006). In view of their structural diversity and extensive plant origin, coumarins have been found to be useful as anti-inflammatory (Fylaktakidou et al. 2004; Hadjipavlou-Litina et al. 2007) and anticancer agents (Irena 2005; Lacy and O'Kennedy 2004). They have also been employed as fluorescent probes in the study of biochemical mechanisms (Katerinopoulos 2004; Khalilah et al. 2012).

Retrosynthetic analysis of the target coumarin skeleton provides two major synthetic routes (Figure 10.1). Route A requires an aromatic *o*-hydroxy carbonyl compound and a two-carbon fragment observed in Knoevenagel and Perkin reactions reflecting the [4+2] approach for the construction of six-membered heterocycles. Route B represents the reaction between phenols and three carbon fragments associated with the Pechmann cyclization. In recent years, newer synthetic methodologies have also been applied to the synthesis of a variety of coumarins that have avoided the use of concentrated sulfuric acid. The present chapter reviews various

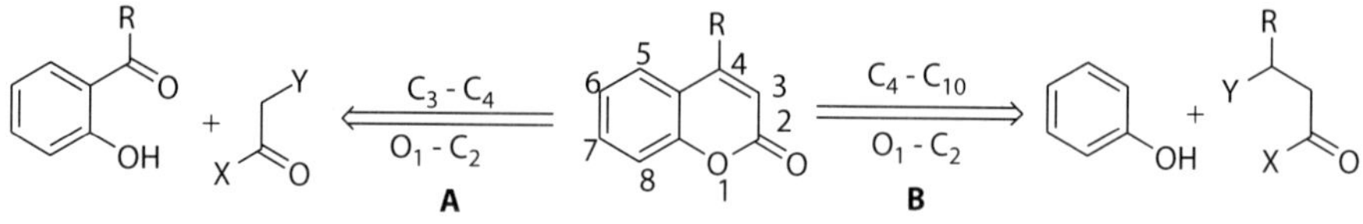

FIGURE 10.1 Numbering and retrosynthetic analysis of chromene[2*H*]-2-ones.

heterogeneous catalysts employed in the synthesis of structurally interesting and biologically active coumarins.

10.2 LEWIS ACIDS AS HETEROGENEOUS CATALYSTS IN PECHMANN CYCLIZATION

One of the earliest attempts in this area was to employ Lewis acid-like anhydrous aluminum chloride in the reaction of phenols and β-keto esters (Sethna and Shah 1945). This idea served as the basis for the application of other metal salts and complexes for this purpose.

The use of anhydrous aluminum chloride and dry HCl gas in the synthesis of 3,4-dihydrocoumarins possessing a hydroxy group at the C-5, C-6, and C-7 positions was reported (Green et al. 1969). This strategy was applied to the synthesis of xanthyletin (Das-Gupta and Das 1969). The application of dipyridine Cu(II) chloride (Rajitha et al. 2006) as an efficient catalyst for the synthesis of 4-alkylcoumarins was demonstrated. The rate of the solvent-free reaction was enhanced under microwave (MW) irradiation, leading to higher yield and less reaction time. The reaction time for the unsubstituted phenols was generally as high as 70–140 min, whereas for meta-substituted phenols it was 30–50 min. Under the influence of MW irradiation the reaction time was reduced to 10–15 min and a yield enhancement of 10%–30% was observed. Other Lewis acids and metal salts employed in the synthesis of coumarins are mentioned in Table 10.1. The general scheme for the Lewis acid–catalyzed synthesis of coumarins is given in Scheme 10.1.

10.3 SOLID ACID CATALYSTS IN SYNTHESIS OF COUMARINS

Application of zeolites, clays, and resins as surrogates for mineral acids and metal chlorides in the field of aromatic electrophilic substitution (Lionel et al. 1993) provided an impetus for their application in the synthesis of coumarins.

Silica gel–supported sulfuric acid has been employed as a catalyst in the reaction of phenols with β-keto esters. Pechmann cyclization under these solvent-free conditions was found to occur in the temperature range of 60°C–120°C with yields in the range of 60%–85%. Wide applicability of this catalyst has been demonstrated with cyclic β-keto esters and even trifluoromethyl ethyl acetoacetate (Benyaram et al. 2009). Silica gel–supported perchloric acid ($HClO_4·SiO_2$) was also employed in the synthesis of coumarins under solvent-free conditions (Muchchintala et al. 2006). Extending the application of metal hydrogen sulfates in organic synthesis (Salehi et al. 2003),

TABLE 10.1
Coumarins Obtained under Lewis Acid–Catalyzed Pechmann Cyclization[a]

Entry	Lewis Acid	Phenols	β-Keto Esters	Structure of the Coumarin	Yield (%)	Author
1	$CuPyCl_2$	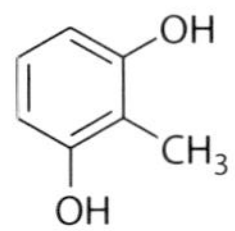	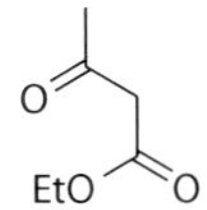	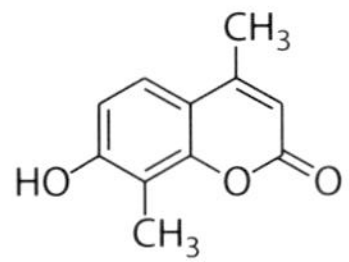	85	Rajitha et al. (2006)
2	$TiCl_4$	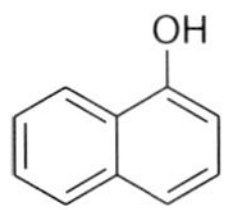	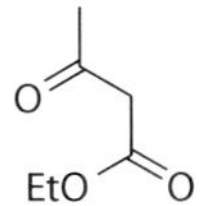	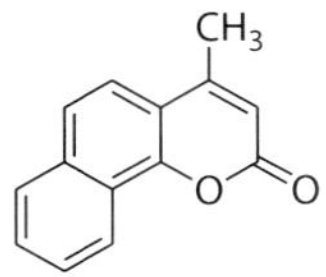	90	Hassan and Abbas (2005)
3	$InCl_3$	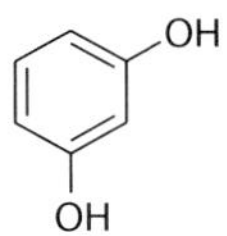	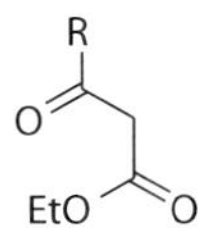	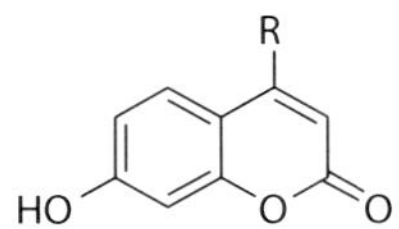 $R = CH_3$ $R = CF_3$ $R = CH_2Cl$	98 96 93	Subhas et al. (2002)
4	$Yb(OTf)_3$	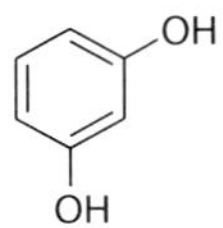	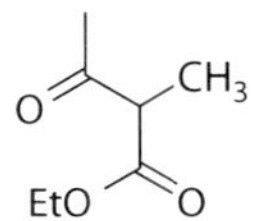	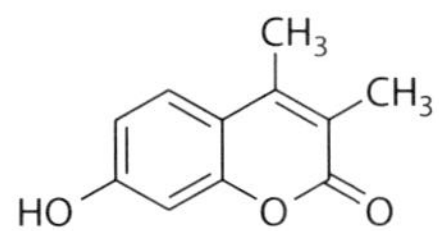	85	Limin et al. (2003)

(continued)

TABLE 10.1 (Continued)
Coumarins Obtained under Lewis Acid–Catalyzed Pechmann Cyclization[a]

Entry	Lewis Acid	Phenols	β-Keto Esters	Structure of the Coumarin	Yield (%)	Author
5	$Sm(NO_3)_3{\cdot}6H_2O$	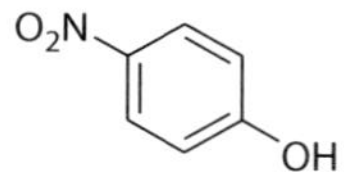	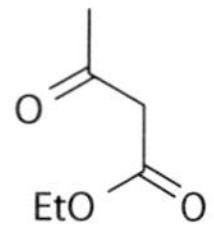	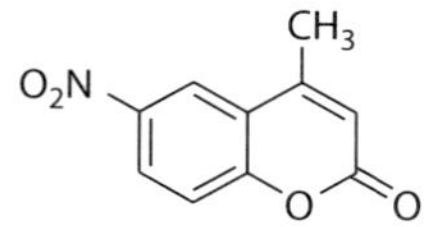	75	Sushilkumar and Devanand (2004)
6	$Bi(NO_3)_3{\cdot}5H_2O$	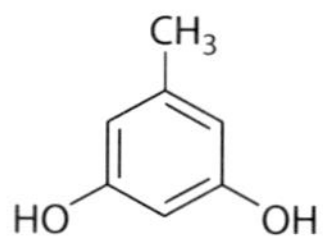	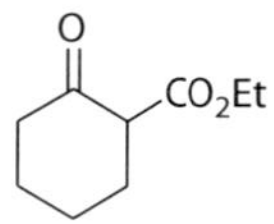	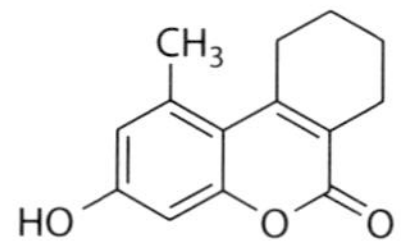	77	Varughese et al. (2005)
7	$ZrCl_4$	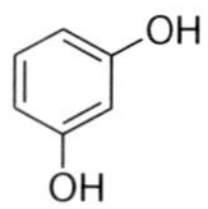	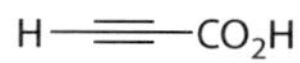	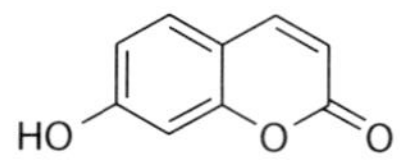	87	Gangadasu et al. (2004)
8	$ZnCl_2$	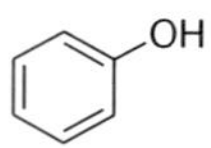	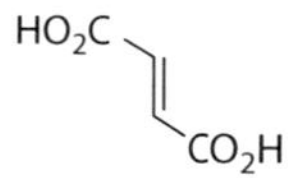	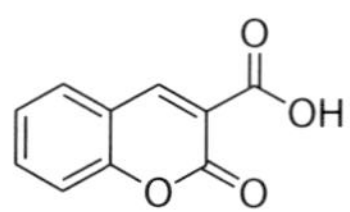	89	Helavi et al. (2003)
9	$Ce(NH_4)_2(NO_3)_6$	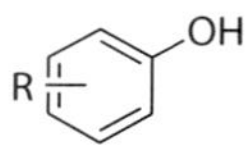	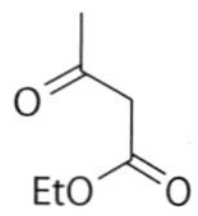	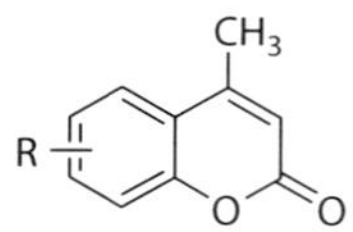	>92	Thirupathi Reddy et al. (2008)

10	BF_3	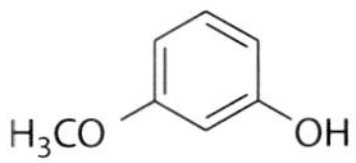	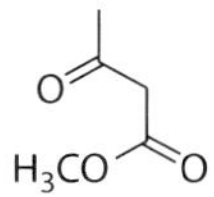	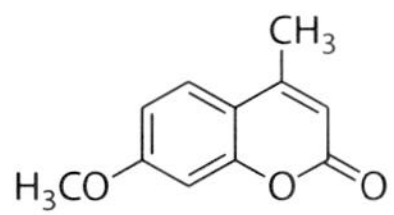	98	Stoyanov and Mezger (2005)
11	NH_4VO_3	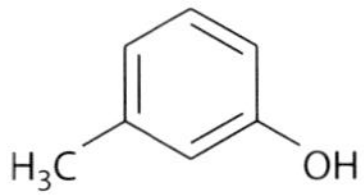	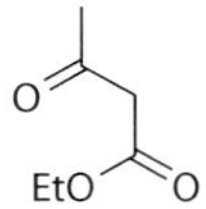	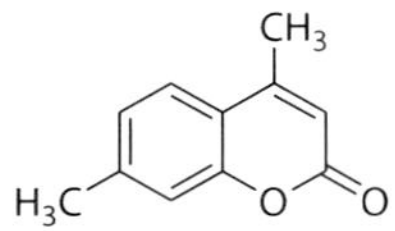	85	Priyanka et al. (2009)
12	$CoPyCl_2$	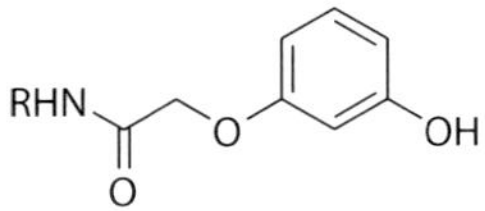	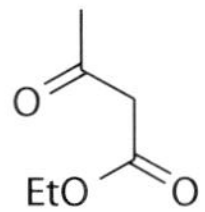	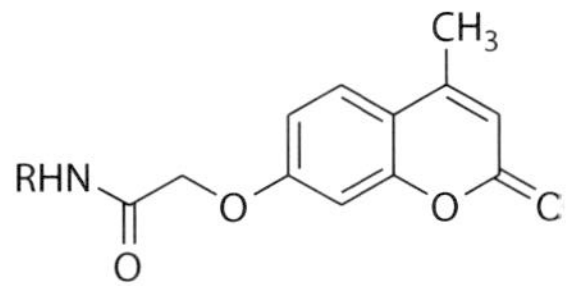	80	Rajanarendra et al. (2010)
12	LiBr	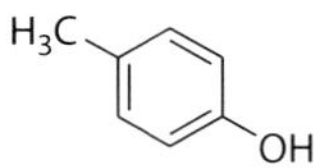	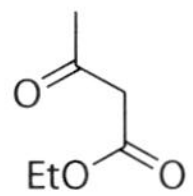	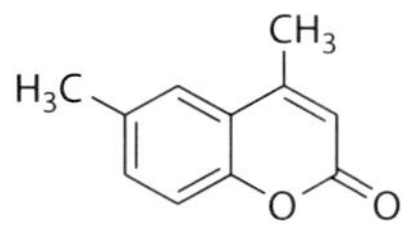	85	Sanjay et al. (2007)

[a] Each paper has reported a number of coumarins. In the table only one derivative has been chosen, randomly.

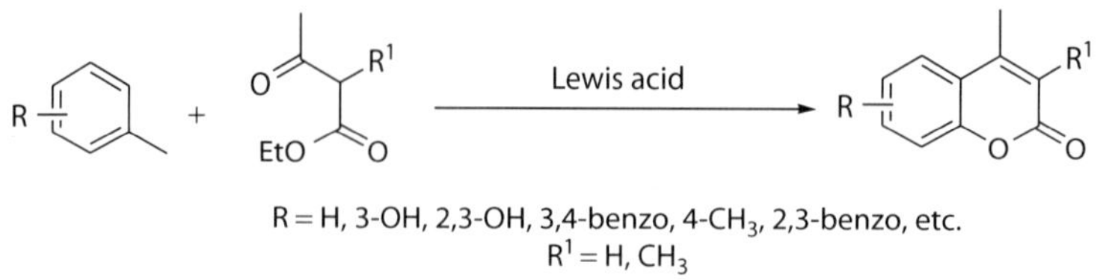

SCHEME 10.1

Shinde et al. (2008) have prepared silica gel–supported sodium hydrogen sulfate and demonstrated its application in the synthesis of coumarins. The reactions were conducted in acetonitrile and the reaction time in most of the cases was 1 h, which included the weakly reactive naphthols. Application of nanocrystalline sulfated zirconia in the reaction of meta-substituted phenols has also been investigated (Tyagi et al. 2007) under different conditions. In view of the slow reaction rates observed in nitrobenzene and toluene, the solvent-free conditions were adopted under MW irradiation to obtain high yields (95%–99%), of 7-amino- and 7-hydroxy-4-methylcoumarins in the reaction of *m*-amino and *m*-hydroxy phenols with ethyl acetoacetate. A low percentage of catalyst is required and its reusability has been claimed as a novel feature of this synthesis.

Environmentally benign synthesis routes for umbelliferone (**I**), 7-hydroxy-4-methylcoumarin (**II**), 3,4-dihydrocoumarin (**III**), and tricyclic coumarin (**IV**) (Figure 10.2) were developed using zeolite-H-β or Amberlyst-15 as solid catalysts. This resulted in the minimization of harmful mineral acid wastes. The reactions of resorcinol with acrylic acid, propynoic acid, and ethyl acetoacetate were investigated in toluene and *p*-chlorotoluene under reflux conditions to obtain coumarins in 60%–80% yield (Hoefnagel et al. 1995; Gunnewegh et al. 1995).

A novel composite solid acid catalyst in the form of SO_4^{-2}/Ce_xZrO_2 was employed in the synthesis of hydroxycoumarins. Low loads (10%), moderate temperature (393 K), and high yields (77%–94%) are the main features of this method (Benyaram et al. 2006).

Commercially available heteropoly acids that are nontoxic have been employed as heterogeneous catalysts in organic synthesis (Izumi et al. 1992). Wells–Dawson heteropoly acid was employed as a catalyst for the reaction between phenols and β-ketoesters under solvent-free conditions at high temperatures (130°C). A special feature of this method lies in employing α-methyl ethyl acetoacetate, thus facilitating the introduction of methyl groups at the C-3 positions in the resulting 4-substituted coumarins (Eromanelli et al. 2004) (Scheme 10.2).

Similarly, application of phosphotungstic acid has also been employed in the reaction of di- and tri-hydroxy benzenes with ethyl acetoacetate. The reactions were

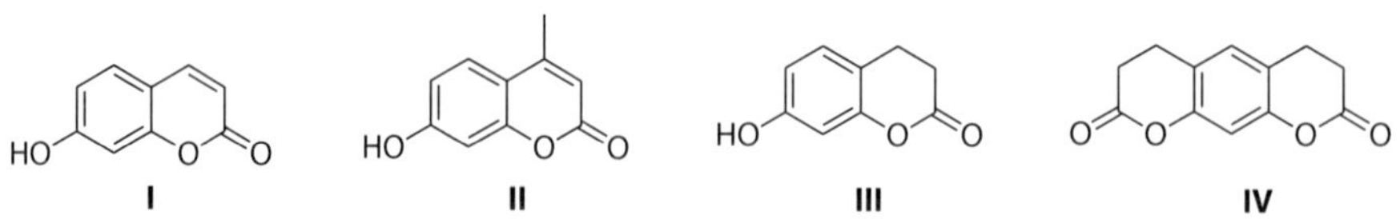

FIGURE 10.2 Umbelliferone (**I**), 7-hydroxy-4-methylcoumarin (**II**), 3,4-dihydrocoumarin (**III**), and tricyclic coumarin (**IV**).

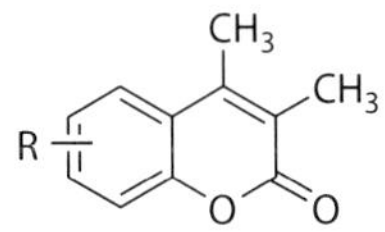

R = 7-OH; 7,8-OH; 7-CH_3; 5,7-OH; 7,8-benzo

SCHEME 10.2

conducted in toluene under reflux conditions to obtain a variety of 4-methylcoumarins (China Raju et al. 2009).

Application of montmorillonite clays for the synthesis of coumarins has been systematically investigated. It has been shown that phenols with electron-withdrawing groups such as NO_2 and CHO at the para-position do not lead to the expected Pechmann-cyclized product. This method offers easy separation of products, consistent yields and reusability of the clay catalyst (Tong-Shuang et al. 1998). Other solid acid catalysts employed are presented in Table 10.2.

Application of MW irradiation in the solid acid-catalyzed reaction of resorcinol and phloroglucinol with ethyl acetoacetate and propenoic and propynoic acid resulted in the formation of corresponding hydroxy coumarins in very high yields, avoiding the acidic waste associated with aqueous and Lewis acids (De la Hoz et al. 1999).

Inorganic ion exchangers such as phosphates and tungstates of metals in higher oxidation states have been found to possess surface acid sites and thermal stability. The M(IV) salts, where M = Zr, Ti, or Sn, have been employed in the reaction of methyl acetoacetate and dihydroxybenzene, leading to the formation of coumarins in moderate yields (Joshi and Chudasama 2008). Recently, alumina sulfuric acid (ASA) has been developed as a reusable catalyst for Pechmann reactions between phenols and β-keto esters (Ali et al. 2013).

Polymer-supported solid acid catalysts have also been found to be useful in a number of organic transformations and are eco-friendly in nature (Benaglia et al. 2003). Poly(4-vinylpyridine)-supported sulfuric acid has been shown to be an efficient catalyst in Pechmann cyclization reactions between phenols and acetoacetic esters (Kalyan and Ruli 2011). These reactions have been conducted under solvent-free conditions using MW irradiation leading to higher yields of 4-alkylcoumarins. Polyaniline salts have been employed as catalysts for the synthesis of 4-methylumbelliferone. It has been shown that the efficiency of the catalyst depends upon the acid strength and the total amount of acid present on the polyamine chain (Srinivasan and Rampally 2004).

TABLE 10.2
Solid Acid Catalysts in Synthesis of Coumarins

Sl. No.	Catalyst	Author
1	Nafion-H	Chaudhari (1983)
2	Zeolite H-BEA	Laufer et al. (2003)
3	Amberlyst-15	Rammohan et al. (2012)
4	W/ZrO_2	Benyaram et al. (2001)
5	Montmorillonite K-10	Stephane et al. (2001)

10.4 SOLID BASE CATALYSTS IN THE SYNTHESIS OF COUMARINS

Unlike solid acid catalysts, there are very few supports on the application of solid base catalysts in this area. Limited application of weakly basic alumina phosphate nitride catalysts in Knoevenagel condensations has been reported (Climent et al. 1996). The ability of mesoporous materials to acquire a basicity by exchange was applied to a few organic reactions (Kloetstra and Van Bekkum 1995). Coupling of triazabicyclodecene (TBD) with mesoporous silicious (MCM-41) support has resulted in a basic catalyst employed in the Michael reaction of cyclopentenone and ethyl cyanoacetate. This was successfully extended to the synthesis of 3-cyanocoumarin by the reaction of salal with ethyl cyanoacetate (Yarlagadda et al. 1997). Subsequently, Mg-Al hydrotalcites were employed in the reaction of *o*-hydroxy carbonyl compounds and active methylene compounds leading to a variety of 3- and 4-substituted coumarins. The reactions were conducted in an argon atmosphere at high temperature (110°C–120°C). High selectivity, low cost, and reusability of the catalyst are the salient features of this method (Ramani et al. 1999).

10.5 METAL-CATALYZED APPROACHES

Improved synthetic routes for functionalized coumarins related to aflatoxins have been reported based on the application of Pd(0) complexes (Trost and Toste 1996) in the reaction of phenols and alkynoates. The synthetic strategy was based on electrophilic palladation of phenols followed by carbometalation of the alkynoate followed by protonolysis, leading to the generation of cinnamate intermediates that would undergo an *in situ* cyclization to coumarin. Compounds reported by this strategy are shown in Figure 10.3.

Palladium-catalyzed cross-coupling of 2-iodophenols with carbon monoxide and alkynes has been applied to the synthesis of a number of 3,4-disubstituted coumarins (Larock and Kadnikov 2000) (Scheme 10.3).

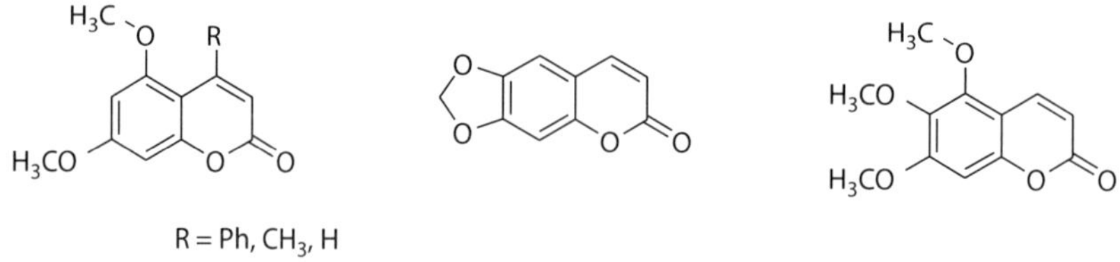

FIGURE 10.3 Alkoxy coumarins obtained by Pd(0) catalysis.

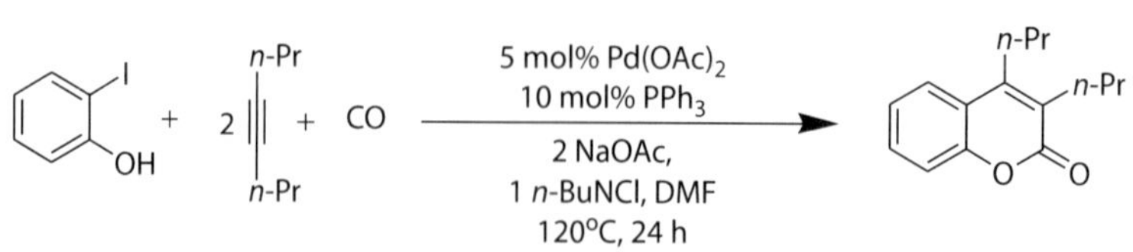

SCHEME 10.3

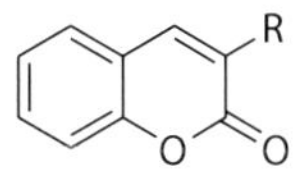

R = CH_3, *t*-butyl, C_6H_5, etc.

SCHEME 10.4

Rhodium-catalyzed-carbonylation of *o*-alkynyl phenols has resulted in the formation of 3-substituted coumarins and 3-substituted benzofuran-2-ones, under the water-gas shift reaction conditions (Takashi et al. 1998). $Rh_6(CO)_{16}$ was employed as the catalyst in the carbonylation. The reaction has been shown to involve oxidative addition and insertion of carbon monoxide followed by reductive elimination. Coumarins were obtained in low yield (20%–40%) (Scheme 10.4).

Zinc-mediated transesterification of β-keto esters has been successfully applied to the synthesis of coumarins by using phenols and iodine (Subhash et al. 2002). The reaction occurred in the presence of iodine in toluene under reflux conditions. An important feature of this technique is that the authors have observed the formation of coumarins in the case of nitrophenols, leading to 6-nitro- and 7-nitro-4-methylcoumarins.

The ring-closing olefin metathesis (RCM) strategy has been applied to the synthesis of benzofuran derivatives via the intermediacy of styrenyl allyl ether dienes (Chang and Grubbs 1998). Later, styrenyl acrylates were subjected to the RCM strategy leading to the formation of coumarins and many 3- and 4-substituted compounds (Chatterjee et al. 2003). Zinc oxide nanoparticles were found to be efficient catalysts in the reaction between various salicyldehydes and 1,3-dicarbonyl compounds leading to the formation of 3-substituted coumarins (Vinay Kumar et al. 2011).

10.6 APPLICATION OF ION-EXCHANGE RESINS TO SYNTHESIS OF COUMARINS

Use of cation-exchange resins as an alternative for concentrated sulfuric acid (Mastalgi and Andric 1958) was demonstrated by the addition of Amberlite IRC-120 as a reagent in the Pechmann reaction, which resulted in very low yield of coumarin. Application of Zeo-karb 225 and Amberlite IR-120 was found to be more useful in this reaction (John and Isra Elstam 1961). In these reactions, the quantity of cation resins was restricted to the range 20%–30%, in view of their ability to hydrolyze the β-keto esters. Due to the thermal stability of the resins, the reactions were conducted at higher temperatures (120°C–130°C). Recently, Amberlyst ion-exchange resins have been shown to be efficient catalysts for the synthesis of 7-hydroxy-4-methylcoumarins (Sabou et al. 2005).

Synthesis of iminocoumarins (Scheme 10.5) was accomplished by using Amberlite IRA 900 in a reaction of salicylaldehydes and nitriles (Mhiri et al. 1999). Subsequently bis-iminocoumarins were obtained by using Amberlyst in cyclohexane for the reaction between arylacetonitriles and salicylaldehydes (Houcine et al. 2003).

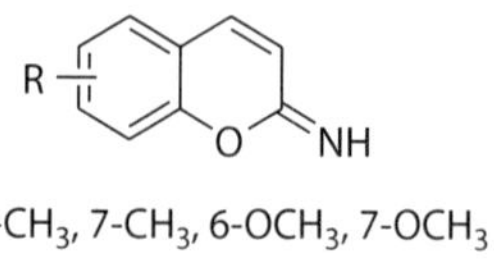

SCHEME 10.5

10.7 IONIC LIQUIDS IN THE SYNTHESIS OF COUMARINS

Ionic liquids have been recognized as designer solvents and are now accepted as environmentally benign media in a number of chemical transformations. Their property of acting as heterogeneous catalysts in aldol condensations (Abello et al. 2004) and transesterification (Liu et al. 2012) has emerged as an area of great interest.

1-Butyl-3-methylimidazolium chloroaluminate was employed in the Pechmann reaction of phenols and ethyl acetoacetate (Potdar et al. 2001). It was shown that the ionic liquid played the dual role of solvent and catalyst in the high-yielding synthesis of coumarins. Overcoming the problems of Lewis acidity and the resulting toxic wastes, 1-butyl-3-methylimidazolium hexafluorophosphate was employed as a neutral ionic liquid (Potdar et al. 2005) in Pechmann cyclization.

Application of MW irradiation using 1-butyl-3-methylimidazolium hydrogen sulfate in the reactions between phenols and β-keto esters resulted in high yields and reduced reaction time (Singh et al. 2005).

The catalytic ability of niobate ionic liquids generated in combination with 1-butyl-3-methylimidazolium chloride and niobium pentachloride was tested in Pechmann cyclizations. The reaction of phenols and ethyl acetoacetate, leading to the formation of coumarins, was conducted in an acidic mixture of 60 mol% (Soares et al. 2007).

Pechmann cyclizations have been reported to occur in water when catalyzed by ionic liquids. These acyclic ionic liquids, possessing a SO_3H group, were able to bring about this cyclization reaction. *N,N,N*-Trimethyl-*N*-propane sulfonic acid ammonium hydrogen sulfate was found to be the most efficient and reusable catalyst (Dong et al. 2008).

The moisture sensitivity of 1,3-dimethylimidazolium methyl sulfate has been applied to bring about Knoevenagel condensation between aromatic aldehydes and active methylene compounds. The introduction of L-proline as a promoter in this reaction conducted with *o*-hydroxybenzaldehydes resulted in high yields (Verdia et al. 2011). Coumarin-3-carboxylic acid was obtained during an ionic liquid–mediated reaction of Meldrum's acid with *o*-hydroxyaryl aldehydes (Darvatkar et al. 2008). Expanding the applicability of ionic liquids, 3-substituted coumarins were reported by a combination of Wittig and Knoevenagel condensations. Reactions of triphenyl phosphine α-chloroesters and *o*-hydroxybenzaldehyde in ionic liquid directly led to the formation of 3-substituted coumarins (Valizadeh and Vaghefi, 2009). A variety of 3-substituted coumarins were obtained from an imidazolium phosphate ionic liquid using salicylaldehydes and active methylene compounds. Reaction times were reduced, and reusability of the catalyst was enhanced (Valizadeh and Gholipour, 2010).

Butyl-methylimidazolium bromide was found to be an efficient medium for the synthesis of 3-acetoacetylcoumarins via the reaction between salicylaldehydes and 4-hydroxy-6-methyl-2*H*-pyran-2-one (Shi et al. 2009).

Knoevenagel condensation between aromatic aldehydes and active methylene compounds has also been induced by Lewis acid ionic liquids leading to the generation of electrophilic alkenes as major products. Extension of this methodology to *o*-hydroxybenzaldehydes resulted in the formation of coumarins (Harjani et al. 2002).

10.8 CONCLUSION

Most of the catalysts have been employed for the synthesis of known hydroxyl- or alkoxycoumarins. Application of heterogeneous catalysts has resulted in environmentally benign, high-yielding experimental techniques, with considerable reduction in acidic and toxic wastes. The potential of these catalysts can be applied to other acid/base-catalyzed reactions and construction of structurally analogous heterocyclic systems. The ability of these catalysts needs to be exploited for the synthesis of commercially available coumarins with the aim of reducing manufacturing costs. Lastly, their application to the synthesis of unknown coumarins will prove their utility in multistep transformations.

REFERENCES

Abello, S., Medina, F., Rodriguez, X., Cesteros, Y., Salagre, P., Sueirase, J. E., Tichit, D., Bernard, C. 2004. Supported choline hydroxide (ionic liquid) as heterogeneous catalyst for aldol condensation reaction. *Chem. Commun.* 1096–1097.

Ali, A., Majid, A., Eskandar, K. 2013. Easy access to coumarin derivatives using alumina sulfuric acid as an efficient and reusable catalyst under solvent free conditions. *J. Chem.* 2013, ID 767825.

Benaglia, M., Puglisi, A., Cozzi, F. 2003. Polymer supported organic catalysts. *Chem. Rev.* 103(9): 3401–3430.

Benyaram, M. R., Boningari, T., Meghshyam, K. P. 2009. One-pot synthesis of substituted coumarins catalyzed by silica gel supported sulfuric acid under solvent-free conditions. *Open Catal. J.* 2: 33–39.

Benyaram, M. R., Meghshyam, K. P., Lakshmanan, P. 2006. Sulfated Ce_xZr_{1-x} solid acid catalyst for the solvent free synthesis of coumarins. *J. Mol. Catal. A Chem.* 256: 290–294.

Benyaram, M. R., Vangala, R. R., Giridhar, D. 2001. Synthesis of coumarins catalyzed by eco-friendly W/ZrO_2 solid acid catalyst. *Synth. Commun.* 31: 3603–3607.

Chang, S., Grubbs, R. H. 1998. A highly efficient and practical synthesis of chromene derivatives using ring-closing olefin metathesis. *J. Org. Chem.* 63: 864–866.

Chatterjee, A. K., Toste, F. D., Goldberg, S. D., Grubbs, R. H. 2003. Synthesis of coumarins by ring-closing metathesis. *Pure Appl. Chem.* 75: 421–425.

Chaudhari, D. D. 1983. Heterogeneous catalysis by solid super acid: Nafion-H catalysed von Pechmann condensation. *Chem. Ind.* 568.

China Raju, B., Hari Babu, T., Madhusudana Rao, J. 2009. $H_3PW_{12}O_{40}$ catalysed efficient synthesis of 4-substituted coumarins. *Indian J. Chem.* 48B: 120–123.

Climent, M. J., Corma, A., Fornes, V., Frau, A., Guil-Lopez, R., Ihorra, S., Primo, J. 1996. Aluminophosphates oxynitrides as base catalysts: Nature of the base sites and their catalytic implications. *J. Catal.* 163: 392–398.

Darvatkar, N. B., Deorukhkar, A. R., Bhilare, S. V., Raut, D. G., Salunkhe, M. M. 2008. Ionic liquid-mediated synthesis of coumarin-3-carboxylic acids via Knoevenagel condensation of Meldrum's acid with *ortho*-hydroxyaryl aldehydes. *Synth. Commun.* 38(20): 3508–3513.

Das-Gupta, A. K, Das, K. R. 1969. Coumarins and related compounds. Part VI. A new approach to xanthyletin. *J. Chem. Soc. C.* 1969: 33–34.

De la Hoz, A., Andres, M., Ester, V. 1999. Use of microwave irradiation and solid acid catalysts in an enhanced and environmentally friendly synthesis of coumarin derivatives. *Synlett* 5: 608–610.

Dong, F., Jain, C., Kai, G., Qunrong, S., Zuliang, L. 2008. Synthesis of coumarins via Pechmann reaction in water catalysed by acyclic acidic ionic liquids. *Catal. Lett.* 121: 255–259.

Eromanelli, G. P., Bennardi, D., Ruiz, D. M., Baronetti, G., Thomas, H. J., Autino, J. C. 2004. A solvent free synthesis of coumarins using a Wells-Dawson heteropolyacid as catalyst. *Tetrahedron Lett.* 45: 8935–8939.

Fylaktakidou, K. C., Hadjipavlou-Litina, D. J., Litinas, K. E., Nicolaides, D. N. 2004. Natural and synthetic coumarin derivatives with anti-inflammatory/antioxidant activities. *Curr. Pharm. Des.* 10: 3813–3833.

Gangadasu, B., Narender, P., Raju, B. C., Rao, V. J. 2004. $ZrCl_4$ catalysed solvent free synthesis of coumarin. *J. Chem. Res.* 7: 480–481.

Green, B., Gupta, A. K. D., Das, K. R., Chatterje, R. M. 1969. Coumarins and related compounds. Part IV. Aluminium chloride-catalysed reaction of phenols with methyl acrylate: A new approach to the synthesis of hydroxycoumarins. *J. Chem. Soc. C.* 1969: 29–33.

Gunnewegh, E. A., Hoefnagel, A. J., VanBekkum, H. 1995. Zeolite catalysed synthesis of coumarin derivatives. *J. Mol. Catal. A Chem.* 100: 87–92.

Hadjipavlou-Litina, D., Christos, K., Eleni, P., Marianna, D., Antonia, A., Haralambos, E. K. 2007. Anti-inflammatory and antioxidant activity of coumarins designed as potential fluorescent zinc sensors. *J. Enzy. Inhib. Med. Chem.* 22: 287–292.

Harjani, J. R., Nara, S. J., Salunkhe, M. M. 2002. Lewis acidic ionic liquids for the synthesis of electrophilic alkenes via the Knoevenagel condensation. *Tetrahedron Lett.* 43: 1127–1130.

Hassan, V., Abbas, S. 2005. An efficient procedure for the synthesis of coumarin derivatives using $TiCl_4$ as catalyst under solvent-free conditions. *Tetrahedron Lett.* 46: 3501–3503.

Helavi, B. V., Solabannavar, S. B., Salunkhe, R. S., Mane, R. B. 2003. Microwave-assisted solventless Pechmann condensation. *J. Chem. Res.* 5: 279–280.

Hoefnagel, A. J., Gunnewegh, E. A., Downing, R. S., Bekkum, H. V. 1995. Synthesis of 7-hydroxycoumarins catalysed by solid acid catalysts. *J. Chem. Soc. Chem. Commun.* 2: 225–226.

Houcine, A., Mehdi, F., Yves Le, B., Rachid El, G. 2003. Synthesis of new bis-iminocoumarins by Schmidt reaction catalysed by ion-exchange resins. *Synth. Commun.* 33: 1821–1828.

Irena, K. 2005. Synthetic and natural coumarins as cytotoxic agents. *Curr. Med. Chem. Anti-Cancer Agents* 5: 29–46.

Izumi, Y., Urabe, K., Onaka, M. 1992. *Zeolite Clay and Heteropolyacids in Organic Reactions.* VCH: Tokyo.

John, E., Isra Elstam, S. 1961. Use of cation exchange resins in organic reactions in the Von Pechmann reaction. *J. Org. Chem.* 26: 240–242.

Joshi, R., Chudasama, U. 2008. Synthesis of coumarins via Pechmann condensation using inorganic ion exchangers as solid acid. *Catal. J. Sci. Ind. Res.* 67: 1092–1097.

Kalyan, J. B., Ruli, B. 2011. Poly(4-vinylpyridine)-supported sulfuric acid: An efficient solid acid catalyst for the synthesis of coumarin derivatives under solvent free conditions. *Monatsh Chem.* 142: 1253–1257.

Katerinopoulos, H. E. 2004. The coumarin moiety as chromophore of fluorescent ion indicators in biological systems. *Curr. Pharm. Des.* 10: 3835–3852.

Khalilah, G. R., William, H. H., Charlo, P. B., Christine, K. P., Melissa, L. K., Niren, M. 2012. Fluorescent coumarin thiols measure biological redox couples. *Org. Lett.* 14: 680–683.

Kloetstra, K. R., Van Bekkum, H. 1995. Mesoporous material containing framework tectosilicate by pore-wall recrystallization. *J. Chem. Soc. Chem. Commun.* 1005–1006.

Kulkarni, M. V., Kulkarni, G. M., Lin, C., Sun, C. M. 2006. Recent advances in coumarins and 1- azacoumarins as versatile biodynamic agents. *Curr. Med. Chem.* 13: 2795–2818.

Lacy, A., O'Kennedy, R. 2004. Studies on coumarins and coumarin-related compounds to determine their therapeutic role in the treatment of cancer. *Curr. Pharm. Des.* 10: 3797–3811.

Larock, R. C., Kadnikov, D. V. 2000. Synthesis of coumarins via palladium-catalyzed carbonylative annulation of internal alkynes by *o*-iodophenols. *Org. Lett.* 2: 3643–3646.

Laufer, M. C., Hausmann, H., Holderich, W. F. 2003. Synthesis of 7-hydroxycoumarins by Pechmann reaction using Nafion resin/silica nanocomposites as catalysts. *J. Catal.* 218(2): 315–320.

Limin, W., Jianjun, X., He, T., Changtao, Q., Yun, M. 2003. Synthesis of coumarin by $Yb(OTf)_3$ catalysed Pechmann reaction under the solvent-free conditions. *Indian J. Chem.* 42B: 2097–2099.

Lionel, D., Pierre, L., Keith, S. 1993. Heightened selectivity in aromatic nitrations and chlorinations by the use of solid supports and catalysts. *Acc. Chem. Res.* 26: 607–613.

Liu, F., Wang, L. Sun, Q., Zhu, L., Meng, X., Xiao, F.S. 2012. Transesterification catalysed by ionic liquids on superhydrophobic mesoporous polymers: Heterogeneous catalysts that are faster than homogeneous catalysts. *J. Am. Chem. Soc.* 134: 16948–16950.

Mastalgi, P., Andric, N. 1958. Catalytic action of ion exchange resins in the Pechmann and Knoevenagel reactions. *Compt. Rend.* 246: 3079–3081.

Mhiri, C., El Gharbi, R., Le Bigot, Y. 1999. Polymer supported reagents: Novel methodology for selective and general synthesis of iminocoumarins. *Synth. Commun.* 29: 3385–3399.

Muchchintala, M., Vidavalur, S., Guri Lakishmi, V. D., Yerra Koteswara, R., Chunduri, V. R. 2006. A solvent-free synthesis of coumarins via Pechmann condensation using heterogeneous catalyst. *J. Mol. Catal. A Chem.* 255: 49–52.

Potdar, M. K., Mohile, S. S., Salunkhe, M. M. 2001. Coumarin synthesis via Pechmann condensation in Lewis acidic chloroaluminate ionic liquid. *Tetrahedron Lett.* 42: 9285–9287.

Potdar, M. K., Rasalkar, M. S., Mohile, S. S., Salunkhe, M. M. 2005. Convenient and efficient protocols for coumarin synthesis via Pechmann condensation in neutral ionic liquids. *J. Mol. Catal. A Chem.* 235: 249–252.

Priyanka, G. M., Ranadeep, S. J., Anant, R. G., Ganesh, R. J., Charansingh, H. G. 2009. Ammonium metavanadate: A mild and efficient catalyst for the synthesis of coumarins. *Bull. Chem. Soc.* 30: 2969–2972.

Rajanarendra, E., Firoz, P. S., Nagi Reddy, M. 2010. Synthesis of new isoxazolyl coumarins by eco-friendly dipyridine cobalt chloride catalyzed Pechmann reaction at ambient temperature. *Indian J. Chem.* 49B: 532–535.

Rajitha, B., Naveen Kumar, V., Someshwar, P., Venu Madhav, J., Narsimha Reddy, P., Thirupathi Reddy, Y. 2006. Dipyridine copper chloride catalyzed coumarin synthesis via Pechmann condensation under conventional heating and microwave irradiation. *ARKIVOC* 2006(12): 23–27.

Ramani, A., Chanda, B. M., Velu, S., Sivasanker, S. 1999. One-pot synthesis of coumarins. Catalysis by the solid base, calcined Mg-Al hydrotalcite. *Green Chem.* 1: 163–165.

Rammohan, P., Taradas, S., Shampa, K. 2012. Amberlyst-15 in organic synthesis. *Arkivoc* 2012(1): 570–609.

Sabou, R., Hoelderich, W. F., Ramprasad, D., Weinand, R. 2005. Synthesis of 7-hydroxy coumarin via the Pechmann reaction with Amberlyst ion-exchange reins as catalysts. *J. Catal.* 232: 34–37.

Salehi, P., Khodaei daje, M. M., Zolfigol, M. A., Zeinoldini, S. 2003. Catalytic Friedel-Crafts acylation of alkoxy benzenes by ferric hydrogensulfate. *Synth. Commun.* 33: 1367–1373.

Sanjay, K., Anil, S., Jagir, S. S. 2007. LiBr-mediated solvent free von Pechmann reaction: Facile and efficient method for the synthesis of 2*H*-chromen-2-ones. *ARKIVOC* 2007(15): 18–23.

Sethna, M. S., Shah, N. M. 1945. The chemistry of coumarins. *Chem. Rev.* 36: 1–62.

Shi, D. Q., Zhou, Y., Rong, S. F. 2009. Ionic liquid [bmim]Br, as efficient promoting medium for synthesis of 3-acetoacetyl coumarin derivatives without the use of any catalyst. *Synth. Commun.* 39: 3500–3508.

Shinde, N., Chavan, F., Madje, B., Bharad, J., Ubale, M., Ware, M., Shingare, M. 2008. Silicagel supported $NaHSO_4$ catalyzed organic reaction: An efficient synthesis of coumarins. *Bull. Catal. Soc. India* 7: 41–45.

Singh, V., Kaur, S., Sapehiyia, V., Singh, J., Kad, G. L. 2005. Microwave accelerated preparation of [bmim][H_2SO_4] ionic liquid: An acid catalyst for improved synthesis of coumarins. *Catal. Commun.* 6: 57–60.

Soares, V. C. D., Alves, M. B., Souza, E. R., Pinto, I. O., Rubim, J. C., Andrade, C. K. Z., Suarez, P. A. 2007. Organo-niobate ionic liquids: Synthesis, characterization and application as acid catalyst in Pechmann reactions. *Int. J. Mol. Sci.* 8: 392–398.

Srinivasan, P., Rampally, C. S. 2004. Synthesis of 7-methylcoumarin using polyaniline supported acid catalyst. *J. Mol. Catal. A Chem.* 209: 117–124.

Stephane, F., Valerie, T., Thierry, B. 2001. Microwave acceleration of the Pechmann reaction on graphite/montmorillonite K10: Application to the preparation of 4-substituted 7-aminocoumarins. *Tetrahedron Lett.* 42: 2791–2794.

Stoyanov, E. V., Mezger, J. 2005. Pechmann reaction promoted by Boron trifluoride dehydrate. *Molecules* 10: 762–766.

Subhas, D. B., Rudradas, A. P., Babu, M. H. 2002. The indium(III) chloride-catalyzed von Pechmann reaction: A simple and effective procedure for the synthesis of 4-substituted coumarins. *Tetrahedron Lett.* 43: 9195–9197.

Subhash, P. C., Shivasankar, K., Sivappa, R., Kale, R. 2002. Zinc mediated trans-esterification of β-ketoesters and coumarin synthesis. *Tetrahedron Lett.* 43: 8583–8586.

Sushilkumar, S. B., Devanand, B. S. 2004. Samarium(III) catalyzed one-pot construction of coumarins. *Tetrahedron Lett.* 45: 7999–8001.

Takashi, S., Eiji, Y., Kojiro, H., Shi-Wei, Z., Shigetoshi, T. 1998. Rhodium-catalysed cyclic carbonylation of 2-alkynylphenols: Synthesis of benzofuranones and coumarins. *J. Chem. Soc. Perkin Trans. 1* 477–483.

Thirupathi Reddy, Y., Vijayakumar, N. S., Peter, A. C., Pavan, K. D., Narsimha Reddy, P., Rajitha, B. 2008. Ceric ammonium nitrate (CAN): An efficient catalyst for the coumarin synthesis via Pechmann condensation using conventional heating and microwave irradiation. *Synth. Commun.* 38: 2082–2088.

Tong-Shuang, L., Zhan-Hui, Z., Feng, Y., Cheng-Gaung, F. 1998. Montmorillonite clay catalysis. Part 7. An environmentally friendly procedure for the synthesis of coumarins via Pechmann condensation of phenols with ethyl acetoacetate. *J. Chem. Res.* 38–39.

Trost, B. M., Toste, D. 1996. A new palladium-catalyzed addition: A mild method for the synthesis of coumarins. *J. Am. Chem. Soc.* 118: 6305–6306.

Tyagi, B., Mishra, M. K., Jasra, R. V. 2007. Synthesis of 7-substituted 4-methyl coumarins by Pechmann reaction using nano-crystalline sulfated-zirconia, *J. Mol. Catal. A Chem.* 276: 47–56.

Valizadeh, H., Gholipour, H. 2010. Imidazolium-based phosphinite ionic liquid (IL-$OPPh_2$) as reusable catalyst and solvent for the Knoevenagel condensation reaction. *Synth. Commun.* 40: 1477–1485.

Valizadeh, H., Vaghefi, S. 2009. One-pot Wittig and Knoevenagel reactions in ionic liquid as convenient methods for the synthesis of coumarin derivatives. *Synth. Commun.* 39: 1666–1678.

Varughese, M. A., Ramakrshna, P. B., Shriniwas, D. S. 2005. Bismuth(III) nitrate pentahydrate a mild and inexpensive reagent for the synthesis of coumarin under mild conditions. *Tetrahedron Lett.* 46: 6957–6959.

Verdia, P., Santamarta, F., Tojo, E. 2011. Knoevenagel reaction in [MMIm][MSO_4] synthesis of coumarins. *Molecules* 16: 4379–4388.

Vinay Kumar, B., Bhojya Naik, S. H., Girija, D., Vijaya Kumar, B. 2011. ZnO nano particles as catalyst for efficient green one-pot synthesis of coumarins through Knoevenagel condensation. *J. Chem. Sci.* 123: 615–621.

Yarlagadda, V. S. R., Dirk, E. D. V., Pierre, A. J. 1997. 1,5,7-Triazabicyclo[4.4.0]dec-5-ene immobilized in MCM-41: A strongly basic porous catalyst. *Angew. Chem. Int. Ed. Engl.* 36: 2661–2663.

11 Silver

A Versatile Heterogeneous Catalyst for Heterocyclic Synthesis

Chetna Ameta and K. L. Ameta

CONTENTS

11.1 INTRODUCTION

Heterogeneous catalysis refers to the form of catalysis where the phase of the catalyst differs from that of the reactants. "Phase" here refers not only to solid, liquid, and gas, but also immiscible liquids, for example, oil and water. The great majority of practical heterogeneous catalysts are solids and the great majority of reactants are gases or liquids (Rothenberg 2008). Heterogeneous catalysis is of paramount importance in many areas of the chemical and energy industries. It attracted Nobel prizes to Fritz Haber and Carl Bosch in 1918, Irving Langmuir in 1932, and Gerhard Ertl in 2007. Adsorption is commonly an essential first step in heterogeneous catalysis.

Adsorption is when a molecule in the gas phase or in solution binds to atoms on a solid or liquid surface. The binding molecule is called the adsorbate, and the surface to which it binds is the adsorbent. The process of the adsorbate binding to the adsorbent is called adsorption. The reverse of this process (the adsorbate splitting from adsorbent) is called desorption. In terms of catalyst support, the catalyst is the adsorbate and the support is the adsorbent.

Two types of adsorption are recognized in heterogeneous catalysis, although many processes fall into an ambiguous range between the two extremes. In the first type, physisorption, only small changes to the electronic structure of the adsorbate are induced. Typical energies for physisorption are from 2 to 10 kcal/mol. The second type is called chemisorption, in which the adsorbate is strongly perturbed, often combined with bond-breaking. Energies for typical chemisorptions range from 15 to 100 kcal/mol. In physisorption, the adsorbate is attracted to the surface atoms by van der Waals forces. A mathematical model for physorbtion was developed to predict the energies of basic physisorption of nonpolar molecules. The analysis of physisorption for polar or ionic species is more complex. Chemisorption results in the sharing of electrons between the adsorbate and the adsorbent. Chemisorption is traditionally described by the Lennard-Jones potential, which considers various cases, two of which are:

- Molecular adsorption, where the adsorbate remains intact. An example is alkene binding by platinum.
- Dissociation adsorption, where one or more bonds break concomitantly with adsorption. In this case, the barrier to dissociation affects the rate of adsorption. An example of this is the binding of H_2, where the H–H bond is broken upon adsorption (Masel, 1996).

In heterogeneous catalysis, the reactants diffuse to the catalyst surface and adsorb onto it, via the formation of chemical bonds. After reaction, the products desorb from the surface and diffuse away. Understanding the transport phenomena and surface chemistry such as dispersion is important. If diffusion rates are not taken into account, the reaction rates for various reactions on surfaces depend solely on the rate constants and reactant concentrations. For solid heterogeneous catalysts, the surface area of the catalyst is critical since it determines the availability of catalytic sites. Surface areas can be large; for example, some mesoporous silicates have areas of 1000 m^2/g. The most common approach to maximizing surface area is the use of catalyst supports, which are the materials over which the catalysts are spread.

11.2 HETEROGENEOUS CATALYSIS IN HETEROCYCLIC SYNTHESIS

Heterocyclic synthesis involving transition-metal complexes has become of common use in the past decade because a transition-metal-catalyzed reaction can directly build complicated molecules from readily accessible starting materials under mild conditions. In comparison with other transition metals, silver(I) complexes have long been believed to have low catalytic efficiency, and most commonly, they are used as either cocatalysts or Lewis acids. Now, Ag-catalyzed reactions have emerged as important synthetic methods for a variety of organic transformations. Ag(I) is known to interact with multiple bonds, such as alkenes, alkynes, and allenes. In addition, the use of silver(I) is more economical than other expensive transition metals. Efforts in studying homogeneous silver-catalyzed organic transformations have mostly focused on asymmetric catalysis. For instance, Yamamoto and others have developed the Yamamoto–Yanagisawa system [BINAP + silver(I) system] for enantioselective allylation reactions (Yanagisawa et al. 1999); some applications of this methodology to the synthesis of heterocycles are beginning to appear (Jimenez-Gonzalez et al. 2006). The purpose of this chapter is to summarize those heterocyclizations in which silver salts play an important role: addition of nucleophiles to alkynes, allenes, and olefins; cycloaddition reactions, with special focus on enantioselective [3 + 2]-cycloaddition of azomethine ylides and nitrilimines; and [4 + 2]-cycloaddition of imines.

11.2.1 Silver-Catalyzed Synthesis of Five-Membered Heterocycles

A mild and direct process for C–C bond formation from propargylic alcohols and olefins has been developed in the presence of a silver catalyst (Ji et al. 2009) (Scheme 11.1). In this reaction, trace amounts of water were necessary and allene alcohols and 1,3-dienes were obtained selectively.

An efficient and one-pot synthesis of multisubstituted pyrroles with high diversity and in a regioselective manner from the reactions of suitably substituted (*Z*)-enynols with amines or sulfonamides under mild reaction conditions has been developed (Lu et al. 2009). This synthesis was realized via a cascade process in the presence of gold/silver (Au/Ag) or boron trifluoride etherate/gold/silver ($BF_3{\cdot}Et_2O$/Au/Ag) catalysts, which could catalyze amination and cycloisomerization reactions in the same vessel. Scheme 11.2 illustrates that either silver trifluoromethanesulfonate or a mixture of

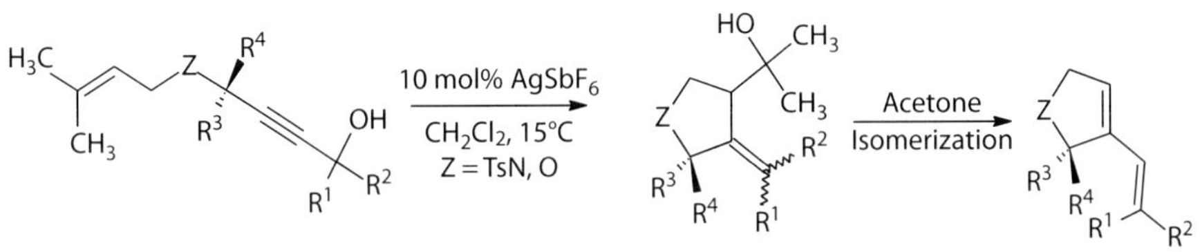

SCHEME 11.1

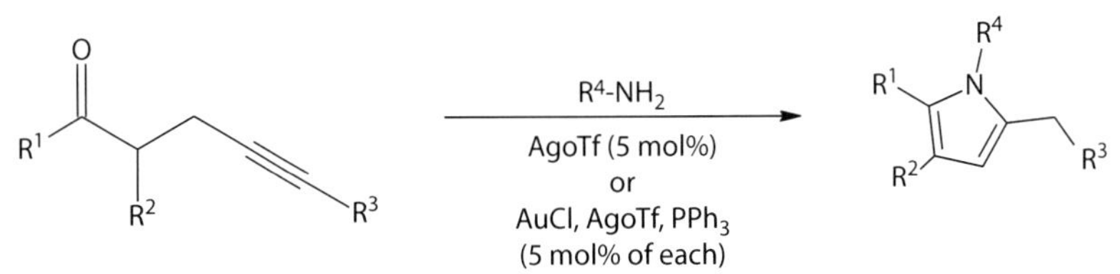

SCHEME 11.2

gold(I) chloride, silver trifluoromethanesulfonate, and triphenylphosphine catalyze the formation of pyrroles from substituted β-alkynyl ketones and amines (Harrison et al. 2006). These reactions proceed when using 5 mol% of catalyst with yields of isolated pyrroles ranging from 13% to 92%.

11.2.2 One-Pot Silver-Catalyzed and PIDA-Mediated Sequential Reactions

The addition/oxidative cyclization of alkynes with amines in the presence of an $AgBF_4$ catalyst and a PIDA oxidant leads to polysubstituted pyrroles (Scheme 11.3). The reaction corresponds to the construction of a pyrrole fragment, which also provides a new method for the formation of C–C bonds (Liu et al. 2010).

11.2.3 Silver-Catalyzed Hydroamination

It has been shown that functionalized pyrroles can be efficiently prepared using a two-step sequence. This sequence involves the propargylation of secondary enaminones using *n*-BuLi and propargyl bromide, followed by intramolecular hydroamination catalyzed by silver nitrate (Robinson et al. 2005a). The hydroamination can be carried out at room temperature (overnight) or in a domestic microwave oven (60 s).

11.2.4 Silver(I)-Promoted Cyclization

In a convenient one-pot process, easily accessed propargyl vinyl ethers and aromatic amines are effectively converted into tetra- and pentasubstituted 5-methylpyrroles, which can further be transformed into 5-formylpyrroles via 2-iodoxybenzoic acid (IBX)-mediated oxidation (Binder and Kirsch 2006). The cascade reaction proceeds through a silver(I)-catalyzed propargyl Claisen rearrangement, an amine condensation, and a gold(I)-catalyzed 5-*exo-dig* heterocyclization, as shown in Scheme 11.4.

11.2.5 Silver-Mediated Addition

A highly efficient intermolecular addition of 1,3-diketones to alkenes catalyzed by $AuCl_3$/AgOTf was developed (Scheme 11.5). A mechanistic rationale for the reaction has been proposed via an α-C–H activation.

Treatment of allenyne-1,6-diols with gold and silver catalysts selectively produced 2,5-dihydrofuran and furan derivatives, respectively, in good to excellent

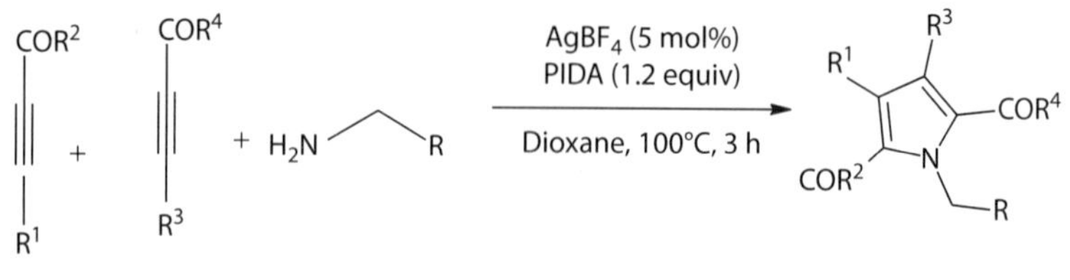

SCHEME 11.3

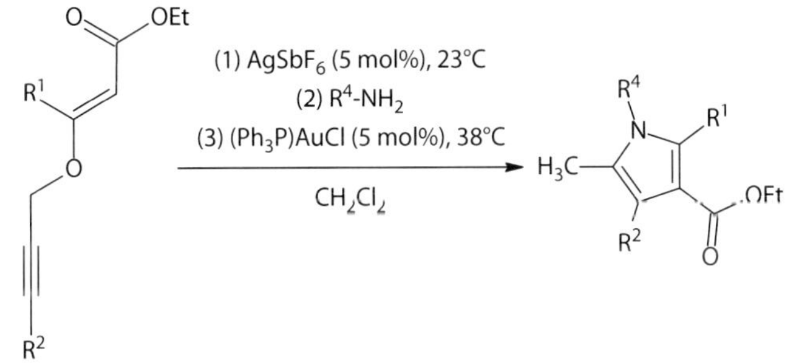

SCHEME 11.4

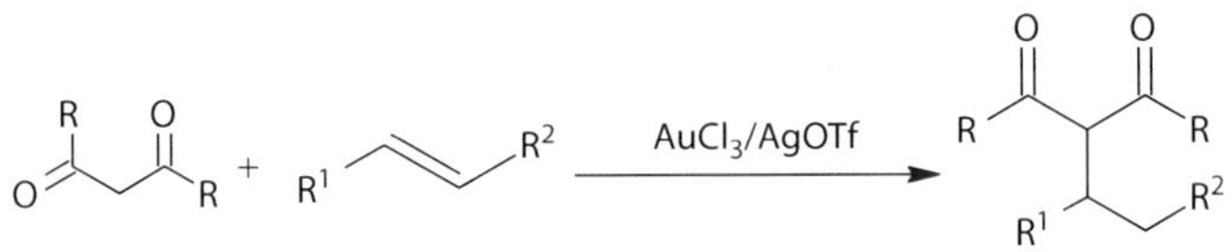

SCHEME 11.5

yields through the selective activation and differentiation of the double and triple bonds in allenyne-1,6-diols. It is noteworthy that the chemoselectivity is clearly switched simply by changing the metal from gold to silver (Kim and Lee 2008). Scheme 11.6 shows that functionalized furans are conveniently formed by a new silver(I)-catalyzed reaction of alk-1-ynyl oxiranes in the presence of *p*-toluenesulfonic acid and methanol (Yao and Li 2004; Blanc et al. 2009).

Novel tetracyclic and pentacyclic indole derivatives can be prepared from readily available *gem*-dibromovinyl substrates in a single step by means of an efficient Pd-catalyzed domino Buchwald–Hartwig amination/direct arylation reaction. Enhanced reactivity and selectivity are obtained by the addition of silver salts (Bryan and Lautens 2008).

11.2.6 Silver-Promoted Domino Pd-Catalyzed Amination/Direct Arylation

Transition-metal-catalyzed domino reactions have been used as powerful tools for the preparation of polysubstituted furans in a one-pot manner. An efficient synthetic

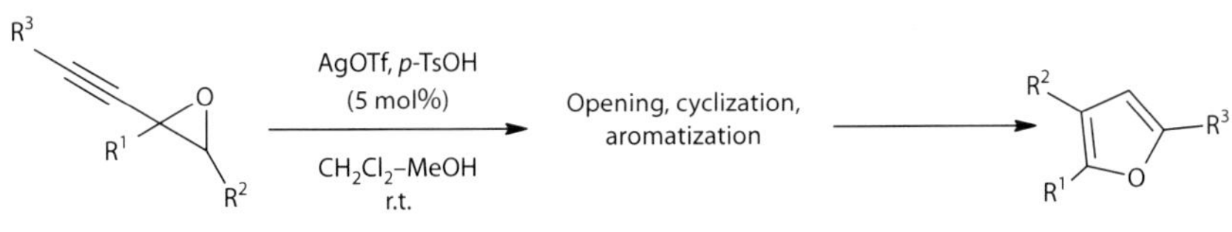

SCHEME 11.6

method has been developed for the construction of tri- or tetrasubstituted furans from electron-deficient alkynes and 2-yn-1-ols by a silver-catalyzed domino reaction (Cao et al. 2010). It is especially noteworthy that a 2,3,5-trisubstituted 4-ynylfuran was formally obtained in an extremely direct manner without tedious stepwise synthesis. In addition, regioisomeric furans were observed when substituted aryl alkynyl ketones were employed. This methodology represents a highly efficient synthetic route to electron-deficient furans for which catalytic approaches are scarce. The reaction proceeds efficiently under mild conditions with commercially available catalysts and materials.

11.2.7 Silver(I)-Catalyzed Cascade

Scheme 11.7 illustrates that gold and silver triflate–catalyzed intramolecular hydroarylation of allenic anilines and phenols offers an efficient route to dihydroquinoline and chromene derivatives under mild reaction conditions (Watanabe et al. 2007). The hydroarylation takes place at the terminal or central allenic carbon, depending on the substrate structure, leading to a highly selective formation of six-membered rings.

A set of cycloisomerization methodologies of alkynyl ketones and imines with concurrent acyloxy, phosphatyloxy, or sulfonyloxy group migration, which allow for the efficient synthesis of multisubstituted furans and N-fused heterocycles, has been developed (Schwier et al. 2007). Investigation of the reaction course employing ^{17}O-labeled substrates allowed for elucidation of the mechanisms behind these diverse transformations. It was found that, while the phosphatyloxy migration in conjugated alkynyl imines in their cycloisomerization to N-fused pyrroles proceeded via a [3,3]-sigmatropic rearrangement, the analogous cycloisomerization of skipped alkynyl ketones proceeds through two consecutive 1,2-migrations, resulting in an apparent 1,3-shift, followed by a subsequent 1,2-migration through competitive oxirenium and dioxolenylium pathways. Investigations of the 1,2-acyloxy migration of conjugated alkynyl ketones to furans demonstrated the involvement of a dioxolenylium intermediate. The mechanism of cycloisomerization of skipped alkynyl ketones containing an acyloxy group was found to be catalyst dependent; Lewis acid catalysts caused an ionization/S_N1' isomerization to the allene, followed by cycloisomerization to the furan, while transition-metal catalysts evoked a Rautenstrauch-type mechanistic pathway. Further synthetic utility of the obtained phosphatyloxy-substituted heterocycles was demonstrated through their efficient employment in the Kumada cross-coupling reaction with various Grignard reagents.

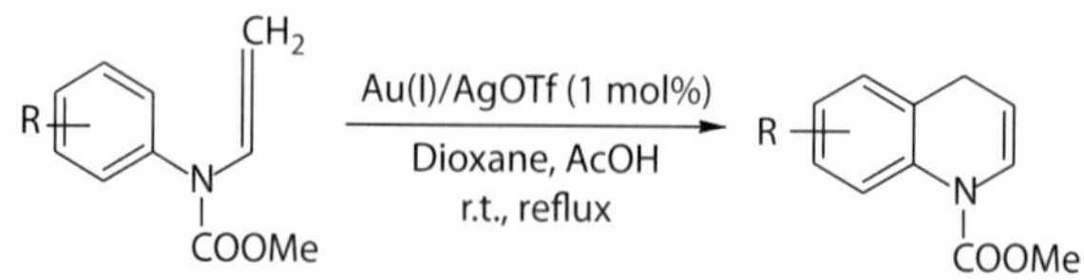

SCHEME 11.7

Two complementary protocols for assembly of multisubstituted N-fused heterocycles have been developed (Seregin et al. 2008). It was demonstrated that 1,3-disubstituted N-fused heterocycles, including indolizines, pyrroloquinoxalines, and pyrrolothiazoles, can easily be synthesized via an exceptionally mild and efficient method involving a novel silver-catalyzed cycloisomerization of propargyl-containing heterocycles. Alternatively, 1,2-disubstituted heterocycles can be accessed through the novel cascade transformation involving an alkyne-vinylidene isomerization with concomitant 1,2-shift of hydrogen, silyl, stannyl, or germyl groups. This mild and simple method allows for selective and highly efficient synthesis of indolizines, pyrroloisoquinolines, pyrroloquinoxalines, pyrrolopyrazines, and pyrrolothiazoles (Scheme 11.8).

Aromatic heterocycles are highly important structural units in a vast number of biological active natural compounds, pharmaceuticals, and materials. They are also important intermediates in organic synthesis, often providing access to other highly desirable structures. Thus, there is a compelling need to develop novel and more general methods for the synthesis of heterocycles. In recent years, the use of transition-metal-catalyzed transformations truly revolutionized the area of heterocyclic chemistry. A particularly attractive approach toward this goal involves the incorporation of molecular rearrangement steps into the transition-metal-catalyzed cycloisomerization cascade reactions. Scheme 11.9 covers the most important recent advances in the Cu-, Ag-, and Au-catalyzed syntheses of five-membered aromatic heterocycles proceeding with 1,2- and 1,3-migrations of various groups during the assembly of the heterocyclic ring (Dudnik et al. 2010).

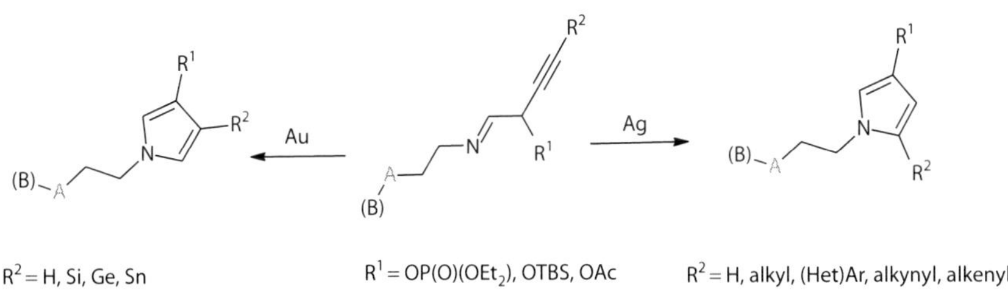

SCHEME 11.8

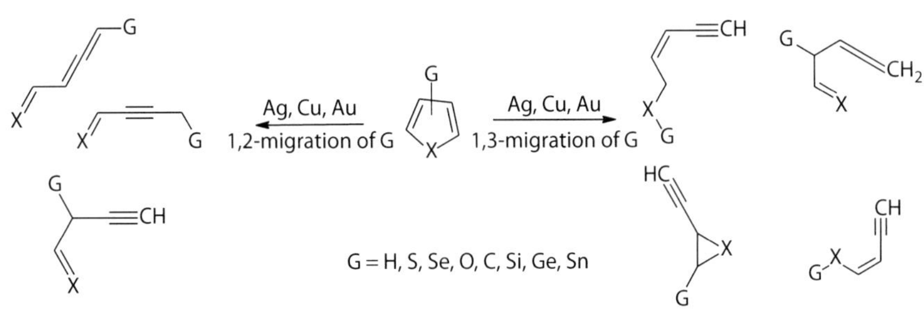

SCHEME 11.9

Cycloisomerization of 1,*n*-enynes and diynes is a powerful method in organic synthesis to access heterocyclic compounds and has drawn increasing attention from organic chemists. The transition-metal-catalyzed cycloisomerization to synthesize five- or six-membered heterocyclic compounds uses 1,*n*-enynes and diynes having a propargylic ester moiety (Zhang et al. 2012). Firstly, 2,3-disubstituted 3-pyrrolines are synthesized via gold-catalyzed cycloisomerization of 1,6-diynes. In addition, a novel silver-catalyzed tandem 1,3-acyloxy migration/Mannich-type addition/elimination of the sulfonyl group of *N*-sulfonylhydrazone-propargylic esters to 5,6-dihydropyridazin-4-one derivatives also occurs. Furthermore, there are three examples of the synthesis of bicyclic compounds via titanium- or rhodium-catalyzed carbocyclization of enynes. In this context, it is shown that 1,*n*-enynes and diynes containing propargylic esters are highly reactive and useful starting materials for the cycloisomerization catalyzed by a transition-metal catalyst. Westling et al. (1986) studied a convergent approach to heterocycle synthesis via silver ion–mediated α-ketoimidoylhalide-arene cyclizations. It is an application for the synthesis of the erythrinane skeleton. The reaction of symmetrical diols and oligo(ethylene glycol)s with a stoichiometric amount of *p*-toluenesulfonyl chloride in the presence of silver(I) oxide and a catalytic amount of potassium iodide led selectively to the monotosylate derivatives in high yields (Bouzide and Sauve 2002). Polysubstituted cyclic ethers were obtained readily upon treatment of the corresponding diols with an excess of silver oxide. The high selectivity was explained on the basis of the difference in acidity between the two hydroxy groups, which undergo an intramolecular hydrogen bonding (Scheme 11.10).

A highly *endo*-selective gold- and silver-catalyzed cycloisomerization of 1,4-diynes has been developed (Wilckens et al. 2010). By employing electron-rich phosphines and *N*-heterocyclic carbenes as ligands, a number of 1,4-diynes could be cyclized in a desymmetrizing fashion to form dihydrodioxepines and enantiomerically enriched tetrahydrooxazepines (Scheme 11.11).

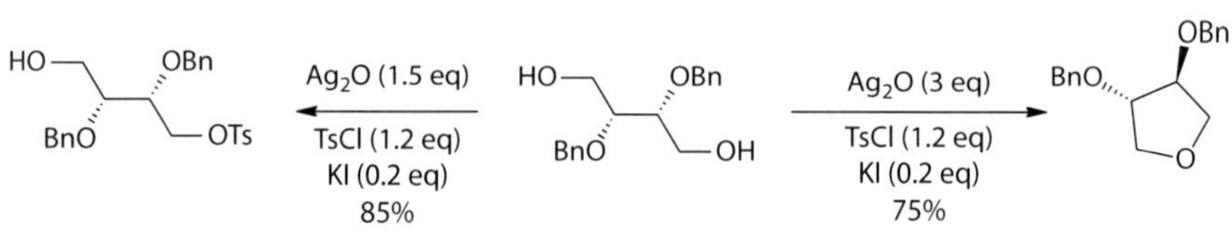

SCHEME 11.10

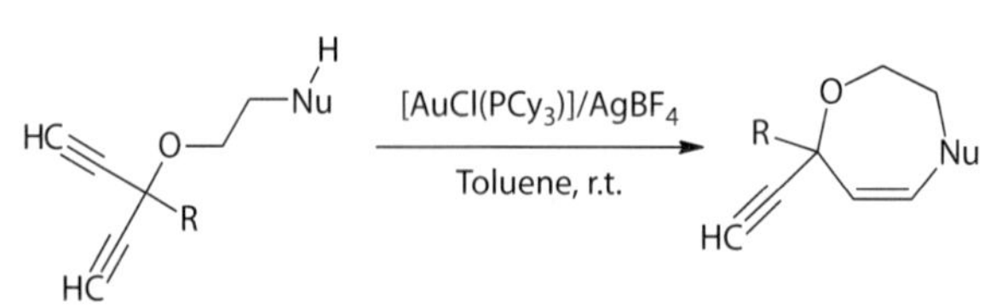

SCHEME 11.11

The heteroannulation catalyzed by silver salts of alkynols or alkynoic acids is considerably enhanced by the presence of a propargylic C–O bond. This method allows for rapid access to highly functionalized heterocycles, such as α-methylene oxolanes or oxanes and γ-methylene pentanolactones (Dalla and Pale 1999).

A novel synthetic approach to the synthesis of enantiomerically pure 2,5-disubstituted pyrrolines has been described (Esseveldt et al. 2005). The methodology involves an Ag-catalyzed 5-*endo-dig* cyclization of enantiopure aryl-substituted acetylene-containing amino acids. It has also been shown that the obtained pyrrolines can be efficiently transformed into the corresponding saturated 5-aryl-substituted proline derivatives. Dalla and Pale (1994) described the total synthesis of the naturally occurring *cis* 2-hexadecyl-3-hydroxy-4-methylene butyrolactone via silver-catalyzed heterocyclization. Marshall and Pinney (1993) studied stereoselective synthesis of 2,5-dihydrofurans by sequential S_N2' cleavage of alkynyloxiranes and silver(I)-catalyzed cyclization of the allenylcarbinol products. Silver salts that have a basic counterion are efficient catalysts for the regiospecific intramolecular addition of various acetylenic alcohols. Silver carbonate in aromatic solvent proved to be the best catalyst (Pale and Chuche 2000). Alkynols in which the two reacting parts of the molecule are relatively close together in space required only catalytic amounts of silver ions, while others cyclize readily with stoichiometric quantities. This heterocyclization reaction provides mild and convenient access to 2-methylene-oxolanes or oxanes. Scheme 11.12 shows that a dinuclear silver(I) compound efficiently catalyzes the intramolecular amidation of saturated C–H bonds of carbamates and sulfamates (Cui and He 2004). This highly regioselective, stereospecific reaction offers a practical method for the construction of cyclic nitrogen-containing organic molecules.

N-Bridgehead pyrroles are efficiently prepared from cyclic secondary vinylogous carbamates using a two-step sequence (Robinson et al. 2005b). This sequence involves *C*-propargylation followed by silver-catalyzed intramolecular hydroamination. Hydroamination is brought about using microwave irradiation and affords the desired *N*-bridgehead pyrroles rapidly and in good yield. Cyclic secondary vinylogous carbamates are prepared using a mild, economical procedure.

11.2.8 Silver-Promoted Cationic Cyclizations

The electrocyclic opening of cyclopropane derivatives containing internal nucleophilic groups provides a new route to vinyl lactones, tetrahydropyrans, and tetrahydrofurans (Danheiser et al. 1981). The silver salt initiates the opening of the ring

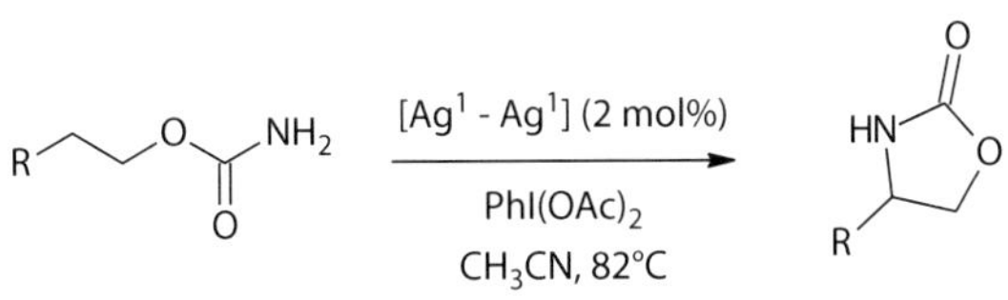

SCHEME 11.12

of *gem*-dibromocyclopropane, a process that leads to the formation of an incipient π-allyl cation, which is intramolecularly captured by the nucleophile to form the desired heterocycle. The method has been extended to the synthesis of pyrrolidines and applied to the total synthesis of (−)-γ-lycorane (Banwell et al. 2000). Scheme 11.13 involves a silver(I)-promoted electrocyclic ring-opening/π-allyl cation cyclization sequence to deliver the hexahydroindole, which participates in a Suzuki cross-coupling reaction with arylboronic acid to give the tetracyclic compound.

1,2-Dioxetanes (Kopecky et al. 1975; Jefford and Deheza 1999) and 1,2-dioxolanes (Bloodworth et al. 1986) can be prepared through silver salt-induced cyclizations of γ-bromoalkyl *t*-butyl peroxides via cyclic trialkylperoxionium intermediates.

2,5-Disubstituted tetrahydrofurans can be obtained with a high degree of stereocontrol by ring contraction of bromotetrahydropyrans induced by silver tetrafluoroborate in acetone (Ting and Bartlett 1984; Bartlett and Chapuis 1986). ω-Chloroalkanohydrazides cyclize when treated with $AgBF_4$ to yield N,N-disubstituted lactone hydrazones (Enders et al. 1985). The treatment of *S*-glycosidic silyl enol ethers with silver triflate can be used to prepare monocyclic *C*-glycosides (Craig et al. 1993, 1998) as single, *cis*-fused diastereomers for both the five- and six-membered templates (Scheme 11.14).

11.2.9 Silver-Induced Macrocyclizations

Novel thiaarenecyclynes, in which two thioether units and two benzene rings are alternately inserted into the single bonds of cyclooctatetrayne are synthesized. The cyclization is considered to proceed by a double zipper (concerted) reaction involving aromatization and successive protodemetalation. Sulfur analogs do not exhibit the same behavior (Kobayashi et al. 2003).

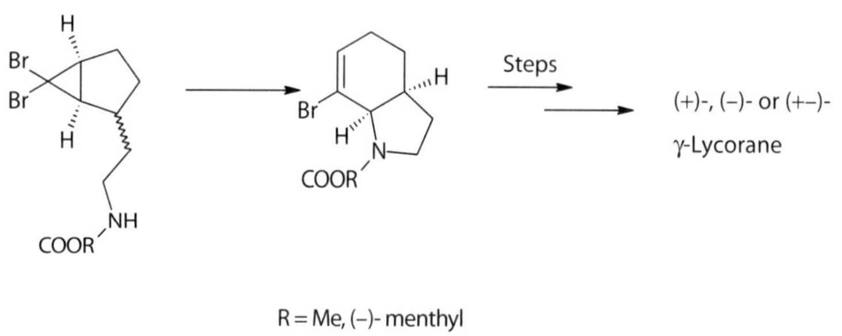

SCHEME 11.13

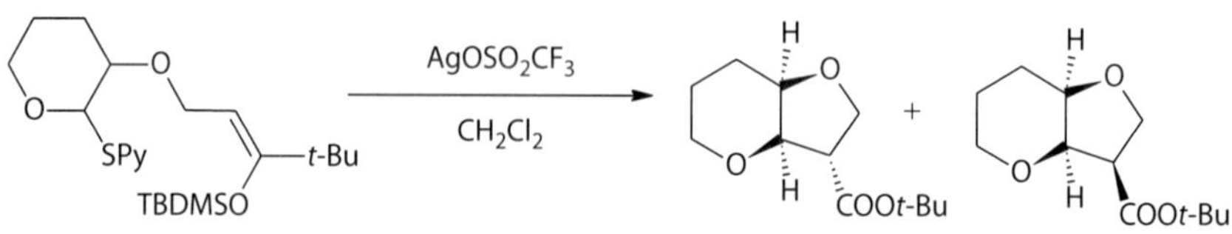

SCHEME 11.14

11.2.10 Silver-Catalyzed Aza-Diels–Alder Cycloadditions

Asymmetric aza-Diels–Alder reactions provide a useful route to optically active nitrogen-containing heterocyclic compounds such as piperidines, tetrahydroquinolines, and so on.

Although successful examples of diastereoselective approaches using chiral auxiliaries have been reported, few examples of enantioselective reactions are known (Kobayashi 2001). The role of Ag(I) has already been described in the azomethine and nitrilimine 1,3-dipolar cycloadditions discussed previously; in addition, other reactions such as aza-Diels–Alder may also be catalyzed by silver salts. Different chiral *P*,*P*-ligands, such as BINAP and Tol-BINAP, were tested. Four Lewis-acidic silver phosphane complexes partnered with $[1\text{-}closo\text{-}CB_{11}H_{12}]^-$ and $[1\text{-}closo\text{-}CB_{11}H_6Br_6]^-$ have been synthesized and studied by solution nuclear magnetic resonance (NMR) spectroscopy and solid-state X-ray diffraction techniques. In the complex $[Ag(PPh_3)(CB_{11}H_{12})]$, the silver is coordinated with the carborane by two stronger 3c–2e B–H–Ag bonds, one weaker B–H–Ag interaction, and a very weak $Ag\cdots C_{arene}$ contact in the solid state (Patmore et al. 2002). The method was extended to a three-component reaction where the unstable imines could be generated in situ. Only a trace amount of the desired cycloadduct was observed when the diene was added in one portion, but under slow addition, the yield improved dramatically (51%–90%). In some cases, yields were improved with the aid of the nonionic surfactant Triton X-100. Aza-Diels–Alder reactions of Danishefsky's diene with imines in water took place smoothly in the presence of a catalytic amount of silver triflate to afford dihydro-4-pyridones in high yields. The silver triflate-catalyzed three-component reactions starting from aldehydes, amines, and Danishefsky's diene were also performed efficiently. In the three-component reactions with benzaldehyde, the addition of a nonionic surfactant was found to be effective (Loncaric et al. 2003). A readily available isoleucine-based phosphine ligand is used to promote Ag-catalyzed Mannich reactions between silylketene acetals and various alkynyl imines (Scheme 11.15). Reactions can be effected in the presence of 5 mol% catalyst, without the need for rigorous exclusion of air, and with commercially available solvents (without purification) to afford the desired β-alkynyl-β-amino esters in 84%–94% and 61%–91% isolated yield (Josephsohn et al. 2005).

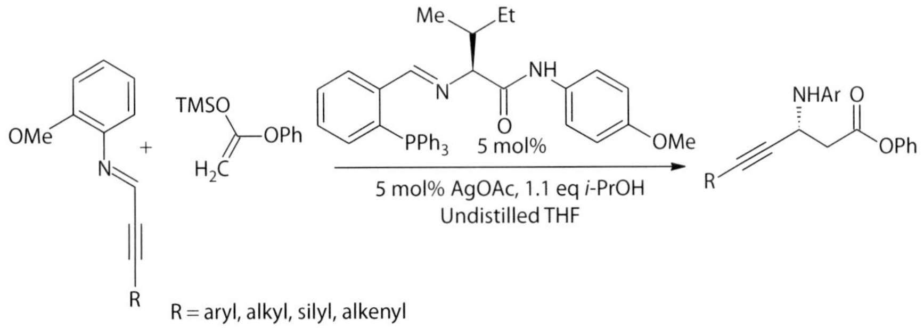

SCHEME 11.15

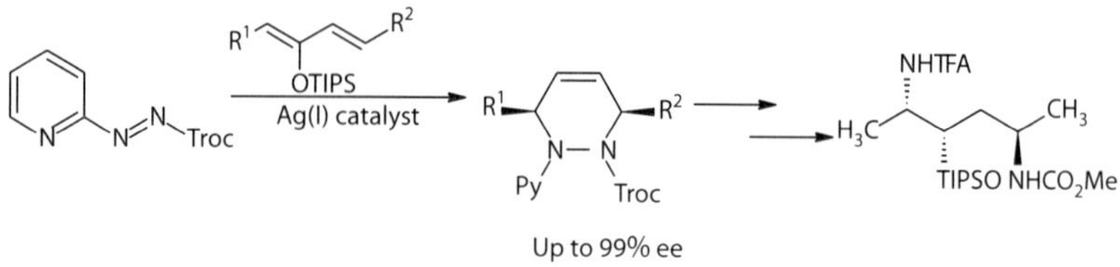

SCHEME 11.16

Kawasaki and Yamamoto (2006) describe studies in which an azo hetero-Diels–Alder adduct was furnished in high regio- and enantioselectivity using azopyridine as a reagent and silver as a catalyst (Scheme 11.16) The obtained hetero-Diels–Alder adduct was easily converted to the corresponding chiral 1,4-diamino alcohol.

11.2.11 Other Types of Cyclization

$Ag_3PW_{12}O_{40}$ is a novel and recyclable heteropoly acid (HPA) for the synthesis of 1,5-benzodiazepines under solvent-free conditions (Yadav et al. 2004a). *o*-Phenylenediamines undergo smooth condensation with ketones having hydrogens at the α-position on the surface of the HPA ($Ag_3PW_{12}O_{40}$) under extremely mild conditions to afford the corresponding 1,5-benzodiazepines in excellent yields with high selectivity. The catalyst can be recovered by simple filtration and can be reused in subsequent reactions. It is also used in synthesis of quinolines (Yadav et al. 2004b) (Friedländer annulation) or 3,4-dihydropyrimidinones (Yadav et al. 2004c). A Biginelli three-component condensation of an aldehyde, β-keto ester, and urea proceeds smoothly on the surface of the silver salt of the HPA ($Ag_3PW_{12}O_{40}$) in water to afford the corresponding 3,4-dihydropyrimidinones in high-to-quantitative yields under mild conditions. The heterogeneous solid acid provides ease of separation of the catalyst and isolation of the products. The recovered catalyst can be recycled in subsequent reactions with consistent activity. Compared to the classical Biginelli reaction conditions, this new method has the advantages of improved yields, reusability of the catalyst, an eco-friendly solvent, ease of isolation of products, and simplicity in the experimental procedure.

11.2.12 Silver as Cocatalyst with Pd, Au, Cu, Pt, Rh, and Ru

Silver salts are also known to be used as additives to other active transition metals. In this way, silver compounds generally react with transition-metal halides to generate catalytically more active cationic species; often they are critical for successful organic transformations. Many examples have been reported in different kinds of reactions. Many other syntheses of heterocycles in which silver salts are combined with other transition-metal complexes have been described (Schemes 11.17 through 11.21), mainly with Au (Reich et al. 2006; Nguyen et al. 2006; Suhre et al. 2005; Liu et al. 2005; Jung and Floreancig 2006; Bender and Widenhoefer 2006), Cu (Towers et al. 2003), Pt (Charruault et al. 2004), Rh (Evans et al. 2005),

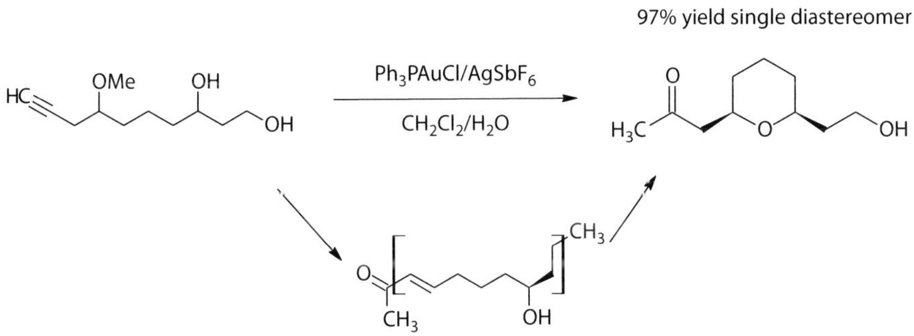

SCHEME 11.17

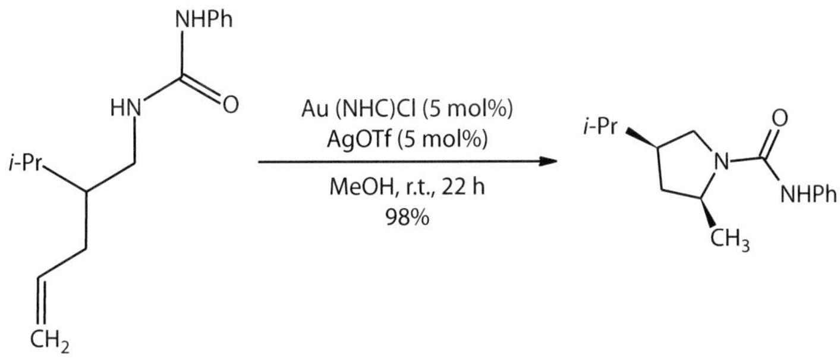

SCHEME 11.18

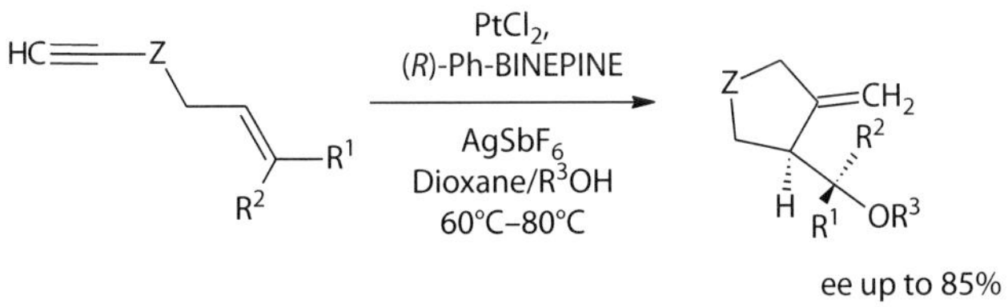

SCHEME 11.19

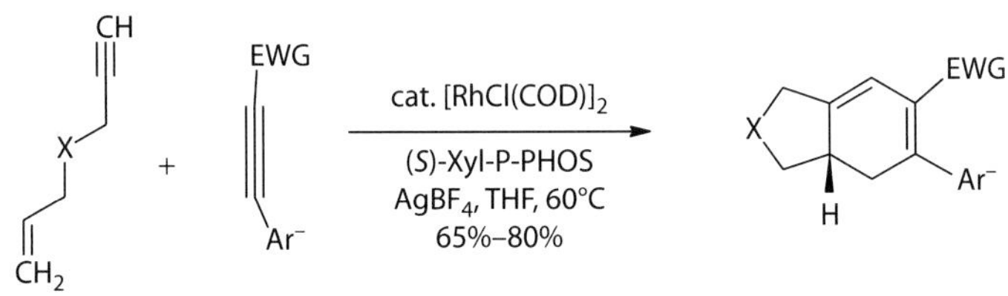

SCHEME 11.20

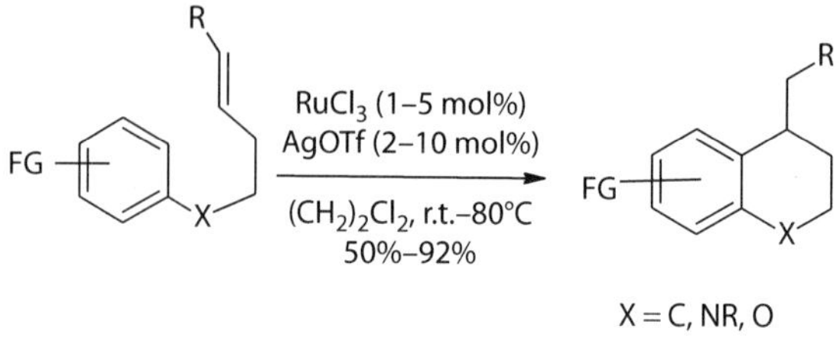

SCHEME 11.21

and Ru (Youn et al. 2004). The role of silver in these processes has scarcely been described. However, for palladium-catalyzed reactions, the influence of silver salts seems to be clearer.

The mixed-metal system $PdCl_2$ and $AgBF_4$ can initiate the cyclization of an indole ring onto an olefin to form six- and seven-membered rings. The role of the mixed-metal catalyst seems to involve enhancement of the electrophilicity of the palladium chloride in the presence of silver ions (Trost and Fortunak, 1982). In the preparation of a variety of tricyclic ring systems by palladium-catalyzed cyclizations of unsaturated arylhalides, the addition of silver salts dramatically reduces double-bond isomerizations of the cyclization products (Abelman et al. 1987). Moreover, depending upon how the HX is scavenged, either enantiomer of the Heck product can be formed with good selectivity using a single enantiomer of a chiral diphosphine ligand. A moderately strong Brønsted base must be present to obtain useful good results. Silver salts having weakly basic counterions (e.g., OTf^-, NO_3, BF_4^-) do not effectively promote Heck cyclization (Ashimori et al. 1998a, 1998b). Sato et al. (1994) have reported the catalytic asymmetric synthesis of the decalins from the prochiral alkenyl iodides. They proposed that the Heck reaction proceeds via a 16-electron Pd+ intermediate in the presence of Ag(I), but via a neutral palladium intermediate in the absence of Ag(I). In addition, the Ag(I) counteranion plays an important role in the reaction, as it largely influences the asymmetric induction. Counteranions that make tight ion pairs with Pd+ interfere with the ideal square planar geometry in the intermediate and give products of low enantiomeric excess (ee). Thus, the use of silver-exchanged zeolite leads to an "anion-free" square planar Pd+ cation intermediate, which affords high asymmetric induction levels. Yang and coworkers have found that cationic palladium complexes exhibit high reactivity toward coordination of alkenes or alkynes to bring about efficient carbonylation reactions. Such cationic complexes can be easily prepared in situ by reaction or silver salts of BF_4^-, ClO_4^-, and BAr_4^- with organopalladium halides in the presence or tertiary phosphine ligands or chelating diamine (diimine) ligands. When trying to convert without silver salts, no 3-aroylbenzo[*b*]furans were detected; only the corresponding phenol ester was obtained (Hu et al. 2002). The authors have proposed a complex as intermediate, generated from organopalladium iodide. This complex contains a cationic metal center with a Lewis acid character and has a stronger tendency to coordinate the unsaturated triple bond.

11.3 CONCLUSION

In the past several years, significant progress has been made in the exploration of silver-based heterocyclizations. Silver complexes have been shown to efficiently catalyze the intramolecular addition of oxygen and nitrogen nucleophiles to alkynes, allenes, and olefins to generate oxygen and nitrogen heterocycles. Moreover, numerous natural and unnatural products have been prepared by synthetic routes that have a silver-mediated 1,3-dipolar cycloaddition of an azomethine ylide or a nitrilimine as a crucial step for the construction of a nitrogen-containing, five-membered heterocycle.

In addition, silver-catalyzed asymmetric aza-Diels–Alder reactions provide a useful route to optically active nitrogen-heterocyclic compounds such as piperidines or pyridazines. Substituted dihydrobenzofurans can also be enantioselectively prepared through silver-promoted allylation of aldehydes. Other types of silver-mediated cyclizations can also be used in the synthesis of tetrahydrofurans, tetrahydropyrans, 1,2-dioxetanes, 1,2-dioxolanes, medium-sized lactones, dihydroisoquinolines, and so on. Silver salts can also be used as cocatalysts with other transition metals. Unique activity was observed for these silver-based systems in several cases. Consequently, the use of silver can enrich several available heterocyclization methods, and further developments in the application of chiral silver complexes will hopefully appear in the near future.

REFERENCES

Abelman, M. M., Oh, T. and Overman, L. E. 1987. Intramolecular alkene arylations for rapid assembly of polycyclic systems containing quaternary centers. A new synthesis of spirooxindoles and other fused and bridged ring systems. *J. Org. Chem.* 52: 4130–4133.

Ashimori, A., Bachand, B., Calter, M. A., Govek, S. P., Overman, L. E. and Poon, D. J. 1998a. Catalytic asymmetric synthesis of quaternary carbon centers. Exploratory studies of intramolecular Heck reactions of (*Z*)-α,β-unsaturated anilides and mechanistic investigations of asymmetric Heck reactions proceeding via neutral intermediates. *J. Am. Chem. Soc.* 120: 6488–6499.

Ashimori, A., Bachand, B., Overman, L. E. and Poon, D. J. 1998b. Catalytic asymmetric synthesis of quaternary carbon centers. Exploratory investigations of intramolecular Heck reactions of (*E*)-α,β-unsaturated 2-Haloanilides and analogues to form enantioenriched spirocyclic products. *J. Am. Chem. Soc.* 120: 6477–6487.

Banwell, M. G., Harvey, J. E. and Hockless, D. C. R. 2000. Electrocyclic ring-opening/π-allyl cation cyclization reaction sequences involving *gem*-dihalocyclopropanes as substrates: Application to syntheses of (±)-, (+)-, and (−)-γ-lycorane. *J. Org. Chem.* 65: 4241–4250.

Bartlett, P. A. and Chapuis, C. 1986. Synthesis of polyether-type tetrahydrofurans via hydroperoxide cyclization. *J. Org. Chem.* 51: 2799–2806.

Bender, C. F. and Widenhoefer, R. A. 2006. Room temperature hydroamination of *N*-alkenyl ureas catalyzed by a gold (I) *N*-heterocyclic carbene complex. *Org. Lett.* 8: 5303–5305.

Binder, J. T. and Kirsch, S. F. 2006. Synthesis of highly substituted pyrroles via a multimetal-catalyzed rearrangement-condensation-cyclization domino approach. *Org. Lett.* 8: 2151–2153.

Blanc, A., Tenbrink, K., Weibel, J.-M. and Pale, P. 2009. Silver(I)-catalyzed cascade: Direct access to furans from alkynyloxiranes. *J. Org. Chem.* 74: 4360–4363.

Bloodworth, A. J., Chan, K. H. and Cooksey, C. J. 1986. Oxymetalation. 20. Conversion of cyclopropanes into 1,2-dioxolanes via tert-butyl peroxymercuration, bromodemercuration, and silver salt induced cyclization. *J. Org. Chem.* 51: 2110–2115.

Bouzide, A. and Sauve, G. 2002. Silver(I) oxide mediated highly selective monotosylation of symmetrical diols. Application to the synthesis of polysubstituted cyclic ethers. *Org. Lett.* 4: 2329–2332.

Bryan, C. S. and Lautens, M. 2008. Silver-promoted domino Pd-catalyzed amination/direct arylation: Access to polycyclic heteroaromatics. *Org. Lett.* 10: 4633–4636.

Cao, H., Jiang, H., Mai, R., Zhu, S. and Qi, C. 2010. Silver-catalyzed one-pot cyclization reaction of electron- deficient alkynes and 2-Yn-1-ols: An efficient domino process to polysubstituted furans. *Adv. Synth. Catal.* 352: 143–152.

Charruault, L., Michelet, V., Taras, R., Gladiali, S. and Genet, J.-P. 2004. Functionalized carbo- and heterocycles via Pt-catalyzed asymmetric alkoxycyclization of 1,6-enynes. *Chem. Commun.* 850–851.

Craig, D., Payne, A. H. and Warner, P. 1998. Template-directed intramolecular *C*-glycosidation stereoselective synthesis of monocyclic *C*-glycosides. *Tetrahedron Lett.* 39: 8325–8328.

Craig, D., Pennington, M. W. and Warner, P. 1993. Stereoselective template-directed *C*-glycosidation. Synthesis of 5-membered oxygen heterocycles via cation-mediated intramolecular cyclization reactions. *Tetrahedron Lett.* 34: 8539–8542.

Cui, Y. and He, C. 2004. A silver-catalyzed intramolecular amidation of saturated C–H bonds. *Angew. Chem. Int. Ed.* 43: 4210–4212.

Dalla, V. and Pale, P. 1994. Silver-catalyzed heterocyclization: First total synthesis of the naturally occurring *cis* 2-hexadecyl-3-hydroxy-4-methylene butyrolactone. *Tetrahedron Lett.* 35: 3525–3528.

Dalla, V. and Pale, P. 1999. Silver-catalyzed cyclization of acetylenic alcohols and acids: A remarkable accelerating effect of a propargylic C–O bond. *New J. Chem.* 23: 803–805.

Danheiser, R. L., Morin, J. M., Yu, M. and Basak, A. 1981. Cationic cyclizations initiated by electrocyclic cleavage of cyclopropanes. Synthesis of lactones, tetrahydropyrans, and tetrahydrofurans. *Tetrahedron Lett.* 22: 4205–4208.

Dudnik, A., Chernyak, N. and Gevorgyan, V. 2010. Copper-, silver-, and gold catalyzed migratory cycloisomerization leading to heterocyclic five-membered rings. *Aldrichim. Acta.* 43: 37–46.

Enders, D., Brauer-Scheib, S. and Fey, P. 1985. A simple and efficient synthesis of lactone *N,N*-dialkylhydrazones and their isomeric *N*-(dialkylamino)-lactams. *Synthesis* 1985(4): 393–396.

Esseveldt, B. C. J., Vervoort, P. W. H., Delft, F. L. and Floris, P. J. T. 2005. Novel approach to 5-substituted proline derivatives using a silver-catalyzed cyclization as the key step. *J. Org. Chem.* 70: 1791–1795.

Evans, P. A., Lai, K. W. and Sawyer, J. R. 2005. Regio- and enantioselective *inter*molecular rhodium-catalyzed [2+2+2] carbocyclization reactions of 1,6-enynes with methyl arylpropiolates. *J. Am. Chem. Soc.* 127: 12466–12467.

Harrison, T. J., Kozak, J. A., Corbella-Pane, M. and Dake, G. R. 2006. Pyrrole synthesis catalyzed by AgOTf or cationic Au(I) complexes. *J. Org. Chem.* 71: 4525–4529.

Hu, Y., Zhang, Y., Yang, Z. and Fathi, R. 2002. Palladium-catalyzed carbonylative annulation of *o*-alkynylphenols: Syntheses of 2-substituted-3-aroyl-benzo[*b*]furans. *J. Org. Chem.* 67: 2365–2368.

Jefford, C. W. and Deheza, M. F. 1999. The efficient synthesis of 1,2-dioxetanes from indene and 1,2-dihydronaphthalene. *Heterocycles* 50: 1025–1031.

Ji, K.-G., Shu, X.-Z., Zhao, S.-C., Zhu, H.-T., Niu, Y.-N., Liu, X.-Y. and Liang, Y.-M. 2009. Novel carbon–carbon bond formation from propargylic alcohols and olefin toward five-membered heterocyclic rings catalyzed by $AgSbF_6$. *Org. Lett.* 11: 3206–3209.

Jimenez-Gonzalez, L., Garcia-Muñoz, S., Alvarez-Corral, M., Muñoz-Dorado, M. and Rodriguez-Garcia, I. 2006. Silver-catalyzed asymmetric synthesis of 2,3-dihydrobenzofurans: A new chiral synthesis of pterocarpans. *Chem. Eur. J.* 12: 8762–8769.

Josephsohn, N. S., Carswell, E. L., Snapper, M. L. and Hoveyda, A. H. 2005. Practical and highly enantioselective synthesis of β-alkynyl-β-amino esters through Ag-catalyzed asymmetric Mannich reactions of silylketene acetals and alkynyl imines. *Org. Lett.* 7: 2711–2713.

Jung, H. H. and Floreancig, P. E. 2006. Gold-catalyzed heterocycle synthesis using homopropargylic ethers as latent electrophiles. *Org. Lett.* 8: 1949–1951.

Kawasaki, M. and Yamamoto, H. 2006. Catalytic enantioselective hetero-Diels–Alder reactions of an Azo compound. *J. Am. Chem. Soc.* 128: 16482–16483.

Kim, S. and Lee, H. 2008. Cyclization of allenyne-1,6-diols catalyzed by gold and silver salts: An efficient selective synthesis of dihydrofuran and furan derivatives. *Adv. Synth. Catal.* 350: 547–551.

Kobayashi, S. 2001. In Kobayashi, S., Jorgensen, K. A. (eds.), Catalytic enantioselective aza-Diels-Alder reactions. *Cycloaddition Reactions in Organic Synthesis*, p. 187. Wiley-VCH: Weinheim, Germany.

Kobayashi, S., Wakumoto, S., Yamaguchi, Y., Wakamiya, T., Sugimoto, K., Matsubara, Y. and Yoshida, Z. 2003. Synthesis and properties of novel thiaarenecyclynes. *Tetrahedron Lett.* 44: 1807–1810.

Kopecky, K. R., Filby, J. E., Mumford, C., Lockwood, P. A. and Ding, J.-Y. 1975. Preparation and thermolysis of some 1,2-dioxetanes. *Canadian J. Chem.* 53: 6–8.

Liu, W., Jiang, H. and Huang, L. 2010. One-pot silver-catalyzed and PIDA-mediated sequential reactions: Synthesis of polysubstituted pyrroles directly from alkynoates and amines. *Org. Lett.* 12: 312–315.

Liu, Y., Song, F., Song, Z., Liu, M. and Yan, B. 2005. Gold-catalyzed cyclization of (*Z*)-2-En-4-yn-1-ols: Highly efficient synthesis of fully substituted dihydrofurans and furans. *Org. Lett.* 7: 5409–5412.

Loncaric, C., Manabe, K. and Kobayashi, S. 2003. AgOTf-catalyzed aza-Diels–Alder reactions of Danishefsky's diene with imines in water. *Adv. Synth. Catal.* 345: 475–477.

Lu, Y., Fu, X., Chen, H., Du, X., Jia, X. and Liu, Y. 2009. An efficient domino approach for the synthesis of multisubstituted pyrroles via gold/silver-catalyzed amination/cycloisomerization of (*Z*)-2-en-4-yn-1-ols. *Adv. Synth. Catal.* 351: 129–134.

Marshall, J. A. and Pinney, K. G. 1993. Stereoselective synthesis of 2,5-dihydrofurans by sequential SN2′ cleavage of alkynyloxiranes and silver(I)-catalyzed cyclization of the allenylcarbinol products. *J. Org. Chem.* 58: 7180–7184.

Masel, R. I. 1996. *Principles of Adsorption and Reaction on Solid Surfaces*, Wiley Series in Chemical Engineering. Wiley-Interscience: New York, 1996.

Nguyen, R.-V., Yao, X. and Li, C.-J. 2006. Highly efficient gold-catalyzed atom-economical annulation of phenols with dienes. *Org. Lett.* 8: 2397–2399.

Pale, P. and Chuche, J. 2000. Silver-catalyzed cyclization of acetylenic alcohols: Synthesis of functionalized 2-methylene-oxolanes. *Eur. J. Org. Chem.* 2000: 1019–1025.

Patmore, N. J., Hague, C., Cotgreave, J. H., Mahon, M. F., Frost, C. G. and Weller, A. S. 2002. Silver phosphanes partnered with carborane monoanions: Synthesis, structures and use as highly active Lewis acid catalysts in a hetero-Diels–Alder reaction. *Chem. Eur. J.* 8: 2088–2098.

Reich, N. W., Yang, C.-G., Shi, Z. and He, C. 2006. Gold (I)-catalyzed synthesis of dihydrobenzofurans from aryl allyl ethers. *Synlett* 8: 1278–1280.

Robinson, R. S., Dovey, M. C. and Gravestock, D. 2005a. Silver-catalysed hydroamination: Synthesis of functionalised pyrroles. *Eur. J. Org. Chem.* 505–511.

Robinson, R. S., Dovey, M. C. and Gravestock, D. 2005b. Silver-catalyzed hydroamination: Synthesis of *N*-bridgehead pyrroles, incorporating a protection-deprotection strategy for preparation of cyclic secondary vinylogous carbamates. *Eur. J. Org. Chem.* 2005: 505–511.

Rothenberg, G. 2008. *Catalysis: Concepts and Green Applications*. Wiley-VCH: Weinheim.

Sato, Y., Nukui, S., Sodeoka, M. and Shibasaki, M. 1994. Asymmetric Heck reaction of alkenyl iodides in the presence of silver salts. Catalytic asymmetric synthesis of decalin and functionalized indolizidine derivatives. *Tetrahedron* 50: 371–382.

Schwier, T., Sromek, A. W., Yap, D. M. L., Chernyak, D. and Gevorgyan, V. 2007. Mechanistically diverse copper-, silver-, and gold-catalyzed acyloxy and phosphatyloxy migrations: Efficient synthesis of heterocycles via cascade migration/cycloisomerization approach. *J. Am. Chem. Soc.* 129: 9868–9878.

Seregin, I., Schammel, A. and Gevorgyan, V. 2008. Multisubstituted N-fused heterocycles via transtion metal-catalyzed cycloisomerization protocols. *Tetrahedron* 64: 6876–6883.

Suhre, M. H., Reif, M. and Kirsch, S. F. 2005. Gold(I)-catalyzed synthesis of highly substituted furan. *Org. Lett.* 7: 3925–3927.

Ting, P. C. and Bartlett, P. A. 1984. Stereocontrolled synthesis of trans-2,5-disubstituted tetrahydrofurans. *J. Am. Chem. Soc.* 106: 2668–2671.

Towers, M. D. K. N., Woodgate, P. D. and Brimble, M. A. 2003. Addition of 2-[(trimethylsilyloxy)]furan to 2-acetyl-1,4-benzoquinone using chiral non-racemic copper(II)-pybox catalysts. *Arkivoc* (i): 43–55.

Trost, B. M. and Fortunak, J. M. D. 1982. Cyclizations initiated by a palladium (2+)-silver(1+) mixed-metal system. *Organometallics* 1: 7–13.

Watanabe, T., Oishi, S., Fujii, N. and Ohno, H. 2007. Gold-catalyzed hydroarylation of allenes: A highly regioselective carbon–carbon bond formation producing six-membered rings. *Org. Lett.* 9: 4821–4824.

Westling, M., Smith, R. and Livinghouse, T. 1986. A convergent approach to heterocycle synthesis via silver ion mediated α-ketoimidoyl halide-arene cyclizations. An application to the synthesis of the erythrinane skeleton. *J. Org. Chem.* 51: 1159–1165.

Wilckens, K., Uhlemann, M. and Czekelius, C. 2010. Gold-catalyzed *endo*-cyclizations of 1,4-diynes to seven-membered ring heterocycles. *Chem. Eur. J.* 15: 13323–13326.

Yadav, J. S., Reddy, B. V. S., Kumar, S. P., Nagaiah, K. and Lingaiah, P. S. 2004a. $Ag_3PW_{12}O_{40}$: A novel and recyclable heteropoly acid for the synthesis of 1,5-benzodiazepines under solvent-free conditions. *Synthesis* 6: 901–904.

Yadav, J. S., Reddy, B. V. S., Sreedhar, P., Rao, R. S. and Nagaiah, K. 2004b. Silver phosphotungstate: A novel and recyclable heteropoly acid for Friedländer quinoline synthesis. *Synthesis* 14: 2381–2385.

Yadav, J. S., Reddy, B. V. S., Sreedhar, P., Reddy, J. S. S., Nagaiah, K., Lingaiah, N. and Saiprasad, P. S. 2004c. Green protocol for the Biginelli three-component reaction: $Ag_3PW_{12}O_{40}$as a novel, water-tolerant heteropolyacid for the synthesis of 3,4-dihydropyrimidinones. *Eur. J. Org. Chem.* 2004: 552–557.

Yanagisawa, A., Kageyama, H., Nakatsuka, Y., Asakawa, K., Matsumoto, Y. and Yamamoto, H. 1999. Enantioselective addition of allylic trimethoxysilanes to aldehydes catalyzed by *p*-Tol-BINAP small middle dot AgF. *Angew. Chem. Int. Ed.* 38: 3701–3703.

Yao, X. and Li, C.-J. 2004. Highly efficient addition of activated methylene compounds to alkenes catalyzed by gold and silver. *J. Am. Chem. Soc.* 126: 6884–6885.

Youn, S. W., Pastine, S. J. and Sames, D. 2004. Ru(III)-catalyzed cyclization of arene-alkene substrates via intramolecular electrophilic hydroarylation. *Org. Lett.* 6: 581–584.

Zhang, D. H., Zhang, Z. and Shi, M. 2012. Transition metal-catalyzed carbocyclization of nitrogen and oxygen-tethered 1,n-enynes and diynes: Synthesis of five or six-membered heterocyclic compounds. *Chem. Commun. (Camb.)* 48(83): 10271–10279.

12 Mesoporous Materials from Novel Silica Source as Heterogeneous Catalyst

Sujitra Wongkasemjit and Rujirat Longloilert

CONTENTS

12.1 INTRODUCTION

Since 1992, when mesoporous molecular sieves in the M41S family were discovered by Mobil's research group (Kresge et al. 1992; Beck et al. 1992), these materials have attracted significant attention due to their large surface area, ordered pore-structure array, and narrow pore-size distribution (Shao et al. 2005). There are three categories of M41S divided by different arrays, namely, hexagonal MCM-41 possessing honeycomb arrays of nonintersecting uniformly sized channels, cubic MCM-48 with *Ia3d* symmetry having a three-dimensional bicontinuous channel system, and unstable lamellar MCM-50, which collapses on template removal (Sayari 1996; Vinu et al. 2003; Hartmann 2005, respectively).

The SBA family has also gained much attention from many scientists owing to its thicker walls, providing a better hydrothermal stability and larger pore size than M41S. It is well known that SBA-1 material is analogous to the cubic assemblage of globular micelles in amphophilic surfactant solutions. Its structure has been described as a cage type with open windows (Tanglumlert et al. 2007). In the same family, SBA-15 has a highly ordered hexagonal topology and larger pore size than MCM-41 (Zhao et al. 1998). The mesoporous silicas are typically achieved by using surfactants or amphiphilic triblock copolymers as a structure-directing agent in basic or acidic media.

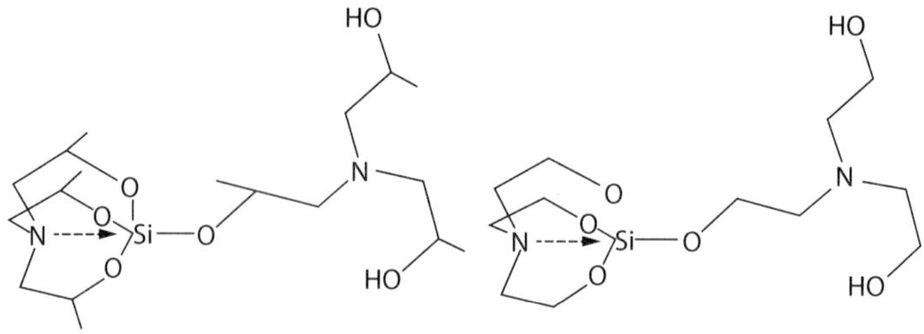

FIGURE 12.1 Structures of home-made silatranes.

In general, the sources of silica for both families are sodium silicate, tetraethyl orthosilicate (TEOS), and tetramethyl orthosilicate (TMOS) (Giraldo et al. 2007). In this chapter, we introduce our home-made organometallic silica sources, known as silatranes (Figure 12.1).

These novel precursors are not only easily prepared from silicon dioxide and trialkanolamine in ethylene glycol solvent, but are also moisture stable. With this remarkable latter property, silatrane precursors have proved to be good candidates to achieve the synthesis of not only mesoporous (Thanabodeekij et al. 2006; Tanglumlert et al. 2007; Samran et al. 2011) (Figure 12.2), but also microporous materials (Sathupunya et al. 2002, 2003; Phiriyawirut et al. 2003; Phonthammachai et al. 2003) (Figure 12.3).

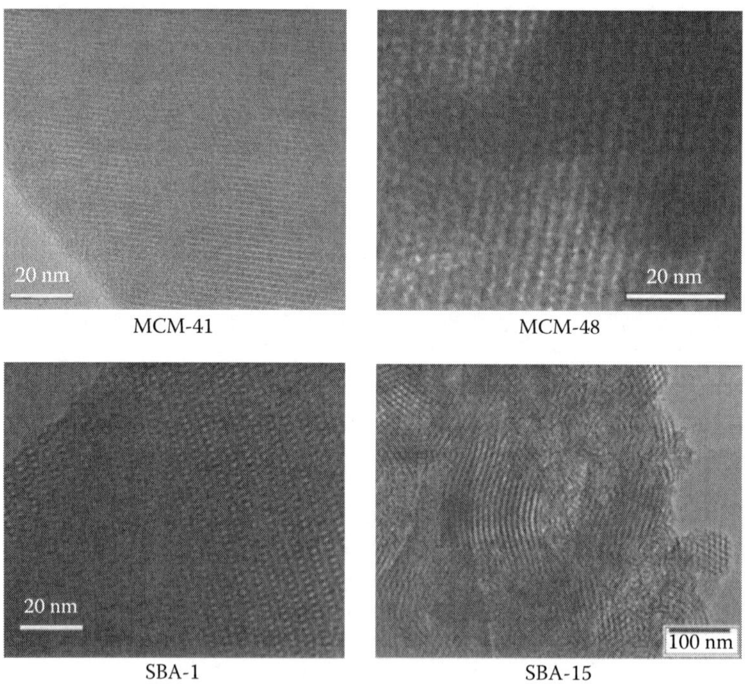

FIGURE 12.2 Examples of mesoporous materials from silatranes.

FIGURE 12.3 Examples of microporous materials from silatranes.

In addition, the applications of these materials, including heterogeneous catalysts, are mentioned with reference to their notable properties.

12.2 SYNTHESIS OF MESOPOROUS SILICA

Typically, either the alkaline route or the acid route is used for synthesizing mesoporous silica with the aid of amphiphiles as templates (Beck et al. 1992; Huo et al. 1994). In the acid route, both the hydrolysis and the condensation rates are promoted mostly at the end of the silica polymers to form linear silicate ions, providing a fuzzier and softer network. In contrast, the alkaline route favors both hydrolysis and condensation, leading to a highly condensed and compact structure (Wen and Wilkes, 1996).

In 1993, Stucky and coworkers synthesized MCM-41 at different temperatures for various reaction times, and studied the structures and compositions of the obtained mesoporous materials. They explained the mechanism of the ordered mesoporous silica formation and proposed the following model: The oligomeric silica polyanions (denoted as I^-) were bound to cationic head groups of surfactants (denoted as S^+); then, the surfactant–silicate complexes (S^+I^-) were formed due to the charge density matching at the interface between silicate anions and surfactant cations. In the early stages of the process, lamellar structures were formed first, due to thermodynamic favor. After the polymerization of the silicate species proceeded, the charge density matching of anion silicates decreased. In order to match the reduced charge density and keep the silicate-to-surfactant ratio constant, the head group area of the surfactant assembly expanded, resulting in the formation of hexagonal mesophases.

However, in 1994, Huo et al. extended the model derived by Stucky and coworkers so that it could be used to explain the formation of various mesostructured nanomaterials. The model could be applied to S^-I^+ by varying $[H^+]$ in solutions and used to explain not only the S^-I^+ system, but also both negative charges (S^-, I^-) and positive charges (S^+, I^+) of surfactant and inorganic moieties. The formation of mesophases and their structures in this case was assisted by the counterions (i.e., X^+ = alkaline metal ions and X^- = halide ions) to form a sandwich-like triple layer ($S^-X^+I^-$ or $S^+X^-I^+$).

12.3 NOVEL SILICA SOURCE: "SILATRANE" AND ITS APPLICATION

Silatranes are a class of chelate compounds of pentacoordinated silicon. They have been known and investigated since 1963, with initial studies by Voronkov and coworkers. Because of both their electron density distribution and their spatial structure, the specific biological activity of silatranes is of wide interest (Voronkov et al. 2012). In addition to their physicochemical properties, the structural properties of silatranes have also been widely considered (Voronkov et al. 1989; Cerveau et al. 1990; Nasim et al. 1991; Hencsei 1991). Typically silatranes are synthesized via the reaction of halo-, hydro-, and alkoxysilanes with triethanolamine or its derivatives (Voronkov et al. 1982). In 1991, Laine and coworkers developed a method for synthesizing organosilicon compounds directly from an economical starting material, namely silica and ethylene glycol (EG), in only one step in the presence of a tetraethylenediamine (TETA) catalyst (Laine et al. 1991; Bickmore and Laine 1996). Their work resulted in the so-called oxide one-pot synthesis (OOPS) process. Moreover, Piboonchaisit et al. (1999) successfully synthesized silatrane complexes from silica and triisopropanolamine (TIS) via the OOPS process with and without TETA (Scheme 12.1), and the products obtained exhibited similar properties. However, the reaction time is two times faster in the presence of TETA, meaning that TETA could be used as an accelerator for this reaction.

Owing to the hydrolytic stability in air for periods up to several weeks, silatranes are good candidates for use as precursors in ceramic processing via the sol-gel technique, as studied by Charoenpinijkarn et al. (2001), who mechanistically investigated the sol-gel processing of silatranes. They found that the silatrane complexes are potential candidates as ceramic precursors through the hydrolytic sol-gel processing method. Not only the silatranes, but also other moisture-stable metal alkoxides (e.g., alumatrane, cerium glycolate, titanium glycolate, etc.) were synthesized by Wongkasemjit's research group (Waldner et al. 1996; Piboonchaisit et al. 1999; Charoenpinijkarn et al. 2001; Jitchum et al. 2001; Opornsawad et al. 2001; Phonthammachai et al. 2003; Ksapabutr et al. 2004) (Figure 12.4). These materials were prepared directly from inexpensive metal oxides using ethylene glycol solvent via the OOPS process and provided highly pure metal alkoxides.

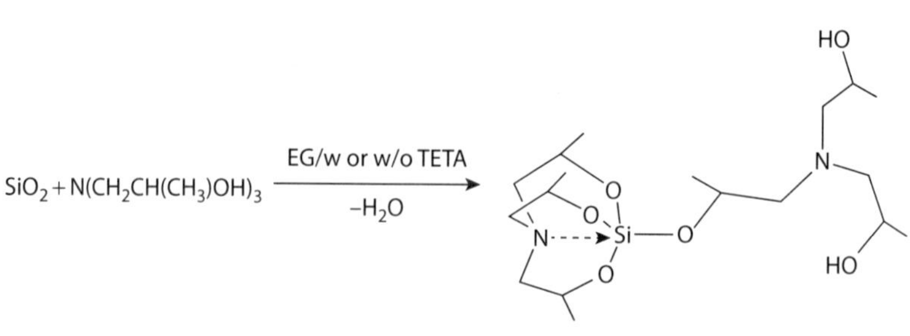

SCHEME 12.1 Synthesis of silatranes.

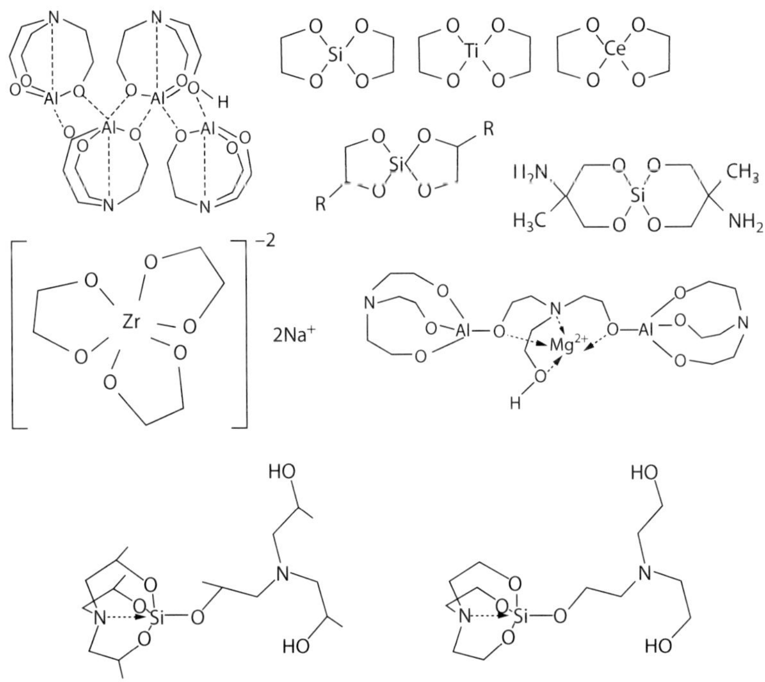

FIGURE 12.4 Metal alkoxides synthesized by Wongkasemjit's research group.

In 2006, the extremely large surface area MCM-41 was successfully synthesized directly from a silatrane precursor by Thanabodeekij and coworkers. They investigated the optimum conditions by studying the effects of surfactant concentration, ion concentration, aging time, and temperature of the system. The results showed that the MCM-41 structure could be formed in a narrow range of ion concentration at a wide range of temperatures. However, the larger pore size was obtained at a higher temperature. In this case, 100°C provided a high-quality MCM-41. The Brunauer–Emmett–Teller (BET) surface area was affected by the surfactant concentration. Interestingly, a surface area as high as 2400 m^2/g was achieved with a surfactant-to-silatrane ratio of 0.6 at 60°C, while 1.72 cm^3/g of pore volume was obtained at a similar surfactant ratio and 100°C (Thanabodeekij et al. 2006).

In 2007, Tanglumlert et al. synthesized cubic SBA-1 at room temperature using a cationic surfactant. This work studied the influence of acidity, alkyl chain length of surfactant, and synthesis temperature on the SBA-1 formation, and the results showed that the shape of SBA-1 crystals relied on the alkyl chain length of the surfactant. Three-dimensionally ordered mesopores were obtained at a high surfactant concentration and elevated synthesis temperature (50°C). The morphology of the obtained materials indicated an octahedron, consistent with six square {100} and 12 hexagonal {110} planes with a cubic symmetry. The surface area was as high as 1000–1500 m^2/g with an adsorption volume of 0.6–1.0 cm^3/g.

In 2011, Samran and coworkers successfully synthesized a well-ordered and stable dimensional mesoporous SBA-15 silica at room temperature using a nonionic triblock copolymer ($EO_{20}PO_{70}EO_{20}$) as a structure-directing agent. In this work, SBA-15 was prepared using two different routes. One experiment was conducted at room temperature while the other was via the conventional microwave-assisted hydrothermal method. The products of the two methods were found to be comparable and provided large surface areas (486–613 m^2/g), pore diameters (45–67 Å), and channel volumes (0.6–0.8 cm^3/g). Additionally, the morphology of all synthesized materials had the form of packed bundle-like silica tubes with approximate *p6mm* symmetry. The first SBA-15 synthesis method provides the advantage of using room temperatures for an economical, energy-saving process in the large-scale production of thermally stable SBA-15.

Recently, MCM-48, with *Ia3d* symmetry and a three-dimensional pore structure, was also synthesized using silatrane as a silica precursor (Longloilert et al. 2011). This research focused on the effects of synthesis parameters, namely crystallization temperature and time, cetyl trimethylammonium bromide (CTAB) surfactant concentration, amount of NaOH, and silica source. The results revealed that all synthesis parameters affected the synthesis of MCM-48. Various techniques used for characterizing MCM-48 showed a long-range ordered structure with a truncated octahedral shape and a surface area as high as 1300 m^2/g with a pore size of 2.86 nm.

Mechanistically, the reaction generated triethanolamine (TEA) molecules, which were the by-product from the hydrolysis of silatrane. TEA can also act as a structure-directing agent for the system, resulting in a lower requirement of CTAB concentration. Moreover, TEA could also improve the surfactant-packing parameter that encourages MCM-48 formation.

12.4 SYNTHESIS OF METAL-LOADED MESOPOROUS SILICA

Both SBA and M41S types of mesoporous silica possess large surface area, high thermal and hydrothermal stability, and a possibly modifiable pore size. Due to their advantageous properties, they have extensively demonstrated promising results as catalyst supports to be used in industrial processes, as adsorbents and as host structures for nanometer-size guest compounds (Barton et al. 1999). Several metal ions, such as titanium, chromium, molybdenum, and iron, have been successfully introduced into the mesoporous silica framework (Zhang et al. 2002; Shao et al. 2008; Dai et al. 2001; Vinu et al. 2007). Wongkasemjit's research group have expanded their work by incorporating heteroatoms into the silica framework (Figure 12.5) and have

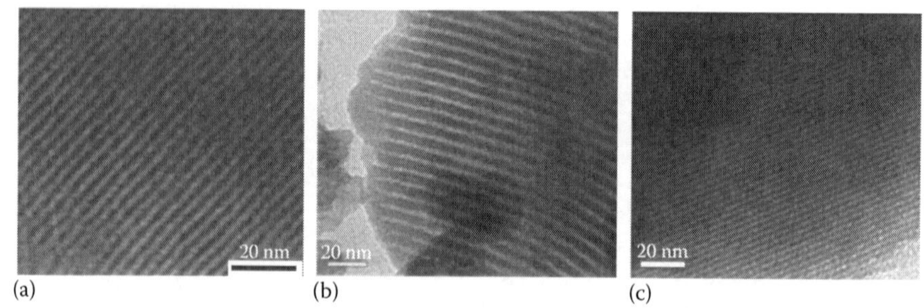

FIGURE 12.5 Titanium-loaded (a) MCM-41, (b) SBA15, and (c) MCM-28.

also studied their activity toward the oxidation of organic molecules (Thanabodeekij et al. 2007; Thitsatarn et al. 2008; Tanglumlert et al. 2008; Wongkasemjit et al. 2009; Longloilert et al. 2012; Maneesuwan et al. 2013).

Iron-containing mesoporous materials have been widely studied due to the unique catalytic performance of selective reduction, hydrocarbon oxidation, and acylation and alkylation reactions (Vinu et al. 2007). Thus, Tanglumlert et al. (2008) were interested in the room temperature synthesis of Fe-SBA-1 using $FeCl_3$ via the sol-gel process. The results illustrated that up to 6 wt% Fe could be contained in the SBA-1 framework without destroying the mesopore order. Nevertheless, extraframework FeO_6 clusters were also found, as suggested by electron spin resonance (ESR) spectroscopy. The BET surface area was 1062 m^2/g, with a pore diameter around 2.1 nm.

Other colleagues (Thitsartarn et al. 2008) in Wongkasemjit's research group synthesized Fe-MCM-41 and investigated the synthesis parameters to determine the optimal conditions. Many parameters were observed, including temperature, aging time, calcination rate, and amount of Fe loaded. They also prepared the materials by both the sol-gel and impregnation methods in order to compare their catalytic activity on the epoxidation of styrene as a catalyst. The results revealed that the optimal conditions to obtain hexagonal Fe-MCM-41 were at 60°C reaction temperature for 7 h and 550°C calcination temperature with 1°C/min. The maximum level of Fe loading while maintaining the MCM-41 structure was 2.5%. However, at higher 2.1% of Fe, the iron species were found within both the framework and extraframework. The in-framework iron content was active for the conversion of styrene with H_2O_2. Moreover, the catalyst prepared from the sol-gel technique performed better than the impregnated catalyst in the epoxidation of styrene. With Fe loading of 1%, levels of 65% selectivity of styrene oxide and 22% styrene conversion were achieved via the sol-gel process.

Generally, a catalyst used at high temperature or in boiling water would result in the loss of hydrothermal stability. This problem can be overcome by loading two different heteroatoms in support, known as bimetallic catalysts. These catalysts usually improve stability and also provide catalytic activity and selectivity. Thus, in 2013, Maneesuwan and coworkers prepared Fe-Ce-MCM-48 from silatrane via the sol-gel technique using cerium glycolate and $FeCl_3$ as metal precursors. The amounts of the two metals loaded on the MCM-48 were varied and characterized, with the results showing that Fe-Ce-MCM-48 was successfully synthesized and a high dispersion of Fe and Ce in the silica framework was obtained with ratios of 0.01 Fe/Si and 0.01–0.07 Ce/Si. All samples possessed a large surface area (approximately 1200 m^2/g) and a narrow pore-size distribution. This material was proposed as a good catalyst for phenol hydroxylation and to improve hydrothermal properties. A similar study by Longloilert et al. (2012) incorporated only Ce or Cr in the mesoporous MCM-48 synthesized from silatrane and studied the MCM-48's properties as influenced by the introduction of Ce as well as Cr. The results showed that the materials still retained the structure of MCM-48, with a long-range ordered structure and large surface area (up to 1500 m^2/g), even loading metals (Ce or Cr) to the structure. The hydrothermal stability test of Ce- or Cr-MCM-48 performed better than that of pure MCM-48, the structure of which was completely destroyed after the test. Interestingly, according to the scanning electron microscope (SEM) results, the MCM-48 structure morphologically changed when loaded with different metal species. Based on the diffuse reflectance ultraviolet-visible (DRUV) spectroscopy results, the metal extraframework was detected at high metal content loadings.

The incorporation of a large transitional metal atom (i.e., Mo) in the support is significant for the catalysis processes in the petroleum industry, namely the hydrodesulfurization and hydrodenitrogenation processes (Reddy et al. 1998). A high catalytic performance of Mo catalysts was found in hydrogenation (Halachev et al. 1998), oxidation (Parvulescu and Su 2001), and photocatalytic reactions. Hence, Thanabodeekij et al. (2007) were interested in the production of Mo-MCM-41 creating a high dispersion of Mo onto MCM-41 support through incipient wetness impregnation using silatrane and molybdenum glycolate precursors. The structure of hexagonal arrays was maintained even when 10% mol of Mo (or 0.265 g MoO_3/g SiO_2) was loaded. The surface area reached 1600 m^2/g with 3 nm pore size. The photocatalytic activity of the synthesized Mo-MCM-41 catalyst was tested with regard to peroxidative bromination, and the results showed that Mo loading onto uncalcined MCM-41 exhibited a high activity than Mo loading onto calcined MCM-41. Additionally, in 2009, Wongkasemjit et al. synthesized Mo-SBA-1 via the sol-gel process at room temperature while Che et al. (2001) did so at low temperature (0°C). The materials obtained through this work maintained a well-ordered mesostructure and a surface area >1000 m^2/g. The maximum loading of tetrahedral-coordinated Mo was around 5 mol% without any extraframework. This material was studied for its activity on the epoxidation reaction of styrene monomers. The optimum conditions for this reaction were found to be a reaction temperature of 70°C for 3 h reaction time, using 0.1 g of catalyst containing 7.2 mol% Mo content. The maximum level of styrene conversion reached 60%.

As is well known, titanium-substituted porous materials are mostly used in the catalysis field, especially in selective oxidation reactions using hydrogen peroxide as an oxidant. Due to their high performance, they can be found in many studies (Rajakovic et al. 2003; Anand et al. 2006; Thanabodeekij et al. 2005), including those of Wongkasemjit's group. Although Ti-substituted microporous materials, such as TS-1 and TS-β, are also of interest, the limitation of using large reactant molecules becomes a major drawback due to the materials' small pores. This limitation could be solved by the incorporation of titanium into the mesoporous materials, such as the M41S or SBA families.

Thanabodeekij et al. prepared Ti-MCM-41 using silatrane and titanium glycolate precursors via the sol-gel method. The obtained materials had a very large surface area (up to 2300 m^2/g) even with titanium loading of 1%–5%. The MCM-41 structure was maintained as the Ti content increased. They also tested these catalysts through peroxidative bromination, and the activity was found to be good. Then, in 2008, Tanglumlert et al. synthesized Ti-SBA-1 at room temperature using the same precursors. The results demonstrated that Ti was accommodated on SBA-1 support up to 10 wt%. The surface area was around 1000 m^2/g with an average pore diameter of 2 nm.

Apart from the oxidation of organic molecules, the catalytic pyrolysis of waste tires to light olefins using mesoporous material containing metals was also performed. Generally, the production of light olefins has been derived mostly from steam crackers and refinery fluid catalytic cracking units. Moreover, their demand has been continuously increasing, and the waste tire material has caused environmental problems. Therefore, the pyrolysis of waste tires to obtain light olefins, such as ethylene and propylene, has been viewed as a promising solution to both issues. Dũng et al. (2009) prepared Ru/MCM-41 via conventional wetness impregnation to use as a catalyst in waste tire pyrolysis. They found that a considerably higher yield of light olefins (four times higher than noncatalytic pyrolysis) was achieved over 2% wt Ru/MCM-41. Furthermore, this catalyst also produced the lightest oil with the highest concentration of single ring aromatics and lowest content of polycyclic aromatics. Two years later, Witpathomwong et al. (2011) investigated the activity of a 0.7% Ru/MCM-48 catalyst on the same reaction. They found that this catalyst produced twice as many light olefins as the noncatalytic pyrolysis. Additionally, it could reduce the poly- and polar-aromatic compounds, as well as the sulfur content in the derived oil.

12.5 APPLICATIONS

It is generally agreed that the most noteworthy properties of mesoporous materials are their large surface area, adjustable pore size, high pore volume, well-ordered structure with uniform pores, and hydrophilic character. These advantages are suitable for the adsorption and transformation of large organic molecules (Corma et al. 1994; Reddy et al. 1994), especially in such biological applications as biosensors and biocatalysts (Hartmann 2005). Furthermore, many attempts have been made to use mesoporous materials as drug delivery agents (Vallet-Regi et al. 2001). Moreover, further uses of the bioactivity behavior of mesoporous silica (i.e., SBA-15, MCM-48, and MCM-41) have also been proposed (Vallet-Regí et al. 2006).

12.6 CONCLUSION

Many types of mesoporous silicas have been synthesized from the moisture-stable silica source known as silatrane. These materials possess extraordinary properties, making them valuable in a wide range of applications. Until now, their potential for biological applications has attracted much attention. However, the activities described in this chapter have focused mainly on the organic reaction. Due to their unusual and useful characteristics, it is recommended that these materials be studied further with regard to their biological applications.

REFERENCES

Anand, R., Hamdy, M. S., Gkourgkoulas, P., Maschmeyer, T., Jansen, J. C., and Hanefeld, U. 2006. Liquid phase oxidation of cyclohexane over transition metal incorporated amorphous 3D-mesoporous silicates M-TUD-1 (M = Ti, Fe, Co and Cr). *Catal. Today.* 117(1–3):279–283.

Barton, T. J., Bull, L. M., Klemperer, W. G., Loy, D. A., McEnaney, B., Misono, M., Monson, P. A., et al. 1999. Tailored porous materials. *Chem. Mater.* 11(10):2633–2656.

Beck, J. S., Vartuli, J. C., Roth, W. J., Leonowicz, M. E., Kresge, C. T., Schmitt, K. D., Chu, C. T. W., et al. 1992. A new family of mesoporous molecular sieves prepared with liquid crystal templates. *J. Am. Chem. Soc.* 114:10834–10843.

Bickmore, C. and Laine, R. 1996. Synthesis of oxynitride powders via fluidized-bed ammonolysis, Part I: Large, porous, silica particles. *J. Am. Ceram. Soc.* 79:2865–2877.

Cerveau, G., Chuit, C., Colomer, E., Corriu, R. J. P., and Reye, C. 1990. Ferrocenyl compounds containing two hypervalent silicon species. *Organometallics.* 9:2415.

Charoenpinijkarn, W., Suwankruhasn, M., Kesapabutr, B., Wongkasemjit, S., and Jamieson, A. M. 2001. Sol–gel processing of silatranes. *Eur. Polym. J.* 37:1441–1448.

Che, S., Sakamoto, Y., Yoshitake, H., Terasaki, O., and Tatsumi, T. 2001. Synthesis and characterization of Mo-SBA-1 cubic mesoporous molecular sieves. *J. Phys. Chem. B.* 105(43):10565–10572.

Corma, A., Navarro, M. T., and Pariente, J. P. 1994. Synthesis of an ultra large pore titanium silicate isomorphous to MCM-41 and its application as a catalyst for selective oxidation of hydrocarbons. *J. Chem. Soc. Chem. Commun.* (2):147–148.

Dai, L.-X., Teng, Y.-H., Tabata, K., Suzuki, E., and Tatsumi, T. 2001. Catalytic application of Mo-incorporated SBA-1 mesoporous molecular sieves to partial oxidation of methane. *Micropor. Mesopor. Mater.* 44–45:573–580.

Dũng, N. A., Klaewkla, R., Wongkasemjit, S., and Jitkarnka, S. 2009. Light olefins and light oil production from catalytic pyrolysis of waste tire. *J. Anal. Appl. Pyrol.* 86(2):281–286.

Giraldo, L. F., López, B. L., Pérez, L., Urrego, S., Sierra, L., and Mesa, M. 2007. Mesoporous silica applications. *Macromol. Sym.* 258(1):129–141.

Halachev, T., Nava, R., and Dimitrov, L. 1998. Catalytic activity of (P)NiMo/Ti-HMS and (P)NiW/Ti-HMS catalysts in the hydrogenation of naphthalene. *Appl. Catal. A Gen.* 169(1):111–117.

Hartmann, M. 2005. Ordered mesoporous materials for bioadsorption and biocatalysis. *Chem. Mater.* 17(18):4577–4593.

Hencsei, P. 1991. Mass-spectrometric study of ring substituted silatranes. *Struct. Chem.* 2:21.

Huo, Q., Margolese, D. I., Ciesla, U., Feng, P., Gier, T. E., Sieger, P., Leon, R., Petroff, P. M., Schüth, F., and Stucky, G. D. 1994. Generalized synthesis of periodic surfactant/inorganic composite materials. *Nature.* 368:317–321.

Jitchum, V., Sun, C., Wongkasemjit, S., and Ishida, H. 2001. Synthesis of spirosilicates directly from silica and ethylene glycol/ethylene glycol derivatives. *Tetrahedron.* 57(18):3997–4003.

Kresge, C. T., Leonowicz, M. E., Roth, W. J., Vartuli, J. C., and Beck, J. S. 1992. Ordered mesoporous molecular sieves synthesized by a liquid-crystal template mechanism. *Nature.* 359:710–712.

Ksapabutr, B., Gulari, E., and Wongkasemjit, S. 2004. One-Pot synthesis and characterization of novel sodium tris(glycozirconate) and cerium glycolate precursors and their pyrolysis. *Mater. Chem. Phys.* 83(1):34–42.

Laine, R., Blohowiak, K., Robinson, T., Hoppe, M., Nardi, P., Kampf, J., and Uhm, J. 1991. Synthesis of pentacoordinate silicon complexes from SiO_2. *Nature.* 353:642–644.

Longloilert, R., Chaisuwan, T., Luengnaruemitchai, A., and Wongkasemjit, S. 2011. Synthesis of MCM-48 from silatrane via sol–gel process. *J. Sol-Gel Sci. Technol.* 58(2):427–435.

Longloilert, R., Chaisuwan, T., Luengnaruemitchai, A., and Wongkasemjit, S. 2012. Synthesis and characterization of M-MCM-48 (M = Cr, Ce) from silatrane via sol–gel process. *J. Sol-Gel Sci. Technol.* 61(1):133–143.

Maneesuwan, H., Longloilert, R., Chaisuwan, T., and Wongkasemjit, S. 2013. Synthesis and characterization of Fe-Ce-MCM-48 from silatrane precursor via sol–gel process. *Mater. Lett.* 94:65–68.

Nasim, M., Livantsova, L. J., Krutko, P. D., Zaitseva, G. S., Lorbertth, J., and Otto, M. 1991. Synthesis of 2-Silatranyl-and 2-(3,7,10-trimethylsilatranyl) acetaldehydes. *J. Organomet. Chem.* 402:313.

Opornsawad, Y., Ksapabutr, B., Wongkasemjit, S., and Laine, R. 2001. Formation and structure of tris(alumatranyloxy-I-propyl)amine directly from alumina and triiospropanolamine. *Eur. Polym.* J. 37(9):1877–1885.

Parvulescu, V. and Su, B. L. 2001. Iron, cobalt or nickel substituted MCM-41 molecular sieves for oxidation of hydrocarbons. *Catal. Today.* 69(1–4):315–322.

Phiriyawirut, P., Magaraphan, R., Jamieson, A. M., and Wongkasemjit, S. 2003. Morphology study of MFI zeolite synthesized directly from silatrane and alumatrane via the sol–gel process and microwave heating. *Micropor. Mesopor. Mater.* 64(1–3):83–93.

Phonthammachai, N., Chairassameewong, T., Gulari, E., Jameison, A. M., and Wongkasemjit S. 2003. Oxide one pot synthesis of a novel titanium glycolate and its pyrolysis. *J. Met. Mater. Min.* 12:23.

Piboonchaisit, P., Wongkasemjit, S., and Laine, R. 1999. A novel route to Tris(silatranyloxy-i-propyl) amine directly from silica and triisopropanolamine, Part I. *Sci. Asia.* 25:113–119.

Rajakovic, V. N., Mintova, S., Senker, J., and Bein, T. 2003. Synthesis and characterization of V- and Ti-substituted mesoporous materials. *Mater. Sci. Eng. C.* 23(6–8):817–821.

Reddy, K. M., Moudrakovski, I., and Sayari, A. 1994. Synthesis of mesoporous vanadium silicate molecular sieves. *J. Chem. Soc. Chem. Commun.* (9):1059–1060.

Reddy, K. M., Wei, B., and Song, C. 1998. Mesoporous molecular sieve MCM-41 supported Co-Mo catalyst for hydrodesulfurization of petroleum resids. *Catal. Today.* 43(3–4):261–272.

Samran, B., Aungkutranont, S., White, T., and Wongkasemjit, S. 2011. Room temperature synthesis of Ti-SBA-15 from silatrane and titanium-glycolate and its catalytic performance towards styrene epoxidation. *J. Sol-Gel Sci. Technol.* 57(2):221–228.

Sathupunya, M., Gulari, E., and Wongkasemjit, S. 2003. Na-A (LTA) Zeolite synthesis directly from alumatrane and silatrane by sol–gel microwave techniques. *J. Eur. Ceram. Soc.* 23:2305–2314.

Sathupunya, M., Gulari, E., and Wongkasemjit, S. 2002. ANA and GIS zeolite synthesis directly from alumatrane and silatrane by sol-gel process and microwave techniques. *J. Eur. Ceram. Soc.* 22:1293–1303.

Sayari, A. 1996. Catalysis by crystalline mesoporous molecular sieves. *Chem. Mater.* 8(8):1840–1852.

Shao, Y., Wang, L., Zhang, J., and Anpo, M. 2005. Novel synthesis of high hydrothermal stability and long-range order MCM-48 with a convenient method. *Micropor. Mesopor. Mater.* 86:314.

Shao, Y., Wang, L., Zhang, J., and Anpo, M. 2008. Synthesis and characterization of high hydrothermally stable Cr-MCM-48. *Micropor. Mesopor. Mater.* 109:271–277.

Tanglumlert, W., Imae, T., White, T. J., and Wongkasemjit, S. 2007. Structural aspects of SBA-1 cubic mesoporous silica synthesized via a sol–gel process using a silatrane precursor. *J. Am. Ceram. Soc.* 90(12):3992–3997.

Tanglumlert, W., Imae, T., White, T. J., and Wongkasemjit, S. 2008. Preparation of highly ordered Fe-SBA-1 and Ti-SBA-1 cubic mesoporous silica via sol-gel processing of silatrane. *Mater. Lett.* 62(30):4545–4548.

Thanabodeekij, N., Gulari, E., and Wongkasemjit, S. 2007. Highly dispersed Mo-MCM-41 produced from silatrane and molybdenum glycolate precursors and its peroxidation activity. *Powder Technol.* 173(3):211–216.

Thanabodeekij, N., Sadthayanon, S., Gulari, E., and Wongkasemjit, S. 2006. Extremely high surface area of ordered mesoporous MCM-41 by atrane route. *Mater. Chem. Phys.* 9:131–137.

Thanabodeekij, N., Tanglumlert, W., Gulari, E., and Wongkasemjit, S. 2005. Synthesis of Ti-MCM-41 directly from silatrane and titanium glycolate and its catalytic activity. *Appl. Organomet. Chem.* 19(9):1047–1054.

Thitsartarn, W., Gulari, E., and Wongkasemjit, S. 2008. Synthesis of Fe-MCM-41 from silatrane and $FeCl_3$ via sol–gel process and its epoxidation activity. *Appl. Organomet. Chem.* 22(2):97–103.

Vallet-Regi, M., Rámila, A., del Real, R. P., and Pérez-Pariente, J. 2001. A new property of MCM-41: Drug delivery system. *Chem. Mater.* 13(2):308–311.

Vallet-Regí, M., Ruiz-González, L., Izquierdo-Barba, I., and González-Calbet, J. M. 2006. Revisiting silica based ordered mesoporous materials: Medical applications. *J. Mater. Chem.* 16(1):26–31.

Vinu, A., Murugesan, V., and Hartmann, M. 2003. Pore size engineering and mechanical stability of the cubic mesoporous molecular sieve SBA-1. *Chem. Mater.* 15:1385–1393.

Voronkov, M. G., Belyaeva, V. V., and Abzaeva, K. A. 2012. Basicity of silatranes (review). *Chem. Heterocycl. Comp.* 47(11):1330–1338.

Voronkov, M. G., Dyakov, V. M., and Kirpichenko, S. V. 1982. Silatranes. *J. Organomet. Chem.* 233:1–147.

Voronkov, M. G., Sorokin, M. S., Klyuchnikov, V. A., Shvetz, G. N., and Pepekin, V. J. 1989. Thermochemistry of organosilicon compounds. *J. Organomet. Chem.* 359:301.

Waldner, K. F., Laine, R. M., Dhumrongvaraporn, S., Tayaniphan, S., and Narayanan, R. 1996. Synthesis of a double alkoxide precursor to spinel ($MgAl_2O_4$) directly from $Al(OH)_3$, MgO, and triethanolamine and its pyrolytic transformation to spinel. *Chem. Mater.* 8:2850–2857.

Wen, J. and Wilkes, G. L. 1996. Organic/Inorganic hybrid network materials by the sol−gel approach. *Chem. Mater.* 8(8):1667–1681.

Witpathomwong, C., Longloilert, R., Wongkasemjit, S., and Jitkarnka, S. 2011. Improving light olefins and light oil production using Ru/MCM-48 in catalytic pyrolysis of waste tire. *Energy Procedia.* 9:245–251.

Wongkasemjit, S., Tamuang, S., Tanglumlert, W., and Imae, T. 2009. Synthesis of Mo-SBA-1 catalyst via sol–gel process and its activity. *Mater. Chem. Phys.* 117(1):301–306.

Zhang, W.-H., Lu, J., Han, B., Li, M., Xiu, J., Ying, P., and Li, C. 2002. Direct synthesis and characterization of titanium-substituted esoporous molecular sieve SBA-15. *Chem. Mater.* 14(8):3413–3421.

Zhao, D., Feng, J., Huo, Q., Melosh, N., Fredrickson, G. H., Chmelka, B. F., and Stucky, G. D. 1998. Triblock copolymer syntheses of mesoporous silica with periodic 50 to 300 angstrom pores. *Science.* 279(5350):548–552.

Index

A

B

C

G

H

N

O

P

T

U

V

W

X

Y

Z